The Benefit/Risk RATIO

A Handbook for the Rational Use
of Potentially Hazardous Drugs

Edited by

Hans C. Korting

Dermatologische Klinik and Poliklinik
Ludwig-Maximilians Universität
Munich, Germany

Monika Schäfer-Korting

Fachbereich Pharmazie
Freie Universität Berlin
Berlin, Germany

CRC Press
Taylor & Francis Group
Boca Raton London New York

CRC Press is an imprint of the
Taylor & Francis Group, an **informa** business

CRC Press
Taylor & Francis Group
6000 Broken Sound Parkway NW, Suite 300
Boca Raton, FL 33487-2742

© 1999 by Taylor & Francis Group, LLC
CRC Press is an imprint of Taylor & Francis Group, an Informa business

First issued in paperback 2019

No claim to original U.S. Government works

ISBN 13: 978-0-367-44781-6 (pbk)
ISBN 13: 978-0-8493-2791-9 (hbk)

Visit the Taylor & Francis Web site at
http://www.taylorandfrancis.com

and the CRC Press Web site at
http://www.crcpress.com

Introduction

The use of drugs as remedies for various types of diseases has a long tradition. As beneficial effects represent the aim of their use, people have for a long time tended mainly to expect wanted effects upon application. Although the existence of unwanted effects being linked to the use of drugs was noted by physicians, it was mainly experts — in particular pharmacologists — who considered side-effect-relevant features when discussing the profile of activity of a given drug. A prominent German pharmacologist, Dr. Gustav Kuschinsky, from Mainz, Germany became well known in some parts of the world for stating that a wanted effect should never be expected from the use of a drug if there is no documented *unwanted* effect.

Experiences with thalidomide, however, fundamentally changed the way drugs are looked at by the general public as well as the medical community. In the 1960s this new type of drug was introduced in Germany and other countries to provide sleep to those suffering from sleeplessness. It was mainly this effect, in the beginning, that was well known to the public. Some time later it became obvious that the drug also had an important potential for embryotoxicity leading to malformations referred to as dysmelia. At the time the new drug was introduced to the German market, there was not even a drug law existing in this country. Faced with this experience, however, the legal situation soon changed, and a German Drug Law (Arzneimittelgesetz) was passed by the German Parliament in 1976. The scientific foundation of this law had been created by Dr. Kuschinsky and pharmacologists of the Freie Universität Berlin including Dr. Hans Herken. The law provided a detailed framework for future handling of drugs. An important principle in the law is that the value of a new — or existing — compound had to be established by weighing benefits and risks together. It was not until 1995, however, that the principle of benefit-to-risk ratio assessment was finally overtly established.

In other countries, the situation has been similar. In the U.S., however —at least from 1962 on (the year amendments to the 1938 drug law were passed) — it was an established principle for the evaluation of a new compound to consist of assessment of both efficacy and toxicity based on animal and human studies. In fact, thalidomide never made its way to approval as a hypnotic in the U.S. In our context, it may look less relevant that this, in fact, was not due to embryotoxicity, which at this time was unknown on both sides of the Atlantic, but to neurotoxicity, which was only detected on clinical application of the drug.

As described above, the assessment of the benefit/risk ratio is most relevant when a new drug is to be approved by drug regulatory authorities, whether it be the Bundesinstitut für Arzneimittel und Medizinprodukte (BfArM) or the Food and Drug Administration (FDA). Benefit-to-risk assessment of drugs, however, is not only relevant in a legal context, but is an ongoing concern of every physician practicing medicine. Although this is undebatable in principle, the idea, so far, has not yet been adequately presented to doctors. Even when examining recent textbooks of pharmacology or clinical pharmacology and therapeutics, one will rarely come across this approach of looking at drugs — if the term is even introduced. This might simply reflect a lack of basic work in the field in the past; however, in recent years, interest in the subject has, fortunately, increased. As an example, at least since 1995, the journal *Drug Safety* regularly contains a section on risk/benefit assessment. Many more articles specializing in this particular subject can be found in a variety of journals and books.

Against this background, it was the intention of the editors to provide the reader with a more comprehensive look at various facets of the subject. In fact, it is the primary aim of this monograph to collect data from various authors and working parties on a variety of general and special aspects — first, to provide the clinician with data allowing optimization of his decisions on drug treatment in a given case, and, second, to provide the interested researcher, whether a physician, pharmacist, or basic scientist, with a sound foundation from which to further increase our knowledge.

Foreword

Old drugs and new drugs share at least one characteristic — their use is associated with an implicit or explicit assessment of their benefit/risk ratio. The ideal therapeutic agent has a risk equal to zero and a benefit equal to one, so that the benefit/risk ratio is infinite, a convenient way to convey the idea that the ideal therapeutic agent does not exist. However we estimate the benefit (a measure of efficacy) and the risk (a measure of toxicity), benefits and risks have to embody an estimate of the benefit of no treatment, i.e., the spontaneous cure rate, and the estimate of the risk of no treatment.

— R. Palminteri, 1988

Since the beginning of the 1990s, we have been working on the improvement of topical glucocorticoids. Since their introduction in 1952, topical glucocorticoids have become the mainstay of topical treatment of skin diseases in general and atopic eczema — one of the most frequent types of skin diseases — in particular. While available congeners of medium strength clearly meet the expectations in terms of efficacy, their safety has become a subject of widespread debate. In fact, nowadays many patients and parents of young patients refuse the application of a topical glucocorticoid even in severe manifestations of inflammatory skin diseases because they fear unwanted effects — in the U.S. mainly suppression of the hypothalamic-pituitary axis, in Europe mainly skin atrophy, i.e., systemic or topical side effects. In severe cases, atopic eczema as a prime indication for topical glucocorticoids can lead to growth retardation. The inability to apply topical glucocorticoids can, thus, have a major impact on the fate of a patient in his childhood or adolescence. For this reason, clear alternatives to conventional medium-potent glucocorticoids were needed. When the introduction of a new chemical class of congeners, i.e., the topical glucocorticoids of the so-called nonhalogenated double-ester type seemed to harbor a potential alternative, the editors of this book, a pharmacologist and a dermatologist, felt challenged to assess activity, efficacy, and safety of this new class of glucocorticoids. In fact, we were fortunate in being able to demonstrate that the benefit/risk ratio of this class of drugs was improved, conflicting with the previous dogma that safety of a given topical glucocorticoid was directly linked to efficacy. Performing several types of investigations it was possible to demonstrate this improved benefit-to-risk ratio not only in qualitative but also in quantitative terms. Moreover, it was possible to base the concept not only on findings *in vivo* and particularly in man, but also on *in vitro* findings providing insight into underlying mechanisms. In this context pharmacokinetic aspects were particularly relevant.

Against this background we became interested in the subject of benefit-to-risk ratio assessment on a larger scale. Looking into other fields of drug treatment we became aware of two facts: On the one hand there was no indication that benefit-to-risk ratio assessment had become a general feature of the judgment of drugs in pharmacology — not even in clinical pharmacology. On the other hand, there was increasing evidence that in various fields of general and special interest, research workers — whether they be physicians or basic scientists —currently try to provide the public with relevant data. These various efforts, however, were widely spread within the scientific literature. This did not make us expect that benefit-to-risk ratio assessment of drugs might soon become a major feature of drug evaluation in clinical medicine or at least within the specialized scientific community. So we felt that we should try to prepare a monograph on the subject, getting experts from all over the world involved who had already become aware of the special relevance of the topic. To make sure that only really dedicated authors would contribute to the book, we decided to invite only those who had previously published on the subject.

As benefit/risk ratio assessment is still a new topic, we hoped to find a publisher for the monograph we had in mind. But we were also concerned that publishing houses might not be convinced that the subject could gain as much interest as needed to provide a sound basis for the publication of a book.

Therefore, we were extremely happy when CRC Press very quickly decided in favor of publishing the book when presented with the idea. In particular, we appreciate the dedicated support of Carol S. Messing. Moreover, we would like to thank all those authors and groups of authors who, despite all their other commitments, dedicated both time and effort to the preparation of their contribution to the book. Finally, we have to thank Mrs. Gabriela Esser and Mrs. Beatrice Sandow, who, as our secretaries, managed the contact with the authors and with CRC Press. Without their most valuable contribution this book might not have been realized.

We are clearly aware of the fact the benefit-to-risk ratio assessment of drugs is still in its infancy. We hope that readers of this book will share our opinion that a lot of work has already been done allowing implementation of the principle not only in registration of drugs by drug authorities, but also by physicians who must make choices in the face of disease of individual patients. To reach maturity, we certainly need further contributions to subjects from various fields of drug treatment, and it is one of the major aims of this book to attract more research workers to look into the field and make their specific contributions. In fact, it would be a particular pleasure to us for this monograph to be widely used enabling the publisher and us to provide another edition filling the gaps we have at the present time.

In order to inprove our current knowledge, we would particularly welcome comments of our readers.

Hans C. Korting and Monika Schäfer-Korting

The Editors

Hans Christian Korting was born in Tübingen, Germany in 1952. He studied medicine at Johannes Gutenberg University in Mainz as a scholar of Studienstiftung des Deutschen Volkes and was approved as a physician in 1977. From 1977 to 1979 he was associated with the Departments of Medicine and Microbiology at Central Medical Facilities of the German Army in Koblenz. From 1979 to the present he has been employed at the Dermatologische Klinik and Poliklinik of the Ludwig Maximilians University in Munich. In 1983 he was approved as a dermatologist, and in 1985 he obtained his Ph.D. He was appointed adjunct professor in 1992, and in 1995 he was appointed lieutenant-colonel of the reserve of the Medical Corps of the German Army.

Dr. Korting has published 150 clinical and experimental original papers primarily on cutaneous infectiology and pharmacology as well as 50 case reports, 50 review articles, 13 contributions to handbooks, and 13 books (as coauthor or coeditor). He has received 5 scientific awards including the Paul Gerson Unna Preis of the Deutsche Dermatologische Gesellschaft (1990) and Forschungsförderungspreis of the Deutschsprachige Mykologische Gesellschaft (1994). He is a member of 12 scientific societies and acts as deputy chairman of the Gesellschaft für Dermopharmazie and Deutschsprachige Mykologische Gesellschaft. He is the head of the Therapy Committee of the Deutsch Dermatologische Gesellschaft and vice chairman of the Committee on Quality Assurance.

Dr. Korting is coeditor or member of the editorial board of 12 scientific journals including *Mycoses*, *Skin Pharmacology*, and *The Journal of Molecular Medicine*.

Monika Schäfer-Korting was born in Gießen, Germany in 1952. She studied pharmacy at Johann Wolfgang Goethe University in Frankfurt and was approved as a pharmacist in 1976. She obtained a Ph.D. from the University of Frankfurt in 1977, and from 1976 to 1979 and 1982 to 1992 was employed by the Institute of Pharmacology of the Faculty of Pharmacy and Food Sciences of the Johann Wolfgang Goethe University in Frankfurt. From 1980 to 1981 she was affiliated with the Institute of Pharmacology of the Faculty of Medicine of Technische Universität in Munich. In 1983 she was approved as an expert in pharmacology. From 1992 to 1994 she was acting chairman of the newly founded Chair of Pharmacology and Toxicology at the Faculty of Pharmacy of Freie Universität in Berlin. In 1993 she was approved as a pharmacist for drug information. In 1994 she was appointed Chairman of the Department of Pharmacology and Toxicology. In 1997 she was appointed Dean of the Faculty of Pharmacy, and in the same year she became Vice President of the Freie Universität.

Major research interests include modern analytical procedures for active compounds of drugs as well as dermatopharmacology. Dr. Schäfer-Korting is the author of more than 50 original publications and 20 review articles as well as 2 books. She is acting as board member of the Gesellschaft für Dermopharmazie and of the Board of Professors of Pharmaceutical University Institutes of the Federal Republic of Germany. She serves on the editorial boards of, e.g., *Pharmazie*, *Skin Pharmacology*, and *Mycoses*.

Contributors

Neal L. Benowitz
San Francisco General Hospital
and
University of California Medical
   Center
San Francisco, California

Hélène Bocquet
Service de Dermatologie
Hôpital Henri Mondor
Université Paris XII
Créteil, France

Giorgio W. Canonica
Allergy and Clinical Immunology
   Service
Department of Internal Medicine
University of Genoa
Genoa, Italy

Christy Chuang-Stein
Clinical Biostatistics II
The Pharmacia & Upjohn Company
Kalamazoo, Michigan

Dario Civalleri
Department of Surgery
University of Genoa School of
   Medicine
Genoa, Italy

Pietro Compagnucci
Division of Internal Medicine
Hospital of Camerino
Camerino, Italy

Franco DeCian
Department of Surgery
University of Genoa School of
   Medicine
Genoa, Italy

Piet De Doncker
Janssen Research Foundation
Beerse, Belgium

Fabien Durand
Vitro-Bio
Le Breuil-sur-Couze, France

Mervyn J. Eadie
Department of Medicine
University of Queensland
Brisbane, Australia

Mauro Esposito
Section of Pharmacotoxicology
National Cancer Institute
Genoa, Italy

Karen T. Ferrer
Department of Pathology
University of Illinois Medical
   Center
Chicago, Illinois

Alberto A. Gabizon
Sharet Institute of Oncology
Hadassah Hebrew University
   Medical Center
Jerusalem, Israel

Rodolphe Garraffo
Clinical Pharmacokinetics Unit
Department of Pharmacology
University Hospital
Nice, France

Aditya K. Gupta
Division of Dermatology
Department of Medicine
Sunnybrook Medical Sciences
   Center and the
University of Toronto
Toronto, Canada

Anja Gysler
Institut für Pharmazie II
Pharmakologie und Toxikologie
Freie Universität Berlin
Berlin, Germany

Joerg Hasford
Department for Medical
   Informatics, Biometry and
   Epidemiology (IBE)
University of Munich
Munich, Germany

Michel Heenen
Service de Dermatologie
Université Libre de Bruxelles
Hôpital Erasme
Brussels, Belgium

J. Michael Kilby
Outpatient Clinic
University of Alabama
Birmingham, Alabama

Gilles Klopman
Department of Chemistry
Case Western Reserve University
Cleveland, Ohio

Hans C. Korting
Dermatologische Klinik and
   Poliklinik
Ludwig Maximilians Universität
Munich, Germany

Laurence Le Cleach
Service de Dermatologie
Hôpital Henri Mondor
Université Paris XII
Créteil, France

Orest T. Macina
Department of Environmental and
   Occupational Health
University of Pittsburgh
Pittsburgh, Pennsylvania

Nicola Magrini
Department of Pharmacology
University of Bologna
Bologna, Italy

ix

Joseph F. Mortola
Division of Reproductive
 Endocrinology
Beth Israel Hospital
Boston, Massachusetts

Gianmauro Numico
Department of Medical Oncology
National Cancer Institute
Genoa, Italy

Giovanni Passalacqua
Allergy and Clinical Immunology
 Service
Department of Internal Medicine
University of Genoa
Genoa, Italy

David J. Peace
Section of Hematology/Oncology
University of Illinois
Department of Medicine
Chicago, Illinois

John Pinney
Pinney Associates
Bethesda, Maryland

Herbert S. Rosenkranz
Department of Environmental and
 Occupational Health
University of Pittsburgh
Pittsburgh, Pennsylvania

Jean-Claude Roujeau
Service de Dermatologie
Hôpital Henri Mondor
Université Paris XII
Créteil, France

Fausto Santeusanio
Dipartimento di Medicina Interna e
 Scienze Endocrine e Metaboliche
 (DIMISEM)
Perugia, Italy

Monika Schäfer-Korting
Institut für Pharmazie II
Pharmakologie und Toxikologie
Freie Universität Berlin
Berlin, Germany

Bodo Schwartzkopff
Department of Cardiology,
 Pneumology and Angiology
Heinrich Heine University
Düsseldorf, Germany

Ravi Shrivastava
Vitro-Bio
Le Breuil-sur-Couze, France

Bodo E. Strauer
Department of Cardiology,
 Pneumology and Angiology
Heinrich Heine University
Düsseldorf, Germany

Petra A. Thürmann
Institut für Klinische Pharmakologic
Klinikum Wuppertal
Wuppertal, Germany

Giuseppe Traversa
Department of Epidemiology and
 Biostatistics
Istituto Superiore de Sanità
Rome, Italy

Warren Wong
Division of Hematology/Oncology
Loyola University Medical Center
Maywood, Illinois

Jianguo Zhi
Department of Clinical
 Pharmacology
Hoffmann-La Roche, Inc.
Nutley, New Jersey

Contents

1

The Role and Analysis of Benefit and Risk in the Development of New Drugs

Christy Chuang-Stein

The Pharmacia and Upjohn Company

1.1   Introduction

With the strong emphasis on quality-of-life evaluation and medical outcome research in the current drug development environment, it is surprising that the effort to conduct a meaningful benefit and risk assessment to support the regulatory approval of a new drug stays at the minimum level. By meaningful benefit and risk assessment, I mean an assessment whereby each patient is considered as a whole with their safety and efficacy experience in a trial combined to obtain an overall outcome measure. The closest measure to achieve this goal is the YUYOSEI[1] instrument commonly used in Japanese clinical trials. YUYOSEI is a global assessment given by the treating physician to reflect, in the physician's opinion, an individual's overall response to a treatment. By its very definition, YUYOSEI has objective and subjective components, since the sense of *well-being* of a patient can be quite subjective. For this reason, YUYOSEI has recently come under attack for not being objective enough and there has been some discussion of discontinuing its use. In the author's opinion, throwing away this global measurement because of its subjective component is foolish, because subjective evaluation exists in many aspects of clinical trials. One example is the reporting of clinical symptoms by individuals in a trial. Therefore, instead of advocating the abolition of YUYOSEI, I want to argue the need to make this seemingly subjective measure as objective as possible when it is used for decision-making purposes.

In an article[2] on the pitfalls in the drug approval process, Miller pointed out three deficiencies that often delay the review of a new drug application (NDA) and subsequently the application's approval. The three deficiencies are: (1) failure to adequately characterize the dose–response relationship, (2) failure to focus the development on issues directly related to the proposed indications, and (3) failure to adequately anticipate the risk–benefit issues that ultimately determine the wording of the package insert. Among the three deficiencies, the first has been addressed extensively in the context of dose–response study designs. Yet, despite the efforts that have been devoted to this subject, the overwhelming emphasis in most clinical development plans to bring a new drug or treatment to market in an expeditious manner has led many pharmaceutical sponsors to bypass some important steps in the drug development process. The most noted one is a rigorous phase II program to select an optimal dose and administration schedule. The second deficiency is frequently found in situations where a clinical development program was initiated with the hope of registering a new chemical entity for multiple indications. Therefore, the lack of focus with respect to the proposed indication at the time of the NDA filing is an inevitable outcome of an overextended drug development plan. Contrasting to the first two, the third deficiency remains largely an outstanding issue because of the lack of generally agreed-upon procedures to conduct a relevant benefit/risk assessment and the lack of regulatory guidelines on the subject. The lack of methodology is mostly due to the lack of attention devoted to this topic in our current regulatory environment, where it is a routine practice to summarize the safety and efficacy data individually and separately without a conscious effort to integrate them in a new drug application.

In this chapter, we will focus on the role of benefit and risk assessment as well as methods to conduct this assessment, pivoting risk against benefit in the evaluation of a new drug. For the latter, we will concentrate on an algorithm-oriented approach to combine risk (measured by adverse reactions) and benefit (measured by treatment efficacy) into a global assessment measure that shares the meaning of YOYUSEĬ. Yet, the use of an algorithm ensures that the process of information aggregation be as objective as possible and not dependent on the individual treating physicians as in the case of YOYUSEĬ. In addition, the use of the algorithm, when automated, guarantees that all factors of concern will be included in the final decision. The latter is an advantage not enjoyed by the human mind due to its known limitations in responding to complicated decision-making needs.

In Section 1.2, we will discuss certain fundamental concepts that underlie the benefit and risk evaluation. Section 1.3 finds some commonly used procedures to conduct such evaluations. In Section 1.4, we will focus on an approach proposed by Chuang-Stein.[3] The approach by Chuang-Stein is called attention to for detailed discussion because of its flexibility and applicability. We will apply the approach to a data set in the same section to help with the exposition. Section 1.5 concludes this chapter by offering some comments and addressing some questions raised by the participants at The Drug Information Association Second Annual Biostatistics Meeting in Tokyo in August 1995 where this approach was presented.

1.2 Concepts Underlying Benefit/Risk Assessment

For convenience, we will use *benefit/risk* to mean benefit to risk in the most general sense and not restrict it to merely the ratio between benefit and risk. In the author's opinion, there are two fundamental concepts that are important to benefit/risk assessment. First, while it is obvious that interventions are given with the hope of curing or controlling existing diseases or palliating discomfort or pain, there is also the danger that the interventions may cause harm or injury. Second, in deciding whether the likely risk can justify the potential benefit, it is a long-accepted notion that the evaluation should be conducted with respect to the underlying disease or symptoms as well as other available therapeutical options.

These fundamental concepts led to the decision to ban the use of diethylstilbestrol (DES) during pregnancy in 1971 when it was concluded that the increased risk of developing clear-cell cervicovaginal cancer among the daughters of women who took the medication during their pregnancies to prevent miscarriages outweighed the medication's potential benefit. These same concepts helped the creation of the accelerated drug approval process by the U.S. Food and Drug Administration for drugs treating life-threatening conditions such as AIDS. Under the accelerated drug approval process, an accelerated approval can be granted based on positive results on markers that are "*reasonably likely to predict clinical efficacy.*" In such cases, the approval is typically contingent on the pharmaceutical sponsor's commitment to conduct post-marketing trials where the clinical benefit of the drug can be evaluated. Full approval will be granted if the clinical benefit can be demonstrated in such trials; otherwise the accelerated approval can be withdrawn.

On August 10, 1995, the Public Citizen's Health Research Group in the U.S. petitioned the U.S. FDA to immediately add warning labels on quinidine and strengthen the warning on other Class I antiarrhythmic drugs because of the increased risk of death due to the proarrhythmic effects of quinidine-containing drugs.[4]

Because of the side effect profile of quinidine-containing drugs, the Group maintained that the quinidine indication should be restricted to life-threatening arrhythmia and not be used on patients with asymptomatic ventricular premature contractions. This was not the first time that the value of quinidine-containing drugs for non-life-threatening arrhythmia was questioned. The same issue will undoubtedly be debated again in the future with benefit/risk argument occupying center stage.

In July of 1995, the Oncology Advisory Committee to the U.S. Food and Drug Administration recommended the approval of Lilly's gemcitabine (Gemzar) as a first line treatment for pancreatic cancer even though the potential benefit of Gemzar over 5-FU was evaluated in only one controlled clinical trial.[5] This recommendation represented a recognized departure from the traditional requirement of replicated results from two independent clinical trials. The Committee members endorsed such a departure because of the seriousness of pancreatic cancer and the general lack of effective alternative regimens for the target patient population. The Committee arrived at this decision after evaluating Gemzar's benefit and risk relative to the indicated disease and the available alternatives.

The deliberation of the Oncology Advisory Committee over the approvability of Gemzar is not unique. It is a well-known fact that it is much easier to gain regulatory approval of a new cancer regimen for patients whose cancer has become refractory to standard treatments than to gain the approval of the regimen as a first-line therapy. The process of weighing risk, both known and unknown, against benefit takes place before any new drug or treatment is allowed to enter the marketplace. This process is needed, or else how can the net "*worthiness*" of a new drug or treatment be determined?

1.3　Measures to Assess Benefit and Risk Simultaneously

Throughout the years, various measures have been used to simultaneously address the benefit and risk of a regimen. The YUYOSEI measure discussed earlier is one of them. But more frequently, benefit/risk consideration has been based on the calculation of a benefit/risk ratio with benefit measured by the response rate and risk determined by the incidence of serious side effects. The interpretation of an acceptable benefit/risk ratio is, therefore, the number of efficacy events needed for each serious adverse event incurred. This ratio statistic was used by Payne and Loken[6] to evaluate the benefit and risk of radiology. Based on 40 million mass chest X-ray exams in Japan in 1968, Payne and Loken came up with a rough benefit/risk ratio of greater than 50 to 1. The 50 to 1 ratio was obtained from 723 somatic and genetic deaths to 39,250 curable pickups of tuberculosis, lung cancer, and leukemia.

A measure to compare the benefits and risks of several treatments is the incremental benefit/risk ratio. Incremental benefit/risk ratios were used by Chang and Fineberg[7] to compare five possible strategies for managing patients afflicted with the polymyalgia rheumatica (PMR) syndrome. It was known to physicians that neither nonsteroidal antiinflammatory drugs (NSAIDs) nor low doses of corticosteroids, two common modalities for PMR, can completely protect PMR patients from blindness, a result from the associated giant cell arteritis. While higher and more toxic doses of corticosteroids can eliminate this risk, medication-induced side effects such as perforated ulcer, gastrointestinal bleeding, septicemia, symptomatic vertebral or long bone fracture, and avascular necrosis of bone can be severe also. The central issue was how one could balance a strategy's effectiveness in preventing blindness with its side effects, which can be severe, with more effective strategies. Chang and Fineberg constructed the incremental benefit/risk ratio for each pair of contiguous strategies with strategies ordered by their probabilities of inducing severe steroid toxicity. In this case, the incremental benefit/risk ratio represents the number of additional cases of severe toxicity for each case of blindness averted in moving from one strategy to another. Therefore, selection among the strategies could be based on the acceptable values for the incremental benefit/risk ratio.

If data are available to allow the estimation of the dose-benefit and dose-injury curves, Andrews[8] proposed using the fitted curves to search for an optimum level of radiotherapeutic dose to treat cancer patients. The analysis underlying the procedure is called the *receiver operating characteristics* (ROC) analysis. Other measures to evaluate cancer treatments include quality-adjusted life expectancy as well as quality-adjusted time without symptoms of disease and toxic effects. The latter measures have been used recently to evaluate the cost-effectiveness of certain medical practices such as elective hysterectomy[9] and the use of hormone replacement therapy to prevent osteoporosis and reduce cardiac diseases among menopausal women.[10-12]

In terms of making treatment selection, Hilden[13] proposed a procedure that would incorporate a patient's own judgment on the value of the potential benefits and risks of the treatments into consideration. Hilden argued that the investigators should present the outcome of the significance test for treatment comparisons in terms of a constant "c", which represents the number of units of complications a patient is willing to accept in order to achieve one unit of benefit. Hilden's approach was later generalized by Chuang-Stein et al.[14] In addition to generalizing Hilden's benefit/risk measure, Chuang-Stein et al.[14] also proposed two additional measures that carry the flavor of benefit/risk ratio. To construct the measures, Chuang-Stein et al. proposed to classify the outcome of each patient in a trial into one of five categories based on the perceived benefit and side effects of the medications. The five categories are: efficacy without side effects, efficacy with side effects, no efficacy and no side effects, no efficacy with side effects, and finally side effects leading to withdrawal from an ongoing treatment. The three measures proposed by Chuang-Stein et al. combine the five potential outcomes either through a weighted average of their chance of occurrence or a ratio between the weighted averages of the more desirable and the not-so-desirable subsets of the outcomes. For applications and interpretation of these three measures, interested readers are referred to the original publication.

One disadvantage of all the measures mentioned above is the loss of information when a patient's safety experience in a trial is dichotomized. The dichotomization can greatly reduce our ability to differentiate between the safety outcomes of different treatments. Our inability to make this differentiation can, in turn, reduce the sensitivity of the procedures to make treatment comparisons. This is especially important when none of the side effects are serious, but those associated with one treatment are much more annoying than those of the other. The same can be said of the efficacy outcome when efficacy is measured on an interval scale. One natural question then is, is it possible to treat benefit and risk as continuous variables? To address this question, Chuang-Stein[3] proposed another benefit/risk measure that simultaneously quantifies the benefit and the risk. This new benefit/risk measure will be the subject of the next section.

1.4   A Continuous Benefit-Less-Risk Measure

To help with the exposition, I will use an example of two trials conducted under the same protocol to compare two drugs, labeled as drug A and drug B, for the treatment of angina pectoris. Since the two trials followed the same protocol, the results from the two trials are pooled as if they came from the same trial. Between the two trials, a total of 86 patients with exertional angina were randomized to receive either an investigational medication or a beta-blocker with 43 patients in each treatment group. The trials included a two-week tapering period and a two-week placebo run-in period before patients received their randomized treatments. During the first 4 weeks of treatment, the medications under comparison were given three times a day and their doses could be escalated if the doses were tolerated and if, in the opinion of the investigators, the escalation was beneficial to the patients. After 4 weeks of treatment, patients receiving the beta-blocker were maintained for two additional weeks with dose at the level established at the end of the fourth week. As for patients receiving the investigational medication, they were taken off the medications and placed on a placebo for 2 weeks to allow for a more accurate evaluation of the effect of the experimental medication. The efficacy of the medication was measured by the change in time to angina during the exercise tolerance test (ETT) at the end of the fourth treatment week from that at baseline. The baseline value was defined to be the time to angina during the ETT conducted at the end of the 2-week placebo run-in period.

With the efficacy measurement well-defined, the issue automatically turns to our ability to summarize and quantify a patient's safety experience in the trials. Unlike efficacy experience, a patient's safety experience as reflected by side effects or adverse drug reaction, typically comes from many different sources and can be quite diverse in nature. It can be the observations by the clinicians (signs), complaints of untoward events from the patients (symptoms), laboratory assay reports, and physiological tests such as ECG and CT scan. In order to summarize an individual's risk associated with a treatment, we need to consolidate this mass amount of data into a more meaningful and manageable format. One option to do so is an approach proposed by Chuang-Stein et al.[15] I will briefly describe the approach here.

The approach has three basic assumptions. First, it assumes that a set of J classes representative of body functions can be selected to cover all areas of safety concerns for a particular situation. Examples of the classes are cardiovascular, hepatic, central nervous system, pulmonary, metabolic, dermatologic, etc. Second, the approach assumes that an overall intensity grade can be determined for each patient within each class using all the safety information relevant to that class. An intensity grade 0 for a class would indicate that there is no safety concern for that class. The assignment of the overall intensity grade should be based on a set of predetermined rules that take into account issues such as multiple occurrences of events and events that existed prior to receiving the study treatments. Depending on the underlying disease and the type of compounds under comparison, the number of grades can vary from class to class. Third, the approach assumes that weight W_{jk} can be assigned to the kth intensity grade within the jth event class with higher weights corresponding to higher intensity grades and thus representing worse safety outcomes within the class. Furthermore, the approach assumes that the weights can also reflect the relative seriousness of the various intensity grades across different classes. An example is to assign the same set of weights to all event classes when the numbers of intensity grades are the same for all the classes. Under this choice of weights, events of the same grades in different classes will have the same weights and the safety analysis will treat all classes with equal importance.

In the example involving the two angina clinical trials, we chose ten classes and used four grades from 0 to 3 to summarize the safety data within each class. The consolidated safety summaries for six patients from each treatment group are given in Table 1.1. Details on rules to determine the grades in these two trials can be found in Chuang-Stein et al.[15] As for the weights for the four intensity grades within each of the ten classes, one possible choice is to consider weights that are products of a class-specific score v_j and the equally spaced scores $\{0,1,2,3\}$. The score v_j reflects the relative seriousness of the ten event classes in the intended patient population. One such set of $\{v_j\}$ is: 6 for cardiovascular, 3.5 for pulmonary, 3 for neurological/psychiatric, 2 for hematologic and musculoskeletal, 1.5 for GI/hepatic, GU/renal, and metabolic, 1 for special senses and dermatologic. Under this weighting scheme, a cardiovascular event is considered to be six times as serious as a dermatologic event of the same grade. With this choice of weights, the worst total safety score an individual can get is 69 when the individual experienced side effects of the worst grade in all ten event classes. An individual's total safety score is the sum of the weights assigned to those intensity grades in the ten classes that describe the individual's overall safety experience in the respective class. We would like to point out that the above is but one of many possible ways to assign weights to the ten event classes. These weights were, however, considered to be reasonable by one of the clinical monitors in the cardiovascular area at The Upjohn Company (now The Pharmacia and Upjohn Company).

Relative to the worst total safety score, we can compute the normalized safety score for each individual. The normalized score is simply the ratio between an individual's total safety score and the maximum total score of 69. This normalized score has a value between 0 and 1 and can be viewed as a summary of risk for an individual. Thus, a risk summary of 1 corresponds to the worst safety experience that one could possibly have in a trial. Using the risk summary for each individual patient, we can compare different treatment groups with respect to their mean risk summary. When we did so in our example, we found that drug A had a slightly worse safety profile than drug B even though the difference was not significant at the 5% level. The p-value was found to be 0.258. The *average* risk (summary) for drug A is 0.113 and for drug B 0.022. The higher average risk for drug A is primarily due to the higher weights assigned to the cardiovascular events, which occurred with a higher frequency in the group receiving drug A. The frequencies of events in the ten classes for the two treatment groups are given in Table 1.2.

As mentioned earlier, the primary objective of the studies was to investigate the beneficial effects of the drugs measured by the increase in the treadmill tolerance time at the end of the fourth treatment week over that at baseline. The average increase time for the two groups was 144.16 seconds for drug A

TABLE 1.1   A Sample of the Overall Intensity Grades Assigned to the 10 Event Classes for Patients in the Two Randomized Trials

Drug	Pt.	CV	Hematology	GI/ Hepatic	GU/ Renal	Neurologic/ Psychiatric	Pulmonary	Special Senses	Metabolic/ Nutrition	Dermatologic	Musculoskeletal
A	1	0	0	0	0	1	0	0	0	0	0
	2	0	0	0	0	0	0	0	0	0	0
	3	1	1	0	0	0	1	0	0	0	0
	4	0	0	1	0	2	0	0	0	0	0
	5	2	0	0	0	0	2	0	0	0	0
	6	2	0	0	0	2	0	2	2	0	0
	...					...	...				
B	1	0	0	0	3	2	0	0	0	0	0
	2	0	0	0	1	0	0	0	0	0	0
	3	0	0	0	1	0	0	2	0	0	0
	4	0	0	1	3	0	0	0	0	0	1
	5	1	0	0	0	0	0	0	0	0	0
	6	0	1	2	1	0	0	0	0	0	2

TABLE 1.2   Results from Consolidating the Safety Data of All Patients in Two Randomized Trials (Chuang-Stein et al.[15]).

Class	Drug	Grade 0	Grade 1	Grade 2	Grade 3
Cardiovascular	A	29	3	8	3
	B	38	4	1	0
Hematologic	A	40	3	0	0
	B	40	3	0	0
Gastrointestinal/hepatic	A	37	3	2	1
	B	36	3	4	0
Genitourinary/renal	A	40	2	1	0
	B	38	2	3	0
Neurological/psychiatric	A	26	4	13	0
	B	15	11	10	7
Pulmonary	A	37	4	2	0
	B	41	1	1	0
Special senses	A	41	1	1	0
	B	41	0	2	0
Metabolic/nutritional	A	40	1	2	0
	B	39	2	2	0
Dermatologic	A	41	1	1	0
	B	41	0	2	0
Musculoskeletal	A	39	3	1	0
	B	37	4	2	0

(standard deviation = 62.81 seconds) and 112.86 seconds for drug B (standard deviation = 79.19 seconds). When applying the two-sample t-test to compare the mean increase time between the two groups, we obtained a p-value of 0.045, providing some evidence that, at the 5% level, the mean increase time for drug A was higher than that for drug B. Recall that drug A also had a slightly worse safety profile. The question is whether the slightly worse safety profile of drug A would offset its superior performance when the performance of the drugs is based on both efficacy and safety.

To adjust the observed benefit by risk, Chuang-Stein[3] proposed to discount the observed benefit by a fraction of the risk summary to obtain a risk-adjusted benefit. Chuang-Stein's approach uses a concept similar to that underlying the adjustment to the life expectancy to obtain the quality-adjusted life expectancy. In other words, if we let r_{ti} represent the risk summary for the ith individual in the tth (t = 1,2) treatment group and e_{ti} the observed benefit for the same individual, Chuang-Stein proposed to obtain a risk-adjusted benefit as

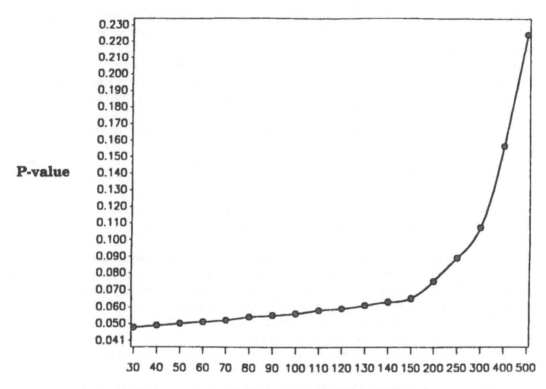

FIGURE 1.1　Plotting the P-value against the discounting factor f.

$$e_{ti}^* = e_{ti} - f \cdot r_{ti} \qquad (1.1)$$

where f controls the amount of discounting that will take place. The choice of f in (1.1) depends on many factors such as other available treatments and the nature of the disease/symptoms. Since benefit is adjusted for risk by subtracting a multiple of the risk from the benefit, the *net* benefit measure on the left-hand side in (1.1) was termed *benefit-less-risk* by Chuang-Stein.[3] Once the benefit is adjusted for risk, Chuang-Stein suggested proceeding with the comparisons between the two treatment groups using the risk-adjusted benefit as if the latter were the data.

In our example, the efficacy measure e_i is the observed increase in the treadmill tolerance time at the end of four treatment weeks over that at baseline. In general, when the efficacy of an intervention is measured on the interval scale, the change in the response from its baseline value is typically used to estimate the benefit. When the efficacy is measured by a binary response variable, Chuang-Stein proposed to let e_i be 1 if a response is obtained and 0 otherwise. If the response is measured on an ordinal scale and there are numeric values associated with the categories, Chuang-Stein proposed to use the appropriate difference in the numeric values for e_i; otherwise she proposed to assign equally spaced scores to the ordinal response categories and let these scores play the role of the numerical values in calculating differences.

We now get back to the question of whether the better efficacy result with drug A was achieved at the expense of a slightly worse safety outcome. To attempt an answer to the question, we applied the approach discussed above and adjusted the efficacy results. We tried a range of values for the discounting factor f. For each f, we computed the adjusted benefit for each individual and proceeded with the comparison between the two treatment groups using the adjusted benefit. P-values obtained under different adjusting factors f are plotted against f in Figure 1.1. As can be seen from Figure 1.1, as f becomes large, the adjustment starts to have more impact on the efficacy comparison and the superiority of drug A starts to disappear. This finding on the *net* benefit of drug A relative to drug B is expected considering the role of the discounting factor in the adjustment process.

A question ensues naturally, What constitutes a reasonable choice for f in this case? One possible choice is based on the rationale that the placebo effect is not worthy of a grade 2 cardiac event. The placebo effect has been quoted in the literature to be around 27 seconds. Since a grade 2 cardiac event carries a normalized score of 12/69, we can equate 27 to the product of f and 12/69 and solve for f. The f satisfying this condition was found to be around 155. With $f = 155$, the p-value based on the risk-adjusted benefit is around 0.065, no longer significant at the 5% level. However, the relatively small p value under this choice of f still provides some evidence for the generally better efficacy result with drug A.

In this example, the effect of adjusting the increase in the ETT time on the treatment comparison amounts to subtracting the quantity 0.022 f from the difference in the observed mean changes, i.e., 31.30, between the two groups (drug A – drug B). Thus, a value of 155 for f resulted in an adjusted difference of 27.89 (= 31.30 – 0.022 × 155) between the two groups. Notice that in this example, the adjustment is not substantial. This is primarily due to the fact that the incidence of side effects was relatively low for both groups in the trials. Nevertheless, one should keep in mind that in general, the conclusion under the above approach depends on how the weights $\{W_{jk}\}$ are chosen and how the discounting factor f is selected.

It should be clear by now that our ability to quantitatively summarize a patient's safety experience in a trial is essential to the construction of the risk-adjusted benefit measure. It should also be clear that the observed benefit can be adjusted not only for the full spectrum of safety experience, but also for selected medical events that are of primary interest. An example for the latter is the serious or life-threatening adverse drug reactions. Furthermore, it should be clear that the risk-adjusted efficacy can be negative if one experienced adverse reaction but not efficacy.

Like any other adjustment methods, the benefit/risk evaluation discussed above has a subjective component. When applied by an individual to make a treatment selection, the weights $\{W_{jk}\}$ as well as the discounting factor f should be chosen to reflect the individual's assessment of the possible side effects and the potential benefit. This is highly desirable because in the end, individuals with the help of their physicians need to decide whether or not to receive a particular treatment. When applying the approach to compare the relative merits of treatments in the indicated population on the average, it would be most helpful if a group of experts in the field can agree on what constitutes a safety concern for the population involved and how benefit can best be balanced with risk in this population. It is possible that for certain situations, no agreement can be reached. The latter reflects a rather diverse decision-making process where the same treatments possess quite different values in the mind of different experts. In this case, treatment selection is less clear and can be influenced by many factors that defy logical programming.

1.5   Discussion

Currently, benefit/risk assessment in the development of a new drug is done on an ad hoc basis. While benefit/risk ratio was frequently mentioned during the regulatory deliberations on the approvability of new drugs, the author has not yet found one case where the user of the phrase was specific about how this assessment should be conducted. Instead, the phrase "benefit/risk ratio" was merely used to reflect the need to examine benefit and risk in one package. In other words, the phrase as it is being used now often does not have any substantive meaning. Admittedly it might be difficult to reach a consensus on what constitutes an acceptable ratio when ratio carries the same meaning as in Payne and Loken,[6] the lack of this consensus, especially from the regulatory perspective, does not help pharmaceutical sponsors evaluate the likelihood of success of their compounds under development.

All benefit/risk assessment methodologies aim to combine benefit and risk into one summary statistic for evaluation. The approach detailed in Section 1.4 is no exception. However, the approach in Section 1.4 expands on the concept of risk by allowing the incorporation of a wide spectrum of safety issues that are of potential concern to the target patient population. The process to consolidate the safety experience mimics what treating physicians typically have to go through in their clinical practice to decide if a treatment is "*safe*" for a particular patient. The process, with the assistance of computer programming of the algorithm, can produce consistent results when all factors of interest are the same. Because of the judgmental nature of the approach, it is possible that a user of the approach cannot come up with a definitive weighting scheme or a definitive discounting factor. Therefore, we recommend conducting the assessment with different choices on the weights and the discounting factor to see how robust the conclusions are to these choices. This process is iterative and the results should be viewed as suggestive instead of confirmatory. It should be apparent that this effort requires extensive input from the medical personnel. In the author's opinion, this effort represents a joint adventure between the medical establishment and their statistical colleagues and should be pursued with urgency and sincerity.

As we said above, results from the benefit/risk assessment outlined in Section 1.4 are suggestive instead of confirmatory. Even though we used p-value in Section 1.4 to illustrate the impact of the discounting factor f on the comparison, the p-value only serves as a reference point and should not be regarded as pertaining to rigorous hypothesis testing. Another reason for the exploratory nature of the p-value in this context is the fact that phase II and III trials are typically designed for efficacy evaluation. In such trials, safety data are collected as secondary end points. In fact, the collection of safety data continues beyond the market entry of a new drug and it is not uncommon that the safety profile of a new drug was not yet fully characterized at the time of market entry. Therefore, it is hard to prespecify hypotheses related to the safety profiles for confirmatory testing in phase II or III trials. This is different from the efficacy end points, which need to be rigorously documented and validated prior to drug approval under the traditional drug approval process.

Despite what we said above, phase III data offer us the best chance to evaluate a drug's safety profile and therefore an opportunity to conduct a meaningful benefit/risk assessment prior to the drug's market entry. Understandably we might still get surprises in terms of adverse events at the phase III stage. However, using this to preclude the conduct of a benefit/risk assessment at this stage is ludicrous. The truth is that we might never know the full spectrum of the safety issues of a drug until many years after the drug has been widely consumed by the public as in the DES case. To say that we therefore are prevented from conducting a meaningful benefit/risk assessment is like an ostrich burying its head in the sand. We need to realize that evaluating the value of a new drug is an ongoing and evolving process. As more information becomes available, we need to update our safety database as well as the benefit/risk assessment of the drug. We need to fully utilize whatever information that is available to us to serve the public. We need to be flexible and fluid in our evaluation and not get bogged down by the semantics of hypothesis testing. Strictly bound by the requirement for rigorous hypothesis testing is not only rigid but throwing away precious information. One ought to remember the lamppost phenomenon. While it is comforting to seek the refuge of things that are of high technicality, it might be more beneficial to society if we do not lose sight of the human aspects of medicine.

There is an increasing emphasis on conducting a pharmacoeconomic evaluation of a new drug to decide the worthiness of a drug in terms of its cost and benefit. While the discussion in this chapter focuses on the benefit and risk of a drug with risk defined solely by adverse reactions to the drug, the concept of the drug's ultimate *"worthiness"* and *"value"* is nevertheless the driving force. The requirement to come up with one summary figure to balance risk against benefit forces one to think hard about the mechanism to balance these two opposing forces and attach a measure of worth to these forces. The same mechanism is employed to varying degree in many other facets of our lives such as the purchase of a car or a house.

Benefit/risk assessment, in the current regulatory atmosphere, remains largely a concept rather than a routine practice. Even though everyone agrees on the need to conduct such an assessment, few can come up with the specifics to actually carry it out. Therefore, even though Miller[2] cited the lack of attention paid to the risk–benefit issues as one pitfall that frequently delays the drug review and approval process, pharmaceutical sponsors rarely receive any guidance from the regulatory agencies regarding the basics and contents of a pertinent benefit/risk assessment package. As a result, benefit/risk assessment remains largely an ad hoc process and has become only a lip service in many situations.

The bottom line is: treatments are being selected for patients every day after *weighing the risks against the benefits* among the available options. What we need is to find a way to formulate this decision-making process so that all factors of concerns are included in the equation. The latter ultimately has to rely on numerically balancing risk against benefit.

References

1. Sakuma, A., Subjectivity in clinical medicine, presented at The Drug Information Association Second Annual Biostatistics Meeting in Tokyo, August 30, 1995.
2. Miller, L. L., Pitfalls in the drug approval process: dose-effect, experimental design, and risk–benefit issues, *Drug Information Journal*, 26, 251, 1992.
3. Chuang-Stein, C., A new proposal for benefit-less-risk analysis in clinical trials, *Controlled Clinical Trials*, 15, 30, 1994.
4. F-D-C Reports, "The Pink Sheet," T&G 4–5, August 14 1995.
5. F-D-C Reports, "The Pink Sheet," T&G 8–9, July 31 1995.
6. Payne, J. T., Loken, M. K., A survey of the benefits and risks in the practice of radiology. *CRC Critical Reviews in Clinical Radiology and Nuclear Medicine*, 6, 425, 1975.
7. Chang, R. W., Fineberg, H. V., Risk-benefit considerations in the management of polymyalgia rheumatica, *Medical Decision Making*, 3, 459, 1983.
8. Andrews, J. R., Benefit, risk and optimization by ROC analysis in cancer radiotherapy, *International Journal of Radiation Oncology, Biology and Physics*, 11, 1557, 1985.
9. Sandberg, S. I., Barnes, B. A., Weinstein, M. C., Braun, P., Elective hysterectomy. Benefits, risks, and costs, *Medical Care*, 23, 1067, 1985.
10. Weinstein, M. C., Estrogen use in postmenopausal women — costs, risks, and benefits, *New England Journal of Medicine*, 303, 308, 1980.
11. Weinstein, M. C., Schiff, I., Cost-effectiveness of hormone replacement therapy in the menopause, *Obstetrical and Gynecological Survey*, 38, 445, 1983.
12. F-D-C Reports, "The Pink Sheet," T&G 6–7, September 11 1995.
13. Hilden, J., Reporting clinical trials from the viewpoint of a patient's choice of treatment, *Statistics in Medicine*, 6, 745, 1987.
14. Chuang-Stein, C., Mohberg, N. R., Sinkula, M. S., Three measures for simultaneously evaluating benefits and risks using categorical data from clinical trials, *Statistics in Medicine*, 10, 1349, 1991.
15. Chuang-Stein, C., Mohberg, N. R., Musselman, D. M., Organization and analysis of safety data using a multivariate approach, *Statistics in Medicine*, 11, 1075, 1992.

2

Prediction of the Benefit/Risk Ratio from *In Vitro* Data

Ravi Shrivastava
Vitro-Bio

Fabien Durand
Vitro-Bio

2.1   Introduction

Assessment of the benefit/risk ratio for drugs and chemicals from *in vitro* data is currently receiving unprecedented attention. Increasingly tough legislation and the consequent development cost increases associated with social and political demands to reduce *in vivo* animal experiments constitute the main incentives for food and drugs, chemical, and cosmetics industries to actively search for suitable *in vitro* procedures to evaluate the activity and toxicity of their molecules.[1]

In this review, an overview will be given of the current status and evolving trends in the use of *in vitro* tests in assessing the safety of drugs and chemicals, the currently available means of predicting the benefit/risk ratio *in vitro*, and the validation and acceptance by regulatory bodies of such methods. Emphasis will be placed on the most rapidly evolving areas in benefit/risk assessment *in vitro*.

2.2   Current Status of *In Vitro* Tests in Safety Assessment

2.2.1   Recent Developments in Safety Assessment of Drugs and Chemicals

Over the last decade, lots of resources, time, and energy have been spent in the search for suitable alternatives to animal testing in all sectors of biomedical research and industry.[2] Today, progress in cell culture technology and biotechnology have broadened the list of potential alternatives to *in vivo* testing.

In 1981, the European Council of Ministers of the European Economic Community (EEC) voted a sixth amendment to the classification, packaging, and labeling of Dangerous Substances Directive

0-8493-2791-1/99/$0.00+$.50
© 1999 by CRC Press LLC

(67/548/EEC), which forced all the manufacturers and importers of new chemicals within the EEC to provide full details of safety data, which include toxicological, ecotoxicological, and physicochemical data before launching any novel substances into the common market for sale or use. Although these safety data requirements are not as strict as the requirements for pharmaceuticals, pesticides, or food additives, they still oblige industry to provide results in accord with the protocols of the Organization for Economic Cooperation and Development (OECD).

The European Inventory of Existing Chemical Substances (EINECS) and the European Core Inventories (ECOIN) consist of approximately 100,000 entries, probably representing 50,000 to 60,000 chemical substances. It has been claimed that among these chemicals, only 8% meet current OECD guideline protocol requirements for safety assessment and that 45% of these studies should be repeated.[3] If such claims were correct, it would require approximatively 600,000 new studies in order to bring the current population of chemicals in line with current regulatory guidelines. The cost of conducting safety studies is estimated at US $150,000 for a minimum package and US $10,000,000 to register a drug, pesticide, or food additive. To fulfill the above recommendations would cost between US$ 5 to 50 billion.

The regulatory requirements for the assessment of chemical safety that industry is now obliged to provide fall broadly into four categories: (1) the classification, packaging, and labeling requirements for transportation and occupational health control, (2) the registration or reregistration of a new drug, food additive, or pesticide, (3) reassessment of a previously developed product for its continued sale and use, and (4) new formulations of known products.

The dossiers required for registering each type of product are clearly defined in the various tests of the national and international guidelines. The majority of safety data requirements are based on *in vivo* studies in experimental animals such as rats, dogs, rabbits, mice, guinea pigs, pigs, and birds.[4] In the early 1980s the Organization for Economic Cooperation and Development (OECD) issued guidelines with regard to the experimental design of these studies in an effort to harmonize what had become standard study types. Rigid adherence to these guidelines and the scientific validity of these studies are often criticized due to the quite large numbers of animals required on one hand and the quantity and quality of scientific information collected on the other.

An analysis of OECD guidelines would indicate that over 2300 animals are required to provide safety data to register a drug (Table 2.1). In reality, these numbers represent a strict minimum, because more than one route of administration or exposure is usually necessary, since additional dose finding studies are often performed.[5]

These guidelines are not confined to rodents or dogs; studies on other mammals may also be required, especially in assessing putative environmental contaminants.

Safety studies in industry are primarily focused on the use of laboratory bred rodents as surrogates for man. Large numbers of animals are incorporated in each group and in each study due to biological animal to animal variation. More groups in a study and a greater range of exposure levels increase the statistical probability of detecting an effect if it exists. A step sequence or tier approach to evaluate safety is used in OECD guidelines considering that short-term studies, such as acute toxicity using few animals, should precede longer-term studies, such as subchronic, chronic, or oncogenicity studies. Each preceding study provides data on which the subsequent more costly and extensive studies can be appropriately designed.

As a result of progress in *in vitro* technology and the increasing political and scientific demand for the reduction, refinement, and replacement of animal experimentation to evaluate the benefit/risk of chemicals, *in vitro* studies are now undergoing a period of rapid growth and development.

The concept of alternative technique was first developed in the beginning of the 19th century with the aim to minimize pain and suffering and reduce unnecessary animal experimentation.

More recently, in 1978, the topic of *in vitro* testing came to the forefront of public conscience, when a campaign directed toward the removal of the Draize test from cosmetic testing was organized by animal rights advocates. The Draize test was developed in 1944 to evaluate ophthalmic safety of chemicals. Other tests such as LD_{50} and the phenol coefficient test for skin damage were developed in the 1930s.[6] The

TABLE 2.1   Average Minimum Number of Rodents and Dogs Required to Perform *In Vivo* Investigations in Order to Register a Drug According to the OECD Protocols

	Type of Study	Rodents/Rabbits	Dogs
Acute effects	LD$_{50}$, MTD	60	10
	(Maximum tolerated dose)		
	Irritancy	6	
	Ocular irritancy	3	
	Sensitization	20	
Subchronic	28 day subacute	80	38
	90 day subacute	160	
Chronic and oncogenicity	12 month	320	48
	18 month mouse	400	
	24 month rat	400	
Development and reproductive	Rat	160	
	Rabbit	96	
	Segment	320	
Genetic toxicity	Micronucleus	45	
	Cytogentics	45	
Toxicokinetics	Metabolism	24	
	Distribution	120	
	SUBTOTAL	2259	134
	TOTAL	2393	

reliability and the reproducibility of the Draize test, and perhaps the lack of other alternative tests have brought it to widespread use in the past five decades in ocular irritancy testing in the cosmetics industry.[7]

Although animal rights advocates accelerated the movement to seek alternative testing, unifying concern for the welfare of animals is a major element that continues to broaden the commitment of industry and academia to seek suitable *in vitro* tests.[8]

2.2.2   The 3 R's Concept

At the University Federation for Animal Welfare's 1957 symposium on Humane Techniques in the Laboratory (UFAW, 1957) the concept of the three R's (reduction, refinement, and replacement) as a means of removing inhumanity from animal experimentation was first discussed in depth.[9] The three R's approach to alternative testing encompasses the entire spectrum of available possibilities. The most important point the concept makes is that alternative testing can be accomplished by means other than replacing *in vivo* tests with *in vitro* tests. It indicates that the goal of alternative testing is to minimize the pain and suffering of research animals (reduction) by decreasing the number of animals used, as well as the severity and replications of tests required (refinement). The ideal but optimistic alternative is the implementation of a validated nonanimal test that is of equal scientific value as its predecessor (replacement).

Russell and Burch[10] distinguished between relative replacement, in which animals would still be required but would not be exposed to any distress in the actual experiment, and absolute replacement, in which animals would not be required at any stage at all. They regarded nonrecovery experiments on animals under anesthesia and the use of spinal and decerebrated animals, i.e., animals whose central nervous systems had reliably been made insensitive, as examples for relative replacement. They also discussed humane killing of animals to provide cells, tissues, and organs for *in vitro* studies. Tissue culture involving human and nonvertebrate tissue was regarded as a bridge between relative and absolute replacement.

The 1960s saw great progress in relation to industrial and academic expansion of scientific activities, and exciting developments in molecular and cell biology led to nonanimal techniques as fundamental to progress in the biomedical sciences. At the same time, however, expectations of greater safety for human beings and demands for greater protection of the environment gradually led to a dramatic expansion of routine safety in animals, including the introduction of new testing requirements every time a new problem was identified. For various reasons, this equilibrated compromise in the debate on animal experimentation between total abolition on one hand and scientific liberty on the other steadily gained acceptance.

This led to the development of many government associated or free associations, such as:

- FRAME (Fund for the Replacement of Animals in Medical Experiments) in 1979 in Britain.[11]
- CAAT (The Johns Hopkins Center for Alternative to Animal Testing) in 1982 in U.S.A.[12]
- MEIC (Multicenter Evaluation of *In vitro* Cytotoxicity) in 1989 in Sweden.[13]
- PAEXA (Pour l'Alternative à l'Expérimentation Animal) in 1989 in France.
- ECVAM (European Center for the Validation of Alternative Methods) in 1991 in Italy.

In line with the recommendations of the council of Europe Convention for the Protection of Vertebrate Animals used for experimental and other scientific purposes, the council of ministers of the EC had meanwhile adopted Directive 86/609/EEC, which also has a three R's basis, suggesting the following major clauses:[9,14]

- An experiment shall not be performed if another scientifically satisfactory method of obtaining the results sought, not entailing the use of an animal, is reasonably and practicably available.
- When an experiment has to be performed, the choice of species shall be carefully considered and, where necessary, explained to the authority. In a choice between experiments, those which use the minimum number of animals, involve animals with lower degree of neurophysiological sensivity, cause the least pain, suffering, distress or lasting harm, and which are most likely to provide satisfactory results, shall be selected.
- All experiments shall be designed to avoid distress and unnecessary pain and suffering to the experimental animals.
- Where it is planned to subject an animal to an experiment in which it will, or may, experience pain which is likely to be prolonged, that experiment must be specifically declared and justified to, or specifically authorized by the authority. The authority shall take appropriate judicial or administrative action, if it is not satisfied that the experiment is of sufficient importance for meeting the essential needs of man or animal.

2.3 Available Means of Predicting Benefit/Risk Ratio from *In Vitro* Data

Evaluating benefit/risk ratios on the basis of *in vitro* data invariably involves the desired pharmacological activity on one hand and toxicity on the other. Traditionally benefit/risk assessments are based on data obtained *in vivo* or from complex systems, and despite various incentives to do so industry is slow in adopting *in vitro* procedures. Indeed it is not yet commonplace to read literature reports dealing with pharmacology and toxicology data *in vitro* in the same manuscript.

2.3.1 Procedures in Assessing Pharmacological Activity

Isolated organ experiments have been in use for over a century and play a key role in forming the basis of the receptor theory in modern pharmacology. They will not be discussed here. Since the adaptation of radioimmunoassay technology to membrane receptors in the early 1970s, the affinity of drugs and chemicals for a wide variety of pharmacological targets, including receptors, ion channels, enzymes, and other proteins can be assessed easily, rapidly, and economically *in vitro*, using cultured cells, cellular homogenates or particulate fractions. Cultured cells are widely used to assess cellular mechanisms of toxicity[15,16] and to understand drug interactions.[17] The advent of gene technology over the last decade has opened new doors in encoding human receptors or other proteins, and expressing the gene encoding for the protein in question in a stable and reliable manner in host cells. This means that drug/chemical affinity and agonist/antagonist or partial agonist activity can now be commonly detected *in vitro* in cells transfected with the gene encoding for the human protein in question. This technology has nevertheless raised several issues. The first is the fact that receptor diversity within a given species is far greater than might have been predicted previously, and in addition that significant pharmacological variability may occur between species, indicating that studies in nonhumans may be misleading. The second is the two-state model of receptor theory[18-20] in which stably expressed receptors following gene transfection, either *in vitro* in host cells, or in transgenic animals *in vivo*, tend to be "overexpressed" and may sometimes exhibit constitutive activity, i.e., activate G proteins and second messengers in the absence of agonist stimulation, which does not generally appear to be the case for naturally expressed receptors.

The third and most important issue is *in vivo* safety assessment of drugs or chemicals with affinity for a given human receptor which is not necessarily expressed in laboratory animals, or possesses slightly different amino acid sequences, leading to reduced or increased affinity at the receptor/protein homolog expressed naturally in laboratory animals. It may be agreed, however, that safety for a desired human receptor/protein that is not expressed or is present in small proportions in laboratory animals is still a valid approach in evaluating the toxicity of the drug/chemical in question, which is not mediated by the targeted protein (i.e., toxicity separated from extended pharmacological effects). Under such circumstances, *in vitro* benefit/risk assessments can usually be performed easily by evaluating both pharmacological (affinity, intrinsic activity, and antagonism) and toxicological action within the same cell population, which may or may not be stably transfected with a human gene product. A trend in this direction is already underway in several major pharmaceutical companies worldwide.

2.3.2 Tissue Culture Technology

In the general sense, tissue culture includes all those studies in which plant or animal cells, tissues, or organs are maintained in a viable state outside the donor for long periods of time (generally more than 24 h). Upon initial observation, it appears that this is not an alternative to animal testing, but merely a technique of animal testing. Tissue culture, however, is considered an alternative to animal testing because it reduces the pain and suffering experienced by the animals required for an experiment. Typically, tissues used in a tissue culture experiment are obtained from relatively few animals, while the quantity of results obtained from the study using tissue cultured from one animal can be substantial compared to that obtained by conducting an *in vivo* study using many animals. The obvious limitation of cell culture techniques is the absence of neurohormonal mechanisms, hence the whole body response is absent and therefore extrapolation with the *in vivo* situation is difficult.

The metabolic and growth response of a given tissue is solely dependent on the *in vitro* environment. Tissue is maintained *in vitro* by media that contain the nutrients necessary for cell maintenance or growth.

Primary cell culture is obtained from the tissue of freshly sacrificed animals. The organ of interest is then dissociated by chemical or mechanical means into single cells or cell clusters. The cells are then plated onto a synthetic or natural matrix and allowed to grow for a variable period of time. The type of cells and the culture environment in which they are maintained is well defined, thus allowing identical test conditions for all the chemicals tested or experiments performed. A tissue culture plate contains multiple chambers of test unit; therefore the control and the test chambers are treated uniformly prior to experimentation. Individual culture plates are homogeneous, since the tissue from the animal was combined prior to dissociation. This reduces the variance between test units compared to the variance obtained using individual animals as the test unit. This makes cell culture systems suitable for standardization because the reagent, culture media, and the test conditions employed are usually standardized for different cell types. The use of sera from different animals and culture handling techniques may differ from laboratory to laboratory, thereby producing variability. This problem can be solved by the use of artificial serum and by using a standard protocol for each type of test.

Cell lines are similar to primary cultures, apart from the fact that cells undergo unlimited cell division. The viability of primary cell cultures is short term (days to weeks), whereas the continuous cell lines can remain alive for longer periods (months). Very often these cells are derived from a tumor cell. These hybrid cells possess the characteristics of the original cell of interest, with the additional ability of unlimited division *in vitro*. The use of cell line cultures has a significant effect on validation, since test to test variability can be minimized.

The correlation between *in vitro and in vivo* data is meaningful because living tissue is employed in each case. The possibilities for the use of human cells is another advantage, as the results obtained *in vitro* can be extrapolated and compared to the effects upon human cells.

Due to these advantages of cell culture technology, primary and cell line cultures are very often used as an *in vitro* system of choice for the assessment of the benefit/risk ratios of a wide range of chemicals and biologicals. Considerable effort has been made to demonstrate a positive correlation between *in vitro* results obtained using cell cultures and the real results obtained *in vivo* in animals or humans.

2.3.3 *In Vitro* Safety Evaluation

2.3.3.1 The LD_{50} Test

Since its introduction in 1927 by Trevan et al.,[6] the LD_{50} test in rats and mice has been widely used to evaluate the acute toxicity of a variety of substances, including drugs, pesticides, industrial chemicals, cosmetics, and food additives. The scientific validity and usefulness of the LD_{50} test was increasingly criticized as an unnecessary waste of resources, particularly with respect to the large number of animals employed and for the reason that the lethal dose of a substance can be determined fairly accurately by other methods that require fewer animals.[21]

Over the past several decades the standard practice of determining LD_{50} doses with 95% confidence limits is now widely criticized by scientists and animal welfare groups alike.[22] This is not surprising, since the LD_{50} of a given chemical often varies at least three- to tenfold between different animal species and strains, and also according to the mode of compound administered, environmental conditions and the statistical method employed to calculate the LD_{50} value.[21] Therefore, the use of this measure as a general index of toxicity, as a guide to additional toxicological studies, or as a guide to human toxicity is inaccurate and involves extensive use of animals.

Many proposals have been made to reduce the number of animals necessary for this test. For example, the use of a fixed dose method[22,23] or even the complete abolition of this test have been proposed if the toxicity of a test chemical can be estimated using alternative *in vitro* methods.[24]

A positive correlation between *in vitro* cytotoxicity and LD_{50} in rats and mice has been shown.[25,26] Attempts have been made to measure the concentration of drugs in human plasma capable of inducing unwanted side effects and to compare these with the 50% cytotoxic concentrations of the same drugs *in vitro*.[27] While establishing an *in vitro–in vivo* correlation, however, the question regarding the validity of these results to reveal human toxicity is not yet fully answered, apart from initial efforts made by Helberg et al.,[28] and Barile et al.,[29] who showed a positive correlation between *in vitro* cytotoxicity and the human acute lethal doses for the first ten chemicals in the Multicenter Evaluation of *In Vitro* Cytotoxicity study (MEIC) list.

The MEIC program has established a list of 50 reference chemicals, and currently more than 140 laboratories worldwide have studied the toxicity of these compounds using their own methods and techniques. The MEIC organization will assess and compare these results from different laboratories with the aim of replacing the classical LD_{50} test and, more than that, comparing the *in vitro* toxicity values with human plasma toxic concentrations of the same substance. The aim is to validate a reliable *in vitro* model or models provided that the predictive validity of results from this study can be demonstrated.[30] We have analyzed the cytotoxicity of 48 of the 50 MEIC selected chemicals in several different cell types.

In vivo LD_{50} in rats and mice results were obtained from the literature and represented the median toxic value for a compound after single oral administration in rats and mice. For *in vitro* studies, rat hepatocyte primary cultures, MDBK (Madin-Darby Bovine Kidney) and McCOY (human epithelial) cell lines were exposed to different concentrations of the test chemicals.

Cell growth, cell morphology, cell viability, and lactate dehydrogenase (LDH) release were measured at each time point. The minimum test compound concentration inducing morphometric changes in

50% cells or 50% of cell death and/or 50% to 100% increase in hepatocyte LDH release compared to control values was determined. The cytotoxicity values obtained for each of three parameters were then averaged and rounded to give the final CT_{50} (50% cytotoxicity) concentration in each cell type for each test compound. Similarly, the minimum test compound concentration inducing marked morphometric changes or >50% cell death along with >100% increases in hepatocyte LDH release compared to control cultures was also determined. Cytotoxicity values were averaged as indicated above to obtain the final CT_{100} (100% cytotoxicity) concentration.

The results obtained by Shrivastava et al.[24] showed a better than 75% correlation between *in vitro* cytotoxicity with the actual *in vivo* LD_{50} values (R = 0.77 to 0.83), indicating that the type of cells used has little or no influence on *in vitro–in vivo* toxicity correlations. In general, the concentration of test compound required to produce 50% to 100% cytotoxicity was less for the hepatocyte cultures than in the McCOY or MDBK cells. This difference was related probably to the fact that hepatocytes contain a plasma membrane that is more fragile to pressure forces than McCOY or MDBK cells or to the fact that hepatocytes have the capacity to metabolize compounds.

A few chemicals were found to be less toxic *in vitro* compared to the *in vivo* LD_{50} estimation. This phenomenon was observed particularly for compounds that did not dissolve well in the culture medium or formed a medium-solvent partition (e.g., alcohols) so that direct contact of the compound with the cells in culture was impaired. This could explain why the cytotoxicity values obtained with such compounds were lower than their potential toxicity *in vivo*. For such chemicals a better *in vitro–in vivo* correlation may be obtained using suspension cultures under agitation to allow direct contact between cells and compound under investigation, although this remains to be proved.

The LD_{50} test *in vivo* generally requires a minimum of 20 rats or mice, which represents about 1000 animals to evaluate *in vivo* LD_{50} values for the 48 test compounds used in this study. On the other hand, hepatocyte cultures prepared from 10 to 20 rats are sufficient to screen the same number of compounds *in vitro*.

These results demonstrate that accurate *in vivo* LD_{50} doses in rats and mice could be predicted for at least 75% of the selected chemicals using cell culture techniques. These results strengthen the initial concept of the MEIC program, namely that *in vitro* cytotoxicity tests can be used to predict approximate *in vivo* LD_{50} values and that a correlation exists between *in vitro* cytotoxicity and acute human toxicity.

2.3.3.2 The Maximum Tolerated Dose (MTD) Test

Classic *in vivo* toxicity tests that involve short-term and long-term administration of compounds to different animal species constitute the major and most important part of toxicological work. For pharmaceuticals, toxicological evaluation in a 4-week MTD study in both a rodent and nonrodent species is currently recommended by EEC countries prior to a 1-week test substance exposure to human beings in phase I clinical trials. Very often, dogs and rats are used for this purpose, but the reliability and necessity of such MTD tests remains an integral part of the ethical debate in toxicology circles.[31] For this reason, a simple *in vitro* test and a method to correlate *in vitro* with *in vivo* toxicity data, even if this could only be used to predict an approximate *in vivo* toxic dose, would provide an extremely useful tool for reducing the number of animals used and the animal suffering caused during MTD experiments.

The reason for using mammalian cells *in vitro* to evaluate toxicity is based on the fact that chemical toxicity *in vivo* ultimately occurs at the cellular level, and can consequently be studied *in vitro*. For this reason, Shrivastava et al.[32] initially attempted to correlate *in vivo* MTD doses for certain development drug candidates with *in vitro* data, and a working hypothesis was established that permitted this extrapolation. The results reported here were subsequent to a single blind trial in order to validate or refute the initial hypothesis.

The MTD of 25 orally well-absorbed compounds with a range of pharmacological and toxicological profiles had already been determined in both rats and dogs. The compounds were coded and then tested blind to the investigator(s) in several different types of carefully selected, relatively "sensitive" and relatively "resistant" cells in culture. The *in vitro* results were then compared with *in vivo* data on the basis of the proposed hypothesis.

The *in vivo* experiments were performed using four male and four female dogs or 24 male and 24 female rats divided in four equal groups. The first group of animals received gelatin (0.5% w/v) and mannitol (5.0% w/v) suspension vehicle, and served as controls. Depending upon the pharmacological class of the drug, animals in group 2 received daily oral administration of the test compound suspended in gelatin–mannitol for a period of 4 weeks. On the basis of the initial toxicity results observed in group 2, the animals in group 3 received either a higher or lower dose, 1 week after the start of dosing in group 2. Similarly, the animals in group 4 were given a higher or lower dose, 1 week after the start of dosing in group 3. An interval of 1 week for dosing in the last two groups served to adjust the dose according to the initial toxicity results observed in the other groups.

All animals were observed for the manifestation of clinical signs throughout the dosing period. Physical, hematological, and blood biochemical examinations were performed before the start and at the end of the dosing period in all animals. Food consumption and body weight were determined twice a week during the study. All animals were killed using an overdose of an anesthetic agent at the end of the 4-week dosing period, for determination of gross abnormalities and organ weight changes. The potential target organs were fixed in formaldehyde for histopathological examination.

A drug dose that induced any change compared to vehicle-treated animals in one or more of the above-mentioned parameters was considered as the threshold toxic dose, while a dose that caused moderate to marked clinical, physical, hematological, biochemical, organ weight, or histological changes was considered to be the toxic dose. The *in vitro* tests were performed in a similar manner to that for the LD_{50} test described previously.

A working hypothesis was established to correlate *in vitro* with *in vivo* toxicity as there exists no established method of extrapolating *in vitro* toxicity findings to predict an *in vivo* toxic dose. This is hardly surprising, since there are multiple factors that must be taken into account, e.g., intraspecies and interspecies differences, drug or metabolite related toxicity, route and duration of drug administration, pharmacological effects. Indeed, biological responsiveness is known to be variable and often requires population analysis to gain even an approximate idea of what the overall species response to a particular compound might be. Bearing in mind these limitations, however, a hypothesis was proposed to predict the *in vivo* MTD using the *in vitro* data in the following manner:

- Correlating *in vitro* drug concentration and *in vivo* toxic dose: In *in vitro* experimentation, the concentration of test compound remains practically the same in the culture medium throughout the experimental time period. In whole-animal experimentation, however, there is continous metabolism and excretion of compounds and a need for drug plasma levels to be maintained by repeated dosing at regular intervals depending on their pharmacokinetic profiles. *In vitro*, cells are directly exposed to the drug and no absorption or diffusion barriers exist, as is the case *in vivo*. Given these marked differences and, as an initial formula to convert the *in vivo* toxic concentration to an *in vivo* toxic dose, the concentration in µg/ml (or mg/l) *in vitro* was taken and empirically expressed as mg/kg of body weight/day.
- Correlating animal species with different types of cells from different organs and species: *In vivo* toxicological studies have demonstrated that a large variation exists in the sensitivity of different animal species to a particular compound. It is known that dogs are more sensitive than rats in this respect. Cultured cells originating from different species and also from different organs of the same species show differences in the susceptibility to the toxic effects of test compounds. After examining a wide range of the cell types, it was observed that rat hepatocytes are more sensitive than MDBK cells, whereas McCOY cells exhibit intermediate sensivity. Since the interspecies variability in response to drugs represents the extremes of variability observed within a species, it was simply considered that hepatocytes might have comparable sensivity to most sensitive species *in vivo* in the dog, whereas MDBK cells might exhibit a sensitivity comparable to that observed in the rat. McCOY cells, having an intermediate sensitivity, served as *in vitro* control cell line.
- Drug-related or metabolite-related toxicity: When a compound is found to be toxic *in vivo*, a simple procedure was carried out *in vitro* to determine whether the toxicity is drug- or metabolite-related, provided that the present drug is metabolized in the liver to give the active toxic molecule.

TABLE 2.2 *In vivo* threshold toxic doses and *in vivo* mild to moderate oral toxic doses (mg/kg/day) of compounds A to Y studied in four-week maximum tolerated dose studies in rats and dogs, and the corresponding *in vitro* predictions (mg/kg/day) determined in MDBK cells for rats and in hepatocytes for dogs.

Test Article Code	Rats				Dogs			
	Threshold Toxic Dose		Mild-Moderate Toxic Dose		Threshold Toxic Dose		Mild-Moderate Toxic Dose	
	In Vivo (mg/kg/d)	Predicted Dose (mg/kg/d)	*In Vivo* (mg/kg/d)	Predicted Dose (mg/kg/d)	*In Vivo* (mg/kg/d)	Predicted Dose (mg/kg/d)	*In Vivo* (mg/kg/d)	Predicted Dose (mg/kg/d)
A	50	300	200	400	85	250	185	400
B	125	100	400	400	65	50	200	200
C	25	10	50	25	10	5	20	10
D	125	150	250	300	50	40	100	100
E	50	40	100	80	10	15	25	30
F	75	60	100	80	25	20	50	40
G	75	50	100	100	10	8	15	20
H	10	10	20	15	6	10	12	15
I	15	10	30	25	4	3	14	10
J	15	20	35	50	5	5	20	10
K	125	100	200	200	30	75	70	100
L	100	80	80	100	80	80	100	100
M	800	500	2000	2000	250	800	350	2000
N	—	30	100	75	20	10	30	20
O	100	100	150	200	15	15	30	25
P	75	80	125	100	80	80	120	100
Q	80	100	300	200	40	50	150	100
R	—	—	—	—	10	10	20	20
S	25	15	50	25	50	40	50	60
T	25	30	50	50	10	10	20	20
U	20	15	25	20	6	6	12	12
V	25	15	25	30	8	10	15	20
W	75	100	100	100	50	70	100	100
X	25	15	50	30	15	15	20	25
Y	100	100	250	200	50	40	80	100

(— = Results not available)

At first, the parent drug was exposed at different concentrations to isolated rat hepatocytes and MDBK cells, and CT_{50} and CT_{100} concentrations were determined in each case. Since the MDBK cells are generally more resistant than rat hepatocytes, a drug concentration that was toxic in hepatocytes, but nontoxic in MDBK cells, was found. Since hepatocytes may metabolize the compound, after 24 hours of drug exposure the supernatant from hepatocytes that was toxic to the latter cells was transferred to fresh MDBK cells and studied for 72 hours. If cytotoxicity occurred in MDBK cells exposed to the supernatant, it was most probable that a metabolite(s) in addition to the parent drug was responsible for the toxicity. On the other hand, if no toxicity was found, then it was most likely that toxicity was due to the parent drug.

The results concerning the *in vivo* threshold toxic and mild to moderately toxic doses of compounds A to Y determined in standard four-week MTD studies in rats and dogs are given in Table 2.2, along with the corresponding *in vitro* cytotoxic concentrations.

Employing this hypothesis, the findings of the present study demonstrated a good correlation, for the large majority (>80%) of compounds tested, between the anticipated *in vivo* toxic drug doses that were derived from the *in vitro* experiments and the actual toxic doses found in the *in vivo* MTD studies.

Out of 25 compounds studied, however, the predictions of the MTD *in vivo* in rats and dogs for two compounds (A and M) were wrong for one or both species. This may be because compound A was found to be toxic *in vivo* as a result of some biochemical changes linked with no apparent clinical abnormalities in the *in vitro* tests. Compound M, on the other hand, was found to be toxic *in vivo*, probably as a result of an extension of its pharmacologic effects.

The initial test in standard safety testing of an ethical pharmaceutical product is the MTD in rats and dogs in which the threshold and toxic doses of compounds are derived. Although the number of animals required for this test is small, it is one of the major animal consuming tests in the entire range of safety studies, as it is the initial experiment of absolute necessity. Being the first safety test with an unknown compound, it may also prove to be the most poorly tolerated by animals compared to successive experiments performed in the light of these initial toxicity results.

Therefore, an *in vitro* test implemented early in drug development, which could predict even approximate *in vivo* MTD doses, would help to eliminate relatively toxic compounds before starting *in vivo* toxicology testing and/or to select drug doses that would be better tolerated by the study animals.

2.3.3.3   *In Vitro* Models for Activity Evaluation

As described previously, many *in vitro* models are already in use for the evaluation of pharmacological activity in different fields. Therefore, it is practically impossible to describe all existing *in vitro* models in pharmacology.

As cardiovascular diseases are the principal cause of concern because they account approximately for half the deaths occurring in industralized nations of the world, we would nevertheless like to give an example of an *in vitro* model recently created in our laboratory for basic research in atherosclerosis.

The accumulation of circulating cholesterol and cholesterol esters in arterial wall cells is an early event in atherogenesis. Platelet- and leukocyte-derived smooth muscle cell (SMC) growth factors are thought to promote medial SMC migration and subsequent proliferation within the intima, but the cause of these events is still not completely understood.[33] Neointima formation related to SMC proliferation is an important and perhaps the primary lesion in the progression of atherosclerotic plaque.[34] Such changes may also be responsible for stenosis in coronary artery bypass grafts and in postangioplasty for a relatively high rate of restenosis.[35]

Currently available *in vivo* models of atherosclerosis are time-consuming, involving feeding of a diet enriched in cholesterol or saturated fat to produce hypercholesterolemia or removing the endothelium of large arteries by a balloon catheter to stimulate SMC proliferation.[36] Neither model is satisfactory, since in rats, which are a commonly used species for atherogenicity studies, most of the cholesterol is transported on high density lipoproteins (HDL). The rabbit, an herbivore with no cholesterol in its natural diet, is unsatisfactory in the cholesterol-enriched diet model because of the great individual variation in response to the cholesterol and because the lesions resemble the fatty streaks seen only in the initial stages of atherosclerosis in humans. In addition, rabbits also require 8 to 10 weeks of 1% to 2% lipid diet to generate sufficient aortic lesions, which themselves appear to be influenced in a seasonal manner.[37]

Although several *in vitro* models of atherosclerosis exist,[38] they are not routinely used in atherosclerosis research mainly because they involve use of human cells from clinical cases providing no homogeneity of the samples collected and, consequently, of the results obtained. Therefore, a new *in vitro* model was created by modifying the existing methods for the evaluation of the effects of new chemical entities on SMC proliferation along with their protective and/or curative effects on intracellular lipid accumulation.[39]

Rat aortic smooth muscle cells were isolated by enzymatic digestion. SMCs were seeded in 24 well tissue culture plates. These cells were exposed to hyperlipidemic rabbit serum (HLRS, 30 μm) either 24 h before exposure to various concentrations of a test compound (curative effects) or simultaneously with the test compounds (preventive effects).

The antiproliferative and the intracellular lipid lowering effects were measured for different test compound concentrations. All observations were made 6 days after test compound exposure. Intracellular lipid accumulation was measured by fixing the cells with Bouin solution followed by Oil Red-O staining. The amount of lipid (stained red) was measured by image analysis, and the mean difference in intracellular lipid compared to controls was determined.

Cell proliferation was measured by counting the total number of cells/well and by quantifying the total DNA content/well using an automated fluorescence multiwell plate scanner (Cytofluor 2300, Millipore, USA). The mean change in these two parameters compared to untreated controls was calculated.

TABLE 2.3   Effects of test compounds on SMC proliferation and their preventive and/or curative effects on intracellular lipid accumulation compared to control (*p<0.05)

Test Compound	Maximum Non-Cytotoxic Concentration (M)	Concentration Tested (M)	% Change in Cell Proliferation	% Decrease in Lipid Accumulation Preventive Effect	Curative Effect
Verapamil	10^{-5}	10^{-5}	$-17 \pm 4^*$	2 ± 2	0
		3×10^{-6}	$+1 \pm 2$	2 ± 2	0
		10^{-6}	$+9 \pm 2^*$	3 ± 3	0
Diltiazem	3×10^{-5}	3×10^{-5}	$-47 \pm 7^*$	0	0
		10^{-5}	$-19 \pm 3^*$	6 ± 3	5 ± 3
		3×10^{-6}	-1 ± 2	0	0
Nifedipine	10^{-5}	10^{-5}	$-38 \pm 1^*$	13 ± 13	0
		3×10^{-6}	$-19 \pm 3^*$	0	0
		10^{-6}	$-14 \pm 5^*$	0	0
Heparin	100 IU	100 IU	-41 ± 4	16 ± 10	4 ± 3
		75 IU	$-36 \pm 5^*$	16 ± 10	17 ± 3
		50 IU	$-26 \pm 2^*$	1 ± 1	2 ± 2
Simvastatin	10^{-6}	10^{-6}	-21 ± 7	8 ± 3	9 ± 6
		3×10^{-7}	-10 ± 3	0	4 ± 4
		10^{-7}	-7 ± 9	0	0
Clofibrate	3×10^{-5}	3×10^{-5}	-17 ± 5	0	1 ± 1
		10^{-5}	-9 ± 9	9 ± 9	15 ± 5
		3×10^{-6}	-3 ± 1	0	0
Nicotinic acid	10^{-5}	10^{-5}	$+2 \pm 6$	0	3 ± 3
		3×10^{-6}	-9 ± 9	6 ± 4	2 ± 2
		10^{-6}	$+8 \pm 6$	3 ± 6	1 ± 1
Gemfibrozil	10^{-5}	10^{-5}	$-19 \pm 6^*$	28 ± 7	15 ± 8
		3×10^{-6}	-9 ± 9	1 ± 1	2 ± 2
		10^{-6}	$+4 \pm 1$	0	0
Probucol	10^{-5}	10^{-5}	$-29 \pm 3^*$	$100 \pm 0^*$	98 ± 2
		3×10^{-6}	-4 ± 5	$100 \pm 0^*$	$94 \pm 4^*$
		10^{-6}	$+3 \pm 5$	$96 \pm 4^*$	$92 \pm 4^*$

The results of the maximum noncytotoxic concentration, the percent change in SMC proliferation for each concentration of different compound tested, and the percent decrease in intracellular lipid accumulation during preventive and curative treatments are summarized in Table 2.3.

The method employed in the present experiment showed that the intracellular lipid changes in the SMCs cultured in the presence of hyperlipidemic rabbit serum (HLRS) resemble those observed in early atherosclerosis. HLRS contains very high proportions of low and very low density lipoproteins, the oxidation of which is prerequisite for the initiation of atherosclerotic lesions *in vivo* and probably also *in vitro.*

Evidence obtained over the past few years has demonstrated an antiatherogenic activity of certain calcium antagonists, but their mechanism of action is unclear. Calcium entry blockers like verapamil, diltiazem, and nifedipine are known to have no effect on intracellular lipids,[40] which is in agreement with the *in vitro* results of this study.

Probucol was found to be the most active compound in inhibiting lipid entry into the cells and in removing intracellularly accumulated cholesterol *in vitro.* Probucol is known to increase the fractional rate of LDL catabolism, thereby reducing plasma LDL and HDL cholesterol *in vivo* and is also known to be an effective antioxidant.[41] Clofibrate, gemfibrozil, and nicotinic acid have been suggested to be slightly active on intracellular lipids *in vivo*, probably by increasing the activity of the lipoprotein lipase and by reducing VLDL, but their mode of action at the cellular level is unknown.

The antiproliferative effect of calcium channel blockers such as verapamil,[42] diltiazem,[43] and nifedipine[44] has been described in the literature, and the results of this *in vitro* experiment also confirm this finding. *In vivo* antiproliferative effects of heparin[45] and probucol[46] reported in the literature were also confirmed in the present *in vitro* experiment.

These results show that SMC cultures may be a useful initial *in vitro* screening model to detect compounds active upon SMC proliferation and intracellular lipid accumulation.

A large number of compounds can be scanned rapidly, economically, and in a robust manner in such procedures, which would otherwise be practically impossible, or at least at the expense of large numbers of animals *in vivo*.

2.3.4   Structure-Activity-Toxicity-Relationships (SATR)

SATR are of interest with respect to animal testing, because fewer *in vivo* experiments may be required if information from a series of compounds can, at least partly, be related to other compounds. Less testing may be needed to evaluate the benefit/risk ratio of a compound, if SATR can predict its activity or safety.

SATR are *in vitro* models based on the assumption that the factors important for biological activity are represented in the physicochemical properties of a chemical.[47,48]

The SATR approach works, provided that the physicochemical model systems contain information that is relevant for the biological system in question. In order to obtain suitable SATR models, a trend in physicochemical properties should be accompanied by a similar trend in biological responses within a given chemical series. Separate SATR are necessary for each chemical class of compounds. It is only when events at the molecular level in the model systems and the predicted system are very similar over a wide range of chemicals that we can expect a "detailed" SATR to be valid.

The SATR models are used for theoretical prediction of risks such as acute toxicity, mutagenicity, carcinogenicity, and pharmacological activity very often within a given chemical series. Considerable effort has been made during the last few years to transfer biological information from the tested compounds to untested compounds of closely related chemical structures having pharmacological profile to the mother molecule.

SATR may prove to be a potentially powerful tool in designing new molecules, especially using computer generated chemical structures, but this technique is presently underemployed by industry. Concrete examples have been obtained with marketed drugs[49,50] in which interrelationships between potency, lipophilicity, cytotoxicity, and chemical class were investigated for calcium antagonists currently in clinical use. These studies revealed direct relationships between cytotoxicity and chemical class[50] and revealed suprisingly that the most cytotoxic compounds were also found to be the most toxic in clinical use.[49]

2.4   The *In Vitro–In Vivo* Correlation

The correlation between the real *in vivo* results and the predictions obtained through *in vitro* data forms the basis for the acceptance or refusal of the *in vitro* tests. Statistically, correlation refers to the relationship between the model chosen to fit the data and the fit of the data to the model. A good correlation is one in which a high percentage of the results of the *in vitro* test are the same as the results obtained in the *in vivo* test, assuming that the official *in vivo* test already is a valid model for predicting human responses.

An *in vivo* test utilizes the most natural test and most complex conditions available: the living animal. It also utilizes the most complex test conditions, i.e., the living animal. The responses, such as clinical behavior, effect on body weight and food consumption; hematological, blood biochemical, and urine parameter changes; gross anatomical, organ weight, and histopathological alterations; effect on blood pressure; irritation, correlation, and pharmacological responses, are utilized to evaluate the benefit or the risk of product *in vivo*. These end points often do not explore the cellular mechanism by which these reactions occur in the animal. *In vitro* tests evaluate a biochemical cellular response or a segment of the response cascade. Therefore, it is very important to consider these basic *in vitro–in vivo* differences before thinking about an *in vitro* to animal or *in vitro* to human response prediction.

The most important problem to correlate *in vitro* data for the estimation of risk to human beings is the absence of reliable human toxicity data for a vast majority of chemicals. *In vitro* experiments are often performed under controlled experimental conditions along with the measurements of reliable adverse effect parameters minimizing test to test differences. In contrast to this, human beings are exposed to chemicals under different circumstances with wide variations in biological responses. The differing predisposition of humans may greatly influence the sensitivity and responses to a given chemical. In addition to the biological variations that are genetically induced, many other factors such as the state of health, body weight, and size, life-style, route of entry, biorhythm, nutrition, physical activity, comedication and exogenous factors such as environmental conditions may cause marked variations in the manifestation of risk phenomena. In spite of these limitations, excellent *in vitro–in vivo* correlations were shown between basal cell cytotoxicity and *in vivo* systemic toxicity,[51] developmental toxicity,[52] ecotoxicity,[53] and ophthalmic risk.[54]

Considering the relatively large number and the importance that endogenous and exogenous factors may exert on the expression of *in vivo* toxicity, the relatively satisfactory prediction of benefit/risk using simple *in vitro* systems needs to be encouraged. Simple *in vitro* models are able to mimic simple *in vivo* phenomena. Over the last few years, both *in vivo* and *in vitro* investigations have rapidly become indispensable for the risk or benefit assessment. Moreover, *in vitro* investigations are indispensable for the risk or benefit assessment, since these models are relevant to assess adverse or benefical effects at the cellular level. *In vivo* models are nevertheless unavoidable to obtain the whole body reaction.

2.5   Validation of *In Vitro* Tests: Past, Present, and Future

The *in vitro* testing, mechanistic investigations, and biotransformation in *in vitro* models can be used as cost effective models in assessing the benefit/risk ratio of new drugs and chemicals. Generally *in vitro* tests are only applied in the screening phase of the development of chemicals and only a few *in vitro* pharmacology test results are used in the documentation for regulatory authorities. Thus, it is easy to adopt an *in vitro* test for in-house development use, because the only requirement is that the method provides information that can be of value when making a decision on further development.

Some of these methods were rapidly adopted by some industries as preliminary screening tools, and either accepted or rejected after evaluation by parallel use of *in vitro* and conventional *in vivo* methods. Thus, many methods for the prediction of acute toxicity, eye irritancy, phototoxicity, and teratogenicity are now in pratical use or on evaluation premises.[55] Cellular tests for local and systemic reactions of plastic devices, implants, etc. are, in fact, recommended as the test of choice by the U.S. Pharmacopia.[8]

For the *in vitro* methods that are accepted by the regulatory authorities such as the Ames and the gene mutation tests, the situation is quite different. These methods have been proven to be reproducible, reliable, and transferable. It is possible to extrapolate the data from these tests to an *in vivo* situation and to integrate them into risk studies. The purpose of validation studies is to identify procedures that are capable of performing according to these demands, i.e., the relevance and reliability of a test method are established for a particular purpose.[56,57]

At present, there is no standard validation method in use, mainly because the scientific requirements, theoretical foundation, and objective of the exercise vary from test to test. Currently, substantial efforts are focused on the validation of *in vitro* or alternative tests for safety evaluation studies, especially in the fields related to toxicology, probably because most animal suffering is incurred during the processes of toxicity testing and also due to the fact that many *in vitro* methods have been developed and primary validation processes have been launched for certain alternative models.[58] The *in vitro* tests for activity evaluation are still used principally as basic research tools with less emphasis on the validation of these tests.[59,60]

Many *in vitro* tests are developed and used by individual industries but very few of them are accepted by the regulatory authorities. This is probably related to presently deficient validation structures, a lack of resources of individual laboratories, and high cost:benefit ratio of currently available *in vitro* tests.[61] Another problem is that no robust and reliable human data exist for comparison with the predictions obtained using *in vitro* techniques, but efforts by the MEIC organization give a big hope for the future.

Therefore, the animal right movements, industry and government agencies, including OECD and EU, are engaged with the possibilities of promoting *in vitro* test procedures by organized, well planned studies. Some organizations such as FRAME, CAAT, and ERGAT (European Research Group for Alternative Testing) have relayed impulses from the different sources mentioned into proposals by FRAME.

In 1990, a document greatly expanding the scope of discussion on the validation of *in vitro* methods was published by Frazier and Goldberg.[59] The authors prepared a report for the OECD that served as a monograph on the scientific criteria for the validation of *in vitro* toxicity tests. The document defines validation as the process by which the credibility of a candidate test is established for a specific purpose with reliability and reproducibility verified. They identified three main steps in the validation process:

- Intralaboratory assessment.
- Interlaboratory assessment with the use of a group of 200–250 chemicals.
- Regulatory validation, the process by which a test becomes accepted by the regulatory authorities.

As a result of the increasing emphasis on the need for formal validation of alternative *in vitro* tests, a number of committees and organizations devoted to the process began to emerge. The Interagency Regulatory Alternatives Group (IRAG) of the United States, composed of representatives from the Environmental Protection Agency, Food and Drug Administration, and the Consumer Product Safety Commission, and the European Commission Centre for Validation of Alternative Methods (ECVAM) were created to evaluate current regulatory guidelines and to facilitate validation activities, respectively. The Johns Hopkins Center for Alternatives to Animal Testing subcommittee on validation and technology transfer has been also working since 1988 to develop a framework for validation that expresses the insights of those who have described the elements of validation over the last five years and incorporates the recommendations of scientists working in academia, industry, and government. It is hoped that the creation of validation infrastructure composed of the elements described above will further facilitate scientific acceptance and utilization of alternative methodologies and speed implementation of *in vitro* techniques in the benefit/risk assessment.

2.6   Conclusion

The development of *in vitro* tests especially due to advances in cell culture technology along with mathematical models and SATR have opened many possibilities to predicting the benefit/risk ratio without the use of living animals. The application of *in vitro* techniques for benefit/risk assessment, however, is not new. It has gained a significance of its own as a consequence of the reliability of these models to predict reactions to human beings and due to the growing debate about alternatives to animal experimentation in the biological sciences. Tissue culture was devised at the beginning of 19th century to study the behavior of cells free of the systemic variations imposed by homeostasis in the living animal.

Such techniques have been used for many years in pharmacology for preselection of chemicals as potent new drugs or pesticides. The assessment of benefit/risk was focused principally on the use of *in vivo* studies due to the absence of relevant and reliable *in vitro* models and due to regulatory requirements. Other than the *in vitro* tests for genotoxicity there are few *in vitro* alternative tests in regulatory guidelines. As the changes in legislation are slow and difficult, irrespective of how many *in vitro* alternatives are developed, they will not reduce the number of *in vivo* studies for regulatory purposes until the regulations include them as alternatives to fully replace *in vivo* studies. On the other hand, in the present competitive world, the use of *in vitro* techniques is a must during the prescreening or preevaluation of benefit/risk for detecting just a few out of thousands of chemicals that possess the desired benefit/risk ratio. It is heartening to note that many chemical companies have already invested considerable resources in the development of *in vitro* techniques as preevaluation screens to select the most active and best tolerated chemicals. Assessment of the benefit/risk ratio from *in vitro* data is clearly destined to gain considerable impact in the near future.

References

1. Grabau, J. H., Animal issues and society, *Toxicology Letters*, 68, 51, 1993.
2. Rhodes, C., Current status of *in vitro* toxicity tests in the industrial setting: a European viewpoint, *In vitro toxicology*, 2(3), 151, 1988/89.
3. Bronstein, D. A., Some ethical issues in toxicology, *Fundamental and Applied Toxicology*, 7, 525, 1986.
4. Fenten, J. H., The APC report on regulatory toxicity testing, *Alternatives to Laboratory Animals*, 22, 285, 1994.
5. Straughan, D. W., First European Commission report on statistics of animal use, *Alternatives to Laboratory Animals*, 22, 289, 1994.
6. Trevan, J. W., The error of determination of toxicity, *Proc. Roy. Soc. (Lond.) Ser. B.*, 101, 483, 1927.
7. Williams, P. D., Alternatives to ocular testing *in vivo*: scientific and regulatory considerations, *Toxicology Methods*, 4(1), 24, 1994.
8. Underhill, L. A., Dabbah, R., Grady, L. T., Rhodes, C. T., Alternatives to animal testing in the USP-NF: Present and future, *Drug Development and Industrial Pharmacy*, 20(2), 165, 1994.
9. Balls, M., Replacement of animal procedures: alternatives in research, education and testing, *Laboratory Animals*, 28, 193, 1994.
10. Russell, W. M. S., Burch, R. L. The principles of human experimental technique, *South Mimms: UFAW*, 238 pp, 1992.
11. Annett, B., The fund for the replacement of animals in medical experiments (FRAME): the first 25 years, *Alternatives to Laboratory Animals*, 23, 19, 1995.
12. Goldberg, A. M., Frazier, J. M., Brusick, D., Dickens, M. S., Flint, O., Gettings, S. D., Hill, R. N., Lipnick, R. L., Renskers, K. J., Bradlaw, J. A., Scala, R. A., Veronesi, B., Green, S., Wilcox, N. L., Curren, R. D., Framework for validation and implementation of *in vitro* toxicity tests: report of the validation and technology transfer committee of the Johns Hopkins center for alternatives to animal testing, *In vitro Toxicology*, 6(1), 47, 1993.
13. Bondesson, I., Ekwall, B., Hellberg, S., Rombert, L., Stenberg, K., Walum, E., MEIC — A new international multicenter project to evaluate the relevance to human toxicity of *in vitro* cytotoxicity tests, *Cell. Biol. Toxicol.*, 5, 331, 1989.
14. Anonymous, Council directive of 24 November 1986 on the approximation of laws, regulations and administrative provisions of the member states regarding the protection of animals used for experimental and other purposes, *Official Journal of the European Communities*, 1, 262, 1986.
15. Shrivastava, R., Ratinaud, M. H., Julien, R., Comparative flow cytometric analysis of bleomycin toxicity on normal and tumour sheep cell kinetics *in vitro*, *Journal of Applied Toxicology*, 10(2), 99, 1990.
16. Shrivastava, R., Ratinaud, M. H., Julien, R., Differential toxicity of cytosine arabinoside to cell kinetics in normal sheep sinus cells and sheep tumour cell line *in vitro*, *Toxic. in Vitro*, 4(2), 143, 1990.
17. Shrivastava, R., John, G., Chevalier, A., Beaughard, M., Rispat, G., Slaoui, M., Massigham, R., Paracetamol potentiates isaxonine toxicity *in vitro*, *Toxicology Letters*, 73, 167, 1994.
18. Bond, R. A., Leff, P., Johnson, T. D., Milano, C. A., Rackman, H. A., McMinn, T. R., Apparsundaram, S., Hyek, M. F., Kenakin, T. P., Allen, L. F., Lefkowitz, R. J. Physiological effects of inverse agonists in transgenic mice with myocardial overexpression of the $\square\square\beta$adrenoceptor, *Nature*, 374, 272, 1995.
19. Kenakin, T., Agonist-receptor efficacy I: mechanisms of efficacy and receptor promiscuity, *Trends Pharmacol. Sci.*, 16, 188, 1995a.
20. Kenakin, T. Agonist-receptor efficacy II: agonist trafficking of receptor signals. *Trends Pharmacol. Sci.*, 16, 232, 1995b.
21. Rowan, A., Shortcomings of LD_{50}-values and acute toxicity testing in animals, *Acta Pharmacol. Toxicol.*, 52, 52, 1983.

22. Lipnick, R. L., Cotruvo, J. A., Hill, R. N., Bruce, R. D., Stitzel, K. A., Walker, A. P., Chu, I., Goddard, M., Segals, L., Springer, J. A., Myers, R. C., Comparison of the up-and-down, conventional LD_{50} and fixed-dose acute toxicity procedures, *Fd. Chem. Toxicol.*, 33(3), 223, 1995.

23. Aldhous, P., Tide turns against LD_{50}, *Nature*, 352, 489, 1991.

24. Shrivastava, R., Delmani, C., Chevalier, A., John, G., Ekwall, B., Walum, E., Massingham, R., Comparison of *in vivo* acute lethal potency and *in vitro* cytotoxicity of 48 chemicals, *Cell Biology and Toxicity*, 8(2), 157, 1992.

25. Clothier, R. H., Hulme, L. M., Smith, M., Balls, M., Comparison of the *in vitro* cytotoxicities and acute *in vivo* toxicities of 59 chemicals, *Mol. Toxicol.*, 1, 571, 1987.

26. Fry, J. R., Garle, M. J., Hammond, H. H., Hatfield, A., Correlation of acute lethal potency with *in vitro* cytotoxicity, *Toxicol. in Vitro*, 4, 175, 1990.

27. Ekwall, B., Bondesson, I., Castell, J. V., Gomez-Lechon, M. J., Högberg, J., Jover, R., Pondosa, X., Rombert, L., Stenberg, K., Walum, E., Cytotoxicity evaluation of the first ten MEIC chemicals: acute lethal toxicity in man predicted by cytotoxicity in five cellular assays and by oral LD_{50} tests in rodents, *Alternatives to Laboratory Animals*, 17, 83, 1989.

28. Helberg, S., Eriksson, L., Jonsson, J., Lindgren, F., Sjöström, M., Wold, S., Ekwall, B., Gomez-Lechon, M. J., Clothier, R., Accamando, N. J., Grimes, A., Barile, F. A., Nordin, M., Tyson, C. A., Dierickx, P., Shrivastava, R., Tingsleff-Skaanild, M., Graza-Ocanas, L., and Fiskesjö, G., Analogy models for prediction of human toxicity, *Alternatives to Laboratory Animals*, 18, 103, 1990.

29. Barile, F. A., Dierickx, P. J., Kristen, U., *In vitro* cytotoxicity testing for prediction of acute human toxicity, *Cell Biology and Toxicology*, 10, 155, 1994.

30. Ekwall, B., Features and prospects of the MEIC cytotoxicity evaluation project, *Alternative Tierexperimente*, 1, 231, 1992.

31. McConnel, E. E., The maximum tolerated dose: the debate, *Journal of the American College of Toxicology*, 8, 1115, 1989.

32. Shrivastava, R., John, G. W., Rispat, G., Rispat, G., Chevalier, A., Massingham, R., Can the *in vivo* maximum tolerated dose be predicted using *in vitro* techniques? A working hypothesis, *Alternatives to Laboratory Animals*, 19, 393, 1991.

33. Gordon, D., Schartz, S., Cell proliferation in human atherosclerosis, *TMC*, 1, 24, 1991.

34. Raines, E. W., Ross, R., Smooth muscle cells and the pathogenesis of the lesions of atherosclerosis, *Brit. Heart J.*, 69, 30, 1993.

35. Cooper, M. M., Reison, D. S., Rose, E. A., Accelerated atherosclerosis, *Ischemic Heart Diseases*, 6, 581, 1991.

36. Clozel, J. P., Hess, P., Michael, C., Schietinger, K., Baumgartner, H. R., Inhibition of converting enzyme and neointima formation after vascular injury in rabbit and guinea pigs, *Hypertension*, 18, 55, 1991.

37. Atkinson, J. B., Hoover, R. L., Berry, K. K., Swift, L. L., Cholesterol fed heterozygous Watanabe heritable hyperlipidemic rabbits: a new model for atherosclerosis, *Atherosclerosis*, 78, 123, 1989.

38. Orekhov, A. N., *In vitro* models of antiatherosclerotic effects of cardiovascular drugs, *Am. J. Cardiol.*, 66, 23, 1990.

39. Shrivastava, R., Chevalier, A., Slaoui, M., John, G. W., An *in vitro* method using vascular smooth muscle cells to study the effect of compounds on cell proliferation and intracellular lipid accumulation, *Meth. Find. Clin. Pharmacol.*, 15(6), 345, 1993.

40. Tilton, G. D., Buja, L. M., Bilheimer, D. W., Failure of a slow channel calcium antagonist, verapamil, to retard atherosclerosis in the Watanabe heritable hyperlipidemic rabbit: an animal model of familial hypercholesterolemia, *J. Am. Coll. Cardiol.*, 6, 141, 1985.

41. Buckley, M. M. T., Goa, K. L., Price, A. H., Brogen, R. N., Probucol: a reappraisal of its pharmacological properties and therapeutic use in hypercholesterolemia, *Drugs*, 37, 761, 1989.

42. Rouleau, J. L., Parmley, W. W., Stevens, J., Verapamil suppresses atherosclerosis in cholesterol-fed rabbit, *J. Am. Coll. Cardiol.*, 1, 1453, 1983.

43. Sugano, M., Nakashima, Y., Matsushima, T., Suppression of atherosclerosis in cholesterol-fed rabbit by diltiazem injection, *Atherosclerosis*, 6, 237, 1986.
44. Henry, P. D., Bentley, K. I., Suppression of atherogenesis in cholesterol-fed rabbits treated with nifedipine, *J. Clin. Invest*, 68, 1366, 1981.
45. Clowes, A. W., Clowes, M. M., Vergel, S. C., Heparin and cilazapril together inhibit injury-induced intimal hyperplasia, *Hypertension*, 18, 65, 1991.
46. Kita, T., Nagano, Y., Yokode, M., Prevention of atherosclerotic progression in Watanabe rabbits by probucol, *Am. J. Cardiol.*, 62, 13B, 1988.
47. Cordier, A. C., *In vitro* strategy for the safety assessment of drugs, *In Vitro Methods in Toxicology*, 2, 21, 1992.
48. Hüttenrauch, R., Speiser, P., *In vitro–in vivo* correlation: an unrealistic problem, *Pharm. Res.*, 97, 1985.
49. John, G. W., Shrivastava, R., Chevalier, A., Pognat, J. F., Massingham, R., An *in vitro* investigation of the relationships between potency, lipophilicity, cytotoxicity and chemical class of representative calcium antagonist drugs, *Pharmacol. Res.*, 27(3), 253, 1993.
50. Shrivastava, R., John, G., Massingham, R., Calcium antagonists can be classified using *in vitro* toxicity and potency indices, *J. Appl. Toxicol.*, 12(5), 329, 1992.
51. Ekwall, B., Ekwall, K., Comments on the use of diverse cell systems in toxicity testing, *Alternatives to Laboratory Animals*, 15, 193, 1988.
52. Reinhart, C. A., Neurodevelopmental toxicity *in vitro*: primary cell culture models for screening and risk assessment, *Reproductive Toxicology*, 7, 165, 1993.
53. Calleja, M. C., Persoone, G., Geladi, P., Comparative acute toxicity of the first 50 multicentre evaluation of *in vitro* cytotoxicity chemicals to aquatic non-vertebrates, *Arch. Environ. Contam. Toxicol*, 26, 69, 1994.
54. Sina, J. F., Galer, R. G., Sussman, R. G., Gautheron, P. D., Sargent, E. V., Leong, B., Shah, P. V., Curren, R. D., Miller K., A collaborative evaluation of seven alternatives to the Draize eye irritation test using pharmaceutical intermediates, *Fundamental and Applied Toxicology*, 26, 20, 1995.
55. Ekwall, B., Validation of *in vitro* cytotoxicity tests, *in vitro* alternatives to animal pharmacotoxicology, Farmaindustria, Castell, J. V., Gomez-Lechon, M. J., Madrid, 1992, chap. 14.
56. Balls, M., Blaauboer, B., and, Brusnick, D., Report and recommendation of the CAAT/ERGATT workshop on the validation of toxicity test procedures, *Alternatives to Laboratory Animals*, 18, 313, 1990a.
57. Balls, M., Botham, P., Cordier, A., Report and recommendations of an international workshop on promotion of the regulatory acceptance of validated non-animal toxicity test procedures, *Alternatives to Laboratory Animals*, 18, 339, 1990b.
58. Walum, E., Clemendson, C., Ekwall, B., Principles for the validation of *in vitro* toxicology test methods, *Toxic. in Vitro*, 8(4), 807, 1994.
59. Frazier, J. F., Goldberg, A. M., Alternatives to and reduction of animals use in biomedical research, education and testing, *Alternatives to Laboratory Animals*, 18, 65, 1990.
60. Siebert, H., Balls, M., Fentem, J. H., Bianchi, V., Clothier, R. H., Ekwall, B., Garle, M. J., Gomez-Lechon, M. J., Gribaldo, L., Gülden, M., Liebsch, M., Rasmussen, E., Rouget, R., Shrivastava, R., Walum, E., Acute toxicity testing *in vitro* and the classification and labelling of chemicals, ECVAM Workshop Report 16, *Alternatives to Laboratory Animals*, 24, 499, 1996.
61. Shrivastava, R., *In vitro* tests in pharmacotoxicology: can we fill the gap between scientific advances and industrial needs, *Alternatives to Laboratory Animals*, 25, 339, 1997.

3

Evaluation of Therapeutic Benefits and Toxicological Risks Using Structure-Activity Relational Expert Systems

Herbert S. Rosenkranz
University of Pittsburgh

Gilles Klopman
Case Western Reserve University

Orest T. Macina
University of Pittsburgh

3.1 Introduction

While traditional Quantitative Structure Activity Relationship (QSAR) approaches have been used in the design of therapeutic agents, the application of newer, knowledge-based expert Structure Activity Relationship (SAR) systems for this purpose has as yet been limited. However, the availability of these recently developed methods has permitted a number of new approaches to the design and/or recognition of novel, therapeutically promising agents. The application of these approaches has allowed:

1. the systematic analysis of noncongeneric data bases,
2. the ability to derive mechanistic information,
3. "coupling" of data bases of therapeutic effectiveness with those of unwanted side effects (e.g., toxicity),
4. the ability to perform more informed risk assessments, and
5. the ability to screen rapidly for molecules endowed with potentially useful therapeutic properties.

These and other promising applications of the new technology are described herein using MULTICASE, a knowledge-based expert system that is entirely automated and devoid of human bias.

3.2   Methodology and Data Bases

3.2.1   The Correlation of Chemical Structure with Biological Activity

The science of Structure Activity Relationships and Quantitative Structure Activity Relationships (SAR/QSAR) encompasses attempts to correlate chemical structure with biological activity. The representation of chemical structure can range from physicochemical properties (hydrophobic, electronic, steric) to parameters derived directly from the structures under consideration (molecular connectivity indices, structural fragments). A successfully derived SAR/QSAR model can be used to predict the biological activity of compounds as well as to investigate mechanisms of action. Recently published monographs provide details of various SAR/QSAR methods and techniques.[1,2]

The successful application of SAR/QSAR methodologies often depends on the hypothesis that a series of compounds exhibiting activity operate via a common mechanism of action. In most situations it is assumed that structurally similar compounds exhibit their biological effects by the same mechanism of action. Indeed, many successful models have been developed for congeneric series of molecules acting at an identical receptor and exhibiting the same pharmacological effect.

Problems arise, however, when a set of structurally diverse compounds exhibiting the same biological effect are under investigation. The activity may be due to the existence of multiple mechanisms of action and/or the presence of a structural commonality not easily recognizable by the investigator. This is particularly true for toxic effects where the activity of structurally dissimilar compounds may result from different biological processes often involving a sequence of events. Moreover, the exact site or mechanism(s) of action are generally unknown for toxicants.

The Multiple Computer Automated Structure Evaluation (MULTICASE) program was designed to overcome some of the difficulties associated with SAR/QSAR studies. The program is dependent on the fundamental hypothesis that a relationship does indeed exist between chemical structure and biological activity. It is not necessary, however, that a single chemical functionality be responsible for all of the observed activity. MULTICASE is completely automatic and learns directly from the available data consisting of chemical structures and the respective biological activity. The program seeks to find molecular fragments that discriminate between the active and inactive classes of compounds. In addition, MULTICASE differentiates between substructures responsible for activity and those that influence or modulate the activity. The program organizes the entire data base into logical congeneric subsets where the commonality between molecules is based on a rational evaluation of their structures rather than on the arbitrary choice of a common structural feature.

3.2.2   MULTICASE

A detailed description of the MULTICASE algorithm has been previously described in published sources.[3] Input is in the form of a data base composed of chemical structure and biological activity (quantitative or qualitative). Congeneric as well as structurally diverse series of compounds can be analyzed by the system. The program attempts to identify descriptors consisting of molecular fragments ranging from two to ten heavy atoms along with their associated hydrogens, which account for the biological activity of the compounds under study. Molecular fragments are generated as a result of decomposing each individual chemical structure within a data base into all possible constituent groupings of atoms. Thus, descriptors (fragments) are generated solely on the basis of the data under consideration, rather than being selected from a preconceived library. Statistically relevant fragments are identified because they will generally exhibit a nonrandom distribution among the active and inactive classes of compounds. In addition to using molecular fragments, MULTICASE has the ability to identify relevant two-dimensional distances between atoms within a chemical structure.

Once a set of statistically significant descriptors (fragment and/or distance) has been identified, the program seeks to find a descriptor (biophore) that has the highest probability of being responsible for activity. Compounds containing the primary biophore are removed from the analysis, and subsequent biophores are selected that explain the activity of the remaining compounds. This iterative process of selection is continued until either all of the active compounds are accounted for or no statistically significant descriptors remain. Thus, the program separates the entire data base into logical subsets based on an unbiased determination of structural commonality. The presence of biophores determines the potential for a compound to exhibit biological activity. A compound is presumed to be inactive if no biophores are identified within the chemical structure.

In addition to the primary biophores, fragments with slight differences (e.g., -CH$_2$-I vs. -CH-I; -CH$_2$-I vs. -CH$_2$-F) are also identified as expanded fragments. This is an attempt to take bioisosterism into account. Compounds containing expanded biophores are grouped together with the ones containing the primary biophore.

Establishment of a QSAR is attempted within each group of compounds containing a particular biophore in order to identify molecular features that modulate the activity/potency. Modulators are selected by stepwise multivariate linear regression from the associated pool of molecular fragments, distance descriptors, calculated electronic indices (molecular orbital energies, charge densities) and calculated transport parameters (oil/water partition coefficient, water solubility). The contribution (positive or negative) of the modulators to the activity is valid only in the context of compounds containing the relevant biophore; i.e., the resulting local QSAR is valid only within the subset of compounds containing an established biophore (common active substructure).

At this point the analyzed data base can be utilized for predicting the activity of untested compounds. Structural fragments present within a compound that have not been encountered in the learning set are flagged by MULTICASE as "warnings." The contribution of these unknown functionalities to the biological activity under consideration is uncertain and therefore must be taken as an indication that the prediction may not be very reliable. In addition to their use for predictive purposes, biophores and their respective modulators may be helpful in defining mechanisms of action.

3.2.3 META

The *in vivo* biological activity of compounds depends on factors such as absorption, distribution, and metabolism. Attempts to model absorption and distribution often rely on the dependence of activity on the logarithm of the oil/water partition coefficient (log P). The log P value of a compound is a reflection of its ability to partition among various components of the biophase. For metabolism, on the other hand, no single parameter exists that measures a compound's susceptibility to metabolic conversion.

Metabolites resulting from a parent compound may possess biological properties that are undesirable (toxicity). Indeed, an innocuous parent compound may become a toxicant by virtue of the metabolism it undergoes.

In order to address the relevance of metabolism within investigations of structure–biological activity relationships, we have developed a system for predicting the possible metabolites arising from a parent compound. Metabolites generated by META may be subsequently submitted to MULTICASE for predictions of possible biological activity. Our combined META/MULTICASE approach allows for the identification of undesirable side effects that may limit therapeutic effectiveness.

A detailed description of META has been previously published.[4,5] The program is based on an expert evaluation of chemical structure and does not require prior knowledge regarding the actual metabolism of the compound under study. The program functions in conjunction with a number of metabolism dictionaries containing coded biotransformations with relevant information about the structural constraints governing the specificity of each metabolic transformation. The objective of the META program and its associated dictionaries is to model a theoretical "average" mammal. Work is in progress toward the development of more specific dictionaries pertaining to individual species and organs. In addition to metabolic transformations, META also contains a dictionary of spontaneous reactions to detect and process unstable intermediates.

Submission of a compound to META results in the identification of a target fragment within the molecular structure. A target fragment is defined as a molecular domain that is recognized by a specific enzyme. The target fragment is subsequently transformed (metabolized) into the product fragment by the relevant enzymatic system. Primary metabolites are screened for the presence of additional target fragments and further transformed as applicable. The parent compound is thus metabolized into its possible metabolites. Each metabolite is screened against the spontaneous reaction dictionary in order to identify and list products resulting from spontaneous decomposition. The META program keeps an archive of all of the metabolic transformations and will identify the enzyme class and the transformations used to generate any of the metabolites.

3.2.4   Available Data Bases

The MULTICASE program (and its predecessor CASE) has been applied to a variety of biological phenomena in order to determine the relationships between chemical structure and activity. In addition to pharmacological end points,[6-9] we have compiled an extensive catalog of data bases related to toxicology. Several publications have resulted from our investigations of the structural basis of various toxicities.[10-15]

A list of representative data bases currently available to us is shown in Table 3.1. The rodent carcinogenicity and *Salmonella* mutagenicity data bases consist of data generated under the aegis of the United States National Toxicology Program (NTP), wherein strict criteria related to experimental protocol and purity of chemicals were followed. These two data bases used in combination allow for the identification of genotoxic and nongenotoxic carcinogens. In addition, other short-term assays such as the induction of sister chromatid exchange (SCE), chromosomal aberration, and of micronuclei are useful in determining the genotoxic liability posed by a compound. The cellular toxicity toward BALB/3T3 cells can be used as a surrogate data base for systemic toxicity.

The availability of these and other toxicological data bases allows one to identify and thereby avoid any possible unwanted side effects in the molecular design of therapeutic agents. In addition, these data bases may be used to identify potential human health hazards in the environment.

3.3   Applications of META/MULTICASE

3.3.1   Fluoroquinolones

The fluoroquinolones are a relatively new class of antibacterial agents in widespread clinical use. They are effective against Gram-positive and Gram-negative bacteria as well as against mycobacteria. The mechanism of action of the fluoroquinolones involves inhibition of bacterial topoisomerase II (DNA gyrase).[16,17] Recent investigations, however, have indicated that mammalian topoisomerases are also sensitive to inhibition by fluoroquinolones.[18] These interactions with mammalian targets may lead to host-associated toxicities.

TABLE 3.1   Some Available Toxicological Data Bases

Salmonella mutagenicity
Rodent carcinogenicity
Structural alerts for DNA reactivity
Developmental toxicity/teratogenicity
Systemic toxicity
Cellular toxicity
Short-term genotoxicity assays
Contact sensitization
Sensory irritation

The search for newer quinolones with improved efficacy relative to ciprofloxacin continues. Our laboratories have had a continual interest in determining the structural requirements for exhibiting antibacterial activity within the fluoroquinolone class of compounds.[9] The reports of host-associated toxicities prompted us to compare the structural features related to therapeutic benefits and toxic risk. Accordingly, we compiled data bases pertaining to Gram-positive and Gram-negative activity, as well as toxicity. The data were obtained from the literature.[18]

MULTICASE data bases were established for 116 compounds evaluated for antibacterial activity against both Gram-positive and Gram-negative bacteria. Furthermore, data were available regarding the cytotoxic potential of the same series of compounds. Cytotoxicity can be assumed to serve as a surrogate for systemic toxicity. The data base regarding activity against Gram-positive bacteria consisted of 60 inactives, 12 marginally active, and 44 actives. Gram-negative data included 74 inactive, 5 marginal, and 37 active compounds. The same series of compounds evaluated for cytotoxicity resulted in 71 inactives, 9 marginals, and 36 actives. In addition to the cytotoxicity data base, we also established an "inverse cytotoxicity" data base in which noncytotoxic compounds are classified as active, while the cytotoxic compounds are classified as inactive. The "inverse cytotoxicity" data base was created in order to determine "beneficial" biophores, i.e., structural features that are associated with noncytotoxicity. The "inverse cytotoxicity" data base can be used as a final screen in the design process.

Each of the respective data bases was submitted to MULTICASE for analysis. Partial lists of biophores generated from each of the data bases are shown in Tables 3.2 to 3.5. For each of the biophores, the total number of occurrences within the data base is listed as well as the distribution among the inactive, marginal, and active classes. The average activity (in CASE units; 10–19 inactive, 20–29 marginal, 30–99 active) of the compounds containing a particular biophore is also listed. Biophores that are "redundant" with each other (fragments 2 and 3 for the Gram-negative; Table 3.3) indicate overlapping structural features.

The biophores derived for Gram-positive activity (Table 3.2) contain two distance descriptors related to the two-dimensional orientation of atomic centers (Biophores 1 and 2). There is, however, no single structural theme (biophore) that accounts for the majority of the active compounds. This may be an indication that many features are responsible for antibacterial activity against Gram-positives.

For the Gram-negative data base, on the other hand, relatively few structural features account for the majority of the activity (Table 3.3). Thus, the structural requirements for activity against Gram-negative bacteria may be more specific than those required for Gram-positives.

The cytotoxicity data base is similar to the Gram-negative data in that few structural themes are related to activity (Table 3.4). For example, "toxicophore" 1 (and its expanded fragment #2) indicates that for N-1 cyclopropyl-substituted quinolones a fluorine or sulfur substituent at position 8 of the quinolone nucleus contributes to cytotoxicity. "Toxicophore" #3 indicates that a chlorine at the 8-position of the quinolone nucleus together with substituents at the 6 and 7 positions also contributes to cytotoxicity. It also appears that primary amines would be cytotoxic.

The biophores derived from the "inverse toxicity" data base (Table 3.5) can be used to guide the incorporation of structural features beneficial for both antibacterial activity and reduced toxicity. This data base can be utilized as a final "check" on a proposed fluoroquinolone.

The information from each of the above data bases can be utilized to design a fluoroquinolone with maximum beneficial properties (Gram-positive and Gram-negative activity) and minimum risk (cyto-toxicity). The process we have utilized to design such a compound is illustrated in Figure 3.1. The starting point involves structure I, which has the quinolone nucleus substituted at N-1 with cyclopropyl and a fluorine at the 6-position. Incorporation of Gram-positive activity results from the introduction of 4-aminopiperidine at the 7-position. The 4-amino-piperidine moiety was chosen after consideration of biophore #3 of the Gram-positive data base (Table 3.2). The resulting structure II can be further modified in order to incorporate Gram-negative antibacterial activity. Consideration of biophore #1 of the Gram-negative data base (Table 3.3) indicates that a chlorine at the 8-position would confer activity. Placement of the chlorine at the 8-position yields structure III. Structure III contains structural features that are

TABLE 3.2   Biophores Associated with Activity Against Gram-positive Bacteria

Fragment	Nr.of	FR.	In	Ma	Ac	Av.Act.	Nr
`1---2---3---4---5---6---7---8---9---10---------------------------------`							
`[NH -]  <-- 3.0A -->  [F -] conj  generic`	9	2	0	7	48.0++	1
`[N* -]  <-- 7.8A -->  [NH -]       generic`	4	0	0	4	52.0++	2
`CH2-CH2-CH -CH2-CH2-N -C =C -C. =     <3-NH2>`	3	0	0	3	47.0	3
`F -C =C -N -CH2-CH2-CH -CH2-CH2-       <3-C =>`	—	—	—	—	—	4
`N -C. =C -C* -N -CH2-CH2-CH -CH2-      <5-CH2>`	—	—	—	—	—	5
`N -C. =C -C* -N -CH2-CH2-CH -CH2-CH2-`	—	—	—	—	—	6
`NH -CH2-CH -CH2-N -C =CH -`	6	1	1	4	36.0	7
`NH -CH2-CH -CH2-CH2-N  -C =CH -`	—	—	—	—	—	8
`F -C =C -N -CH2-CH -CH2-NH -           <3-CH=>`	—	—	—	—	—	9
`F -C =C -N -CH2-CH2-CH -CH2-NH -       <3-CH=>`	—	—	—	—	—	10
`NH -CH2-C -CH2-N -C =CH -`	1	0	0	1	68.0	11-
`NH2-CH2-C -CH2-N -C =CH -`	1	0	0	1	68.0	12-
`F -C -C =C -N -CH2-CH -CH2-`	2	0	0	2	68.0	13
`N -C. =C -C* -N -CH2-CH -CH2-          <3-C >`	—	—	—	—	—	14
`F -C -C =C -N -CH2-CH2-CH -CH2-`	—	—	—	—	—	15
`N -C. =C -C* -N -CH2-CH2-CH -CH2-      <3-C >`	—	—	—	—	—	16
`CH -N -C. =C -C* -N -CH2-CH2-CH -      <4-O >`	4	1	0	3	53.0	17
`CH -N -C. =C -C* -N -CH2-CH2-CH -      <4-S >`	1	0	0	1	44.0	18-
`F -C =C -C."-CO -C* -CO -OH            <3-Cl >`	2	0	1	1	44.0	19
`Cl -C =C -C* -N -                      <3-F >`	—	—	—	—	—	20
`Cl -C =C -C =C -Cl                     <3-N >`	—	—	—	—	—	21
`Cl -C =C -C =C -C. =                   <4-N >`	—	—	—	—	—	22
`Cl -C =C -C =C -                       <3-F >`	—	—	—	—	—	23
`Cl -C =C -C =C -C."-N -                <5-Cl >`	—	—	—	—	—	24
`Cl -C =C. -CO -C =CH -N -              <5-CO >`	—	—	—	—	—	25
`OH -CO -C* =CO -C. =C -C =C -          <6-Cl >`	—	—	—	—	—	26
`OH -CO -C =CH -N -C. =N -              <5-C =>`	1	0	0	1	68.0	27
`F -C =C -N =C. -N -C =                 <3-N >`	—	—	—	—	—	28
`N =C. -N -C =C -CH =C -                <5-F >`	—	—	—	—	—	29
`CH =C -C =N -C."-N -C =                <3-N >`	—	—	—	—	—	30
`C* -N -C. =N -C =C -CH =C. -           <6-F >`	—	—	—	—	—	31
`N =C. -N -C =CH -CH =C -CH =C -        <7-F >`	—	—	—	—	—	32
`CH2-CH2-CH -CH2-N -C =N -C."-N -C =    <3-NH2>`	—	—	—	—	—	33
`N =C -C =CH -C."-CO -C =CH -N -C =     <3-F >`	—	—	—	—	—	34
`N =C -C =CH -C."-CO -C =CH -N -C =     <7-CO >`	—	—	—	—	—	35
`OH -CO -C* -CO -C. =C -C =C -CH =      <6-CH3>`	2	0	0	2	68.0	36
`N -C. =CH -C =C -C* -CH3               <5-F >`	—	—	—	—	—	37
`CO -C. =C -C =C -CH =C. -N -           <3-CH3>`	—	—	—	—	—	38
`CH3-C =C. -CO -C =CH -N -C. =CH -      <5-CO >`	—	—	—	—	—	39
`N -CH =C -CO -C. =C -C =C -CH =        <6-CH3>`	—	—	—	—	—	40
`N -C. =CH -C =C -C* -CH3               <4-N >`	—	—	—	—	—	41
`C."-N -CH =C -CO -C. =C -C =C -        <7-CH3>`	—	—	—	—	—	42
`CO -C =CH -N -C. =CH -C =C -C* -CH3`	—	—	—	—	—	43
`CO -C =CH -N -C. =CH -C =C -C* -CH3`	—	—	—	—	—	44
`F -C =C -C."-CO -C =CH -N -C. =CH -    <3-CH3>`	—	—	—	—	—	45

C. indicates a carbon common to two rings.

Fragments with a dashed line overlap with the primary biophore (e.g., fragments 4, 5, and 6 overlap with 3).

Fragments 11 and 12 are expanded versions of 7.

Fragment 18 is an expanded version of 17.

conducive to both Gram-positive and Gram-negative antibacterial activity. The compound also contains, however, cytotoxic fragment #3, indicating that 6,7,8 trisubstitution with the 5-position unsubstituted poses a risk (Table 3.4). In order to avoid introducing cytotoxicity into our designed fluoroquinolone, substituents were placed at position 5 of the quinolone nucleus to yield structure IV. Specifically, it was observed that introduction of fluorine or an amino group at this position decreased the cytotoxic potential

TABLE 3.3 Biophores Associated with Activity Against Gram-negative Bacteria

Fragment	Nr.of FR.	In	Ma	Ac	Av.Act.	Nr
1---2---3---4---5---6---7---8---9---10						
CH2-CH2-N -C =C -C."-N -CH -	<5-Cl >	14	3	0	11	35.0	1
CH2-NH -CH2-CH2-N -C =C. -CH =C. -	<7-F >	8	0	1	7	35.0	2
NH -CH2-CH2-N -CH2-CH2-	-------------------					3
CH =C -C =C -C."-N -CH -	<4-F >	10	2	1	7	39.0	4
F -C =C. -N -CH =C -CO -C. =CH -	<4-CH >	-------------------					5
CH -N -C. =C -C =C -C =	<4-O >	1	0	0	1	30.0	6-
NH2-CH -CH2-N -C =C -C."-N -CH -	<6-F >	6	1	1	4	46.0	7
CH -N -C. =C -C" -N -CH2-CH2-CH -	<4-O >	4	1	0	3	28.0	8
F -C =C -C."-CO -C" -CO -OH	<3-OH >	2	0	0	2	68.0	9
OH -C =C -C" -N -	<3-F >	-------------------					10
OH -C =C. -CO -C =CH -N -	<5-CO >	-------------------					11
OH -CO -C" -CO -C. =C -C =C -	<6-OH >	-------------------					12
OH -C =C -C" -N -CH2-CH -CH2-CH2-	<7-NH2>	-------------------					13
OH -CO -C =CH -N -C. =N -	<5-C =>	1	0	0	1	44.0	14
F -C =C -N =C. -N -C =	<3-N >	-------------------					15
N =C. -N -C -CH =C -	<5-F >	-------------------					16
CH =C -C =N -C."-N -C =	<3-N >	-------------------					17
C" -N -C. =N -C =C -CH =C. -	<6-F >	-------------------					18
N =C. -N -C =CH -CH =C -CH =C -	<7-F >	-------------------					19
CH2-CH2-CH -CH2-N -C =N -C."-N -C =	<3-NH2>	-------------------					20
N =C -C =CH -C."-CO -C =CH -N -C =	<3-F >	-------------------					21
N =C -C =CH -C."-CO -C =CH -N -C =	<7-CO >	-------------------					22
N -CH2-CH2-N -C =CH -		1	0	0	1	44.0	23
F -C =C -N -CH2-CH2-N -	<3-CH=>	-------------------					24
CO -C. =CH -C =C -N -CH2-CH2-N -	<5-CH=>	-------------------					25
NH -CH -CH2-N -C =CH -		1	0	0	1	44.0	26-
NH2-CH -CH2-N -C =C -C" -CH3	<5-CH=>	1	0	0	1	30.0	27
NH2-CH -CH2-CH2-N -C =C -C" -CH3	<6-CH=>	-------------------					28
NH2-CH -CH2-N -C =C -C" -NH2	<5-CH=>	1	0	0	1	68.0	29
NH2-CH -CH2-CH2-N -C =C -C" -NH2	<6-CH=>	-------------------					30

C. indicates a carbon common to two rings.
Fragments with a dashed line overlap with the primary biophore (e.g., fragment 3 overlaps with 2).
Fragment 6 is an expanded version of 4.
Fragment 26 is an expanded version of 23.

while retaining the Gram-positive and Gram-negative antibacterial activity. Introduction of a methyl group at this position, however, resulted in the introduction of cytotoxicity (biophore #5 of Table 3.4).

MULTICASE predictions for the final product (structure IV) of the design process with regard to Gram-positive, Gram-negative, cytotoxic, and inverse cytotoxic activities are shown in Figures 3.2, 3.3, 3.4, and 3.5, respectively. In addition to the presence of the relevant biophores, modulators are also utilized in predictions of the potency or extent of activity.

Predictions regarding Gram-positive activity (Figure 3.2) indicate that for the 5-F substituted compound, biophore #3 (Table 3.2) is responsible for the activity. For the structure incorporating a $-NH_2$ at the 5-position, on the other hand, it is a combination of biophores (#1, #3, #6 of Table 3.2) that is responsible for imparting activity. All of the biophores identified result from the introduction of the 4-aminopiperidine at position 7 of the quinolone nucleus.

Predictions for Gram-negative activity for both structures ($X = F$, NH_2) rely on the presence of biophore #1 (Table 3.3). This biophore arises from the introduction of a chlorine at position 8 of the quinolone nucleus.

In evaluations of the cytotoxic potential, MULTICASE does not find any "toxicophores" within the structure for both $X = F$ and $X = NH_2$. The compounds are therefore presumed to be inactive, i.e., noncytotoxic. In addition, both compounds are predicted to be active in the "inverse cytotoxicity" data base, i.e., they are noncytotoxic (Figure 3.4). Both structures contain biophores that are beneficial, i.e., do not contribute to cytotoxicity.

TABLE 3.4   "Toxicophores" Associated with Cytotoxicity

Fragment		Nr.of FR.	In	Ma	Ac	Av.Act.	Nr
1---2---3---4---5---6---7---8---9---10-------------------------------							
CH2-CH2-N -C =C -C."-N -CH -	<5-F >	17	3	2	12	48.0	1
CH2-CH2-N -C =C -C."-N -CH -	<5-S >	1	0	0	1	44.0	2-
CH =C -C =C -C."-N -CH -	<4-Cl >	11	2	0	9	50.0	3
NH2-CH2-		12	2	1	9	52.0	4
[C -] <-- 2.6A --> [CO -] conj generic		6	1	0	5	35.0+++	5
CH -N -C. =C -C" -N -CH2-CH2-CH -	<4-O >	4	1	0	3	43.0	6
CH2-CH2-CH -CH2-CH2-N -C =C -C -	<3-NH2>	1	0	0	1	60.0	7
CH2-CH -CH2-CH2-N -C =C -C -	<5-CH2>	----------------------					8
F -C =C -N -CH2-CH2-CH2-CH2-	<3-CH=>	1	0	0	1	32.0	9
N -C. =CH -C" -N -CH2-CH2-CH2-CH2-		----------------------					10
C" -N -C. =C -C" -N -CH2-C -	<4-C >	1	0	0	1	68.0	11
C" -N -C. =C -C" -N -CH2-CH2-C -	<4-C >	----------------------					12

C. indicates a carbon common to two rings.
Fragments with a dashed line overlap with the primary biophore (e.g., fragment 8 overlaps with 7).
Fragment 2 is an expanded version of 1.

TABLE 3.5   Biophores Associated with Lack of Cytotoxicity

Fragment		Nr.of FR.	In	Ma	Ac	Av.Act.	Nr
1---2---3---4---5---6---7---8---9---10-----------------------------------							
C."-N -CH =C -CO -C. =CH -C =C -	<4-CO >	27	3	2	22	56.0	1
[NH -] <-- 9.0A --> [F -] generic		20	3	0	17	54.0+++	2
OH -CO -C =CH -N -CH2-	<3-CO >	10	0	1	9	75.0	3
CH2-N -CH =C -CO -C. =	<4-CO >	----------------------					4
N -C =C -CH =C. -CO -C =CH -N -CH2-	<3-F >	----------------------					5
OH -CO -C =CH -N -C -	<3-CO >	1	0	0	1	57.0	6-
OH -CO -C =CH -N -C =	<3-CO >	11	1	0	10	55.0	7
C."-CO -C =CH -N -C =	<3-CO >	----------------------					8
F -C =C -CH =CH -	<3-N >	----------------------					9
F -C =CH -C =C -CH =	<4-F >	----------------------					10
F -C =CH -C =CH -CH =	<4-F >	----------------------					11
F -C =CH -CH =C -C" -F		----------------------					12
N -C =CH -CH =C -CH =C -	<5-F >	----------------------					13
F -C =CH -C =C -CH =CH -	<5-N >	----------------------					14
F -C =CH -CH =C -C =CH -	<5-N >	----------------------					15
F -C =CH -C =CH -CH =C -N -		----------------------					16
OH -CO -C =CH -N -C =C -CH =C -	<7-F >	----------------------					17
OH -CO -C =CH -N -CH =	<3-CO >	1	0	0	1	80.0	18-
F -C =C -N -CH2-CH2-CH2-	<3-C =>	8	1	1	6	48.0	19
N -C. =C -C" -N -CH2-CH2-CH2-		----------------------					20
CO -C. =CH -C =C -N -CH2-CH2-CH2-	<5-C =>	----------------------					21

C. indicates a carbon common to two rings.
Fragments with a dashed line overlap with the primary biophore (e.g., fragments 4 and 5 overlap with 3).
Fragment 6 is an expanded version of 3.
Fragment 18 is an expanded version of 7.

3.3.2   Approach to Screening New Drugs

The empirical screening of a wide variety of fermentation products for the presence of new antibiotics significantly different from existing ones (using for example streptomycin-resistant microorganisms) led to the identification of a wide variety of antimicrobial agents, a fair number of which eventually found their way into our antimicrobial armamentarium.[19] Gradually, this empirical approach gave way to the use of semisynthetic and synthetic agents. This was made possible as mechanistic studies identified specific

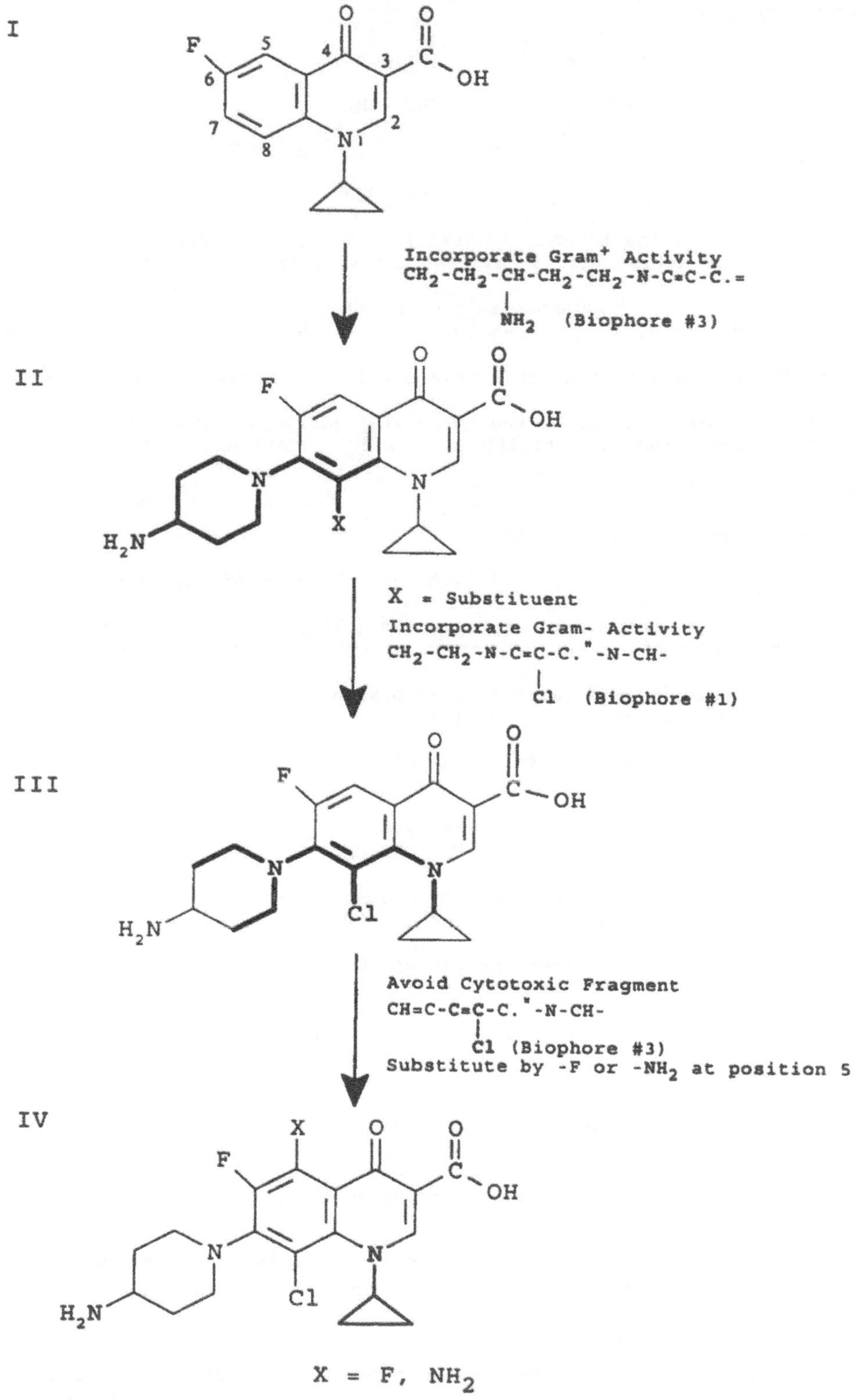

I

Incorporate Gram$^+$ Activity
CH$_2$-CH$_2$-CH-CH$_2$-CH$_2$-N-C=C-C.=
|
NH$_2$ (Biophore #3)

II

X = Substituent
Incorporate Gram- Activity
CH$_2$-CH$_2$-N-C=C-C."-N-CH-
|
Cl (Biophore #1)

III

Avoid Cytotoxic Fragment
CH=C-C=C-C."-N-CH-
|
Cl (Biophore #3)
Substitute by -F or -NH$_2$ at position 5

IV

X = F, NH$_2$

FIGURE 3.1 Design of a nontoxic broad spectrum fluoroquinolone.

STRUCTURE IV (X=F)

The molecule contains the Biophore   (nr.occ.= 2):

```
        C." -C                     CH2 -CH2
              \\                            \
                C   -N                        CH  -NH2
                       \                    /
                         CH2 -CH2
```

***   3 out of the known   3 molecules (100%) containing such Biophore
        are Gm⁺ active with an average activity of   47. (conf.level= 87%)
                                                    Constant is   856.0

** The following Modulator is also present:
    Ln Nr.Bi/Mol.Wt.   = -5.29 ;   Nr.Bioph/MW contrib.is -813.7

** The probability that this molecule is GM⁺ active is 80.0% **

** The compound is predicted to be VERY active, (act.= 42) **
** The projected Gm⁺ activity is   42.0   CASE units **

STRUCTURE IV (X=NH₂)

The molecule contains the Biophore

        2D fragment : [NH2-] <-- 3.0A --> [F -] conj generic

***   7 out of the known   9 molecules (78%) containing such Biophore
        are Gm⁺ active with an average activity of   48. (conf.level= 95%)
                                                    Constant is   49.3
** The following Modulator is also present:
    Log partition coeff.= 0.81 ;   LogP contribution is   1.5

    The molecule also contains the Biophore   :
       CH2-CH2-CH -CH2-CH2-N -C =C -C. =   <3-NH2>

    The molecule also contains the Biophore   :
       F -C =C -N -CH2-CH2-CH -CH2-CH2-   <3-C =>

    The molecule also contains the Biophore   :
       N -C. =C -C" -N -CH2-CH2-CH -CH2-   <5-CH2>

    The molecule also contains the Biophore   :
       N -C. =C -C" -N -CH2-CH2-CH -CH2-CH2-

** The probability that this molecule is Gm⁺ active is 72.7% **

+   increased to 96.9% due to the presence of the extra Biophore

** The compound is predicted to be VERY active, (act.= 51) **
** The projected Gm⁺ activity is   51.0   CASE units **

FIGURE 3.2   Predictions of activity against Gram-positive bacteria.

microbial targets such as DNA gyrases, β-lactamases, or dihydrofolate reductases. The recognition of such targets allowed medicinal chemists to synthesize a plethora of congeneric chemicals that could then be tested. These interactive processes enabled medicinal chemists to become SAR experts with special knowledge of specific classes of agents.

The availability of MULTICASE allowed the development of a highly efficient method based on the approaches described above. It had the added advantage that it could be used to identify potentially

STRUCTURE IV (X=F or NH₂)

The molecule contains the Biophore   (nr.occ.= 2):

```
            CH2 -CH2
                    \
                 N    -C
                   \\        //
                    C    -C.
                   /          \
                 Cl            N
                               /
                              CH
```

*** 11 out of the known 14 molecules (79%) containing such Biophore
    are Gm⁻ active with an average activity of 35. (conf.level= 98%)
                                        Constant is   36.4

** The probability that this molecule is Gm⁻ is 75.0% **

** The activity is predicted to be MODERATE, activity = 36**

** The projected Gm activity is 36.0 CASE units **

FIGURE 3.3   Predictions of activity against Gram-negative bacteria.

effective molecules even without a complete *a priori* understanding of their mechanism of action. This was because MULTICASE became the expert on the basis of the action of specific biological phenomena.

The approach is illustrated herein using the phenomenon of multiple drug resistance (MDR).[20] MDR is an important clinical phenomenon with profound implications on the therapeutic regimens used in cancer chemotherapy. MDR is manifested when, following a course of cancer chemotherapy, patients no longer respond to the agent or agents used in the initial course of treatment. But even more ominous is the observation that the patients, or the cancer cells derived from them, are no longer responsive toward a broad spectrum of unrelated anticancer agents, including some to which the patient had not been exposed previously. The prognosis for such patients is of course a matter of major concern.[21-23]

The basis of MDR appears to be an alteration in a specific glycoprotein that, in its abnormal form, actively exports therapeutic agents.[24-26] A number of empirical observations led to the recognition that *in vitro*, the MDR phenomenon was reversible by a number of seemingly unrelated chemicals (e.g., verapamil, cyclosporin).[27]

Even though the exact mechanism responsible for the reversal of MDR remains largely unknown, the phenomenon is amenable to analysis by MULTICASE. This then may possibly lead to the elucidation of the mechanism involved in this phenomenon. If no biophores are identified, then there may be no structural basis for the phenomenon, or a multitude of mechanisms — each perhaps unique for a specific chemical. Given the reproducibility of the reversal of MDR phenomenon, the second possibility is unlikely.

An examination of the published literature of reversers of MDR resulted in the development of a data base suitable for analysis by MULTICASE.[28] The analysis resulted in the identification of a series of biophores associated with the ability to reverse MDR (Table 3.6). It is to be noted that in spite of overall differences in the apparent nature of the chemicals, the biophores pointed to similarities between them (Table 3.7), suggesting that they shared common mechanisms of actions that can be the subject of further experimentation.

The identification of biophores associated with the reversal of MDR enabled us to search for new molecules with the potential of being clinically useful. In order to accomplish this, we assembled a test set consisting of approximately 46,000 existing, and therefore available, chemicals, i.e., no need to engage

```
STRUCTURE IV (X=F)

The molecule contains the Biophore    (nr.occ.= 1):


                OH   -CO
                        \
                     C    -CO
                    //         \
                          C.   -C
                         //         \\
                                      C
                                     /
                          C    =C
                         /          \
                       Cl            N
```

*** 3 out of the known   4 molecules (75%) containing such Biophore
     are non-toxic with an average activity of   67. (conf.level= 87%)
                                              Constant is   351.7
** The following Modulator is also present:
     LUMO coef.on   O4 is = 0.01 ;   Its contribution is -220.7

     The molecule also contains the Biophore   :
       F -C =C -C" -N -CH2-CH2-CH -NH2   <3-F >

     The molecule also contains the Biophore   :
       CO -C. =C -C =C -N -CH2-CH2-CH -NH2 <3-F >

** The probability that this molecule is non-toxic is 80.0% **

      increased to   91.7% due to the presence of the extra Biophore

** The compound is predicted to be EXTREMELY non-toxic (131) **
** The projected Inverse non-toxic activity is   131.   CASE units **

The molecule contains the Biophore   (nr.occ.= 2):

STRUCTURE IV (X=NH2)

The molecule contains the Biophore   (nr.occ.= 1):

```
                OH   -CO
                        \
                     C    -CO
                    //         \
                          C.   -C
                         //         \\
                                      C
                                     /
                          C    =C
                         /          \
                       Cl            N
```

*** 3 out of the known   4 molecules (75%) containing such Biophore
     are inverse toxic with an average activity of   67. (conf.level= 87%)
                                              Constant is   351.7
** The following Modulator is also present:
     LUMO coef.on   O4 is = 0.01 ;   Its contribution is -325.7

** The probability that this molecule is a non-toxic is 80.0% **

** The compound is predicted to be MARGINALLY non-toxic (26) **
** The projected Inverse non-toxic activity is   26.0   CASE units **

FIGURE 3.4   Predictions of lack of cytotoxic activity.

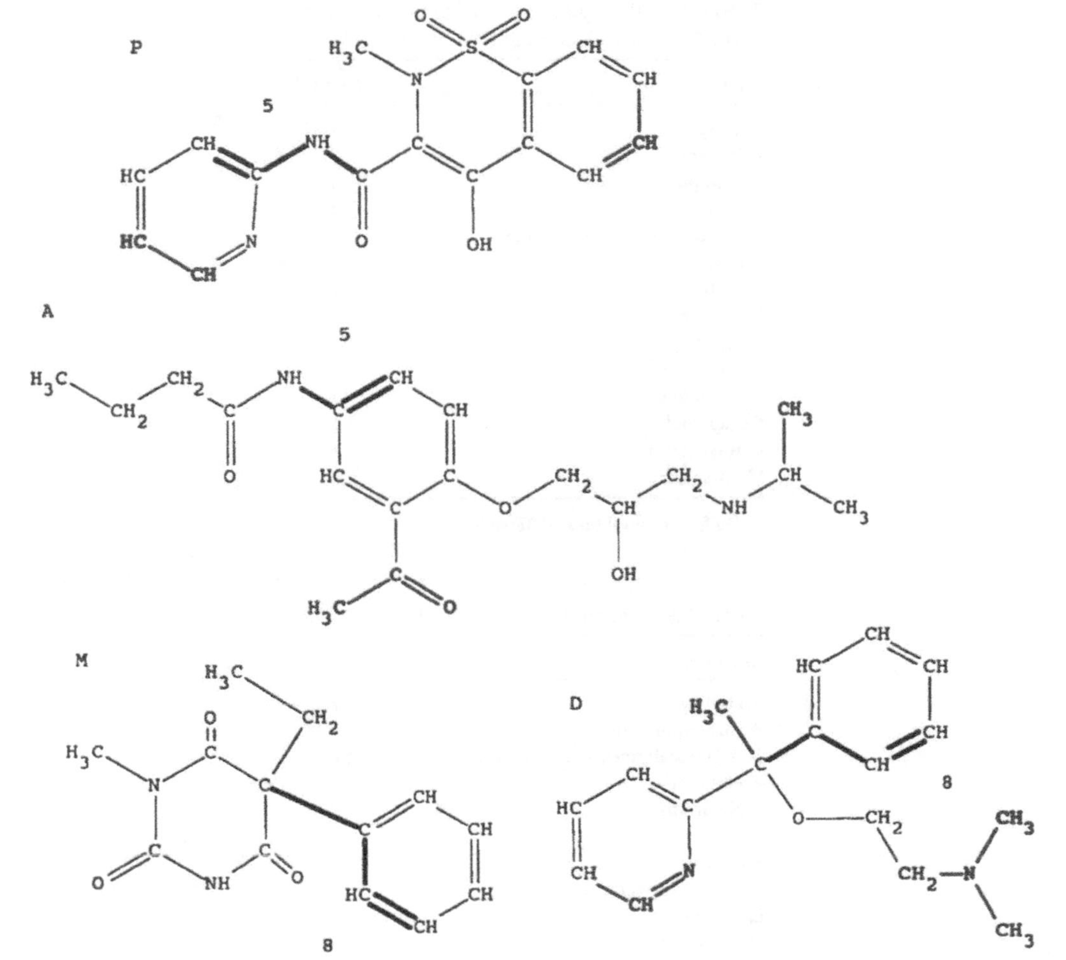

FIGURE 3.5 Some chemicals predicted to reverse MDR: piroxicam (P), acebutolol (A), mephobarbital (M), and doxylamine (D). The putative activities of P and A are due to biophore 5, whereas those of M and D are due to biophore 8. The biophores are shown in bold.

TABLE 3.6 Some Biophores Associated with Reversal of Multiple Drug Resistance

Fragment 1---2---3---4---5---6	Size		Nr. of Fr.	Nr.
[N –] <--11.7A-->	[N –]		28	1
CH2–N –CH2–CH2–			39	2
O –CH =C –CH –C =		<3–O >	12	3
S –CH =C –CH –C =		<3–S >	4	4
NH –C =CH –			4	5
[C –] <--48.1A -->	[NH –]		5	6
N =C. –CH =C –			3	7
CH =CH–C" –C –			6	8

<3–O > indicates the presence of an oxygen atom on the third atom from the left.
Biophores 5 and 8 are shown in Figure 3.5.

TABLE 3.7   Examples of the Origins of Biophores in the Data Base Associated with Reversal of MDR

Chemical	Biophore*
Cyclosporin	1
Diltiazem	1
Nicardipine	1
Monensin	1
Trimethoxybenzoylyohimbine	1
Amiodarone	2
Quinacrine	2
Reserpine	2
Verapamil	2,8
Primaquine	5
Etomidoline	5
Decaprenol	6
Retinyl acetate	8
Fluphenazine	8

*Refers to biophores of Table 3.6.

TABLE 3.8   Chemicals Predicted to Reverse MDR

Chemical	Biophore*
Vindoline	2
7-Fluoroquinoline	7
7-O-Docosahexaenoylocadaic acid	1,6
Acebutolol	5
Alphaprodine	8
Alverine	2
Butacaine	1
α-Dihydroergocryptine	1
Laudanosine	1
Doxylamine	8
Mephobarbital	8
Rapamycin	1
Piroxicam	5
Acranil dihydrochloride	7
Dextromoramide	2,8

* Refers to biophores of Table 3.6.
The structures of piroxicam, acebutolol, mephobarbital, and doxylamine are shown in Figure 3.5.

in synthetic activities. Subgroups of these chemicals were screened by MULTICASE for their potential to reverse MDR and some new ones were identified (Table 3.8). Arrangements are currently being made to test the identified molecules experimentally. The results of these assays will be added to the data base and this should result in the identification of more refined and therefore more predictive biophores. Of course, the availability of our other toxicity data bases (Table 3.1) should help eliminate candidate molecules predicted to possess unacceptable toxicological properties. Thus, while both piroxicam (Figure 3.5) and acebutolol (Figure 3.5) are predicted to reverse multiple drug resistance by virtue of the presence of biophore 5 (Tables 3.6 and 3.8), piroxicam would appear to be a more promising candidate for further testing; unlike acebutolol (Figure 3.6), it is not predicted to be a carcinogen in any of the carcinogenicity data bases available to us. Similarly, both mephobarbital and doxylamine (Figure 3.5) are predicted to revert MDR by virtue of the presence of biophore 8; doxylamine would be the preferred candidate, as the former is associated with a probability of carcinogenicity in rodents. It should be mentioned, however, that neither of the two potential carcinogens (mephobarbital, acebutolol) are

```
The molecule contains the Biophore    (nr.occ.= 1):

  (1)        O    -CH2
                      \
                       CH   -CH2

  ***        6 out of the known    6 molecules (100%) containing such Biophore
             are Mouse carcinogens with an average activity of  38. (conf.level= 98%)
                                                       Constant is        7.8
  **         The following Modulators are also present:
             Ln Nr.Bi/Mol.Wt.    = -5.82 ;        Nr.Bioph/MW contrib.is   39.4
             HOMO coef.on  O1 is =  0.04 ;             Its contribution is   0.4

  (2)        The molecule also contains the Biophore   :
                     CH =CH -C  =C  -CH =                     <3-0 >

  **         The probability that this molecule is a Mouse carcinogen is  87.5%  **
             increased to 93.1% due to the presence of the extra Biophore

  **         The compound is predicted to be VERY active **
  **         The projected Mouse carcinogenic activity is 48.0 CASE units **
```

1

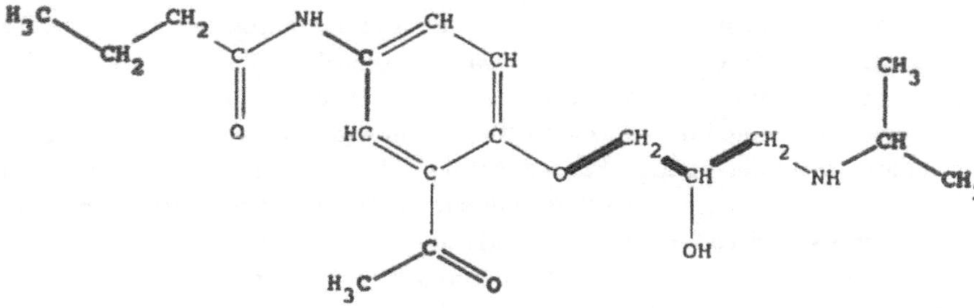

2

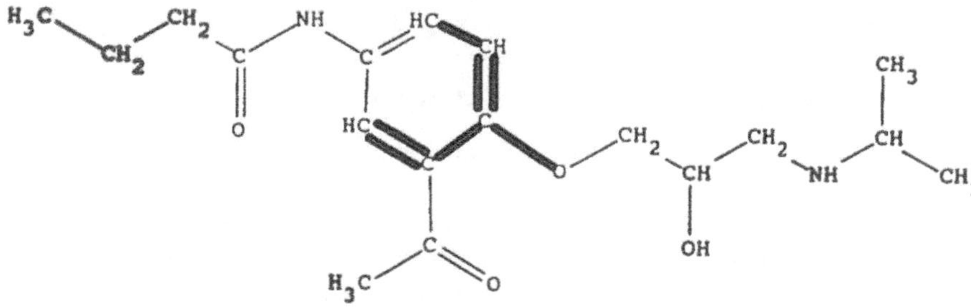

FIGURE 3.6   Prediction of the carcinogenicity of acebutolol. The prediction is associated with the two biophores shown in bold. The predicted potency (48 CASE units) corresponds to a TD_{50} value of 179.2 mg/kg/day.

predicted to be *Salmonella* mutagen carcinogens. Since human carcinogenic risk is usually associated with mutagenic carcinogens, the relevance of these findings will need to be evaluated further. Still, it would seem prudent in a screening program to identify new therapeutic agents to first investigate chemicals not associated with any potential for carcinogenicity.

Alternatively, as with the fluoroquinolones, the promising molecules can be altered synthetically so as to eliminate or greatly decrease the unwanted side effects while retaining and even enhancing their therapeutic potential.

TABLE 3.9   Distribution of 6Å Descriptor Among a Group of Estrogens and Antiestrogens

Chemical	Presence of 6 Å Descriptor
Diethylstilbestrol	+
Estradiol	+
Benzestrol	+
Coumestrol	-
Indenestrol A	+
Megestrol	-
Norgestrel	-
Tamoxifene metabolites*	+
Allenolic Acid	+
LY 117018	-
Ethamoxytriphetol (MER 25)	-
2,6-Bis ((3-methoxy-4-hydroxyphenyl) methylene) cyclohexanone	-
p-Hydroxy-3-phenylacetylamino-2,6-piperidinedione	-

*See, for example, Figures 3.7 and 3.8.

3.3.3   Antiestrogens

In the course of a series of studies on the structural basis of nonmutagenic carcinogens (i.e., "nongeno-toxic" carcinogens), MULTICASE identified a two-dimensional lipophilic descriptor (6.0 Å) associated with carcinogenicity (Figure 3.7) that is also present in some estrogens and some antiestrogens.[29] Our findings suggested a common mechanism for the carcinogenicity of some estrogens, antiestrogens, and their metabolites. Moreover, we hypothesized that the 6 Å descriptor described a portion of the ligand recognized by an estrogen receptor. Further studies revealed that not all antiestrogens possessed the 6 Å descriptor (Table 3.9). Indeed further analysis indicated that the antiestrogens lacking the 6 Å descriptor bound to a different estrogen receptor.[30-32] Additionally, a number of antiestrogens lacking the 6 Å descriptor were predicted by MULTICASE to be devoid of rodent carcinogenicity (e.g., p-hydroxy-3-phenylacetylamino-2, 6-piperidinedione). These observations lead us to suggest that antiestrogens that lack the 6 Å descriptor might be good therapeutic candidates presumably devoid of carcinogenicity. In that connection it is of interest to note that we have developed, in parallel, another expert system, META, capable of predicting metabolism (Figure 3.8) and thereby predict not only the properties of the parent molecules but also of their metabolites.

3.3.4   Informed Risk Assessment: Praziquantel

Schistosomiasis is a widely distributed parasite afflicting millions of people. In the recent past, the antischistosomal agents of choice were niridazole and hycanthone. However, because these agents are mutagens and suspected rodent carcinogens,[33] consideration was given to find a replacement for them. This was because mutagenic carcinogens are considered to present a much greater carcinogenic risk to humans than nonmutagens.[34] In fact, the vast majority of recognized human carcinogens are also mutagens/genotoxicants.[35-37]

The agent chosen to replace the above is praziquantel (PZ, Figure 3.9). The agent is reportedly devoid of mutagenicity[38] and hence, even if it was found to be a rodent carcinogen, being nonmutagenic it would presumably represent a lesser risk to humans.[34] The availability of a number of MULTICASE toxicological models allowed us to explore the potential adverse effects of praziquantel.

Thus, PZ was predicted to be nonmutagenic in *Salmonella*, confirming the previous reports.[38] Additionally, PZ is devoid of structural alerts associated with DNA reactivity. Thus, by these commonly accepted criteria,[39] PZ would be considered "nongenotoxic." Additionally, with the use of the expert system META a further 180 putative metabolites of PZ (Figure 3.10) were generated. These were also submitted to MULTICASE, though none of them were predicted to be mutagenic.

```
The molecule contains the Biophore:

        2D fragment : [C  -]   <-- 6.0A -->   [OH -]

    ***     45 out of the known  47 molecules  (96%) containing such Biophore
            are mouse carcinogens with an average activity of  76. (conf.level=100%)
                                                    Constant is      47.5
    **      The following Modulator is also present:
            Log partition coeff.= 7.23 ;           LogP contribution is   42.2


The molecule also contains the Biophore    (nr.occ.= 2):
```

```
                        //
(B)            CH2 -C
                        \
                        C   =CH
                         `   \
                    CH        CH
                    \\        //
                     CH  -C
                            \
                            OH
```

```
    ***     4 out of the known   4 molecules (100%) containing such Biophore
            are mouse carcinogens with an average activity of   92. (conf.level= 94%)


    **      The probability that this molecule is mouse carcinogen  93.9%  **
            increased to   96.8% due to the presence of the extra Biophore


    **      The compound is predicted to be EXTREMELY active **
    **      The projected carcinogenic potency is   90.0      CASE units **
```

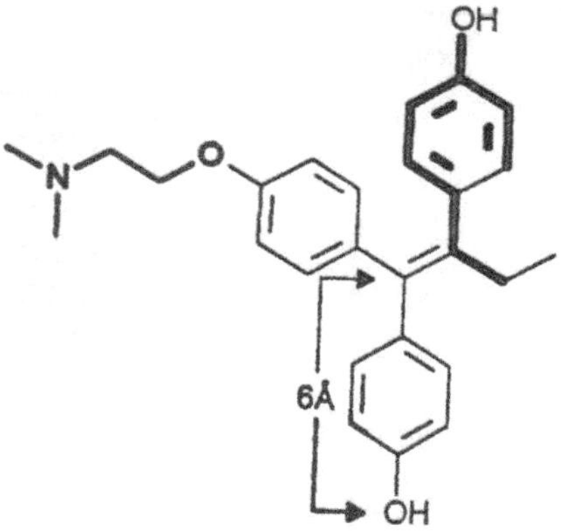

FIGURE 3.7   Predicted carcinogenicity in mice of a metabolite (4,4'-dihydroxytamoxifen) of tamoxifen. This prediction is based upon the presence of the 6 Å 2-D descriptor and biophore B (shown in bold).

Even though PZ is reported to be noncarcinogenic to the rat[40,41] examination of PZ for the presence of structural features associated with carcinogenicity in rodents (CPDB) resulted in the prediction, based upon MULTICASE, that PZ had the potential for being a carcinogen (Figure 3.8). This prediction is based upon the presence in PZ of the biophore $CO–CH_2–N$ which is present in five molecules in CPDB, four of which are carcinogens (Table 3.10).

It is to be noted that while the biophore in ICRF-159 (Figure 3.11) resembles that of PZ, the similarity of the conformation of the biophore in the other molecules to PZ may be more problematic. Moreover,

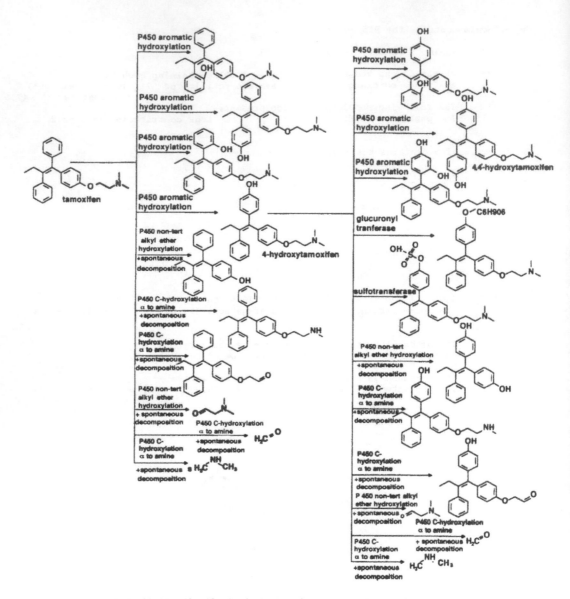

FIGURE 3.8   Identification by META of some tamoxifen metabolites.

the carcinogenicity of 5-nitrofurantoin (Figure 3.11) may, in fact, be derived from the presence of the 5-nitro moiety, a functionality associated with DNA reactivity.[42]

In addition to being associated with an 80% probability of carcinogenicity, the biophore $CO–CH_2–N$ is also predicted to contribute 45.5 CASE units of activity (potency) to the final carcinogenicity of PZ. Based upon QSAR regression analysis it was found by MULTICASE that this biophore is associated with the following modulators: (1) $(–80.1)*((HOMO + LUMO)/2)$ and (2) $(97.2)*(HOMO$ energy on the oxygen atom). Since the calculated electronegativity (i.e., $(HOMO + LUMO)/2)$ of PZ is $–0.12$ (Figure 3.9), the contribution of the electronegativity is $[(–80.1)(–0.12)]$ or 9.6 CASE units. However, because the HOMO on the oxygen is negligible, its contribution to the total activity is insignificant. Hence, a projected potency of 56 CASE units is obtained. This corresponds to a TD_{50} value of 0.31 mmoles/kg/day or 95 mg/kg/day. This is an unexpectedly high projected carcinogenic potency projected for PZ and is in the same range as that of safrole and o-toluidine hydrochloride.

The molecule contains the Biophore (nr. occ.=1):

```
CO -CH2
   \
    N
```

***   4 out of the known 5 molecules (80%) containing such Biophore are Rodent carcinogens with
        an average activity of 38. (conf. level=89%). Constant is 45.5

**   The following Modulators are also present:
        Electronegativity = -0.12; Its contribution is 9.6
        HOMO coef. on O1 is = 0.00; Its contribution is 0.3

**   The probability that this molecule is a Rodent carcinogen is 71.4% **

**   The compound is predicted to be VERY active **
**   The projected Rodent carcinogenic activity is 56 CASE units **

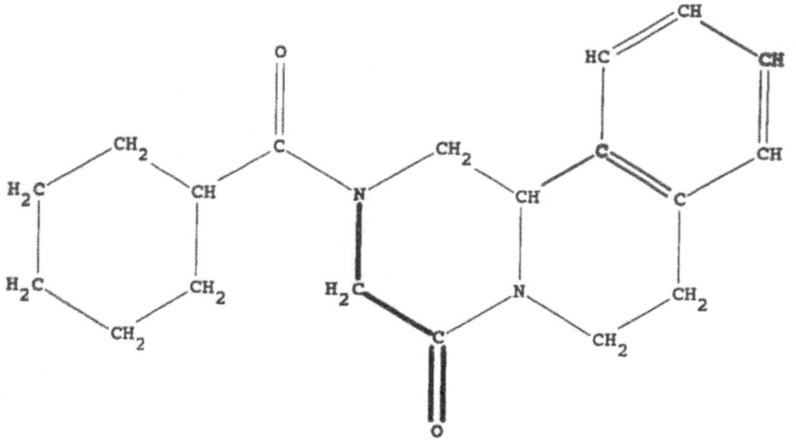

FIGURE 3.9   MULTICASE prediction of the rodent carcinogenicity of praziquantel. This prediction is based upon CPDB. A potency of 56 CASE units corresponds to a TD_{50} value of 0.31 mmoles/kg/day. The derivation of the biophore from the molecules in the data base is shown in Table 3.10.

Examination of PZ using the NTP rodent carcinogenicity data base also results in the prediction based upon CASE of carcinogenicity that is restricted to a single gender of a single species. This prediction is based on the three biophores shown in Figure 3.12. While it may be argued that the molecules contributing biophores to the prediction of carcinogenicity in the NTP data base may represent a different environment than PZ (see Table 3.11 for the origin of the biophores), still the prediction confirms that based upon CPDB. Additionally, it is noteworthy that the biophores are derived primarily from molecules that are also nonmutagenic and non-structurally alerting.

Thus, by the criteria used to characterize "nongenotoxic" carcinogens (see above), PZ is predicted to be a "nongenotoxic" carcinogen. While for "genotoxic" carcinogens there is a uniformity of action, i.e., electrophilic attack on the DNA, there is no such unique action for "nongenotoxic" carcinogens.[34,43] However, cellular and systemic toxicity followed by mitogenesis and cellular proliferation have been suggested as mechanisms of "nongenotoxic" carcinogenicity.[44,45] They could result from receptor-mediated phenomena (e.g., TCDD-, estrogen-, or peroxisome-specific receptors). In that context it should be noted that di(2-ethylhexyl)adipate and di(2-ethylhexyl)phthalate, which share biophores with PZ (Table 3.11), are inducers of peroxisome proliferation and activate specific receptors.[46,47] With respect to receptor-mediated mechanisms, these are thought to occur in a lipophilic environment. Indeed, the molecules associated with the biophores are primarily lipophilic as evidenced by their high log P values

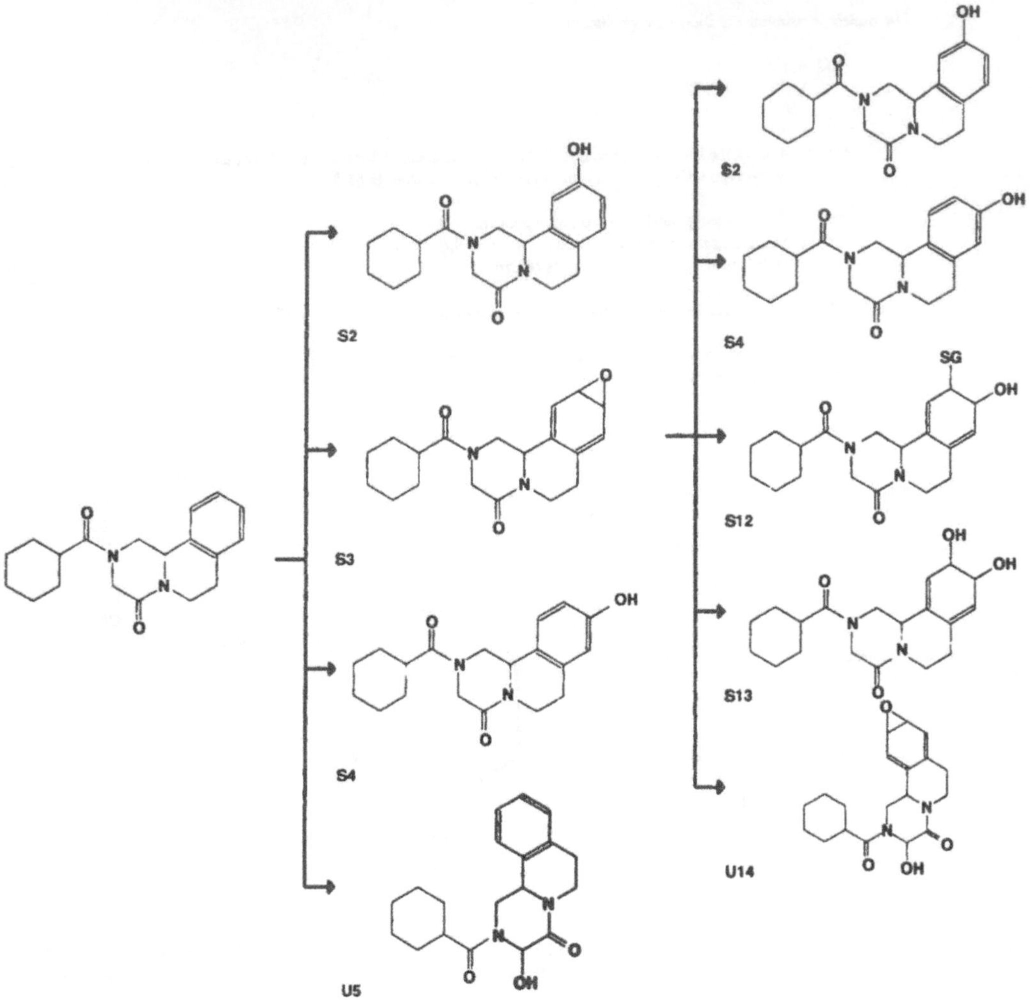

FIGURE 3.10 Example of META-projected metabolism of PZ (S1) and of the further metabolism of S3. S2 and S4 were derived from S1 by cytochrome P450 hydroxylation; S3 by cytochrome P450-epoxidation. U5 and U14 by P450-mediated C-hydroxylation. S12 and S14 were derived from S3 by spontaneous cleavage of the epoxide; S12 by glutathione transferase and S13 by epoxide hydratase. META identified a total of 180 putative metabolites of PZ.

TABLE 3.10   Origin of Biophore CO-CH$_2$-N

Chemical	Case No.	Carcinogenicity TD$_{50}$	NTP**	Mutagenicity
Nitrilotriacetic acid	139139	1450	60	–
Nitrofurantoin	67209	698	60	+
Trisodium nitrilotriacetate	18662538	1530	60	–
ICRF-159	21416875	11		(–)*
EDTA, Trisodium trihydrate	150389	NC	NC	–

Derivation of the biophore present in PZ (see Figure 3.8) from chemicals present in the CPDB. The biophore is shown embedded in ICRF-159, nitrofurantoin, EDTA, and nitrilotriacetic acid in Figure 3.11. The TD$_{50}$ values are expressed in gavage equivalents as mg/kg/day.

*The mutagenicity in *Salmonella* of ICRF-159 is not reported in the NTP data base. However, this chemical lacks a "structural alert" for DNA-reactivity.[56]

**For chemicals also listed in the NTP data base, the carcinogenic spectrum (see text) is given.

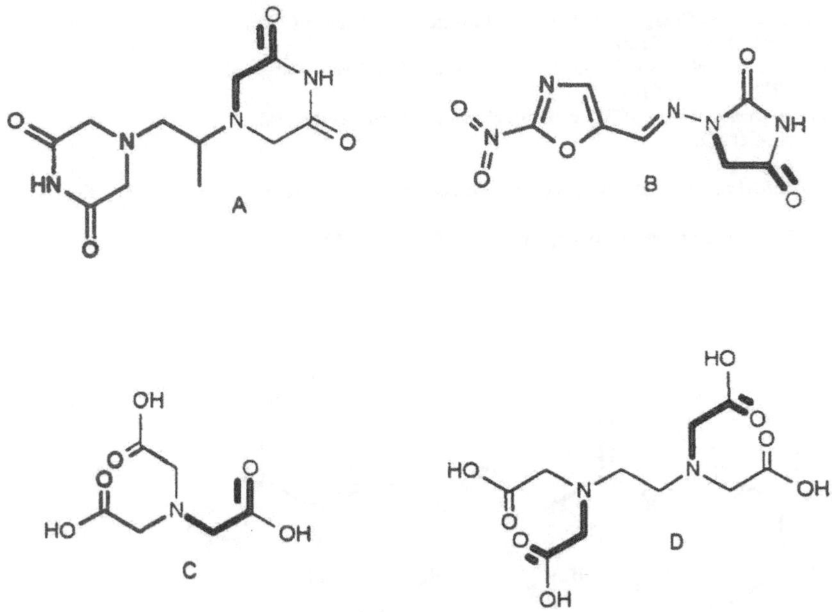

FIGURE 3.11 Presence of biophore CO-CH$_2$-N in ICRF-159 (A) nitrofurantoin (B), nitrilotriacetic acid (C), and EDTA (D). The carcinogenicity and mutagenicity of these chemicals are summarized in Table 3.10.

TABLE 3.11 Origin of Biophores Contributing to Carcinogenicity of Praziquantel

Chemical	Biophore A	Biophore B	Biophore C	Salmonella mutagenicity	Carcinogenicity* (NTP)	TD$_{50}$**	Log P
Di(2-ethylhexyl)phthalate		+		−	60	499	6.0511
Phenestrin		+		−	60	0.21	9.999
Zearalenone		+		−	50	22	3.222
Di(2-ethylhexyl)adipate	+	+		−	40	3050	5.410
Phenylbutazone		+	+	−	60		3.058
Tris(2-ethylhexyl)phosphate		+		−	30	2560	5.997
11-Aminoundecanoic acid	+			−	50	833	1.933
Piperonyl sulfoxide	+			−	30	62.2	2.925
Captan			+	+	40	86.1	2.243

The biophores derived from the NTP data base are shown embedded in PZ in Figure 3.12.
*The carcinogenicity (NTP data case) is expressed as the "carcinogenic spectrum" (see text).
**For chemicals also listed in CPDB, the TD$_{50}$ values, in gavage equivalents, is given in mg/kg/day.
The log P (octanol/water partition coefficient) was calculated as described previously.[57]

(Table 3.11). Thus, it is conceivable that the projected carcinogenicity of PZ may derive from such a receptor mediated activity.

In view of the fact that "nongenotoxic" rodent carcinogens are thought not to present a major risk to humans (see above) and because of the proven therapeutic properties of PZ, it would seem that the beneficial effects of PZ far outweigh the potential risk for carcinogenicity. Still, in view of the widespread use of PZ, a surveillance of its potential for causing cancer in humans appears appropriate.

3.3.5 Mechanisms of Action

The ability of a chemical to induce micronuclei in the bone marrow of rodents is taken as an indication of a genotoxic potential.[48] Indeed, the determination of the ability of a chemical to induce micronuclei

(A) 63 % chance of being ACTIVE due to substructure (Conf. level=87%):
    CH2-CH2-CH2-CH2-
(B) 66 % chance of being ACTIVE due to substructure (Conf. level=98%):
    CH2-CH2-CH2-CH -
(C) 75 % chance of being ACTIVE due to substructure (Conf. level=75%):
    N -CO -CH -

*** OVERALL, the probability of being a Rodent carcinogen is 90.8% ***

** The predicted carcinogenic spectrum is 32 CASE units **

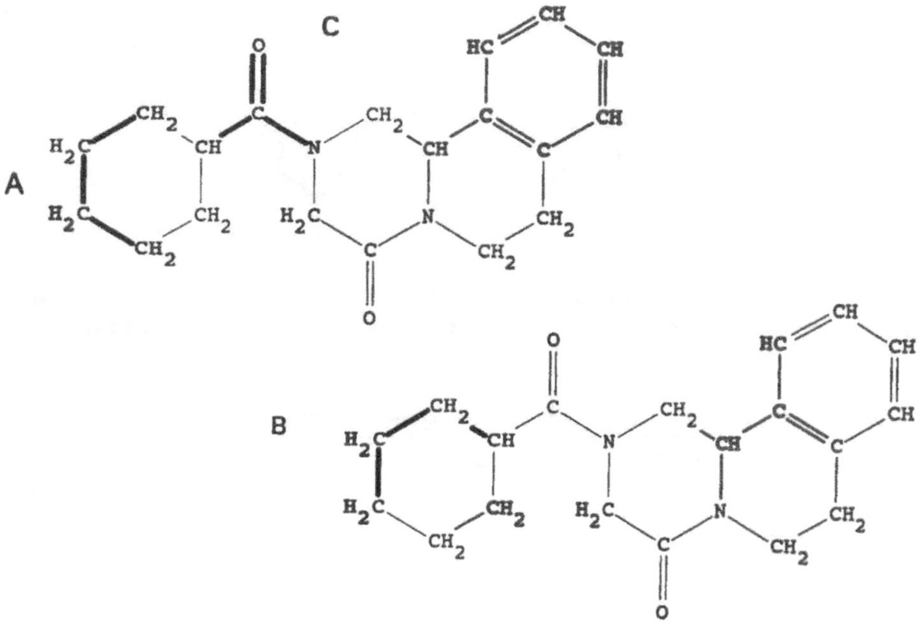

FIGURE 3.12   CASE prediction of the rodent carcinogenicity of praziquantel. This prediction is based on the NTP rodent carcinogenicity data base. The derivation of the biophores from the molecules in the data base is given in Table 3.11.

has been incorporated into testing protocol, especially as a second-tier test for chemicals to confirm the *in vivo* genotoxicity of chemicals that respond positively in an *in vitro* assay such as the *Salmonella* mutagenicity assays.[34,49]

However, recent studies in our laboratory have cast doubt on this possible scheme. Thus, we have processed two data bases: the *in vivo* induction of micronuclei[50] and the ability of chemicals to inhibit tubulin polymerization. As mentioned above, the ability to induce micronuclei is taken as a measure of *in vivo* genotoxicity and therefore of a potential for carcinogenicity. Moreover, genotoxic carcinogens are taken to present a greater risk to humans than nongenotoxic ones.[34] On the other hand inhibition of tubulin polymerization is taken to represent a cancer chemotherapeutic target.

In a study to explore the possible mechanistic relationship between the induction of micronuclei and inhibition of tubulin polymerization,[51] we discovered a significant overlap between the biophores associated with inhibition of tubulin polymerization and those associated with the induction of micronuclei but not with mutagenicity in *Salmonella* or structural alerts for DNA reactivity (Tables 3.12 and 3.13). This suggested to us that the two phenomena were mechanistically related and that a major mechanism for the induction of micronuclei may result from inhibition of tubulin polymerization. If indeed this is

TABLE 3.12 Comparison of Biophores Associated with Tubulin Polymerization Perturbation and Other Biological Phenomena

Tubulin Polymerization Perturbation Fragments	MN	CTox	Salm	SA
Br –C =		X		
NH –C. =CH –		X		
CH″–CH =C. –C. =			X	X	X
CO –O –C =CH –		X	X	
N –C =CH –CH =C –		X	X	X
CH =CH –C =CH –C. =	<3-F >	X	X		X
N –C =CH –CH =C –CH =	X	X	X	
O –C =C –CH =C –CH2-	<3-OH >	X	X	
CH2-C =CH –C =C –CH =	<4-OH >	X		
OH –C =C –CH =CH –C″ –CH2-	<3-O >	X	X	
CH″–CH =C. –C. =C –C =C –	<5-O >	X	X		X
CH2-O –C. =CH –C =C –CH =	<5-O >	X	X	
CH =CH –CH =C –C″ –NH –C. =CH –		X	X	
O –C =C –CH =C. –CH2-CH2-CH –		X	X	
O –C. =CH –C =C –CH =C. –O –	<4-O >	X	X	
OH –C =CH –C =CH –CH =C –O –	<4-CH2 >	X	X		X
CH2-C. =CH –C =C –C =C. –C. =	<5-O >	X	X	
O –CH2-O –C. =CH –C =C –CH =	<6-O >	X	X	
CH3-O –C =C –CH =C –CH =CH –C =	<3-C=>	X	X		X
O –C =C –C =CH –C.″–CH2-CH2-CH –	<3-O >	X	X	
CH3-O –C =C –C =CH –C.″–CH2-CH2-	<4-O >	X	X	
CH2-CH2-C. =CH –C =C –C =C. –C. =	<6-O >	X	X	
O –C =C –C″ –CO –CH =CH –C =CH –	<8-CH=>	X	X	
O –C =C –C =CH –C.″–CH2-CH2-CH –NH –	<3-O >	X	X	
NH –CH –CH2-CH2-C. =CH –C =C –C =C. –	<8-O >	X	X	
O –C =C –CH =C. –CH2-CH2-CH –C. =CH –	<8-NH >	X	X	
O –C =C –C =C –CO –CH =CH –C =CH –	<9-CH=>	X	X	

MN: Micronuclei induction   Salm: *Salmonella* mutagenicity
CTox: Cellular toxicity   SA: Structural alerts

C. indicates a carbon shared by two rings. C″ indicates a carbon atom connected by a double bond to another atom. <3-F> indicates a fluorine atom substituted on the third atom (carbon) from the left. When an atom is shown as unsubstituted (e.g., the C atom of the first biophore shown above) it means that this carbon may be substituted by any atom or functionality (e.g., halogen, CH₃-, etc.) except a hydrogen.

Overlaps between data bases are indicated by X.

TABLE 3.13 Overlap of Biophores Associated with Perturbation of Tubulin Polymerization and Other Biological Phenomena

Data base	Overlap with tubulin polymerization
Induction of micronuclei	71%
Cellular toxicity	71%
Mutagenicity in *Salmonella*	9%
Structural alerts for DNA reactivity	14%

a correct interpretation of the data, it has important consequences regarding the safety or lack thereof of therapeutic agents.

Thus paclitaxel, a promising cancer chemotherapeutic agent,[52] is a powerful perturber of tubulin polymerization,[53] which, in fact, may be the basis of its beneficial effects. Recently, however, the ability of paclitaxel to induce micronuclei was reported.[54,55] This, in turn, led to a warning that paclitaxel may be genotoxic and therefore induce secondary cancers in treated patients.[54] However, unlike other genotoxic agents, paclitaxel is devoid of "structural alerts" for DNA-reactivity, mutagenicity in *Salmonella*, and is nonelectrophilic. Thus it may well be that the *in vivo* induction of micronuclei by paclitaxel is not the result of a genotoxic event but rather of its inhibition of tubulin polymerization, which may be the basis of therapeutic activity. If this is indeed the correct interpretation, then it can be predicted that paclitaxel is not a genotoxic carcinogen and hence does not present an increased risk to patients.

3.4   Conclusions

Application of expert SAR systems to the design of therapeutic agents as well as the evaluation of toxicological hazards affords a useful tool to balance the possible risks posed by such agents with their benefits. It should be emphasized, however, that these expert systems are designed to supplement the armamentarium available to human experts and not to supplant human expertise. In fact, such knowledge-based systems are of maximal efficacy when used by humans with the appropriate expertise. Additionally, with respect to identifying health risks, these systems are not meant to eliminate the experimental determination of toxicological effects. Rather these methods should be used for guidance in the research and development phase, to prioritize toxicological concerns regarding candidate chemicals and to allocate resources appropriately. In the context of these parameters, the expert systems described herein can play a pivotal role in the development of safe and efficacious therapeutic agents.

Acknowledgments

Funding provided by Concurrent Technologies Corporation/National Defense Center for Environmental Excellence in support of the U.S. Department of Defense (Contract No. DAAA21-93-C-0046), the Center for Indoor Air Research and the Center for Alternatives to Animal Testing. All of the studies described in this article were made utilizing the MULTICASE and META programs, available from MULTICASE Inc., 25825 Science Park Dr., #100, Cleveland, OH 44122.

References

1. Hansch, C., Leo, A., Exploring QSAR fundamentals and applications in chemistry biology, American Chemical Society, Washington, D.C., 1995.

2. Kubinyi, H., QSAR: Hansch analysis and related approaches, *Methods and Principles in Medicinal Chemistry,* Volume 1, VCH Publishers, New York, 1993.

3. Klopman, G., MULTICASE 1, A hierarchical Computer Automated Structure Evaluation program, *Quantitative Structure-Activity Relationships,* 11, 176, 1992.

4. Klopman, G., Dimayuga, M., Talafons, J., META 1, a program for the evaluation of metabolic transformation of chemicals, *J. Chem. Inf. Comput. Sci.,* 34, 1320, 1994.

5. Talafous, J., Sayre, L. M., Mieyal, J. J., Klopman, G., META 2, a dictionary model of mammalian xenobiotic metabolism, *J. Chem. Inf. Comput. Sci.,* 34, 1326, 1994.

6. Lee, S. J., Konishi, Y., Yu, D. T., Miskowski, T. A., Riviello, C. M., Macina, O. T., Frierson, M. R., Kondo, K., Sugitani, M., Sircar, J. C., Blazejewski, K. M., Discovery of potent cyclic GMP phosphodiesterase inhibitors, 2-pyridyl and 2-imidazolyl quinazolines possessing cyclic GMP phosphodiesterase and thromboxane synthesis inhibitory activities, *J. Med. Chem.,* 38, 3547, 1995.

7. Klopman, G., Macina, O. T., Drug design based on an artificial intelligence approach. In *Computer Aided Innovation of New Materials II,* Elsevier Science Publishers, Amsterdam, pp. 1135–1140, 1993.

8. Macina, O. T., Rigby, B. S., Computer automated structure evaluation of antifungal 1-vinylimidazoles, 1, 2-disubstituted propenones, and azolylpropanolones, *J. Pharm. Sci.,* 79, 725, 1990.

9. Klopman, G., Macina, O. T., Levinson, M. E., Rosenkranz, H. S., Computer automated structure evaluation of quinolone antibacterials, *Antimicrob. Agents Chemother.,* 31, 1831, 1987.

10. Klopman, G., Pichelintsev, D., Frierson, M., Pennisi, S., Renskers, K., Dickens, M., Multiple computer automated structure evaluation methodology as an alternative to *in vivo* eye irritation testing, *ATLA*, 21, 14, 1993.

11. Rosenkranz, H. S., Ennever, F. K., Dimayuga, M., Klopman, G., Significant differences in the structural basis of the induction of sister chromatid exchanges and chromosomal aberrations in Chinese hamster ovary cells, *Environ. Mol. Mutagen.*, 16, 149, 1990.

12. Rosenkranz, H. S., Klopman, G., The structural basis of the mutagenicity of chemicals in *Salmonella* typhimurium: the national toxicology program data base, *Mutation Res.*, 228, 51, 1990a.

13. Rosenkranz, H. S., Klopman, G., Structural basis of carcinogenicity in rodents of genotoxicants and nongenotoxicants, *Mutation Res.*, 228, 105, 1990b.

14. Rosenkranz, H. S., Takihi, N., Klopman, G., Structure activity-based predictive toxicology: an efficient and economical methods for generating non-congeneric data bases, *Mutagenesis*, 6, 391, 1991.

15. Rosenkranz, H. S., Klopman, G., The application of structural concepts to the prediction of the carcinogenicity of therapeutic agents. In: *Burger's Medicinal Chemistry and Drug Discovery*, 5th Ed., Vol. 1, *Principles and Practice*, Wolff, M. E., Ed., John Wiley & Sons, New York, pp. 223, 1995.

16. Fernandes, P. B., Mode of action, *in vitro* and *in vivo* activities of the fluoroquinolones, *J. Clin. Pharmacol.*, 28, 156, 1988.

17. Domagala, J. M., Hanna, L. D., Heifetz, C. M., Hutt, M. P., Mich, T. F., Sanchez, J. P., Solomon, M., New structure-activity relationships of the quinolone antibacterials using the target enzyme. The development and application of a DNA gyrase assay, *J. Med. Chem.*, 29, 394, 1986.

18. Suto, M. J., Domagala, M. M., Roland, G. E., Mailloux, G. B., Cohen, M. A., Fluoroquinolones: relationships between structural variations, mammalian cell cytotoxicity, and antimicrobial activity, *J. Med. Chem.*, 35, 4745, 1992.

19. Stanier, R. Y., Doudoroff, M., Adelberg, E. A., *The Microbial World*, 2nd Ed., Prentice-Hall, Englewood Cliffs, NJ, pp. 637, 1965.

20. Gottesman, M. M., How cancer cells evade chemotherapy, Sixteenth Richard and Hinda Rosenthal Foundation Award Lecture, *Cancer Res.*, 53, 747, 1993.

21. Bellamy, W. T., Dalton, W. S., Dorr, R. T., The clinical relevance of multidrug resistance, *Cancer Invest.*, 8, 545, 1990.

22. Dalton, W. S., Drug resistance modulation in the laboratory and the clinic, *Semin. Oncol.*, 20, 64, 1993.

23. Bénard, J., Bourhis, J., Riou, G., Clinical significance of multiple drug resistance in human cancers, *Anticancer Res.*, 10, 1297, 1990.

24. Juranka, P. F., Zastawny, R. L., Ling, V., P-glycoprotein: multidrug-resistance and a superfamily of membrane-associated transport proteins, *FASEB J.*, 3, 2583, 1989.

25. Gottesman, M. M., Pastan, I., The multidrug transporter, a double-edged sword, *J. Biol. Chem.*, 263, 12163, 1988.

26. West, I. C., What determines the substrate specificity of the multi-drug-resistance pump? *Trends Biochem. Sci.*, 15, 42, 1990.

27. Beck, W. T., Multidrug resistance and its circumvention, *Eur. J. Cancer*, 26, 513, 1990.

28. Klopman, G., Srivastava, S., Kolossvary, I., Epand, R. F., Ahmed, N., Epand, R.M., Structure-activity study and design of multidrug-resistant reversal compounds by a computer automated structure evaluation methodology, *Cancer Res.*, 52, 4121, 1992.

29. Rosenkranz, H. S., Cunningham, A., Klopman, G., Identification of a 2-D geometric descriptor associated with nongenotoxic carcinogens and some estrogens and antiestrogens, *Mutagenesis*, 11, 95, 1996.

30. Black, L. J., Goode, R. L., Evidence for biological action of the antiestrogens LY117018 and tamoxifen by different mechanisms, *Endocrinology*, 109, 987, 1989.

31. Scholl, S. M., Huff, K. F., Lippman, M. E., Antiestrogenic effects of LY 117018 in MCF-7 cells, *Endocrinology*, 113, 611, 1993.

32. Coradini, D., Biffi, A., Cappelletti, V., Di Fronzo, G., Activity of tamoxifen and new antiestrogens on estrogen receptor positive and negative breast cancer cells, *Anticancer Res.*, 14, 1059, 1994.
33. IARC, Monographs on the Evaluation of Carcinogenic Risks to Humans: Genetic and Related Effects, Supplement 7. Overall Evaluation of Carcinogenicity: An Updating of IARC Monographs Volume 1 to 42. International Agency for Research on Cancer, Lyon, France, 1987.
34. Ashby, J., Morrod, R. S., Detection of human carcinogens, *Nature*, 352, 185, 1991.
35. Ennever, F. K. Noonan, T. J., Rosenkranz, H. S., The predicitivity of animal bioassays and short-term genotoxicity tests for carcinogenicity and non-carcinogenicity to humans, *Mutagenesis*, 2,73, 1987.
36. Bartsch, H., Malaveille, C., Prevalence of genotoxic chemicals among animal and human carcinogens evaluated in the IARC Monograph Series, *Cell Biol. Toxicol.*, 5, 115, 1989.
37. Shelby, M. D., The genetic toxicity of human carcinogens and its implications, *Mutation Res.*, 204, 3, 1988.
38. Bartsch, H., Kuroki, T., Malaveille, C., Loprieno, N., Barale, R., Abbondandolo, A., Bonatti, S., Rainaldi, G., Vogel, E., Davis, A., Absence of mutagenicity of praziquantel, a new, effective, anti-schistosomal drug, in bacteria, yeasts, insects and mammalian cells, *Mutation Res.*, 58, 133, 1978.
39. Ashby, J., Tennant, R. W., Chemical structure, *Salmonella* mutagenicity and extent of carcinogenicity as indicators of genotoxic carcinogenesis among 222 chemicals tested in rodents by the U.S. NCI/NTP, *Mutation Res.*, 204, 17, 1988.
40. Gold, L. S., deVeciana, M., Backman, G. M., Magaw, R., Lopipero, P., Smith, M., Blumenthal, M., Levinson, R., Bernstein, L., Ames, B. N., Chronological supplement to the Carcinogenic Potency Database: Standardized results of animal bioassays published through December 1982, *Environ. Health Perspect.*, 67, 161, 1986.
41. Gold, L. S., Manley, N. B., Slone, T. H., Garfinkel, G. B., Rohrbach, L., Ames, B. N., The fifth plot of the Carcinogenic Potency Database: results of animal bioassays published in the general literature through 1988 and by the National Toxicology Program through 1989, *Environmental Health Perspectives*, 100, 65, 1993.
42. Ashby, J., Tennant, R. W., Definitive relationships among chemical structure, carcinogenicity and mutagenicity for 301 chemicals tested by the U.S. National Toxicology Program, *Mutation Res.*, 257, 229, 1991.
43. IARC, *Mechanisms of Carcinogenesis in Risk Identification*, Vainio, H., Magee, P., McGregor, D., McMichael, A. J., Eds., IARC Scientific Publication No. 116, International Agency for Research on Cancer, Lyon, France, 1992.
44. Cohen, S. M., Ellwein, L. B., Genetic errors, cell proliferation and carcinogenesis, *Cancer Res.*, 51, 6493, 1991.
45. Ames, B. N., Gold, L. S., Chemical carcinogenesis: too many rodent carcinogens, *Proc. Natl. Acad. Sci. U.S.A.*, 87, 7772, 1990.
46. Lewis, D. F. V., Lake, B. G., Interaction of some peroxisome proliferators with the mouse liver peroxisome proliferator-activated receptor (PPAR): a molecular modelling and quantitative structure-activity relationship (QSAR) study, *Xenobiotica*, 23, 79, 1993.
47. Muerhoff, A. S., Griffin, K. J., Johnson, E. F., The peroxisome proliferator-activated receptor mediates the induction of *CYP4A6*, a cytochrome P450 fatty acid ω-hydroxylase, by clofibric acid, *J. Biol. Chem.*, 267, 19051, 1992.
48. Heddle, J. A., Cimino, M. C., Hayashi, M., Romagna, F., Shelby, M. D., Tucker, J. D., Vanparys, Ph., MacGregor, J. T., Micronuclei as an index of cytogenetic damage: past, present, and future, *Environ. Molec. Mutagen*, 18, 277, 1991.
49. Tinwell, H., Ashby, J., Comparative activity of human carcinogens and NTP rodent carcinogens in the mouse bone marrow micronucleus assay: an integrative approach to genetic toxicity data assessment, *Environ. Health Perspect.*, 102, 758, 1994a.
50. Yang, W.-L., Klopman, G., Rosenkranz, H. S., Structural basis of the *in vivo* induction of micronuclei, *Mutation Res.*, 272, 111, 1992.

51. ter Haar, E., Day, B. W., Rosenkranz, H. S., Direct tubulin polymerization perturbation contributes significantly to the induction of micronuclei *in vivo, Mutation Res.*, 350, 331, 1996.
52. Rowinsky, E. K., Onetto, N., Canetta, R. M., Arbuck, S. G., Taxol: the first of the taxanes, an important new class of antitumor agents, *Semin. Oncol.*, 19, 646, 1992.
53. Horwitz, S. B., Taxol (paclitaxel): mechanisms of action, *Ann. Oncol.*, 5 (Suppl. 6):S3–6, 1994.
54. Tinwell, H., Ashby, J., Genetic toxicity and potential carcinogenicity of taxol, *Carcinogenesis*, 15, 1499, 1994b.
55. Long, B. H., Paclitaxel induces the formation of micronuclei and cytotoxicity in human carcinoma cells and mouse fibroblasts in culture, independent of its ability to arrest cells in mitosis, *Proc. Am. Assoc. Cancer Res.*, 36, 1995.
56. Ashby, J., Paton, D., The influence of chemical structure on the extent and sites of carcinogenesis for 522 rodent carcinogens and 52 different chemical carcinogen exposures, *Mutation Res.*, 286, 3, 1993.
57. Klopman, G., Wang, S., A Computer Automated Structure Evaluation (CASE) approach to calculation of partition coefficient, *J. Computational Chem.*, 12, 1025, 1991.

4

Improving the Therapeutic Index through Unusual Routes of Application: Intraarterial Treatment of Liver Metastases

Dario Civalleri
University of Genoa School of Medicine

Gianmauro Numico
National Cancer Institute, Genoa

Franco DeCian, MD
University of Genoa School of Medicine

Mauro Esposito, PhD
National Cancer Institute, Genoa

4.1   Introduction

Many efforts have been made to influence the kinetic behavior of drugs and to direct them to specific targets, thus increasing the therapeutic index. The story has been as old as pharmacology, beginning with different ways of drug administration: oral, subcutaneous, intramuscular, and intravenous. Pharmaceutical technology has subsequently been exploited, changing preparations and molecular forms, and using drug linkage and prodrugs.

In cases of refractory tumors, regional administration of drugs has also been employed. This type of access has the theoretical advantage of increasing local concentrations of drugs with limited systemic exposure. Indeed, many tumors present phases of regional progression that can be treated with administration of cytotoxic drugs to third spaces (including peritoneal and pleural cavities) or vessels afferent to the involved organs or regions. Tumors confined to the liver, limbs, pelvis, head and neck, peritoneum, and even other districts have been treated regionally.

The local advantage of intraarterial access can be expressed according to this simplified model:[1,2]

$$Ra = 1 + \frac{CL}{Q(1-E)}$$

where Ra = regional advantage; CL = total body clearance; E = extraction rate of the drug by the target organ; and Q = blood flow of the perfused vessel.

The liver, in particular, is a target of high clinical interest due to the high incidence of both primary and secondary tumors. Moreover, most drugs are metabolized or inactivated by the liver, thus favoring local advantage with a high first-pass effect when they are given via the hepatic artery or the portal vein. Liver extraction is inversely correlated with blood flow and, for several drugs, is rate-dependent.

Liver tumors also represent an ideal target for regional chemotherapy due to its double vascularization. Indeed the total blood flow to the liver exceeds 1 L/min, albeit one third and two thirds are of arterial and portal origin, respectively, with limited blood flow in the vessel used for access. As a rule, liver parenchyma as well as metastatic emboli and very small tumors are mainly nourished by portal blood, whereas the arterial fraction of tumoral flow becomes predominant as tumor volume increases. This is due to arterial hypervascularization of the proliferating peripheral ring, while hypovascularization, down to ischemia and necrosis, develops in central areas of larger tumors. Consequently, both large and small tumors can have differential perfusion patterns, reflecting the irregular nature of tumor angiogenesis and its relationships with the dual vascular supply of the liver. The arterial route, however, is generally thought to be the most valuable single access for regional treatment of liver tumors in terms of blood flow and of the consequent distribution to the tumor of a drug given regionally. In the tumor, higher concentrations of drugs should be achieved than in the parenchyma, and systemic toxicity should be reduced due to the first pass extraction.

Table 4.1 summarizes the kinetic parameters and the regional advantage after arterial infusion of the most widely employed drugs, compared to intravenous administration. According to the model, the drugs to be chosen should possess, besides a specific activity against the tumor to be treated, the kinetic characteristics of a high liver extraction and of a short half-life. This prompts long-term continuous infusion (CL) schedules with those agents, such as fluoropyrimidines, that exhibit the best single agent activity against gastrointestinal tumors and a cytotoxicity directly correlated with the duration of exposure.

4.2   Implantable Infusion Devices

Regional infusion chemotherapy has actually emerged from a long pioneering stage owing to a rational approach stimulated by the advent of totally implantable infusion devices. Two kinds of systems are presently available: (1) totally implantable pumps for continuous infusion; (2) implantable access ports, specifically intended for bolus infusions.

The first available system has been the Infusaid-400 pump[6,7] (Shiley Infusaid, Norwood, MA, Figure 4.1). It consists of a stainless steel cylinder (8.5 cm wide × 2.8 cm high) containing a 50 mL metal bellows as a reservoir and allowing a simple subcontinuous infusion of about 3 mL/day. A built-in sideport allows additional bolus infusions. The reservoir can be refilled by percutaneous puncture of a self-sealing silicone rubber septum that also automatically recharges its energy source consisting of a double phase

TABLE 4.1   Half-life, First-Pass Liver Extraction, and
Calculated Advantage in Tumor Exposure after
Arterial Infusion of the Most Commonly Used Drugs

	Half-life (min)	% Liver Extraction	Increased Tumor Exposure
5-FU	10	22–45	5–10
FUDR	<10	69–92	100–400
BCNU	<5	—	6–7
MMC	≤10	7–18	3–4
ADM	60	45–50	2
DDP	20–30	8–50	2–7

Adapted from Ensminger, W.D. et al.[2]

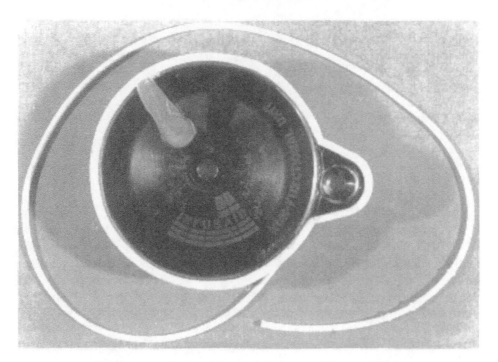

FIGURE 4.1   Infusaid-400 pump (Shiley Infusaid, Norwood, MA).

fluorocarbon gas sealed between the metal bellows and the external structure. The pump is placed in a subcutaneous pocket and, in case of intraarterial (ia) chemotherapy to the liver, the silicone rubber infusion catheter is inserted and fixed into the gastroduodenal artery up to but not into the hepatic artery. A cholecystectomy can be performed, and all collaterals of the proper hepatic artery are divided in such a way that a complete regional perfusion is performed without undue perfusion of extrahepatic organs. In case of aberrant anatomy, the same result can be obtained with modified surgical techniques[8,9] and/or double catheter pumps/ports. A recent advance has been the development of totally implantable and telemetrically programmable pumps such as the Infusaid 1000 and Syncromed DAD (Medtronic, Minneapolis, MN), allowing complex infusions that broaden their possible applications in both systemic and regional chemotherapy.[10]

Improved quality of life with longer outpatient treatments have been made possible by these devices compared to conventional infusion methods.[11,12] Since a major problem with pumps is cost, continuous infusions have also been carried out, with the aid of external pumps, via separate implantable ports[13] (Figure 4.2). The main problems and complications reported with all these systems consist of seroma, hematoma, necrosis, and infection of the subcutaneous pocket; disruption, malfunction and rotation of the device; clotting, rupture, and dislocation of the catheter, both intravascular and extravascular, including asymptomatic catheter erosion of the duodenum; thrombosis and pseudoaneurysm of the artery.[11,14] Though the prevalence of device-related complications is rather high, they appear to be either mild or late in the course of treatment, so that adequate trials of intraarterial infusion can usually be performed.

On this subject, an Italian study was completed (unpublished data) by the National Register of Implantable Systems (RNSI), an open cooperative group aimed at producing controlled trials in regional treatment of tumors. After a standard treatment with 5-fluoro-2'-deoxyuridine (FUdR) continuous infusion in 80 cases (14 days every 28), the 12-month device duration was 92% for Infusaid-400 pumps and 24% for ports fed by external pumps (median 9 months). The median durations of ports patency, employed for bolus ia injections, amounted to 17 and 18 months respectively in 57 cases treated with cisplatin (DDP, 40 mg/mq, days 1 to 3 every 21) and in 22 cases undergoing weekly epirubicin (EPIdx, 30 mg), with 12-month durations of 65% and 78%, respectively. These data indicate that totally implantable pumps are

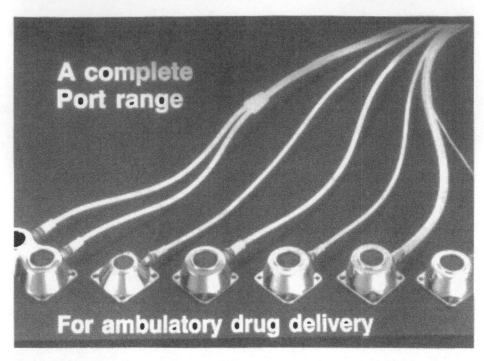

FIGURE 4.2   Implantable ports with single and double catheters, for intravenous, intraarterial, intraperitoneal, and spinal use (Pharmacia Deltec, Uppsala, Sweden).

far better than ports for continuous infusions. Most pumps end functioning for arterial instead of catheter problems, mainly due to the irritating action of drugs. On the other hand, bolus injection appears to be the preferred schedule for ports, as all mechanical and septic problems due to the external pumps are avoided. Even with continuous infusions, however, the ports allow adequate treatments in terms of quality of life and duration, proving two to three times longer compared to conventional access systems. Finally, external pumps connected to ports or pump sideports are the only means to allow ambulatory continuous complex combination chemotherapy.

4.3   Palliative Treatment of Liver Metastases of Colorectal Cancer

4.3.1   Conventional Intraarterial Treatment

4.3.1.1   FUdR Continuous Infusion

Adequate trials have been performed with FUdR continuous infusion (0.2–0.3 mg/m^2/day, days 1 to 14 every 28). The ever growing body of information coming from initial phase II studies[15-27] is confirmed and better defined by the results of randomized phase III trials (Table 4.2). Particularly, three studies performed by the NCI,[28] NCOG,[29] and MSK[30] compared ia vs. intravenous (iv) FUdR infusion. Significantly higher therapeutic response rates have been reported after ia as opposed to iv treatment. Thus, regional treatment improves local control by a factor of 2.5 to 4.2, as further shown by data of the MSK and NCOG. In these studies, patients in the iv arm with liver progression and without evidence of extrahepatic spread were crossed over to ia treatment with an additional response rate of 25% and 11%, respectively. These data are substantially in keeping with the results of the numerous phase II trials of FUdR infusion[15-27] that show response rates ranging from 29% to 67%, as evaluated according to WHO criteria, and median times to progression of up to 12 months. Furthermore, after ia treatment, the randomized studies showed significantly increased times to liver progression, ranging from 9 to 20

TABLE 4.2   Results of the Available Randomized Trials Comparing Intraarterial (ia) vs. Intravenous (iv) or Combined (ia+iv) FUdR Infusion or vs. Best Standard (BS) for Isolated Colorectal Liver Metastases

	NCI (87)[28]		MSK (87)[29]		NCOG (89)[30]		NCCTG (89)[31]		ROUSSY (92)[33]		LONDON (94)[34]		ULM (89)[44]	
	ia	iv	ia	iv	ia	iv	ia	BS	ia	BS	ia	BS	ia	ia+iv
Cases														
Evaluable (n)	24	29	48	51	50	65	33	36	82	84	51	49	23	21
Pretreated (n)					≤5	≤6								
Regional toxicity (%)														
Hepatitis	79	7	42	24	52	0			35				65	43
Jaundice	33	0	19	4	6	0	26							
Biliary sclerosis	21	0	8	0	10*	0	3		23		0**		26	24
Nausea/vomiting	21	35			45	14	13		7					
Chemical gastritis	21	7	8	2					15					
G.D. ulcer	17	0	17	6					9					
Gastritis/G.D. ulcer											2		26	23
Systemic toxicity (%)														
Diarrhea	13	59	2	70	0	58*		18					4	10
Mucositis	0	10			0	24		30					17	19
Cutaneous					0	9								
Leucopenia								45						
Responses (%)	62	17	50	20(25)	42	10(11)	48	21	43	9			52	48
Progressions														
Liver (%)			37	82	10	66					32	78		
Systemic (%)	55	20	56	16(58)	36	21			54	34	68	22	61	33
Time to progr. (months)														
Liver					13	7(6)	15.7	6	14.5	5.5				
Overall	7	9	11	7			6	5						14
Overall survival (median)														
Months	17	13	17	12(18)	17	16(24)	12.6	10.5	15	11			0	43
Days											405	226		
Time to liver failure					p<0.009		p<0.0001							
Survival:														
Overall									p<0.02		p<0.03			
Negative portal nodes	p<0.03													
ia+iv->ia vs. iv			p<0.05											

( ) after cross over
* toxic death
** clinically judged

months, and a trend toward increased survival. In the MSK study, patients treated with ia infusion either initially or after crossover survived longer than those treated with iv FUdR only. In the NCI study, patients with negative portal nodes showed a significantly higher survival after ia as opposed to iv treatment. No significant differences in survival and time to overall progression have been shown in a small NCCTG study[31] (Table 4.2) among pretreated patients randomized to receive either continuous ia FUdR (0.3 mg/kg/day, day 1 to 14 every 28) or bolus iv 5-fluorouracil (5-FU, 500 mg/m²/day, day 1 to 5 every 35). Additional findings of the latter study are significantly increased response rate and time to hepatic progression.

In a recent metaanalysis of four studies by Graf et al.,[32] however, patients with advanced colorectal cancer (59% to 71% with liver metastases) treated with regional chemotherapy demonstrated a positive correlation between tumor response and survival. This was statistically different in patients with complete and partial tumor response as compared with stable and progressive disease (p<0.001). Recent randomized studies confirm this hypothesis. A study by the Gustave Roussy Institute[33] (Table 4.2) showed a significant advantage in responses and survival in patients treated with FUdR continuous infusion, as compared to the "best standard," including no treatment or "placebo" iv 5-FU. One- and two-year survival was 64% and 23% in the treatment group and 44% and 13% in the control group, with 15- and 11-month median, respectively (p<0.02). This was recently confirmed by a further randomized study of Allen-Mersh et al.[34] showing significantly better quality of life (p<0.04) and survival (p<0.03) in patients treated with FUdR continuous infusion compared to conventional palliative treatment. Median survival rose from 226 days in the controls to 405 days in the regionally treated group. An additional finding of these two studies was a limited toxicity, probably due to the lower starting doses of FUdR (0.2 mg/m²/day).

A common feature of all these studies, however, is the limited number of cases, based both on accrual-related problems and even on the assumption that only large improvements of survival could justify a worsening in quality of life that medical oncologists suppose to be connected to regional treatment. As shown by Allen-Mersh et al.,[34] the latter assumption does appear questionable, especially when comparisons were made with "effective" systemic treatments. Indeed, significant toxicity has been reported after both ia and iv infusion, whereas in all reported studies a longer survival was observed after ia FUdR, and at least two studies showed a clearcut survival advantage after ia FUdR. The reported increases in survival after ia treatment, however, did not exceed a few months, and so they still have to be considered clinically unsatisfactory.

A major factor limiting the possible benefits of regional treatment is the high rate of extrahepatic progressions, whose incidence in the phase III trials (Table 4.2) was about 2.5 to 3.5 times higher in the ia as opposed to iv arm. Indeed, the majority of patients with iv treatment show liver progressions. Even higher figures have been reported in the phase II trials,[15,16,22] indicating an important working hypothesis for prevention of extrahepatic progressions.

An additional problem of regional infusion of FUdR lies in the high rate of hepatic and gastroduodenal toxicity. In phase II studies[14,15-27,35] the incidence of gastroduodenal ulcer and chemical gastroduodenitis, mainly related to undue perfusion of the duodeno-pancreatic area, widely ranged from very low figures up to 47% and 59%, respectively. Liver damage is the most relevant toxicity of ia FUdR, with rather variable incidence, depending on diagnostic criteria, drug doses, and duration of treatment. Chemical hepatitis, variously defined as a rise of liver enzymes, amounted to 24% to 76%, while the incidence of transient jaundice appears less variable (18% to 24%). Sclerosing cholangitis appears to be largely related to the different reduction criteria (3% to 29%). The pathogenesis of this severe complication is still undefined.[35] It is refractory to medical treatments and, in its more severe forms, can cause the patients' death despite the interruption of treatment. For these reasons, the starting or maintenance doses of FUdR were reduced to 0.15–0.2 mg/kg/day in most recent trials. Similar figures were reported in phase III trials (Table 4.2) including chemical hepatitis, biliary sclerosis, chemical gastritis, and gastroduodenal ulcer. However, a high incidence of toxicity, including toxic deaths, has also been shown after iv infusion (0.075–0.15 mg/kg/day, days 1 to 14 every 28). While liver damage is nearly absent, toxicity is indicated by diarrhea, mucositis, and colitis.

4.3.1.2   Modified FUdR-Based Regimens, Alternative Drugs and Combinations

Following the obvious suggestions of the randomized trials comparing ia vs. iv FUdR, several studies combining iv 5-FU or FUdR to ia FUdR have recently been initiated, as well as trials of pharmacologic modulation of FUdR with dexamethasone and dipyridamole or biochemical modulation with leucovorin (LV).

With the aim of reducing the local toxicity of FUdR, the NCOG[37] initiated a pilot study alternating ia infusion of low dose FUdR (0.1 mg/kg/day, days 1 to 7 every 35) followed by bolus 5-FU (15 mg/kg) via the pump sideport on days 15, 22, and 29. In 64 evaluable cases (30 pretreated) major responses were observed in 50% of the cases with additional 15% minor responses, and 22.4 months median survival from pump implantation. Liver toxicity was mild, with only 24% of patients requiring dose modifications, and, most important, no treatment terminations due to toxicity. Interestingly, the liver became the initial site of failure in up to 60% of patients, suggesting a reduced local efficacy of the treatment.

Trials of biochemical modulation of fluoropyrimidines has already been started with various combinations of ia and iv FUdR and 5-FU. Warren et al.[38] showed a 48% response rate with acceptable toxicity in 31 patients given weekly ia 5-FU (1.5 mg/m² in 24 h) combined with iv LV (400 mg/m²). Kemeny et al.[39] combined ia FUdR and iv 5-FU plus LV with promising results. The same authors[40,41] studied 63 patients treated according to six schedules of combined LV (15 to 30 mg/m²/day) and FUdR (0.2 to 0.3 mg/kg/day), with administration times ranging from 7 consecutive days over a 14-day period to 14 consecutive days over a 42 day period. The higher dose-intensity schedules showed higher toxicity and higher response rates than after FUdR alone with 1- and 2-year survivals of 86% and 62%, respectively. The low dose LV (15 mg) combined with full dose FUdR (given in 14 consecutive days over a 28 day period) showed a 69% response rate and a relatively low toxicity, with 24% incidence of transient jaundice and 6% cholangitis out of 16 patients.

A total dose of 20 mg ia dexamethasone for 14 days per cycle was employed to reduce FUdR toxicity. In a randomized study,[42] patients treated with FUdR + dexamethasone showed a trend toward reduced liver toxicity (9% vs. 30%) with higher FUdR doses, and significantly higher response rates (71% vs. 40%) compared to FUdR alone. Similar encouraging results have been reported in a phase II study combining dexamethasone to FUdR plus LV.[43] In 33 previously untreated patients, a 78% response rate, with a median survival of 24.8 months and respective 1- and 2-year survival rates of 91% and 57%, was observed. Even previously treated patients showed a 54% response rate. The overall incidence of sclerosing cholangitis was as low as 3%, thus encouraging further application of this schedule in larger trials.

To investigate the possibility of delaying systemic progressions, a randomized study was initiated at the University of Ulm (Table 4.2), comparing simple ia to ia+iv infusion of FUdR.[44] The preliminary results on 44 evaluable cases suggest superimposable toxicities and responses in the two arms with a 3-year survival of 0% and 43%, respectively. Such a difference could actually be due to delayed systemic progressions. However, it has to be taken with extreme caution due to the limited number of cases followed up at that time. Indeed, a similar comparison has been reported by Lorenz et al.[45] giving negative results so far. Several randomized studies have currently been initiated on this subject, including a three-arm Italian study (GISCAD) comparing standard iv 5-FU + LV vs. ia FUdR vs. the combined iv + ia treatment.

Modified FUdR schedules have also been tried, including bolus, pulse, short, and continuous complex infusion,[46] whose first results suggest a fascinating chronobiological inference.[47] Indeed, as biological functions in many tumors and host tissues have shown periodic patterns, time optimization of drug schedules might play an important role in improving the therapeutic index of a given treatment. Preclinical and clinical data make it clear that the efficacy of anticancer drugs depends to a substantial extent upon three strictly related temporal variables: drug sequence, interval between doses, and circadian stage of administration. Circadian stage dependency of toxicity has been shown *in vivo* for many anticancer agents, including cisplatin, adriamycin, methotrexate, and fluoropyrimidines.[47,48] A circadian rhythm in drug plasma concentration has been shown by Petit et al.[49] in patients receiving 5-FU as a continuous venous infusion at a constant rate for 5 days. In turn, recent studies have shown a circadian rhythm in tissue concentrations of dihydropyrimidine dehydrogenase, an enzyme involved in 5-FU metabolism.[50] Hence, a circadian modulation of the infusion rates might optimize the therapeutic index of fluoropyrimidines. Von Römeling et al.[51] reported lower toxicity in mice when FUdR was given intraperitoneally during their mid-to-late activity phase. Cosinor analysis confirmed that the circadian timing of peak rather than the intermittency of drug delivery determined toxicity. A randomized clinical trial was then carried out by the same group in patients with metastatic adenocarcinomas, comparing ia FUdR CI with circadian sinusoidal continuous infusion (Sin-CI). Sixty-eight percent of the daily dose was given either iv or ia from 3 to 9 p.m. (2% from 3 to 9 a.m.), using a DAD pump.[51,52] A less frequent and less severe toxicity occurred after Sin-CI than after CI, allowing an increase of dose intensity from 0.46 (CI) to 0.79 mg/kg/week (Sin-CI). Three times as many patients tolerated ia Sin-CI infusion without evidence of liver toxicity. Most studies, however, have been carried out with systemic infusion of various combinations of time-modulated drugs, including 5-FU plus either LV and interferon-α.[53,54] In a randomized study by Levi et al.,[55] using 5-FU + LV + Oxaliplatin, circadian chronomodulation allowed five times less toxicity (P<0.001) with 40% higher doses (P<0.0001) compared to constant infusion, with response rates of 53% and 32% (P = 0.038), progression free survival of 11 and 8 months (n.s.), and overall survival of 19 and 15 months (P = 0.03), respectively.

As mentioned above, in most recent trials with FUdR monochemotherapy, a lower starting dose (0.15–0.2 mg/kg/day) reduced the incidence of sclerosing cholangitis, thus prompting combinations. Mitomycin C (MMC), as well as carmustine and dichloromethotrexate, has been associated in most occasions to prevent or to overcome resistance to FUdR alone.[6,19,56] Indeed, a randomized study comparing hepatic arterial FUdR + MMC + BCNU vs. FUdR alone showed superimposable results in the two arms. The only exception was the significantly higher response rate with similar toxicity after the combination schedule in a subset of patients who had previous systemic chemotherapy[57] (47% vs. 23%, P = 0.035). An experience of the MD Anderson Hospital is interesting.[58] The authors sequentially tested the efficacy of FUdR (100 mg/m^2) 5-day continuous ia infusion every 28 days, given alone or in combination with ia MMC (15 mg/m^2) and/or DDP (100 mg/m^2) bolus infusion on day 1. The observed response rates rose from 30% with FUdR alone to approximately 50% with combinations, with a parallel increase of response duration peaking at 17 months after FUdR + MMC + DDP. Toxicities were acceptable, mostly mild and transient, with all regimens, essentially consisting of nausea and vomiting (100%), transient liver damage (100%), some degree of chronic liver damage (10%), and ototoxicity (3%). A similar experience was carried out by the SAKK[59] with a combination of ia 5-FU and MMC, resulting in a 50% response rate, crude median survival of 19.5 months and up to grade I–II toxicity, including nausea (46%), leucopenia (32%), thrombocytopenia (21%), and abdominal discomfort (25%).

MMC and DDP have lower first-pass liver extractions and longer half-lives[2,60] compared to fluoropyrimidines, particularly to FUdR. In a portion of cases, the ia infusion of these drugs might then lead to delayed systemic progressions due to high drug levels in the peripheral blood. Tumor exposures, however, have been estimated three to four and two to seven times higher when MMC and DDP, respectively, are given by arterial compared to intravenous route. These values suggest that both ia MMC and DDP applications might have a potentially increased efficacy on liver disease with limited local toxicity. In order to pursue a more effective control of the subsequent extrahepatic phase of the disease a rationale exists for coadministration of fluoropyrimidines as the most effective single agents against colorectal cancer. In addition, an additive activity of the two drugs in several tumors, including colorectal cancer,[61] has been suggested.

In this view, an RNSI study has been carried out with the aim of both reducing extrahepatic relapses and liver toxicity.[62] The regimen included ia bolus infusion of DDP (40 mg/m^2/day, days 1 to 3 up to a maximum dose of 1200 mg) combined with iv bolus 5-FU (200 mg/m^2/day, days 1 to 5) every 3 weeks. Patient enrollment has now been stopped with 73 registered cases. An interim analysis on 59 evaluable cases showed a 47% response rate (18% complete, 29% partial, and 8% minor). Once more, however, parallel to the reduction of liver toxicity, the pattern of recurrences proved completely opposite to the full dose ia FUdR. Liver toxicity was virtually absent, whereas late progressions occurred mainly in the liver (69%) with an overall median time of 10 months. Significant toxicities consisted of nausea and vomiting (76%), bone marrow suppression (51%), and peripheral sensitive neuropathies (56%) starting at cumulative DDP dosages of 600 to 720 mg/m^2 and proving chronic in 12% of cases. With the aim of delaying liver progressions and reducing neurotoxicity, a new randomized phase II trial has recently been activated by the RNSI. This compares continuous vs. bolus ia DDP infusion (24 mg/m²/day, days 1 to 5 every 28 up to a maximum dose of 720 mg/m²) associated in both arms with iv bolus infusion of 5-FU (250 mg/m²/day, days 1 to 5 at the first cycle, escalated to 375 at the second, and to 500 at subsequent cycles). After six cycles or in case of liver progression, patients are switched on second line treatment with ia bolus infusion of FUdR. Preliminary results suggest that neurotoxicity was actually reduced, with an overall 12% clinical incidence and less frequent and severe electroneurographic signs. Finally, adequate antiemetic regimens have substantially reduced the incidence of nausea and vomiting.

4.3.2 Innovative Regional Approaches

4.3.2.1 Chemofiltration and Isolated Perfusion

It must be kept in mind, however, that any treatment should be aimed essentially at the cure of patients. In this view, partial responses to treatment should be considered a failure. While waiting for the magic bullet, attention must be paid to those treatment modalities that could substantially improve the activity of the available drugs.

Chemofiltration,[63-65] a technique popularized by Aigner,[63] involves the ia administration of extremely high doses of hydrosoluble drugs followed by the extracorporeal hemofiltration and reinfusion of blood drawn from the vena cava at the level of the hepatic veins. Early pharmacokinetic data suggest that when using drugs with low liver extraction an increase of local concentrations can actually be obtained. Though some encouraging results have been reported,[67] this technique promises little effect, if any, with drugs showing higher liver extraction rates. More promises are coming from a recent technical modification of chemofiltration that involves a double balloon catheter aimed at occluding the vena cava in order to draw and detoxify all blood coming from hepatic veins without any mixing with systemic blood. Ravikumar et al.[67] employed this technique, which involves only an overnight hospital stay, to treat 23 patients with escalating doses of 5-FU or adriamycin (ADM). They showed limited toxicity and no mortality with extraction efficiencies from the hepatic veins ranging from 64% to 91%.

A further technique is the so-called stopflow perfusion[68] that involves double balloon–catheter occlusion of both aorta and vena cava cranially to the hepatic veins and celiac axis, followed by short-term (15 min) low-flow (300 ml/min) extracorporeal circulation of unoxygenated blood after injection of drug in the arterial line.[68,69] Due to low-flow and recirculation high local extraction of drugs can be obtained, also increasing the cytotoxicity of selected drugs. Indeed the drugs to be chosen appear to be the alkylating agents, such as MMC, which, moreover, is known to be preferentially toxic to cells under hypoxic conditions.[70] Isolated perfusion of the organ allows further combinations, as in the case of isolated hyperthermic-antiblastic perfusion that was initially applied to limb tumors.[71]

Hyperthermia is cytotoxic in itself and potentiates the activity of most anticancer drugs. The procedure has also been applied for liver tumors.[72-74] The surgical technique requires the establishment of an extracorporeal circulation through occlusion of supra- and infrahepatic vena cava on a bypass, hepatopetal cannulation of the hepatic artery, and cannulation of the portal vein in both directions. After liver temperatures have been raised over 40°C by means of a heat exchanger, drugs are injected into the circuit and circulated for 60 min. Isolated perfusion has usually been performed only once with either 5-FU alone or combined with MMC (total doses of 1000 mg and 15 to 50 mg, respectively) and followed by ia infusion. Though this procedure appears very demanding in a palliative setting, with an acute lethality of 8%, tumor responses, evaluated as a reduction of either CEA or tumor mass, appear to be excellent. In a consecutive series of 50 cases, complete and partial responsivity, respectively, were as high as 16% and 79% after 5-FU alone and 62% and 28.5% after 5-FU plus MMC. Median survival after 5-FU plus MMC was 22 months, with three cases surviving more than 5 years. Hafström et al.[75] also showed 5 out of 29 patients surviving more than 3 years after the procedure. Hospital lethality, however, were as high as 14% (all dying patients had more than 50% liver involvement), whereas partial tumor regression was registered in 20% of cases. A less invasive technique involves radiofrequencies, as a means of heat transfer, associated with plain arterial infusion.[76]

4.3.2.2   Chemoembolization

An additional interesting modality of ia access consists of chemoembolization, a technique that promises to improve clinical results through a further optimization of regional tumor targeting. Various materials have been used,[46,77-80] including absorbable gelatin (Gelfoam, GF), glutaraldehyde-stabilized collagen (Angiostat), polyvinyl alcohol particles (Ivalon), iodized oil (Lipiodol, LPD), liposomes, ethylcellulose microcapsules, albumin, and starch microspheres (DSM), mixed or loaded with selected cytotoxic agents. These materials differ from each other in size and biodegradability, which determines the level and duration of vascular occlusion. For instance, GF is degraded by collagenase in 48 hours (either capillary or arterial embolization can be obtained with GF powder or larger blocks); Angiostat persists in the capillaries for about 7 days; Ivalon, and perhaps ethylcellulose microcapsules, cause permanent occlusion at the arteriolar level.

A prolonged vascular occlusion has the advantage of associating the cytotoxic effects of drugs and arterial ischemia. Both have selective effects on the tumors, as the parenchyma is mainly nourished by the portal vein. A study by the EORTC GI Group[81] randomized patients to receive hepatic arterial ligation alone or associated with intraportal infusion of 5-FU with negative results in terms of both treatment response and survival. Hafström et al.[82] showed increased survival ($P = 0.0039$) in those patients randomized to receive temporary arterial occlusion with slings, intraportal 5-FU infusion, and oral allopurinol (17 months median) compared to patients receiving no regional or systemic treatment (8 months median).

A possible disadvantage of prolonged occlusion, however, is the induction of selective arterial thrombosis, precluding continuation of treatment. We preferred degradable starch microspheres (Kabi Pharmacia AB, Uppsala, Sweden), as their small size (40 ± 5 μM) and short half-life (15–30 min) may allow repeated distal chemoembolization with limited risk of permanent vascular occlusion. Once injected intraarterially, they provoke a reduction of blood flow to the target organ due to a transient arteriolar–capillary blockage. During degradation by serum amylases, DSM maintain their dimensions and shape until they finally collapse, allowing resumption of flow. They are nontoxic and nonthrombogenic even after repeated injections. According to the models, the first effect of DSM is to improve local extraction and to reduce systemic spillover of coadministered drugs.[83] A further effect is to divert residual blood flow toward the hypovascular tumors (Fig. 4.3),[77,84,85] which are known to be more resistant to chemotherapy than hypervascular ones,[85,86] the respective response rates being generally lower than 20% and higher than 55%. This induces not only a higher drug extraction by the liver as a whole but also higher drug concentrations in virtually unresponsive tumors.[87] Hence, arterial chemoembolization with DSM should optimize the therapeutic index of ia drugs through a reduction of systemic toxicity and an increase of local activity. As the short-lasting effect of DSM can be better exploited with bolus coadministration of the drug, the drugs of choice seem to be the alkylating agents, such as MMC, that, in addition, are known to be preferentially toxic under the hypoxic conditions occurring after DSM embolization.[70]

A pilot study has been carried out by the GI group of the EORTC,[88] including arterial and portal administration of MMC (10 mg/m², day 1) followed by continuous 5-FU infusion (500 mg/m²/day, days 1 to 5 every 28). The ia fraction of MMC was given mixed with the required dosage of DSM as determined by a computerized monitoring method based on the measurement of the effects of DSM on liver passage to the systemic circulation of a radiolabeled substance devoid of hepatic metabolism.[89] Indeed, the degree of flow blockage is highly variable in individuals, depending on total capillarity of the organ, tumor vascularity, and DSM dosages. As compared to other methods of regional treatment, preliminary data show a high incidence of access problems, due to the double cannulation of portal vein and hepatic artery. Though the response rate was in the range of the international experience (43.5%) similar figures were obtained in cases with hypovascular and hypervascular tumors, and a high incidence of complete responses (20%) was also observed. This could be ascribed to the above-mentioned pharmacodynamic effects of DSM-induced blood flow redistribution. Hence, the experience should be continued as a phase III study including simple ia access.

Additional evaluations of DSM have been carried out by British and Japanese groups. A randomized study was reported by Hunt et al.[90] comparing untreated controls (19 cases) vs. plain DSM embolization (20 cases) vs. 5-FU+DSM chemoembolization (22 cases). Once more, due to the limited number of cases, only a trend toward improved survival was observed in patients receiving chemoembolization (13.0

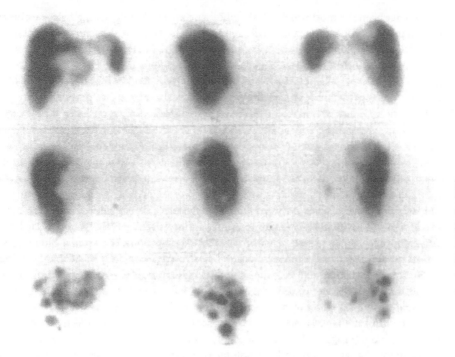

FIGURE 4.3   A large lesion located in the left lobe and several smaller lesions located in the right lobe of the liver were identified with the sulphur colloid scan (*top*). They were classified as grossly hypovascular according to the first ^{99m}Tc-macroaggregated albumin (MAA) perfusion scan (*middle*). Compared with the parenchyma, a redistribution of flow toward all tumor areas is apparent in the second MAA scan, performed immediately after treatment with degradable starch microspheres (*bottom*). The relative residual hypoperfusion of the large mass, compared with the smaller ones, conforms to a substantial degree of central necrosis detected by ultrasound examination.[85]

months median) as compared to controls (9.6 months) and plain embolization (8.7 months). The greatest observed benefit, though not significant, was achieved in the subgroup with <50% hepatic replacement, with 23.6, 10.0, and 10.2 months, respectively. In a pilot study Goldberg et al.[91] employed angiotensin II associated with ia 5-FU+DSM in an attempt to optimize drug targeting to the tumors. Indeed, as tumor circulation is lacunalike and tumor vessels have small and uneven walls, only normal vessels react to vasoconstrictors, thus diverting blood flow to the tumors.[92] Though 5-FU does not appear to be the most suitable drug for chemoembolization, a trend toward an improved survival was reported in 21 cases compared to historical controls. Taguchi et al.[93] randomized 20 patients (15 suffering from colorectal carcinoma) receiving hepatic arterial MMC every 2 weeks compared with 22 cases (10 colorectal) given MMC plus an average of 600 mg of angiographically monitored DSM. The DSM groups showed a higher incidence of transient pain and fever and significantly higher response rates in the whole series (55% vs. 20%) and in those with colorectal carcinoma (40% vs. 20%).

4.4   Chemoembolization of Primary Liver Tumors

Chemoembolization has been widely used for the palliative treatment of primary liver tumors, employing mostly the same drug as in colorectal metastases.[78,80,94] However, some essential differences, related to the histotypes and the frequent association with liver cirrhosis, influence treatment modalities. Indeed, liver cirrhosis influences both complications and survival-related prognoses, and increases the operative risk linked to the possible surgical implant of arterial access systems. Hence, most authors respond best to percutaneous arterial access according to the Seldinger method which almost necessarily involves 1-day cycles.

LPD and GF have been the most widely used embolizing materials. LPD is an ethylester of poppyseed oil that has been used as a contrast medium and can easily be emulsified with most hydrosoluble cytotoxic drugs using mechanical action and/or stabilizer substances. When administered into the hepatic artery, it is trapped in arterioles with a diameter of about 25 µm and mainly concentrates in liver tumor tissue, perhaps because of tumor vascular abnormalities.[95,96] As LPD has a weak influence on arterial blood flow and has mainly to be regarded as a sort of drug carrier to the tumor, it has generally been associated with GF embolization. GF has been mixed with drugs and infused alone; more often it has been infused as a plain embolizing material after infusion of an emulsion containing LPD + drugs. Indeed, GF occludes larger arteries for a longer time in comparison with DSM, possibly inducing an effective ischemic damage of the tumor. The combined treatment modality has two potentially useful effects, consisting of a very slow clearance of the LPD taken up by the tumor and a reversible vascular embolization, allowing repeated cycles.

A water-in-oil emulsion is usually prepared immediately before starting injection with a two-syringe method consisting of up to 20 mL LPD and the required dose of hydrosoluble drug injected over a few minutes into the proper hepatic artery, if possible, until slowing down the arterial blood flow and/or complete filling of the neoplastic nodules. To stabilize the emulsion the use of 2 mL of iopamidol (Iopamiro 300, Bracco, Milano, Italy) or other chemical is allowed. The volume of emulsion was programmed in advance with the aim of allowing infusion of the entire dose of drug. In case the infusion had to be stopped before the entire amount of emulsion had been administered, the injection of the remaining dose of drug can be completed as a water solution. GF is usually injected at the end of the infusion of the emulsion. It is generally available as a micronized material or as cubes to be cut or grated, obtaining small pieces that are soaked in hydrosoluble contrast medium and injected. Especially in cases with cirrhosis, it has often been infused superselectively into the feeding arteries of the obvious tumor masses instead into the proper hepatic artery. This theoretically results in a suboptimal treatment of the disease, as sublinical tumor can be present in untreated portions of the liver. Moreover, the need of percutaneous arterial access, the frequently impaired liver function, and the resumption of a complete arterial patency that usually take a few weeks to be completed make the recycling times rather long, thus inducing a possibly suboptimal 1-day treatment not less than every 4 to 6 weeks.

Although a complete description of the results of lipiodolization of primary liver tumors lies outside our task, it has to be pointed out thousands of cases have been treated worldwide with the only solid evidence of increased tumor responses. Indeed, a major problem of transcatheter chemoembolization also consists of the high variability of interoperator skill and clinical attitude matched with complex diagnostic and operative problems. It involves immediate decisions, taking into account liver function, arterial and portal anatomy and flow, arteriovenous shunts, local and systemic tumor growth, and miscellaneous technical difficulties that concur to deeply influence the risk of toxicity and complications and, ultimately, the efficacy of treatment. This hampers designing and conducting well-controlled response and survival studies. In this view, well-designed randomized controlled studies are still lacking. Some of them involved large numbers of patients to answer hyperfragmented[97] or marginal[98,99] or misleading questions[100] as the comparisons of LPD + ADM vs. LPD + EPIdx or LPD as single agent vs. LPD + ADM. Even misuse of sound statistical techniques is involved. Indeed, in a recent "negative" randomized study of the Groupe d'Etude et de Traitement du Carcinome Hépatocellulaire[101] (with estimated 62% 1-year survival in the chemoembolization group compared to 43.5% in the control group) a sequential design was adopted, the power was set to 90%, and the expected increase in 1-year survival in the treatment group was 25%. Kasugai et al.[96] reported that chemoembolization with LPD + ADM does not seem superior to arterial injection of ADM alone, whereas chemoembolization with LPD + DDP has been a far superior treatment, even though not compared to DDP alone. On the other hand, Chang et al.[102] showed no difference in tumor response and survival in patients treated with LPD + DDP + GF compared to LPD + GF alone. Chemoembolization with LPD + GF appeared superior to LPD alone in a Chinese trial.[103] Two small, though well designed, Western studies comparing supportive treatment vs. chemoembolization with ADM and EPIdx, respectively,[104,105] failed to show any survival advantage even though a considerable response rate has been reported in patients treated with LPD +

EPIdx. This was confirmed by Yoshikawa et al.[106] who showed a 42% response rate in patients receiving chemoembolization with LPD + EPIdx compared to 12% in those receiving arterial EPIdx alone. There also was a tendency toward longer survival in this small study. It has to be pointed out that this technique has already been used in colorectal liver metastases,[107-110] showing similar results.

Finally, arterial embolization with ^{131}I-LPD[111] or ceramic microspheres[112,113] incorporating [92]Y or ^{32}P could also be useful to administer internal radiotherapy to the liver. This method seems to be superior to external radiotherapy that is limited by the radiosensitivity of liver parenchyma (15–20 grays) and has been employed with little benefit in association with FUdR.[114,115] Indeed, two- to sixfold higher tumor doses can be reached with internal radiotherapy due to the preferential targeting to the capillary bed of liver tumors, especially in conjunction with the short-ranged beta emitters that can deliver large local radiation doses with little irradiation to neighboring organs. Vasoactive and/or radiosensitizing drugs, such as misonidazole,[116] bromodeoxyuridine,[117] and DDP itself have also been employed in conjunction with internal radiotherapy. Furthermore, radioactive isotopes as well as cytotoxic drugs[118] can be coupled with monoclonal antibodies. Unfortunately, high systemic radioactivity has been observed after arterial infusion of ^{131}I-LPD, whereas [92]Y seems to interfere with targeting of the Fab end of the molecule. The ^{131}I has then been linked with monoclonal antibodies against alphafetoprotein (AFP), CEA, ferritin,[119,120] etc., being particularly useful in AFP negative primary cancers. Since iv administration is usually employed, these methods lie well outside our present task. Antibody coupled liposomes[121] and albumin microspheres,[122] and even adoptive immunotherapy with interleukin-2[123] and lymphokine-activated killer cells,[124] have already been administered via the hepatic artery, as well as internal radiotherapy with ^{131}I using ia LPD as a carrier. A novel field of intraarterial infusion involves gene therapy. Retroviral mediated transfer of a suicide gene into experimental metastasis has proven efficient in reducing tumor size. Some clinical phase I studies with ia administration are now in progress and first results can be expected in a few years.[125]

4.5   Conclusions

To define the clinical role of regional treatment of liver tumors, the nontumoral prognostic factors that appear to significantly influence clinical outcome have to be considered.[126] Obvious negative factors are pretreatment, poor performance status, weight loss, and poor liver function. In primary tumors, reported poor risk factors[127-129] are jaundice, refractory ascites, portal thrombosis, high-risk esophageal varices, and, generally, advanced cirrhosis. In colorectal patients without liver decompensation, low serum albumin and high lactic dehydrogenase values negatively influence the outcome.[130] Important tumor-related factors are the extent of liver involvement and extrahepatic disease, whose absence (negative nodes) in colorectal patients treated at NCI allowed a significant gain in survival in the arm receiving ia infusion.[28] These factors concur to define the Gennari IV[131] and Okuda III[132] stages, in which survival after ia chemotherapy clearly appears superimposable to the natural history of the disease. Moreover, tumor vascularity is a good predictor of response, whereas responsive patients generally show an increased survival. The value of tumor grading, ploidy patterns, and labeling index,[133] and chemosensitivity[134] should be further investigated.

Though several studies suggest that regional treatment could modify the natural history of the disease, it still remains an investigational procedure that should be applied within controlled clinical trials in dedicated centers. Regional treatment, however, is the preferred investigational standard for palliation of liver tumors. In this setting, only good risk patients should be treated, since prolonged survival with good quality of life can be obtained in those cases. Efforts should be made to delay systemic progressions with associated systemic treatments. It is important to note that survival longer than 5 years has occasionally been obtained with mere regional treatment, whereas additional long-term survivals have been reported in cases submitted to radical liver resection after a response had been obtained with "palliative" regional treatment. The latter suggests the possible use of the regional access for "adjuvant" treatment after radical liver resection. In addition, most responding cases benefit from an improved performance status and can return to normal life and work while on treatment. Indeed, pump infusions can be carried out with minimal inconvenience for the patient. On the contrary, as poor risk patients do not take advantage of conventional regimens, the only opportunity for them could be given by early phase experimental treatments, possibly followed by conventional regimens in selected cases.

Glossary

5-FU	5-fluorouracil
ADM	adriamycin
BCNU	carmustine
CEA	carcinoembryonic antigen
DDP	cisplatin
DSM	degradable starch microsphers (Spherex, Pharmacia AB, Sweden)
EORTC GI	Gastrointestinal Group of the European Organization for Research on Treatment of Cancer
EPIdx	epirubicin
FUdR	5-fluoro-2'-deoxyuridine
GF	gelfoam
GISCAD	Italian Group for the Study of Gastrointestinal Cancer
LPD	lipiodol
LV	leucovorin
MMC	mitomycin C
MSK	Memorial Sloan Kettering Cancer Center
NCCTG	North Central Cancer Treatment Group
NCI	National Cancer Institute
NCOG	Northern California Oncology Group
RNSI	National (Italian) Register of Implantable Systems
SAKK	Swiss Group for Clinical Cancer Research

References

1. Collins, J., Pharmacologic rationale for regional drug delivery, *Journal of Clinical Oncology*, 2, 498, 1984.
2. Ensminger, W. D., Gyves, J. M., Regional chemotherapy of neoplastic diseases, *Pharmacology and Therapeutics*, 21, 277, 1983.
3. Ackerman, N. B., Experimental studies on the circulatory dynamics of intrahepatic blood flow supply, *Cancer*, 29, 435, 1972.
4. Ackerman, N. B., The blood supply of experimental liver metastases. IV. Changes in vascularity with increasing tumor growth, *Surgery*, 75, 589, 1974.
5. Healey, J. E., Jr., Vascular patterns in human metastatic liver tumors, *Surgery Gynecology and Obstetrics*, 129, 1187, 1965.
6. Balch, C. M., Urist, M. M., McGregor, M. L., Continuous regional chemotherapy for metastatic colorectal cancer using a totally implantable infusion pump, *American Journal of Surgery*, 145, 285, 1983.
7. Blackshear, P. J., Dorman, F. D., Blackshear, P. L., Varco, R. L., Buchwald, H., The design and initial testing of an implantable infusion pump, *Surgery Gynecology and Obstetrics*, 134, 51, 1972.
8. Eckhauser, F. E., Knol J. A., Strodel W. E., Shellito J. L., Complicated access for regional infusion chemotherapy, *Archives of Surgery*, 119, 1195, 1984.
9. Watkins, E., Jr., Khazei, A. M., Nahra, K. S., Surgical basis for arterial infusion chemotherapy of disseminated carcinoma of the liver, *Surgery Gynecology and Obstetrics*, 130, 581, 1970.
10. Vogelzang, N. J., Ruane, M., DeMeester, T. R., Phase I trials of an implantable battery-powered programmable drug delivery system for continuous doxorubicin administration, *Journal of Clinical Oncology*, 3, 407, 985.

11. Civalleri D., Cafiero, F., Cosimelli, M., Craus, W., Doci, R., Repetto, M., Simoni, G. A., Regional arterial chemotherapy of liver tumors. I. Performance comparison between a totally implantable pump and a conventional access system, *European Journal of Surgical Oncology*, 12, 277, 1986.

12. Sterchi, J. M., Hepatic artery infusion for metastatic neoplastic disease, *Surgery Gynecology and Obstetrics*, 160, 477, 1985.

13. Niederhuber, J. E, Esminger, W., Gyves, J. W., Liepman, M., Doan, K., Cozzi, R. N., Totally implanted venous and arterial access system to replace external catheters in cancer treatment, *Surgery*, 92, 706, 1982.

14. Hohn, D. C., Rayner, A. A., Economou, J. S., Ignoffo, R. J., Lewis, B. J., Stagg, R. J., Toxicities and complications of implanted pump hepatic arterial and intravenous floxuridine infusion, *Cancer*, 57, 465, 1986.

15. Balch, C.M., Levin, B., Regional and systemic chemotherapy for colorectal metastases to the liver, *World Journal of Surgery*, 11, 521, 1987.

16. Doci, R., Bignami, P., Quagliuolo, V., Civalleri, D., Simoni, G. A., Cosimelli, M., Craus, W., Gennari, L., Continuous arterial infusion of 5-fluorodeoxyuridine for treatment of colorectal liver metastases, *Regional Cancer Treatment*, 3, 13, 1990.

17. Kemeny, M. M., Goldberg, D., Beatty, J. D., Blayney, D., Browning, S., Doroshow, J., Ganteaume, L., Hill, R. L., Kokal, W. A., Riihimaki, D. U., Results of a prospective randomized trial of continuous regional chemotherapy and hepatic resection as treatment of hepatic metastases from colorectal primaries, *Cancer*, 57, 492, 1986.

18. Kemeny, N., Daly, J., Oderman, P., Shike, M., Chun, H., Petroni, G., Geller, N., Hepatic artery pump infusion: toxicity and results in patients with metastatic colorectal carcinoma, *Journal of Clinical Oncology*, 2, 595, 1984.

19. Niederhuber, J.E., Ensminger, W., Gyives, J., Thrall, J., Walker, S., Cozzi, E., Regional chemotherapy of colorectal cancer metastatic to the liver, *Cancer*, 53, 1336, 1984.

20. Ramming, K.P., O'Toole, K., The use of the implantable chemoinfusion pump in the treatment of the hepatic metastases of colorectal cancer, *Archives of Surgery*, 121, 1440, 1986.

21. Rougier, P., Lasser, P., Elias, D., Ghosn, M., Siobide, S., Theodore, C., Lumbroso, J., Intra-arterial hepatic chemotherapy (IAHC) for liver metastasis (LM) from colorectal (CR) origin, *Proceedings of the American Society of Clinical Oncology*, 6, A27, 1987.

22. Schwartz, S.I., Jones, L. S., McCune, C. S., Assessment of intrahepatic malignancies using chemotherapy via an implantable pump, *Annals of Surgery*, 201, 560, 1985.

23. Shepard, K. V., Levin, B., Karl, R. C., Faintuch, J., DuBrow, R. A., Hagle, M., Cooper, R. M., Beschorner, J., Stablein, D., Therapy for metastatic colorectal cancer with hepatic artery infusion chemotherapy using a subcutaneous implanted pump, *Journal of Clinical Oncology*, 3, 161, 1985.

24. Stagg, R. J., Lewis B. J., Friedman M. A., Ignoffo R. J., Hohn D. C., Hepatic arterial chemotherapy for colorectal cancer metastatic to the liver, *Annals of Internal Medicine*, 100, 736, 1984.

25. Trivisionno, D. P., Riether, R. D., Sheets, J. A. Stasik, J. S., Rosen, L., Khubhandani, I. T., Follow-up on a prospective study of continuous hepatic perfusion with implantable pump, *Diseases of the Colon and Rectum*, 29, 691, 1986.

26. Weiss, G. R., Garnick, M. B., Osteen, R. T., Steele, G. D., Wilson, R. E., Schade, G. D., Jr., Wilson, R. E., Schade, D., Kaplan, W. D., Boxt, L., M., Long-term infusion of 5-fluorodeoxyuridine for liver metastases using an implantable pump, *Journal of Clinical Oncology*, 1, 337, 1983.

27. Winton, T. L., Ghent, W. R., Hepatic artery infusion chemotherapy with the Infusaid pump: a Canadian experience, *Canadian Journal of Surgery*, 29, 379, 1986.

28. Chang, A. E., Schneider, P. D., Sugarbaker, P. H., Simpson, C., Culnane, M., Steinberg, S. M., A prospective randomized trial of regional vs. systemic continuous 5-fluorodeoxyuridine chemotherapy in the treatment of colorectal liver metastases, *Annals of Surgery*, 206, 685, 1987.

29. Hohn, D. C., Stagg, J. R., Friedman, M. A., Hammigan, J. F., Rayner, A., Ignoffo, R. J., Acord, P., Lewis B. J., A randomized trial of continuous intravenous vs. hepatic arterial floxuridine in patients with colorectal cancer metastatic to the liver: the Northern California Oncology Group Trial, *Journal of Clinical Oncology*, 7, 1646, 1989.

30. Kemeny, N., Daly, J., Reichman, B., Geller, N., Botet, J., Oderman, P., Intrahepatic or systemic infusion of fluorodeoxyuridine in patients with liver metastases from colorectal carcinoma, *Annals of Internal Medicine*, 107, 459, 1987.

31. Martin, J. K., Jr., O'Connell, M. J., Wieand, H. S., Fitzgibbons, R. J., Jr., Maillard, J. A., Rubin, J., Nagorney, D. M., Tschetter L. K., Krook, J. E., Intra-arterial floxuridine vs. systemic fluorouracil for hepatic metastases from colorectal cancer. A randomized trial, *Archives of Surgery*, 125, 1022, 1990

32. Graf, W., Pahlman, L, Bergstrom, R., Glimelius, B., The relationship between an objective response to chemotherapy and survival in advanced colorectal cancer, *British Journal of Cancer*, 70, 559, 1994.

33. Rougier, P., Laplanche, A., Huguier, M., Hay, J. M., Olliver, J. M., Escat, J., Salmon, R., Julien, M., Roullet Audy, J. C., Gallot, D., Gouzi, J. L., Pailler, J. L., Elisa, D., Lacaine, F., Roos, S., Rotman, N., Luboinski, M., Lasser, P., Hepatic arterial infusion of floxuridine in patients with liver metastases from colorectal carcinoma: long-term results of a prospective randomized trial, *Journal of Clinical Oncology*, 10, 1112, 1992.

34. Allen-Mersh, T. G., Earlam, S., Fordy, C., Abrams, K., Houghton, J., Quality of life and survival with continuous hepatic-artery floxuridine infusion for colorectal liver metastases, *Lancet*, 2, 1255, 1994.

35. Doria, M. I., Shepard K. V., Levin, B., Riddel, R. H., Liver pathology following hepatic arterial infusion chemotherapy. Hepatic toxicity with FUDR, *Cancer*, 58, 855, 1985.

36. Shepard, K. V., Levin, B., Faintuch, J., Doria, M. I., DuBrow, R. A., Riddell, R. H., Hepatitis in patients receiving intra-arterial chemotherapy for metastatic colorectal carcinoma, *American Journal of Clinical Oncology*, 10, 36, 1987.

37. Stagg, R. J., Venook, A. P., Chase, J. L., Lewis, B. J., Warren, R. S., Roh, M., Mulvihill, S. J., Grobman, B. J., Rayner, A. A., Hohn, D. C., Alternating hepatic intra-arterial floxuridine and fluorouracil: a less toxic regimen for treatment of liver metastases from colorectal cancer, *Journal of National Cancer Institute*, 83, 423, 1991.

38. Warren, H. W., Anderson, J. H., O'Gorman, P., Kane, E., Kerr, D. J., Cooke, T. G.,Mc Ardle, C. S., A phase II study of regional 5-fluorouracil infusion with intravenous folinic acid for colorectal metastases, *British Journal of Cancer*, 70, 677, 1994.

39. Kemeny, N., Conti, J. A., Sigurdson, E., Cohen, A., Seiter, K., Lincer, R., Niedzwiecki, D., Botet, J., Chapman, D., Costa, P., Budd, A., et al., A pilot study of hepatic artery floxuridine combined with systemic 5-fluorouracil and leucovorin. A potential adjuvant program after resection of colorectal hepatic metastases, *Cancer*, 71(6), 1964, 1993.

40. Kemeny N., Cohen, A., Bertino, J. R., Sigurdson, E. R., Botet, J., Oderman, P., Continuous intra-hepatic infusion of floxuridine and leucovorin through an implantable pump for the treatment of hepatic metastases from colorectal carcinoma, *Cancer*, 65, 2446, 1990.

41. Kemeny, N., Seiter, K., Conti, J. A., Cohen, A., Bertino, J. R., Sigurdson, E. R., Botet, J., Chapman, D., Mazumdar, M., Budd, A. J., Hepatic arterial floxuridine and leucovorin for unresectable liver metastases from colorectal carcinoma. New doses, schedules and survival update, *Cancer*, 73, 1134, 1994.

42. Kemeny, N., Seiter, K., Niedzwiecki, D., Chapman, D., Sigurdson, E., Cohen, A., Botet, J., Oderman, P., Murray, P., A randomized trial of intrahepatic infusion of fluorodeoxyuridine with dexamethasone vs. fluorodeoxyuridine alone in the treatment of metastatic colorectal cancer, *Cancer*, 69, 327, 1992.

43. Kemeny, N., Conti, J. A., Cohen, A., Campana, P., Huang, Y., Shi, W. J., Botet, J., Pulliam, S., Bertino, J. R., Phase II study of hepatic arterial floxuridine, leucovorin, and dexamethasone for unresectable liver metastases from colorectal origin, *Journal of Clinical Oncology*, 12, 2288, 1994.

44. Safi, F., Bittner, R., Roscher, R., Schuhmacher, K., Gaus, W., Beger, G. H., Regional chemotherapy for hepatic metastases of colorectal carcinoma (continuous intraarterial vs. continuous intraarterial/intravenous therapy). Results of a controlled clinical trial, *Cancer,* 64, 379, 1989.

45. Lorenz, M., Hottenrott, C., Inglis, R., Kirkowa-Reimann, M., Prevention of extrahepatic disease during intraarterial floxuridine of colorectal liver metastases by simultaneous systemic 5-fluorouracil treatment? A prospective multicentric study, *Gan To Kagaku Ryoho,* 16, 3662, 1989.

46. Venook, A., Novel treatments for unresectable liver tumors, in: *American Society Clinical Oncology Educational Booklet,* Twenty-sixth Annual Meeting of the American Society of Clinical Oncology, Washington, D.C., May 20–22, 1990, pp. 81–85.

47. Hrushesky, W. J. M., Circadian timing of cancer chemotherapy, *Science,* 228, 73, 1985.

48. Peleg, L., Ashkenazi, I. E., Carlebach, R., Chaitchick, S., Time-dependent toxicity of drugs used in cancer chemotherapy: separate and combined administration, *International Journal of Cancer,* 44, 273, 1989.

49. Petit, E., Milano, G., Levi, F., Thyss, A., Bailleul, F., Schneider, M., Circadian rhythm-varying plasma concentration of 5-fluorouracil during a five day continuous venous infusion at a constant rate in cancer patients, *Cancer Research,* 48, 167, 1988.

50. Harris, B. E., Song, R., Soong, S. J., Diasio, R. B., Relationship between dihydropyrimidine dehydrogenase activity and plasma 5-fluorouracil levels with evidence for circadian variation of enzyme activity and plasma drug levels in cancer patients receiving 5-fluorouracil by protracted continuous infusion, *Cancer Research,* 50, 197, 1990.

51. Von Römeling, R., Hrushesky, W. J. M., Circadian patterning of continuous floxuridine reduces toxicity and allows higher dose intensity in patients with widespread cancer, *Journal of Clinical Oncology,* 7, 1710, 1989.

52. Wesen C., Olson, G., von Römeling, R., Grage, T., Hrushesky, W. S. M., Circadian modified intraarterial treatment of colo-rectal carcinoma metastatic to the liver allows higher dose intensity to be safely given, *Proceedings of the American Society of Clinical Oncology,* 9, A406, 1989.

53. Adler, S., Lang, S., Langenmayer, I., Eibl-Eibesfeldt, B., Rump, W., Emmerich, B., Hallek, M., Chronotherapy with 5-fluorouracil and folinic acid in advanced colorectal carcinoma. Results of a chronopharmacologic phase I trial, *Cancer,* 73, 2905, 1994.

54. Sparano, J. A., Wadler, S., Diasio, R. B., Zhang, R., Lu, Z., Schwartz, E. L., Einzig, A., Wiernik, P. H., Phase I trial of low-dose, prolonged continuous infusion fluorouracil plus interferon-alpha: evidence for enhanced fluorouracil toxicity without pharmacokinetic perturbation, *Journal of Clinical Oncology,* 11, 1609, 1993.

55. Levi, F. F., Zidani, R., Vannetzel, J. M., Perpoint, B., Focan, C., Faggiuolo, R., Chollet, P., Garufi, C., Itzhaki, M., Dogliotti, L., Iacobelli, S., Adam, R., Kunstlinger, F., Gastiaburu, J., Bismuth, H., Jasmin, C., Misset, J. L., Chronomodulated vs. fixed-infusion-rate delivery of ambulatory chemotherapy with oxaliplatin, fluorouracil, and folinic acid (leucovorin) in patients with colorectal cancer metastases: a randomized multi-institutional trial, *Journal National Cancer Institute,* 86, 1608, 1994.

56. Arisawa, Y., Sutanto-Ward, E., Dalton, R. R., Sigurdson, E.R., Short-term intrahepatic FUdR infusion combined with bolus mitomycin C: reduced risk for developing drug resistance, *Journal of Surgical Oncology,* 56, 75, 1994.

57. Kemeny, N., Randomized trial of hepatic arterial floxuridine, mitomycin, and carmustine vs. floxuridine alone in previously treated patients with liver metastases from colorectal cancer, *Journal of Clinical Oncology,* 11, 330, 1993.

58. Patt, Y. Z., Hepatic arterial infusion of chemotherapy for metastatic colorectal cancer in the liver: why, how and what?, *Antibiotics and Chemotherapy,* 40, 1, 1988.

59. Borner, M., Laffer, U., Ludwig, C., Obrist, R., Metzger, U., Aeberhard, P., Weber, W., Castiglione, M., Obrecht, J. P., Brunner, K., Gloor, F., Cavalli, F., Senn, H. J., Effectiveness and low toxicity of hepatic artery infusion with fluorouracil and mitomycin for metastatic colorectal cancer confined to the liver, *Annals of Oncology,* 1, 227, 1990.

60. Campbell, T. N., Howell, S. B., Pfeifle, C. E., Wung, W. E., Bookstein, J., Clinical pharmacokinetics of intraarterial cisplatin in humans, *Journal of Surgical Oncology*, 21, 755, 1983.

61. Loehrer, P. J., Einhorn, L. H., Williams, S. D., Hui, S. L., Estes, N. C., Pennington, K., Cisplatin plus 5-FU for the treatment of adenocarcinoma of the colon, *Cancer Treatment Reports*, 69, 1359, 1985.

62. Cosimelli, M., Mannella, E., Anza, M., Civalleri, D., Balletto, N., Di-Tora, P., Durante, F., Porcellana, M., Cavaliere, P., Anfossi, A., Two consecutive trials on cisplatin (CDDP) hepatic arterial infusion, and IV 5-fluorouracil (5-FU) chemotherapy for unresectable colorectal liver metastases: an alternative to FUdR base regimens? *Journal of Surgical Oncology*, Suppl 2, 63, 1991.

63. Aigner, K. R., High dose hepatic artery infusion and venous hemofiltration of cytostatic drugs in liver metastases, in *Proc. 3rd Eur. Conf. Clin. Oncol. Cancer Nurs.*, Stockholm, June 16–20, 1985, p. 151.

64. Graham, R. A., Siddik, Z. H., Hohn, D. C., Extracorporeal hemofiltration: a model for decreasing systemic drug exposure with intra-arterial chemotherapy, *Cancer Chemotherapy and Pharmacology*, 26, 210, 1990.

65. Muchmore, J. H., Management of advanced intra-abdominal malignancy using high dose intra-arterial chemotherapy with concomitant chemofiltration, *Regional Cancer Treatment*, 3/4, 211, 1990.

66. Dazzi C., Fiorentini, G., Davitti, B., Priori, T., Cantore, M., Poddie, D., Carosi, V., Marangolo, M., Degli-Albizi, S., Cruciani, G., High-dose intra-arterial plus intraperitoneal chemotherapy combined with hemofiltration in liver metastases from colorectal carcinoma, *Tumori*, 80, 204, 1994.

67. Ravikumar, T. S., Pizzorno, G., Bodden, W., Marsh, J., Strair, R., Pollack, J., Hendler, R., Hanna, J., D'Andrea, E., Percutaneous hepatic vein isolation and high-dose hepatic arterial infusion chemotherapy for unresectable liver tumors, *Journal of Clinical Oncology*, 12, 2723, 1994.

68. Aigner K. R., Pelvic stopflow infusion (PSI) and hypoxic pelvic perfusion (HPP) with mitomycin and melphalan for recurrent rectal cancer, *Regional Cancer Treatment*, 1, 6–11, 1994.

69. Cantore, M., Fiorentini, G., Riitano, G., Davitti, B., Smerieri, F., Dipietro, G., Tona, G., Vaglini, M., Poddie, D., Hypoxic abdominal perfusion (HAP): rationale and evaluation of a new technique, *Regional Cancer Treatment*, 8, 64, 1995.

70. Teicher, B. A., Lazo, J. S., Sartorelli, A. C., Classification of antineoplastic agents by their selective toxicities toward oxygenated and hypoxic tumor cells, *Cancer Research*, 41, 73, 1981.

71. DiFilippo, F., Calabro, A. M., Cavallari, A., Carlini, S., Buttini, G. L., Moscarelli, F., Cavaliere, F., Piarulli, L., Cavaliere, R., The role of hyperthermic perfusion as a first step in the treatment of soft tissue sarcoma of the extremities, *World Journal of Surgery*, 12, 332, 1988.

72. Aigner, K. R., Isolated liver perfusion — long-term results, in *Fourth Int. Conf. Reg. Cancer Treat.*, Berchtesgaden, June 5–7, 1989, L6.

73. Aigner, K. R., Walther, H., Tonn, J., Wenzl, A., Hechtel, R., Merker, G., Schwemmle, K., First experimental and clinical results of isolated liver perfusion with cytotoxics in metastases from colorectal primary, *Recent Results in Cancer Research*, 86, 99, 1983.

74. Schwemmle K., Link, K. H., Rieck, B., Rationale and indications for perfusion in liver tumors: current data, *World Journal of Surgery*, 11, 534, 1987.

75. Hafström, L. R., Holmberg, S. B., Naredi, P. L., Lindner, P. G., Bengtsson, A., Tidebrant, G., Schersten, T. S., Isolated hyperthermic liver perfusion with chemotherapy for liver malignancies, *Surgical Oncology*, 3, 103, 1994.

76. Maeta, M., Kaibara, N., Nakashima, K., Kobayashi, M., Yoshikawa, T., Okamoto, A., Sugiyama, A., A case-matched control study of intrahepatoarterial chemotherapy in combination with or without regional hyperthermia for treatment of primary and metastatic hepatic tumors, *International Journal of Hyperthermia*, 10, 51, 1994.

77. Civalleri, D., Scopinaro, G., Simoni, G., Claudiani, F., Repetto, M., DeCian, F., Bonalumi, U., Starch microsphere-induced arterial flow redistribution after occlusion of replaced hepatic arteries in patients with liver metastases, *Cancer*, 58, 2151, 1986.

78. Lee, Y. T. N. M., Primary carcinoma of the liver: diagnosis, prognosis, and management, *Journal of Surgical Oncology*, 22, 17, 1983

79. *An Update on Regional Treatment of Liver Tumours. The Role of Vascular Occlusion*, 2nd ed.,. Kemeny, N., Carr, B., Hakansson, L., Lindberg, B., Gunnarsson, K., Nilsson, B., Taguchi, T., Khayat, D., Aigner, K. R., eds., Wells Medical, Royal Turnbridge Wells, Kent, U.K., 1995, chap. 3, pp. 27–45.

80. Willmott, N., Chemoembolisation in regional cancer chemotherapy: a rationale, *Cancer Treatment Reviews*, 14, 143, 1987.

81. Gerard, A., Buyse, M., Pector, J. C., Bleiberg, H., Arnaud, J. P., Willems, G., Delvaux, G., Lise, M., Nitti, D., Depadt, G., Hepatic artery ligation with and without portal infusion of 5-FU. A randomized study in patients with unresectable liver metastases from colorectal carcinoma. The E.O.R.T.C. Gastrointestinal Cooperative Group (G.I. Group), *European Journal of Surgical Oncology*, 17, 289, 1991.

82. Hafström, L. R., Engaras, B., Holmberg, S. B., Gustavsson, B., Jonsson, P. E., Lindner, P., Naredi, P., Tidebrant, G., Treatment of liver metastases from colorectal cancer with hepatic artery occlusion, intraportal 5-fluorouracil infusion, and oral allopurinol, *Cancer*, 74, 2749, 1994.

83. Dakhil, S., Esminger, W., Cho, K., Niederhuber, J., Doan, K., Wheeler, R., Improved regional selectivity of hepatic arterial BCNU with degradable microspheres, *Cancer*, 50, 631, 1982.

84. Civalleri, D., Rollandi, G., Simoni, G., Mallarini, G., Repetto, M., Bonalumi, U., Redistribution of arterial blood flow in metastases-bearing livers after infusion of degradable starch microspheres, *Acta Chirurgica Scandinavica*, 151, 613, 1985.

85. Civalleri, D., Scopinaro, G., Balletto, N., Claudiani, F., DeCian, F., Camerini, G., DePaoli, M., Bonalumi, U., Changes in vascularity of liver tumors after hepatic arterial embolization with degradable starch microspheres, *British Journal of Surgery*, 76, 699, 1989.

86. Daly, J. M., Butler, J., Kemeny, N., Yeh, S. D., Ridge, J. A., Botet, J., Bading, J. R., DeCosse, J. J., Benua, R. S., Predicting tumor response in patients with colorectal hepatic metastases, *Annals of Surgery*, 202, 384, 1985.

87. Civalleri, D., Esposito, M., Fulco, R. A., Vannozzi, M., Balletto, N., DeCian, F., Percivale, P. L., Merlo, F., Liver and tumor uptake and plasma pharmacokinetics of arterial cisplatin administered with and without starch microspheres in patients with liver metastases, *Cancer*, 68, 988, 1991.

88. Civalleri, D., Pector, J. C., Hakansson, L., Arnaud, J. P., Duez, N., Buyse, M., Treatment of patients with unresectable liver metastases from colorectal cancer by chemo-occlusion with degradable starch microspheres, *British Journal of Surgery*, 1994, 81, 1338.

89. Starkhammar, H., Hakansson, L., Morales, O., Svedberg, J., Effect of microspheres in intra-arterial chemotherapy. A study of arterio-venous shunting and passage of a labeled marker, *Medical Oncology and Tumor Pharmacotherapy*, 4, 87, 1987.

90. Hunt, T. M., Flowerdew, A. D., Birch, S. J., Williams, J. D., Mullee, M. A., Taylor, L., Prospective randomized controlled trial of hepatic arterial embolization or infusion chemotherapy with 5-fluorouracil and degradable starch microspheres for colorectal liver metastases, *British Journal of Surgery*, 77, 779, 1990.

91. Goldberg, J. A., Kerr, D. J., Wilmott, N., McKillop, J. H., McArdle, C. S., Regional chemotherapy for colorectal liver metastases: a phase II evaluation of targeted hepatic arterial 5-fluorouracil for colorectal liver metastases, *British Journal of Surgery*, 77, 1238, 1990.

92. Sasaki, Y., Imaoka S., Hasegawa, Y., Nakano, S., Ishikawa, O., Ohigashi, H., Taniguchi, K., Koyama, H., Iwanaga, T., Terasawa, T., Changes in distribution of hepatic blood flow induced by intra-arterial infusion of angiotensin II in human hepatic cancer, *Cancer*, 55, 311, 1985.

93. Taguchi, T., Ogawa, N., Bunke, B., Nilsson, B., and DSM Study Group of Japan, The use of degradable starch microspheres (Spherex) with intra-arterial chemotherapy for the treatment of primary and secondary liver tumors — results of a phase III clinical trial, *Regional Cancer Treatment*, 4, 161, 1992.

94. Farmer, D. G., Rosove, M. H., Shaked, A., Busuttil, R. W., Current treatment modalities for hepatocellular carcinoma, *Annals of Surgery*, 219, 236, 1994.

95. Nakakuma, K., Tashiro, S., Hiraoka, T., Ogata, K., Ootsuka, K., Hepatocellular carcinoma and metastatic cancer detected by iodized oil, *Radiology*, 154, 15, 1985.

96. Kasugai, H., Kojima, J., Tatsuta, M., Okuda, S., Sasaki, Y., Imaoka, S., Fujita, M., Ishiguro, S., Treatment of hepatocellular carcinoma by transcatheter arterial embolization combined with intraarterial infusion of a mixture of cisplatin and ethiozided oil, *Gastroenterology*, 97, 965, 1989.

97. Yamashita, Y., Takahashi, M., Koga, Y., Saito, R., Nanakawa, S., Hatanaka, Y., Sato, N., Nakashima, K., Urata, J., Yoshizumi, K., Ito, K., Sumi, S., Kan, M., Prognostic factors in the treatment of hepatocellular carcinoma with transcatheter arterial embolization and arterial infusion, *Cancer*, 67, 385, 1991.

98. Kawai, S., Tani M., Okamura, J., Ogawa, M., Ohashi, Y., Manden, M., Hayashi, S., Inoue, J., Kawarada, Y., Kusano, M., Kubo, Y., Kuroda, C., Sakata, Y., Shimamura, Y., Jinno, K., Takahashi, A., Takayasu, K., Tamura, K., Nagasue, N., Nakanishi, Y., Makino, M., Masuzawa, M., Yumoto, Y., Mori, T., Oda, T., and the Cooperative Study Group for Liver Cancer Treatment of Japan. Prospective and randomized clinical trial for the treatment of hepatocellular carcinoma. A comparison between L-TAE with farmorubicin and L-TAE with adriamycin: Preliminary results (second cooperative study), *Cancer Chemotherapy and Pharmacology*, 33 Suppl, S97, 1994.

99. Watanabe, S., Nishioka, M., Ohta, Y., Ogawa, N., Ito, S., Yamamoto, Y., Prospective and randomized controlled study of chemoembolization therapy in patients with advanced hepatocellular carcinoma, *Cancer Chemotherapy and Pharmacology*, 33 Suppl, S93, 1994.

100. Kawai, S., Tani, M., Okamura, J., Ogawa, M., Ohashi, Y., Monden, M., Hayaashi, S., Inoue, J., Kawarada, Y., Kusano, M., Kubo, Y., Kuroda, C., Sakata, Y., Shimamura, Y., Jinno, K., Takahashi, A., Takayasu, K., Tamura, T., Nagasue, N., Nakanishi, Y., Makino, M., Masuzawa, M., Yumoto, Y., Mori, T., Tishitsugu, Oda., The Cooperative Study Group for Liver Cancer Treatment of Japan, Prospective and randomized clinical trial for the treatment of hepatocellular carcinoma. A comparison of lipiodol-transcatheter arterial embolization with and without adriamycin (first cooperative study), *Cancer Chemotherapy and Pharmacology*, 33 Suppl, S1, 1994.

101. Groupe d'Etude et de Traitement du Carcinome Hépatocellulaire, A comparison of lipiodol chemoembolization and conservative treatment for unresectable hepatocellular carcinoma, *New England Journal of Medicine*, 332, 1256, 1995.

102. Chang, J. M., Tzeng, W. S., Pan, H. B., Yang, C. F., Lai, K. H., Transcatheter arterial embolization with or without cisplatin treatment of hepatocellular carcinoma. A randomized controlled study, *Cancer*, 74, 2449, 1994.

103. Lu, C. D., Qi, Y. G., Peng, S. Y., Lipiodolization with or without gelatin sponge in hepatic arterial chemoembolization for hepatocellular carcinoma, *Chinese Medical Journal (Beijing)*, 107, 209, 1994.

104. Pelletier, G., Roche, A., Ink, O., Anciaux, M. L., Derhy, S., Rougier, P., Lenoir, C., Attali, P., Etienne, J. P., A randomized trial of hepatic arterial chemoembolization in patients with unresectable hepatocellular carcinoma, *Journal of Hepatology*, 11, 181, 1990.

105. Madden, M. V., Krige, J. E., Bailey, S., Beningfield, S. J., Geddes, C., Werner, I. D., Terblanche, J., Randomised trial of targeted chemotherapy with lipiodol and 5-epidoxorubicin compared with symptomatic treatment for hepatoma, *Gut*, 34, 1598, 1993.

106. Yoshikawa, M., Saisho, H., Ebara, M., Iijima, T., Iwama, S., Endo, F., Kimura, M., Shimamura, Y., Suzuki, Y., Nakano, T., Fukuyama, Y., Fujise, K., Nambu, M., Ohto, M., A randomized trial of intrahepatic arterial infusion of 4'-epidoxorubicin with lipiodol vs. 4'-epidoxorubicin alone in the treatment of hepatocellular carcinoma, *Cancer Chemotherapy and Pharmacology*, 33 Suppl, S149, 1994.

107. Kameyama, M., Imaoka, S., Fukuda, I., Nakamori, S., Sasaki, Y., Fujita, M., Hasegawa, Y., Iwanaga, T., Delayed washout of intratumor blood flow is associated with good response to intraarterial chemoembolization for liver metastasis of colorectal cancer, *Surgery*, 114, 97, 1993.

108. Lang, E. K., Brown, C. L., Jr., Colorectal metastases to the liver: selective chemoembolization, *Radiology*, 189, 417, 1993.

109. Tarazov, P. G., Iodized oil in the portal vein after arterial chemoembolization of liver metastases. A caution regarding hepatic necrosis, *Acta Radiologica*, 35, 143, 1994.

110. Yamashita, Y., Prognostic factors in liver metastases after transcatheter arterial embolization or arterial infusion, *Acta Radiologica*, 31, 269, 1990.

111. Nakajo, M., Kobayashi, H., Shimabukuro, K., Shirono, K., Sakata, H., Taguchi, M., Uchiyama, N., Sonoda, T., Shinohara, S., Biodistribution and *in vivo* kinetics of iodine-131 lipiodol infused via the hepatic artery of patients with hepatic cancer, *Journal of Nuclear Medicine*, 29, 1066, 1988.

112. Burton, M. A., Gray, B. N., Kelleher D. K., Klemp. P. F., Selective internal radiation therapy: validation of intraoperative dosimetry, *Radiology*, 175, 253, 1990.

113. Ehrhardt, G. J., Day, D., E., Therapeutic use of ^{90}Y microspheres, *Nuclear Medicine and Biology*, 14, 233, 1987.

114. Ajlouni, M. I., Merrick, H. W., Skeel, R. T., Dobelbower, R. R., Concomitant radiation therapy and constant infusion FUdR for unresectable hepatic metastases, *American Journal of Clinical Oncology*, 13, 532, 1990.

115. Lawrence, W. D., Dworzanin, L. M., Walker-Andrews, S. C., Andrews, J. C., Ten-Haken, R. K., Wollner, I. S., Lichter, A. S., Esminger, W. D., Treatment of cancers involving the liver and porta hepatis with external beam irradiation and intraarterial hepatic fluorodeoxyuridine, *International Journal of Radiation Oncology Biology Physics*, 20, 555, 1991.

116. Minsky, B. D., Leibel, S. A., The treatment of hepatic metastases from colorectal cancer with radiation therapy alone or combined with chemotherapy or misonidazole, *Cancer Treatment Reviews*, 16, 213, 1989.

117. Ensminger, W. D., Hepatic arterial chemotherapy for primary and metastatic liver cancer, *Cancer Chemotherapy and Pharmacology*, 23 Suppl, S68, 1989.

118. Galun, E., Shouval, D., Adler, R., Shahaar, M., Wilchek, M., Hurwitz, E., Sela, M., The effect of anti-α-fetoprotein-adriamycin conjugate on a human hepatoma, *Hepatology*, 11, 578, 1990.

119. Order, S. E., Stillwagon, G. B., Klein, J. L., Leicher, P. K., Siegelman, S. S., Fishman, E. K., Ettinger, D. S., Haulk, T., Kopher, K., Finney, K., Iodine-131 antiferritin, a new treatment modality in hepatoma: a radiation therapy oncology group study, *Journal of Clinical Oncology*, 3, 1573, 1985.

120. Tang, Z. Y., Liu, K. D., Bao, Y. M., Lu, J. Z., Yu, Y. Q., Ma, Z. C., Zhou, X. D., Yang, R., Gan, Y. H., Lin, Z. Y., Radioimmunotherapy in the multimodality treatment of hepatocellular carcinoma with reference to second look resection, *Cancer*, 65, 211, 1990.

121. Matzku, S., Krempel, H., Weckernmann, H. P., Schirrmacher, V., Sinn, H., Stricker, H., Tumour targeting with antibody-coupled liposomes: failure to achieve accumulation in xenografts and spontaneous liver metastases, *Cancer Immunology and Immunotherapy*, 31, 285, 1990.

122. Akasaka, Y., Ueda, H., Takayama, K., Machida, Y., Nagai, T., Preparation and evaluation of bovine serum albumin nanospheres coated with monoclonal antibodies, *Drug Design Discovery*, 3, 85, 1988.

123. Mavligit, G. M., Zukiwski, A. A., Gutterman, J. U., Salem, P., Charnsangavej, C., Wallace, S., Splenic vs. hepatic artery infusion of interleukin-2 in patients with liver metastases, *Journal of Clinical Oncology*, 8, 319, 1990.

124. Komatsu, T., Yamauchi, K., Furukawa, T., Obata, H., Transcatheter arterial injection of autologous lymphokine-activated killer (LAK) cells into patients with liver cancers, *Journal of Clinical Immunology*, 10, 167, 1990.

125. Caruso, M., Panis, Y., Gagandeep, S., Houssin, D., Salzmann, J. L., Klatzmann, D., Regression of established macroscopic liver metastases after *in situ* transduction of a suicide gene, *Proceedings of the National Academy of Sciences U.S.A.*, 90, 7024, 1993.

126. Kim, B. S., Yoo, H. S., Park, Y. S., Lohj, J. K., Current status and future aspects on treatment of liver cancer, *Cancer Chemotherapy and Pharmacology*, 23 Suppl, S118, 1989.

127. Child, C. G., Turcotte, J. G., Surgery and portal hypertension. In *The liver and portal hypertension*, Child, C. G., ed., W. B. Saunders, Philadelphia, 1964, p. 50.

128. Pugh, R. N. H., Murray-Lyon, I. M., Dawson, J. L., Pietroni, M. E., Williams, R., Transection of the oesophagus for bleeding oesophageal varices, *British Journal of Surgery*, 60, 646, 1973.

129. Beppu, K., Inokuchi, K., Koyanagi, N., Nakayama, S., Sakata, H., Kitano, S., Kobayashi, M., Prediction of esophageal hemorrhage by esophageal endoscopy, *Gastrointestinal Endoscopy*, 27, 213, 1981.

130. Kemeny, N., Niedzwiecki, D., Shurgot, B., Oderman, P., Prognostic variables in patients with hepatic metastases from colorectal cancer. Importance of medical assessment of liver involvement, *Cancer*, 63, 742, 1989.

131. Gennari, L., Doci, R., Bignami, P., Bozzetti, F., Surgical treatment of hepatic metastases from colorectal cancer, *Annals of Surgery*, 203, 49, 1986.

132. Okuda, K., Ohtsuki, T., Obata, H., Tomimatsu, M., Okazaki, N., Hasegawa, H., Nakajima, Y., Ohinishi, H., Natural history of hepatocellular carcinoma and prognosis in relation to treatment: study of 850 patients, *Cancer*, 56, 918, 1985.

133. Silvestrini, R., Costa, A., Gennari, L., Doci, R., Bombardieri, E., Bombelli L., Cell kinetics of hepatic metastases as a prognostic marker in patients with advanced colorectal carcinoma, *HPB Surgery*, 2, 135, 1990.

134. Maehara, Y., Sakaguchi, Y., Emi, Y., Kusumoto, T., Kohnoe S., Mori, M., Sugimachi, K., Primary and metastatic liver lesions of clinical colorectal cancer differ in chemosensitivity, *International Journal of Colorectal Diseases*, 5, 87, 1990.

5

Establishing the Benefit/Risk Ratio with Combination Drugs

Petra A. Thürmann
University Hospital Frankfurt

Joerg Hasford
University of Munich

5.1 Introduction

Combination drugs are drugs that contain two or more different pharmacological agents in one preparation. The dosage of the different ingredients is fixed; thus, the term *fixed combination* or *fixed dose combination* is common for this combination of drugs.[1] Among the major advantages physicians claim for combination drugs are: improved effectiveness with a reduced adverse reaction profile, better patient compliance, better protection against drug abuse, and fewer direct costs. A summary of the various rationales for fixed combinations is shown in Table 5.1. Pharmacologists, however, routinely emphasize disadvantages like the lack of a well-proven synergistic beneficial action of the different ingredients, impossibility to dose the different active agents individually according to the patients' responses, and different half-lives of the two or more agents. Despite these differing views fixed combinations present in many countries a relevant share of the drug market. The purpose of this chapter is to outline regulatory, clinical-pharmacological, and methodological requirements for the assessment of the benefits and risks of combination drugs and to present appropriate clinical examples. We conclude with the problems of establishing the benefit/risk ratio.

5.2 Regulatory Requirements

The *United States Food and Drug Administration* was among the first that presented rules for the proper evaluation of combination drugs. The FDA stated as early as 1971:[2] "Two or more medicines may be combined in a single dosage form when each component makes a contribution to the claimed effects and the dosage of each component (amount, frequency, duration) is such that the combination is safe and effective for a significant patient population requiring such concurrent therapy as defined in the

TABLE 5.1   Rationales for Fixed Combination Drugs

Rationales	Examples
Enhancement of efficacy by combining two or more agents with different pharmacological actions	Trimethoprim and sulfamethoxazole; β-blocker plus diuretic for treatment of hypertension
Smaller doses of each individual agent can be used in combination drugs, thereby diminishing the risk of adverse effects	β-blocker plus diuretic in hypertension therapy
Combination drugs reduce the number of pills to be taken, probably increasing compliance	Antihypertensive combination drugs
One agent may ameliorate side effects of the other agent	Thiazide diuretic plus potassium-sparing diuretic; L-dopa plus decarboxylase inhibitor
Increased drug concentration at the site of action	L-dopa plus decarboxylase inhibitor
Addition of one agent may decrease the potential for abuse	Pentazocin plus naloxone
Combination drugs may be more adapted to the physiological situation and reduce adverse effects	Estrogen plus gestagen in oral contraceptives

labeling for the drug." This is still regarded as a relevant requirement. The recent Note for Guidance "Fixed Combination Medicinal Products" of the Committee for Proprietary Medicinal Products of the European Agency for the Evaluation of Medicinal Products (CPMP/EWP/240/95), released in April 1996, offers quite detailed advice.[3] This document will certainly become standard in this field and thus shall be considered here more thoroughly. The potential advantages of fixed combinations, which may justify a particular one, are explicitly mentioned: "a) an improvement of the benefit/risk assessment due to i) addition or potentiation of therapeutic activities of their substances, which result in a level of efficacy similar to the one achievable by each active substance used alone at higher doses than in combination but associated with a better safety profile, or a level of efficacy above the one achievable by a single substance with an acceptable safety profile; ii) the counteracting by one substance of an adverse reaction produced by another one; b) a simplification of therapy, which improves patient compliance." The general rules of this Note for Guidance say that fixed combinations "may not be considered rational if the duration of action of the substances differ significantly" and that "each substance must have documented contribution" to the claimed effect within the combination.[3] "Substances having a critical dosage range or a narrow therapeutic index are unlikely to be suitable for inclusion in fixed combinations." It is noteworthy that the inclusion of a substance intended to produce unpleasant adverse effects as a means of preventing abuse is considered undesirable. The proposed dosage of each substance must be justified by providing substantial evidence that "the combination is safe and effective for a significant population subgroup and the benefit/risk ratio of the fixed combination is equal or exceeds the one of each of its substances taken alone."

Randomized trials with parallel groups comparing placebo, the individual substances given alone, and the fixed combination are — if practicable — the method of choice. This design allows for the evaluation of the type of interaction (e.g., additive, multiplicative, or antagonistic) between the individual components. These trials have to be done in accordance with the current version of the Good Clinical Practice Guideline. With regard to safety, in the case of long-term use, safety data on 300 to 600 patients for 6 months or longer will be called for.[3]

5.3   Clinical Pharmacological Considerations

Drug approval authorities require that each active ingredient of a combination preparation must contribute to its overall safety and/or efficacy; thus, a combination of two or more drugs should be rational with regard to pharmacokinetic and pharmacodynamic characteristics as well as from the clinical point of view.

Combination preparations are developed for several reasons, such as enhancement of efficacy or amelioration of side effects (see Table 5.1). In order to achieve the desired goals, one takes advantage of either a pharmacokinetic or pharmacodynamic "interaction" between two or more drugs.

5.3.1 Pharmacokinetic Aspects of Combination Drugs

It seems noteworthy that pharmacokinetic drug interactions are usually considered undesired risks. Furthermore, a pharmacokinetic basis for a combination preparation is rarely the case. The earlier applied combination of penicillin and probenecid, the latter inhibiting the renal elimination of the former, is no longer used, since a simple increase of the penicillin dose leads to the same effect without adding a further risk. A more important example is the combination of L-dopa and a decarboxylase inhibitor, usually carbidopa or benserazide. Both substances inhibit the peripheral transformation of L-dopa to dopamine, thereby diminishing the peripheral (cardiovascular, gastrointestinal) effects of dopamine. Since carbidopa and benserazide do not penetrate into the CNS, L-dopa can be transformed to dopamine in the CNS and exert its wanted pharmacological action.

In order to increase the duration of action controlled release preparations are usually preferred, but the combination of benzylpenicillin with benzathine penicillin or unmodified insulin plus insulin zinc suspension represent examples of currently prescribed combination drugs. For these combination drugs the benefit in terms of prolongation of the dosing interval is evident.

5.3.2 Pharmacodynamic Aspects of Combination Drugs

Establishing a benefit/risk ratio depends on our knowledge of benefits and risks of a certain drug or combination. Even when a drug is approved for a certain indication, our knowledge of the benefits is limited, since most drugs — especially those for chronic diseases — receive their approval by the responsible authorities based on data on surrogate end points. "A surrogate end point of a clinical trial is a laboratory measurement or a physical sign used as a substitute for a clinical meaningful end point that measures directly how a patient feels, functions or survives. Changes induced by a therapy on a surrogate end point are expected to reflect changes in a clinical meaningful end point. It is implicit in this definition that a surrogate end point is by itself of no value to the patient.[4]" For example, antihypertensive drugs are approved if there is sufficient evidence for a blood pressure–lowering efficacy and a "favorable" safety profile, although the aim of antihypertensive treatment is to prevent end-organ damage and to reduce cardiovascular and total mortality. Some further examples for commonly accepted surrogate end points are given in Table 5.2. Most of these parameters are well characterized risk markers, but their validity as clinical end points has never been proven; e.g., it is not known if the reduction of left ventricular hypertrophy in essential hypertensive patients automatically translates into an improvement of prognosis. The example of the CAST-Study shows the invalidity of a surrogate end point — i.e., the reduction of ventricular arrhythmiae — and the importance of considering side effects — i.e., pro-arrhythmogenic actions.[5] Having these pitfalls in mind when using surrogate end points, benefit cannot be defined, for example, as blood pressure reduction. The wording benefit is clearly aimed at the end point, like cardiovascular morbidity or overall mortality.

TABLE 5.2 Examples of Surrogate and Clinical End Points

Disease/Indication	Surrogate End Points	Clinical End Points
Hypertension	Blood pressure reduction in mmHg, regression of left ventricular hypertrophy	Cardiovascular morbidity and mortality, total mortality
Diabetes mellitus	Blood glucose levels, insulin secretion, HbA_{1C}	Diabetes-associated end-organ damage, survival time
Hyperlipoproteinaemia	LDL/HDL serum lipoproteins	Myocardial infarction, overall mortality
Thromboembolic disorders	Partial thromboplastin time (INR)	Pulmonary embolism, mortality
Ventricular arrhythmia	Reduction of ventricular arrhythmia in the ECG	Sudden cardiac death, overall mortality

5.3.3   Value of Surrogate End Points in Benefit/Risk Assessment

The following example is given to highlight the problem of surrogate end points in the benefit/risk assessment of a widely used combination drug.

The combination of the non-potassium–sparing diuretic hydrochlorothiazide (HCT) and a potassium-sparing diuretic (e.g., triamterene, amiloride) seems to be justified. Both drugs contribute to the overall efficacy and one drug ameliorates the adverse effects of the other.[6] Several large-scale trials demonstrated that non-potassium–sparing diuretics given alone to patients with essential hypertension may lead to an increased risk of sudden cardiac death,[7,8] probably caused by hypokalemia. However, the administration of this fixed combination does not result in a normalization of serum potassium in all patients; moreover, hyperkalemia and hyponatremia may occur in a considerable percentage of patients.[9,10] Furthermore, the addition of triamterene leads to an increased incidence of nephrolithiasis, thereby introducing an additional risk. In this respect the addition of a potassium-sparing diuretic to HCT is claimed to reduce diuretic-induced electrolyte disturbances and by this means reduces the risk of sudden cardiac death. The relevant question to be answered is: Do combination preparations of a thiazide plus a potassium-sparing diuretic really decrease morbidity and mortality of hypertensive patients to a larger extent than thiazides alone? So far, this question has not been addressed by any of the large-scale hypertension trials. In contrast to the above-mentioned MRFIT trial where diuretic therapy with thiazides alone was associated with an increased mortality compared to antiadrenergic drugs,[7] two other trials showed that combination treatment with thiazide and a potassium-sparing drug leads to a decrease in mortality — but not in comparison to thiazide alone.[11,12] The only evidence for a direct comparison is supplied by two recent case-control studies where patients being treated with thiazides had an approximately twofold increased risk of sudden cardiac death in comparison to patients receiving either additional potassium-sparing diuretics or other antihypertensive drugs.[8,12] Siscovick and co-workers were also able to show a clear dose–response risk-relationship for thiazides without potassium-sparing agents where the lowest risk could be demonstrated for 25 mg HCT plus amiloride or triamterene.[13] It should be considered in this context that the addition of triamterene to HCT may reduce the absorption of the latter by almost 50%,[14] suggesting that a decrease in bioavailability rather than avoidance of hypokalemia may be involved in the beneficial effect of combining these drugs. Moreover, a completely different situation may occur for patients with congestive heart failure receiving diuretics. Apparently, in these patients total body content of potassium is not depleted even after long-term therapy. After all, it is surprising that for a combination designed to reduce the risk of monotherapy only "surrogate end-point (serum potassium) trials" were conducted or uncontrolled postmarketing surveillance data are shown,[15] but no prospective data are available for the direct comparison of HCT with and without potassium-sparing drugs with respect to the clinical end point overall mortality.

Questions may also arise when different antihypertensive drugs, such as a β-blocker and a diuretic are combined in order to reduce the maximal doses of the individual medicines and to increase compliance. A benefit in terms of convenience for the patients and of compliance seems to be evident. Diuretic therapy leads to an activation of the renin angiotensin system, which can be partially deactivated by β-blockers; otherwise, β-blockers may induce fluid retention, which can be counterregulated by diuretic therapy. Furthermore, the potentially arrhythmogenic action of diuretics via hypokalemia may also be opposed by β-blockers. In most trials investigating the effect of antihypertensive drugs on morbidity and mortality, β-blockers and diuretics given alone showed a clear benefit, but with the limitation of the potential hazards induced by thiazide monotherapy. If the individual drugs combined are comparable with regard to duration of action, a combination may be assumed as rational. However, if the combination bears a slightly enhanced potential for serious side effects, the convenience gain is clearly overcome. Interestingly, both of the aforementioned case-control studies showed an increased risk of mortality in patients treated with β-blockers when compared with patients not receiving β-blockers, thus questioning the value of β-blockers in hypertensive patients.[8,13] With respect to these findings, the combination of a diuretic with a β-blocker should be reevaluated, especially since β-blockers merely ameliorate the risk of diuretic-induced hazards.

Both examples — the combination of a β-blocker plus a diuretic as well as the combination of hydrochlorothiazide and triamterene — clearly show that the words risk and benefit have to be used and interpreted cautiously.

5.3.4 Benefit/Risk Evaluation of OTC Preparations

A more rigid approach to the establishment of a benefit/risk ratio should be made in OTC preparations. The knowledge of consumers on efficacy of OTC drugs as well as their risks is rather limited. Consumers cannot estimate the risks of a combination drug and also are not able to decide on the real need of a combination drug, e.g., for symptomatic treatment of a common cold. There is no convincing pharmacological justification for the various combinations with aspirin, paracetamol, codeine, caffeine, and vitamin C, but there might be an increased risk for adverse reactions and abuse with fixed combinations of analgesics plus caffeine or codeine. Even multivitamin combinations bear a potential risk if they contain up to 10,000 IU vitamin A per capsule. Multivitamin combinations are often bought by pregnant women, and 10,000 IU vitamin A has been reported to be a teratogenic dose in certain phases of pregnancy.[16] In this case, the pharmacist is the only one who might possess a control function; therefore, a thorough benefit/risk evaluation must have taken place before the consumer can buy such a preparation.

5.4 Biometric Methodology

5.4.1 Measuring Efficacy or Benefit

The clinical preregistration investigation of medicines is usually organized in the partly overlapping phases I, II, and III. Phase I includes the first administration of the new compound to humans — most often healthy young male volunteers — and aims at the pharmacokinetic profile and the tolerable dosage range. With few exceptions, e.g., anticancer drugs, data on efficacy are not collected but only data on safety. In phase II highly selected patients are treated to assess efficacy, efficacious dosages, and necessary treatment duration. Many phase II studies are already randomized trials and may count as pivotal studies. In phase III the knowledge about efficacy, safety, and dosage range is amplified by larger randomized trials, including a broader range of patient populations. In phase II and in many disease areas, e.g., chronic diseases, also in phase III, surrogate end points are used as parameters of efficacy. In case of favorable results approval for marketing will be applied for at the regulatory authorities. After marketing of the product, additional studies on safety and/or efficacy are usually neither required nor regulated. Thus, clinical end point and precise safety data in larger and different patient populations are often missing.

With fixed combinations regulatory authorities expect that the benefit of each of the different agents separately and of their fixed combination has been shown in randomized trials. The factorial design allows the fulfillment of this requirement within one trial, as long as there are not too many different active agents. In the most common case when two different agents are to be combined, the 2 × 2 factorial design is simple and appropriate: one group is treated with agent A, one with agent B, one with the fixed combination AB, and one with placebo (Figure 5.1). The allocation to the groups is random. This design allows for the comparative assessment of the contribution of the individual agents and of the quality and quantity of their interaction in the fixed combination AB. The interaction between A and B is expected to be synergistic but can be antagonistic, too. A great advantage of this design is that half of the patients receive A and half receive B. The statistical analysis first looks for the main effects of A and B. The main treatment effect of A is estimated by $(d_3 + d_4)$: 2 with $d_3 = (x_{11} - x_{21})$ and $d_4 = (x_{12} - x_{22})$. Accordingly, the main treatment effect of B is estimated by $(d_1 + d_2)$: 2 with $d_1 = (x_{11} - x_{12})$ and $d_2 = (x_{21} - x_{22})$. The interaction between A and B is calculated by the difference $(d_2 - d_1)$: 2.[17] Factorial designs can be extended for different dosages of A and B and are useful both for surrogate and clinical end points. Examples for the latter are the U.S. Physicians Health Study[18] and the ISIS-2.[19]

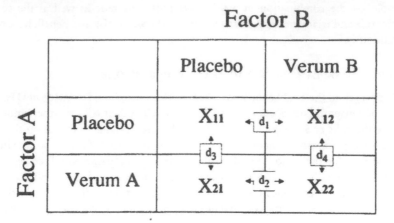

FIGURE 5.1   Design of a 2 × 2 factorial design.

5.4.2   Measuring Risk

Compared to the assessment of the benefits a comprehensive evaluation of the safety profile presents much greater problems. There is no typical, exactly defined clinical symptomatology of an adverse drug reaction to look for. Adverse drug reactions can manifest themselves in all the different body tissues and organs, can alter their functions, and may even impair the offspring. Adverse drug reactions can mimic common diseases and their complications, may be very rare, and may take a long latency period. Drugs can adversely affect individuals, directly and indirectly, even uninvolved individuals and public health, for example, through selection of antibiotic resistant bacterial strains in hospitals. Thus, it may be doubted whether the whole spectrum of the adverse reactions of a drug will ever be known.[20] Essentials of a solid detection, recording, collection, and evaluation of adverse events in the clinical investigation of drugs have been published.[21] Most guidelines concentrate, however, on the proper identification and reporting of beneficial effects. Thus, even in randomized trials, the investigation of adverse effects is not always carried out with the necessary systematic approach and care.

Adverse drug reactions cannot be observed directly; this is only possible for events and changes in a particular patient. Usually, it is no problem to decide whether a particular event is adverse or not. But the assessment whether an adverse event has been really caused by a drug and is to be regarded as an adverse drug reaction is the result of a causality assessment. In a randomized controlled trial the comparison of the frequencies and severities of the adverse events between the groups helps to find the truth. For rare adverse reactions single case causality assessment is needed, where considerable interobserver variability is common. Special decision algorithms have been developed for this purpose.[22,23]

But even if the assessment of adverse reactions is done according to the state of the art during phases I to III the real test phase begins only with the marketing of the drug. The generalizability of the safety profile at this time is rather limited due to the following reason: clinical trials usually admit highly selected, often hospitalized patients. Women with childbearing potential, children, old and/or multimorbid patients are excluded, although they represent major drug consumer groups. Treatment and observation phases are limited, even for chronic diseases. Sample sizes studied are fairly small. The Note for Guidance requires safety data on 300 to 600 patients for 6 months or longer only in the case of long-term use.[3] This allows the identification of very frequent adverse reactions with an incidence of at least 1% only.

The designs of phase IV research are more varied than in phase I to III. There is good evidence that cohort-studies with comparative groups, like, for example, the Prescription Event Monitoring System run by the Drug Surveillance Research Unit, provide reliable results.[24] Company-sponsored observational studies on the contrary usually did not provide valuable knowledge about type, incidence, severity, and outcome of adverse drug reactions.[25,26] In recent years also large randomized trials with a lean protocol have shown to be a valuable design option.[27,28] As this topic cannot be presented in more detail in this chapter, some textbooks are cited.[22,29]

In summary, we think that the safety data base of many drugs currently do not provide the data needed for the solid establishment of a benefit/risk ratio.

5.5 Benefit/Risk Assessment

Any benefit/risk analysis is — like the proof of efficacy — related to the indication of the drug or fixed combination under investigation. The results obtained according to each indication may vary. The establishment of the benefit/risk ratio meets a variety of problems that usually prevent a straightforward and widely accepted procedure:

In the case of a fixed combination drug the following aspects have to be considered:

1. the benefit of each combination partner given alone,
2. the risks of each combination partner given alone,
3. the additional benefit of the combination,
4. the possibly occurring additional risks when both drugs are combined,
5. benefit and risks in the natural history of disease,
6. benefit and risks of therapeutic alternatives.

Ideally the data for the first and third point are available. Benefit can be substantiated by efficacy of effectiveness data, quality of life, costs, or the impact on public health. Even at a rather fundamental level, however, the data needed to aggregate the benefits are quite often missing. As mentioned above the benefit has often been shown only for surrogate end points in highly selected patient samples. Trials admitting the actual users after market authorization, like women, senior people, and polymorbid patients, might never have been done.

Risk is not easy to be measured, as it entails three critical elements:[30] the losses itself, in the form of adverse drug reactions; the significance of losses, i.e., the severity and the impact on a particular patient's quality of life; and the uncertainty associated with these losses, which means the odds that these losses would actually occur. Thus, there is a strong subjective component inherent in this construct and there might be interactions if more than one loss, i.e., adverse reaction, occurs. Similarly, benefit can be regarded as a construct. too. The total drug risk can therefore be formalized as a function:[31]

$$u(Dj) = w_1 adr_1 (f_{1j}) + w_2 adr_2 (f_{2j}) + ... + w_n adr_n (f_{nj})$$

with $u(Dj)$ = total risk of drug Dj, w_i = weight or impact on well-being, adr_i = individual adverse drug reaction and f_i = relative frequency. According to our experience most of the necessary safety data are missing. Even for drugs used widely and for a long time precise data on incidence of serious adverse reactions in real life situations are missing, not to mention a stratification for risk groups and dosage. With combination drugs it is extremely difficult to find out which of the different ingredients is responsible for a particular adverse reaction or if it might be due to an interaction.

Many diseases do not lead to serious or even fatal complications but are resolved without any treatment. For a benefit/risk analysis, quantitative and qualitative data on this natural history, analogous to those for the efficacy and adverse drug reactions, are necessary. Therefore, for example, the incidence of adverse drug reactions must be compared to the spontaneous occurrence of adverse events in the population of patients following the natural history of a disease.[20] In many cases, for ethical reasons the details relating to a natural history may not be ascertainable as the benefit/risk analysis already favors treatment and argues against allowing the disease to take a natural course. Benefits and risks must be ascertained for all therapies of the indications under investigation. Preventive standard therapies can also be included in the analysis.

Benefit and risks are rarely measured in the same dimensions or units, respectively. How to balance a 10 mmHg blood pressure reduction with the occurrence of dry eyes, cold extremities, or occasional nightmares? This problem of lack of commensurability of benefit and risk is crucial as there is no common

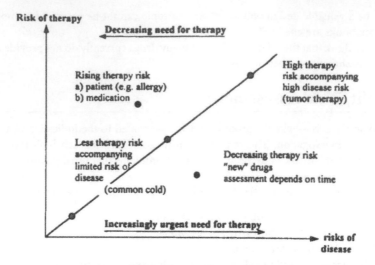

FIGURE 5.2   General considerations for a benefit/risk analysis, after Spilker.[33]

denominator. In some situations, however, it may be feasible to specify the benefit as the expected benefit (e.g., probability of cure according to the results of clinical studies) and to substract the product of the number and severity of adverse events. An example of this is the calculated "time without symptoms and toxic effects" (TWiST) or even a quality-adjusted TWiST.[32] Such a combined "end point" resembles the utility parameters, e.g., the quality-adjusted life-year (QALY), which is routinely used for pharmaco-economic studies. A benefit can then be calculated according to the following equation:

$$\text{treatment benefit} = \text{treatment effects } (= \text{efficacy}) - \text{treatment failures} - \text{adverse events}$$

In principle, all known adverse drug reactions must be simultaneously included in the benefit/risk analysis. This demands an evaluation of the degree of severity of the different adverse drug reactions. For very serious adverse events a partial solution might be to weight the incidence with the fatality rate.

A benefit/risk assessment must also specify its target, a particular patient, a group of patients, or the whole population of patients in a health care system.[33] Although mathematically exact benefit/risk ratios are not currently achievable in most cases, we think less methodological approaches might provide useful results. Figure 5.2 shows the customary, less formalized view of benefit/risk analysis.[34] As the risks of a disease increase, the necessity for therapy becomes increasingly urgent. The more urgent the therapy, the higher the consequent risks that can be accepted.

Establishing the benefit/risk ratio is an important step for rational decisions at practically all levels of health care. Reliable rules for benefit/risk analyses have therefore to be developed. Methodologically sound phase IV research is urgently needed to provide the necessary data.

References

1. Spilker, B., Combination medicine trials, in *Guide to Clinical Trials*, Spilker, B., ed., Raven Press, New York, 1991, chap. 50.
2. Crout, J. R., Fixed Combination Prescription Drugs: FDA Policy, *J. Clin. Pharmacol.*, 14, 249, 1974.
3. Committee for Propriety Medicinal Products (CPMP), Fixed combination medicinal products. Note for Guidance, CPMP/EWP/240/95, *Pharm. Ind.*, 58, 504, 1996.
4. Temple, R. J., A regulatory authority's opinion about surrogate end points, in *Clinical Measurement in Drug Evaluation*, Nimmo, W., Tucker, G., eds., John Wiley, Chichester, 1995, chap. 1.
5. Akiyama, T., Pawitan, Y., Greenberg, H., and the CAST Investigators, Increased risk of death and cardiac arrest from encainide and flecainide in patients after non-Q-wave acute myocardial infarction in the cardiac arrhythmia suppression trial, *Am. J. Cardiol.*, 68, 1551, 1991.

6. Papademetriou, V., Burris, J., Kukich, S., Freis, E. D., Effectiveness of potassium chloride or triamterene in thiazide hypokalemia, *Arch. Intern. Med.*, 145, 1986, 1985.
7. Multiple Risk Factor Intervention Trial Research Group, Multiple Risk Factor Intervention Trial. Risk factor changes and mortality results, *J. Am. Med. Assoc.*, 248, 1465, 1982.
8. Hoes, A. W., Grobbee, D. E., Lubsen, J., Man in't Veld, A. J., van der Does, E., and Hofman A., Diuretics, betablockers, and the risk for sudden cardiac death in hypertensive patients, *Ann. Intern. Med.*, 123, 481, 1995.
9. McMahon, F. G., Okun, R., Vaicatis, J. S., Multicenter study of amiloride/hydrochlorothiazide once daily and triamterene/hydrochlorothiazide twice daily: antihypertensive and potassium-sparing effects, *Curr. Ther. Res.*, 34, 357, 1983.
10. Bayer, A. J., Farag, R., Pathy, M. S. J., Plasma electrolytes in elderly patients taking fixed combination diuretics, *Postgrad. Med. J.*, 62, 159, 1986.
11. Dahlöf, B., Lindholm, L. H., Hansson, L., Schersten, B., Ekbom, T., and Wester, P.O., Morbidity and mortality in the Swedish Trial in Old Patients with Hypertension (STOP-Hypertension), *Lancet*, 338, 1281, 1991.
12. Medical Research Council Working Party, MRC trial of treatment of mild hypertension: principal results, *Br. Med. J.*, 291, 97, 1985.
13. Siscovick, D. S., Raghunathan, T. E., Psaty, B. M., Koepsell, T. D., Wicklund, K. G., Lin, X., Cobb, L., Rautaharju, P. M., Copass, M. K., and Wagner, E. H., Diuretic therapy for hypertension and the risk of primary cardiac arrest, *N. Engl. J. Med.*, 330(26), 1852, 1994.
14. Blume, C. D., Williams, R. L., Upton, R. A., Lin, E. T., and Benet, L. Z., Bioequivalence study of a new tablet formulation of triamterene and hydrochlorothiazide, *Am. J. Med.*, 77 (Suppl 5A), 59, 1084.
15. Hollenberg, N. K., Mickiewicz, C. W., Postmarketing surveillance in 70,898 patients treated with a triamterene/hydrochlorothiazide combination (Maxzide), *Am. J. Cardiol.*, 63, 37B, 1989.
16. Rothman, K. J., Moore, L. L., Singer, M. R., Nguyen, U. D. T., Mannino, S., Milunsky, A., Teratogenicity of high vitamin A intake, *N. Engl. J. Med.*, 333, 1369, 1995.
17. Ansari, H., Studiendesigns, in *Arzneimittelprüfungen und Good Clinical Practice*, Hasford, J., Staib, A. H., eds., MMV Medizin Verlag, München, 1994, chap. 16.
18. Stampfer, M. J., Buring, J. E., Willett, W., Rosner, B., Eberlein, K., Hennekens, C. H., The 2×2 factorial design, its application to a randomized trial of aspirin and carotene in U.S. physicians, *Statistics in Medicine*, 4, 111, 1985.
19. ISIS-2 (Second International Study of Infarct Survival) Collaborative Group, Randomized trial of intravenous streptokinase, oral aspirin, both, or neither among 17,187 cases of suspected acute myocardial infarction, ISIS-2, *Lancet*, 2, 349, 1988.
20. Hasford, J., Victor, N., Risk-benefit analysis of drugs: fundamental considerations and requirements from the point of view of the biometrician. Problems in the assessment of the combination of trimethoprim with sulfamethoxazole, *Infection*, 15 (Suppl 5), S236, 1987.
21. Bethge, H., Czechanowski, B., Gundert-Remy, U., Hasford, J., Kleinsorge, H., Kreutz, G., Letzel, H., Müller, A. A., Selbmann, H. K., Weber, E., Recommendations for the Detection, Recording, Collection and Evaluation of Adverse Events in the Clinical Investigation of Drugs, *Drugs Made in Germany* 34, 10, 1991.
22. Stephens, M. D. B., *Detection of New Adverse Drug Reactions*, 3rd ed., Macmillan Publishers, Basingstoke, 1992.
23. Jones, J. K., Determining Causation from Case Reports, in *Pharmacoepidemiology*, 2nd ed., Strom, B. L., Ed., John Wiley, Chichester, 1994, chap. 26.
24. Wallander, M. A., The way towards adverse event monitoring in clinical trials, *Drug Saf.*, 8(3), 251, 1993.
25. Waller, P. C., Wood, S. M., Langman, M. J. S., Breckenridge, A. M., Rawlins, M. D., Review of company postmarketing surveillance studies, *Br. Med. J.*, 304, 1470, 1992.

26. Hasford, J., Observational Postmarketing Studies — A Contribution to Drug Research in Special Populations?, *Eur. J. Clin. Pharmacol.*, 49, A152, 1995.

27. Hasford, J., Drug risk assessment. A case for large trials with lean protocol, *Pharmacoepidemiology and Drug Safety*, 3, 321, 1994.

28. Mitchell, A. A., Lesko, S. M., When a randomized controlled trial is needed to assess drug safety, *Drug Safety*, 13, 15, 1995.

29. Strom, B. L., *Pharmacoepidemiology*, 2nd ed., John Wiley, Chichester, 1994.

30. Yates, J. F., Stone, E. R., The risk construct, in *Risk-taking Behavior*, Yates, E. F., Ed., John Wiley, Chichester, 1992, chap. 1.

31. Aschenbrenner, K. M., Kasubek, W., Challenging the Cushing syndrome: multiattribute evaluation of cortisone drugs, *Org. Behav. Hum. Perform.*, 22, 216, 1978.

32. Chuang-Stein, C., A new proposal for benefit-less-risk analysis in clinical trials, *Controlled Clin. Trials* 15, 30, 1994.

33. Spilker, B., Benefit-to-risk assessments and comparisons, in *Guide to Clinical Trials*, Spilker, B., Ed., Raven Press, New York, 1991, chap. 99.

34. Dölle, W., Müller-Oerlinghausen, B., Differentialtherapie, in *Grundlagen der Arzneimitteltherapie*, Dölle, W., Müller-Oerlinghausen, B., Schwabe, U., eds., BI-Wissenschaftsverlag, Mannheim, 1986, chap. 4.1.

6

Improving the Therapeutic Index of Antineoplastic Agents through Additional Drugs

Gianmauro Numico
National Cancer Institute, Genoa

Dario Civalleri
University of Genoa School of Medicine

Mauro Esposito
National Cancer Institute, Genoa

List of Abbreviations

CT:	Chemotherapy
MDR:	Multidrug resistance
CSF:	Colony Stimulating Factor
ABMT:	Autologous Bone Marrow Transplantation
EPO:	Erythropoietin
SCF:	Stem Cell Factor
G-CSF:	Granulocyte Colony Stimulating Factor
GM-CSF:	Granulocyte-Macrophage Colony Stimulating Factor
IL3:	Interleukin 3
FA:	Folinic Acid
5FU:	5-Fluorouracil
CH_2-THF:	5,10-Methylene-Tetrahydrofolate
FdUMP:	Fluoro-Deoxiuridine Monophosphate
TS:	Thymidilate Synthetase
IFN:	Interferon
HCL:	Hairy-Cell Leukemia
NK:	Natural Killer Cells

6.1   Introduction

6.1.1   Therapeutic Index

Important results in the field of cancer treatment and cancer-related symptom palliation have been achieved with antineoplastic chemotherapy (CT). However, the resistance of tumors to the currently available cytotoxic drugs is a common clinical experience: in some cases, after a first tumor response, progressive disease or recurrence shows a nonresponsive, progressive growth. In other tumors, drug resistance is present at the beginning of treatment, responses are infrequent, and do not endure. Moreover, antitumoral agents have among the lowest therapeutic indices when compared with any other drug prescribed for the treatment of human diseases inasmuch as patients undergoing CT are often exposed to serious toxic reactions. Acute toxicities such as myelosuppression or mucositis, although reversible, can be life threatening, and chronic toxicities (e.g., neurological, cardiac, renal, gonadal) strongly affect patient quality of life. It is clear that increasing the therapeutic index is a major end point in clinical and basic oncologic research.

While looking for new anticancer agents with high selectivity against tumor cells or innovative therapeutic modalities, it is mainly through the concomitant administration of multiple drugs that CT can be safely employed in clinical practice.

Two strategies have been used to increase the therapeutic index of cytotoxic agents:

1. *Protection against toxicity* with the addition of nonantineoplastic drugs and the dose optimization/escalation of single or associated cytotoxic agents. Examples of such an approach are the hematologic protection afforded by hemopoietic growth factors and renal protection with forced diuresis.
2. Administration of multiple drugs (antineoplastic and nonantineoplastic) to *prevent or overcome the onset of resistance:* multiple agent CT, extensively used in the treatment of solid malignancies, is the most evident example, although many noncytotoxic drugs have been shown to be useful when combined with CT agents.

Both therapeutic approaches have been subject to criticism: single agent CT, at both conventional or high doses, is likely to be inadequate when cancer cells are or become resistant. On the other hand, the association of multiple drugs with different mechanisms of action may not enable reaching the active single agent dose. Clinical evidence, however, suggests that if used in the appropriate setting, both strategies can be of value in increasing the cure rate of human tumors.

In this section, the authors will enter in some detail and will discuss the general mechanisms of resistance to antineoplastic drugs, the dose-escalating approach, and some of the most commonly employed combinations of drugs.

6.1.2   Antineoplastic Agents and DNA Damage

Chemotherapeutic agents act essentially by causing neoplastic cell death through a selective interaction with intracellular targets that are crucial for cellular replication or energy production. Neoplastic cell DNA damage is considered the most important aspect of the cytotoxic activity of antineoplastic drugs. Several types of DNA lesions have been described: base alterations, interstrand or intrastrand DNA-crosslinks, protein-associated DNA cleavage, and DNA single- or double-strand breaks. Furthermore, DNA damage can result not only from drug–DNA interactions but frequently represents the fall-out from a cascade of responses to interactions with non-DNA targets. DNA alterations are induced, for instance, by tubulin binding vinca alkaloids,[1] by topoisomerase II inhibitors,[2] or by the antifolate methotrexate.[3] Though the action of cytotoxic agents is mediated by multiple interactions, it seems likely that DNA damage per se is the central factor in determining antitumoral activity.[4]

6.1.3   Pattern of Tumor Growth and Cytotoxic Activity

Response to CT is deeply affected by the biology of tumor growth. Two models exist that describe the pattern of tumor growth. The first was advanced by Skipper and colleagues[52] in the early 1960s using a cultured tumor cell line, the transplantable leukemia L1210 in BDF or BDA mice. The growth fraction, cell loss fraction, and mitotic cycle duration of this cancer are remarkably stable, from a population size of one cell to its lethal cell number of 10^9 cells. Consequently, L1210 tumors grow exponentially, increasing by a constant percentage per unit of time regardless of the number of cells present. Skipper and co-workers have shown that when such laboratory tumors are treated with anticancer drugs, the fraction of cells killed is always the same regardless of their initial number. This fraction is termed the "log-kill" and increases with the dose of the drug. If a particular dose of an individual drug reduces the number of neoplastic cells from 10^5 to 10^3 (2 logs), the same dose used at a tumor mass consisting of 10^3 cells will result in a 10^1 cells tumor mass. Hence, enough cycles of enough drugs at high enough individual dose levels should be able to kill a very high percentage of the cells. Failure of CT in this model is justified by the excessively high initial tumor cell burden.

Although some neoplasms have a growth pattern that can be defined as exponential, most solid human neoplasms grow exponentially only in the preclinical phase of their natural history, when the tumor burden is low, or after important cytoreduction (e.g., after surgery or radiotherapy). In these phases, CT exerts its greatest activity. Indeed, most solid tumors do not show an exponential growth: their growth fraction decreases with time and can be well described by the Gompertzian curve. When a tumor follows such a "Gompertzian growth model," the killing effect of a cytotoxic drug is not constant and is strictly related to the initial number of cells exposed to the drug. A large tumor burden has in most instances a low growth fraction, and a low cell number is likely affected by the drug: thus response to CT depends on where the tumor lies in its growth curve. Experimental observations imply that there are kinetic reasons for the failure of CT to cure large tumors.

Although the mechanism of action of many cytotoxic drugs is known, the basis of their selective action remains obscure. While it is a common thought that the therapeutic advantage of CT is due to a kinetic difference between normal and neoplastic cells, the assumption that cell killing activity is related to the tumor growth rate has not always been confirmed: for example, the chemosensitivity of breast cancer compared to colon cancer or malignant melanoma does not reside in a difference in their growth rates or S phase fractions.[5] In other words, it is questionable whether factors other than growth rate may account for the differences in sensitivity between normal and neoplastic cells. Specific biochemical targets are another possible explanation of such a difference. In light of this, chemosensitivity and chemoresistance are the complex, final expression of several aspects, which include, in addition to kinetic factors, pharmacokinetic and biochemical factors.

6.1.4   Chemoresistance

In 1950, Burchenal et al.[6] published their observation on the selection of a methotrexate-resistant subline of the L1210 murine leukemia cells. The finding of an acquired cellular resistance to a cytotoxic drug was then reported by many other authors and for almost all the antineoplastic agents known. It is common clinical experience that some tumors are refractory to chemotherapeutic agents, even when diagnosed with apparently minimal tumor volume, or that they acquire a permanent resistance after a few chemotherapy courses. It became increasingly apparent that drug resistant cancer cells continually arise from random mutations and that the selection of clones with the resistant phenotype is induced by the administration of the drug itself. This kind of resistance is "specific" and "permanent" and must be distinguished from the resistance of resting cells. Resting cells are resistant to every kind of cytotoxic action (nonspecific resistance) and can be reverted to cycle and drug sensitivity (nonpermanent resistance). In 1979, Goldie and Coldman[7] developed a model that described the onset of resistance to anticancer drugs. Neoplastic cells are characterized by a high rate of spontaneous mutations caused probably by disturbances in DNA repair systems: this phenomenon has been named genetic instability.

According to the Goldie and Coldman model, tumor cells mutate to drug resistance at a rate dependent on the degree of genetic instability of a particular tumor. Inherently resistant tumors are made up mostly of clones that have been selected to resistance and have become the dominant cell line in the tumor population.

Many specific mechanisms of drug resistance have been revealed that enable the cell to circumvent a well-defined lesion induced by the drug. Mechanisms of drug resistance include decreased uptake caused by alterations in drug transport systems across the cell layer, decreased activation of prodrugs, modification or amplification of the target enzyme of the drug, increased DNA repair, increased inactivation of the drug. Tumor cells can be intrinsically resistant to cytotoxic drugs (intrinsic resistance) or can become resistant after exposure to specific drugs (acquired resistance): in this latter case selective pressure induced by CT is thought to accelerate the rising of mutated clones provided by the biochemical mechanisms necessary for drug inactivation. In some instances, cells become resistant to a single drug or at least to the drugs sharing the same mechanism of action, but in most cases clones with multiple agent resistant phenotypes are selected. The term Multidrug Resistance (MDR) is usually employed to indicate a nonspecific type of resistance that involves structurally unrelated cytotoxic compounds. MDR is classified according to the mechanism underlying resistance: overexpression of the putative drug efflux pump p-glycoprotein (p-170) and of its encoding genes; alterations in topoisomerase enzymes; and alterations in drug metabolizing pathways. These mechanisms are not mutually exclusive.

6.2   Colony Stimulating Factors and Escalating Drug Delivery in Cancer Chemotherapy

CSFs have recently been introduced in clinical settings in order to improve the therapeutic index of antineoplastic drugs. The advent of CSFs has afforded dose optimization and dose escalation of cytotoxic drugs. The therapeutic advantage of such practice is documented only in part, despite accumulating evidence of its protection against the consequences of myelosuppression, and on the consequent improvement of patient quality of life and compliance to CT.[8]

Myelosuppression is the dose-limiting toxicity of the majority of the cytotoxic regimens and frequently results in the need for dose reductions or delays between treatment cycles. Furthermore, hematological toxicity has for a long time limited the delivery of higher CT doses due to the risks of infection related to neutropenia, and the risk of bleeding related to thrombocytopenia. Fifty percent of patients with granulocyte counts less than $500/mm^3$ will develop severe infections, particularly when the neutropenia lasts longer than 3 to 5 days. Anemia rarely represents an acute side effect of CT: it usually becomes a clinical problem after multiple drug administrations and the long-lasting recovery of red cells strongly affects patient quality of life. Schedule optimization and limited dose escalation appear to be possible with CSF support.

6.2.1   Experimental Evidence

For certain tumors, proper drug dosing is the limiting factor for the "capacity to cure" and some investigators have hypothesized that increased CT dose delivery may lead to a higher cure rate.[9]

The Goldie and Coldman model assumes that inherent drug resistance occurs through spontaneous mutations during the growth of the neoplasm with a constant and defined frequency that depends on factors such as histologic type, growth rate, and degree of differentiation. According to this model, escalating drug delivery may reduce the likelihood that chemoresistant cells will emerge and subsequently cause treatment failure.[10] The somatic mutation model predicts a direct relationship between dose delivery and the speed of eradication of drug-sensitive cells, and an inverse relationship between dose delivery and the development of tumor resistance.

The relationship between dose delivery and cell kill is usually depicted as the dose–response curve. The dose–response curve in biological systems is sigmoidal in shape, with a threshold, a lag phase, a linear phase, and a plateau phase. In cancer CT, the difference between the curves of normal and tumor

TABLE 6.1   Cytotoxic Agents Used in High-Dose CT Regimens for Solid Tumors

Drug	Standard Dose (mg/m²)	Dose Administered in High-Dose Regimens (mg/m²)	Organ Toxicity in High-Dose Regimens
Cyclophosphamide	750	7000	Pancarditis, hemorrhagic cystitis
Ifosfamide	5000–9000	20,000	Renal failure, hemorrhagic cysitis
Cisplatin	100	200	Renal failure, neurooototoxicity
Carboplatin	400	2400	Neurooototoxicity, renal failure
Thiotepa	30–50	1300	Neurotoxicity (CNS), mucositis
Carmustine	200	1500	Pneumonitis, VOD
Mitoxantrone	14–20	60–100	Mucositis, heart failure
Melphalan	30	245	Mucositis
Busulfan	1.5 (mg/kg)	16 (mg/kg)	VOD
Etoposide	300–500	4200	Mucositis
Mitomycin C	10–15	50	VOD, pneumonitis, heart failure

CNS: Central Nervous System; VOD: Veno-occlusive disease.

tissues is exploited. When the cytotoxic agent does not show a certain degree of tumor selectivity, increasing its dose may not lead to a higher therapeutic index because of increased toxicity.

There is much evidence indicating that cellular resistance of cultured tumor cells can be overcome by increasing the concentration of certain cytotoxic drugs: the slope of the linear phase of the dose–response curve can be modified by the dose of the agent chosen and is steep for certain drugs. In animal models a reduction of the drug dose in this linear phase leads to the loss of the capacity to cure rodents bearing transplantable tumors, while increasing the dose results in a higher cure rate.

Both *in vitro* and animal model data show that the slope of the dose–response curve is critically determined by the drug chosen: a steep linear dose–response curve is a typical feature of alkylating agents, intercalating agents, platinum-derived agents, anthracyclines, and some antimetabolites. Drugs that typically exert their cytotoxic activity during a specific cell cycle phase, such as vinca alkaloids, have a flat dose–response curve, such that above a particular dose no further cell kill is seen.

Antineoplastic agents suitable for dose escalation (Table 6.1) must have another property: myelosuppression has to be the dose-limiting toxicity, and no major organ injuries must occur after exposure to standard doses. Most alkylating agents share these properties, and since the dosage of many of these can be escalated with protection against hematological toxicity before organ toxicities appear, multiple alkylator regimens are currently under clinical evaluation. Dose escalation is limited by the impairment of organs such as kidney, heart, lungs, CNS, that is often cumulative and dose-related. All of these side effects are not easily manageable, require patient hospitalization and intensive supportive care, and can seriously compromise the quality of life.

Tumor type is another significant variable, given that inherent resistance to antineoplastic drugs is hardly overcome by increasing drug dosage. Von Hoff et al.[11] assessed cell kill of various antineoplastic drugs using short-term cultures of human tumors. These authors showed that drugs such as melphalan, doxorubicin, cisplatin, and mitoxantrone had steep dose–response curves in "sensitive" tumors but most agents had a flat dose–response curve for "resistant" tumors. Generally, the apparent drug sensitivity of the cultured tumor cells correlated with their activity against these tumors in the clinical setting.

In conclusion, experimental studies suggest that there may be an advantage in increasing the dose of selected drugs for the treatment of sensitive tumors.

6.2.2   Clinical Evidence

The log linear relationship between tumor cell kill and administered dose found in laboratory studies is not readily extrapolated to the clinical setting. Data from clinical studies indicating that dose may be important in improving response rate and survival are more controversial.

A first consideration relates to the concept of "dose optimization." The most frequent cause of treatment delays and interruptions and of ad hoc dose reductions is myelosuppression. A few retrospective studies have suggested that lower dose administration has a noticeable clinical impact in terms of response rate. In patients receiving curative, as opposed to palliative, treatment, such dose reductions may adversely affect disease control and long-term survival.[12] These findings have resulted from studies on lymphomas,[13] breast cancer,[14] and small-cell lung cancer.[15] Although prospective randomized trials that clearly support this view are lacking, the use of CSFs when therapy is significantly compromised by myelosuppression seems reasonable.

The positive correlation between drug delivery and outcome seen in the therapeutic dose range and with lower-than-therapeutic doses cannot be applied to CSF-facilitated dose escalation above the current standard dose regimens. Evidence exists that a significant increase in the dose of a drug can cause a response in patients whose tumors were resistant to conventional doses of the same CT.[16,17] Unfortunately, the general impression emerging from studies designed with this end point is that the maximum dose intensification possible with combination CT and CSF support is in the order of 20% to 40%.[18] Consequently, any improvement in response is relatively small and hardly evaluable in prospective studies. Although there are studies suggesting benefits in ovarian, breast, testicular, and small-cell lung cancer,[12] contradictory results have been observed.[19]

The advent of recombinant hemopoietic growth factors has currently made feasible the development of ABMT and of peripheral blood stem cell transplantation. With the support of these techniques, dose-intensities were increased not by 20%, 30%, or 40%, but by 200%, 300%, and even 2000% (Table 6.1). The administration of CSFs in patients undergoing high-dose CT with ABMT has shortened the duration of severe neutropenia and reduced the incidence of infections, the use of antibiotics, and the duration of hospitalization. Moreover, protection from myelosuppression has been made possible also by the reinfusion of circulating hemopoietic progenitor cells collected by leukapheresis during treatment with CSFs. The introduction of higher dose intensities that expanded the range of doses available for clinical applications raised expectations that the outcome of therapy would be improved substantially. Several randomized phase III trials are in progress to study the role of high-dose regimens in solid tumor management. Recent results from a large randomized study confirm the advantage of this approach over standard CT in relapsed non-Hodgkin's lymphoma patients.[20] Another field of interest is breast cancer, both in the adjuvant setting and in the metastatic one. Historical comparisons between standard and high-dose CT suggest some improvement in long-term survival for high-risk breast cancer patients treated with high-dose adjuvant CT; the only randomized trial performed to date confirms these results.[21]

6.2.3   Clinical Pharmacology of Colony Stimulating Factors

The hemopoietic system can be broadly described as a three-level proliferative structure: the compartment of hemopoietic stem cells that are mostly nonproliferating cells (type 1), the compartment of highly proliferative cells that are committed to the various differentiated lineages (type 2), and the compartment of mature cells that are leaving bone marrow (type 3).

The proliferative and differentiative pathways are physiologically regulated by the paracrine action of growth factors. The only exception to the paracrine production of hemopoietins is represented by renal delivery of EPO in response to low hematogenic O_2 pressure.

CT acts by decreasing the number of type 2 (proliferating) cells, and these in turn will (within a few days) lead to a decrease of type 3 cells. Repeated drug administrations may induce severe damage over prolonged periods because they also involve nonproliferating hemopoietic cells. Quiescent stem cells (type 1) are not a primary target of cycle-specific cytostatic drugs, although a marginal involvement has been shown both *in vitro* and *in vivo*. After the damage induced upon type 2 cell populations, these receive unidentified signals triggering them into active proliferation until complete recovery. A different pathogenesis is described for chronic anemia due to nephrotoxic anticancer drugs, especially for platinum derivatives: in addition to the bone marrow directed cytotoxic activity, these agents lower EPO plasma levels because of their action on the renal site of production. Anemia results from these two combined actions, which also explains its long duration.

Leukopenia (mainly neutropenia) is the most obvious clinical sign of bone marrow drug-induced toxicity. The recent availability of recombinant hemopoietic growth factors (SCF, G-CSF, GM-CSF, IL3, EPO, etc.) allows accelerated recovery from cytopenia even after severe myelodepression by stimulating the surviving type 2 cells.

Currently, two hemopoietic growth factors, granulocyte colony-stimulating factor (G-CSF) and granulocyte-macrophage colony-stimulating factor (GM-CSF) are available for use in patients receiving myelosuppressive CT. These are glycosilated monomeric glycoproteins produced in mammalian cells through recombinant DNA technology. Because of the stimulation of proliferation and differentiation of the precursor cells of the neutrophil-granulocyte lineage, subcutaneous and intravenous administration of G-CSF and GM-CSF cause a dose-dependent increase in circulating neutrophils, while GM-CSF also raises circulating eosinophils and monocytes. In addition, CSFs cause progenitor and mature cells to move into the peripheral blood, and are involved in neutrophil activation (GM-CSF also activates eosinophils and monocytes).

The administration of a CSF after CT may be either prophylactic (to prevent infectious complications) or therapeutic (to cure neutropenic fever). The recommended starting dose of G-CSF is 5 μg/kg/day, and 250 μg/m²/day of GM-CSF, to be administered subcutaneously or intravenously. Although optimal scheduling of CSF's prophylactic administration has not been established for standard or moderately escalating dose CT, the most widely used schedule is to initiate treatment 1 day after the last CT administration. The period of administration varies in practice, however, eight to fifteen days from the CT administration or until the neutrophil count exceeding 10,000/mm³ is feasible. Early discontinuation resulting in a 50% decrease in granulocytes after 24 hours may be dangerous. Adverse effects are dose-related and generally mild in the conventional dose range; they include bone pain, fever, erythematous lesions at the site of injection (GM-CSF), and minor biochemical changes.[19]

CSFs administration to patients with severe neutropenia (therapeutic use) has consistently demonstrated the ability to reduce the duration and degree of neutropenia.[22] If this result implies improved clinical outcomes, including a decrease in the number of episodes of fever, in use of antibiotics, or in days of hospitalization is still under debate.

Use of CSFs to prevent neutrophil reduction in patients undergoing dose-intensive regimens (prophylactic use) is another field of active research: the available evidence indicates that CSF administration can facilitate dose maintenance, although improvement in disease control resulting from dose-intensive CT regimens remains to be clearly demonstrated. It is likely, however, that subsets of patients at high risk of developing CT-induced infectious complications may benefit from the prophylactic use of CSFs.[19]

An important limitation of the use of CSFs is that neither G-CSF nor GM-CSF has made an impact on platelet recovery, and at more myelosuppressive CT doses, thrombocytopenia becomes progressively more severe and protracted in relation to neutropenia. Several ongoing clinical trials are evaluating the usefulness of molecules able to increase platelet count, such as SCF, IL3, and thrombopoietin. New molecules will have a major impact in a short time on many aspects of oncologic practice and will probably afford complete protection from bone marrow toxicity. Whether this will result in an improvement in the cure rate of solid human tumors is, however, still an open question.

6.3   Drug Combinations

Although the delivery of high single agent doses has a sound theoretical basis and has given rise to promising clinical results, the use of drug combinations has historically been the most widely used means to increase CT therapeutic index. The model of Goldie and Coldman provides both the theoretical basis and the model for the use of combination CT.

Goldie and Coldman predicted that at a tenable mutation rate of one mutation per million mitoses, the probability of finding no mutants resistant to any one drug in a total population of 10^5 cells is about 90%, and the probability of finding no mutants in 10^7 cells is only 0.0045%. They hence theorized that as many effective drugs as possible should be applied as soon as possible, so that cells that are already resistant to one drug could be killed before they have a chance to mutate to resistance to other drugs. They concluded that the best approach would be true combination CT, giving several drugs simultaneously at full dosages. Their work has been indicated as the theoretical basis for the use of combination CT. The Goldie and Coldman model supports the use of noncross-resistant drugs as the first line CT of solid neoplasms. In general, agents with different mechanisms of action are empirically associated, such as alkylating agents and antimetabolites, anthracyclines, and platinum derived agents in order to achieve an additive cytotoxic action upon multidrug-resistant cell lines. In some cases a true synergistic effect has been demonstrated: cisplatin enhances fluorouracil cytotoxicity probably through a metabolic effect consisting of the increase of the reduced folate intracellular pool.[23,24] In some studies the etoposide/cisplatin combination has been shown to produce a cell kill 6 to 7 logs higher than did either drug used alone, although the mechanism of synergy between these two drugs is still unexplained.[25] In most instances, however, the combined action of cytotoxic agents must be defined as additive.

An emerging issue is the use of drugs that do not share cytotoxic activity with anticancer agents. Biochemical mechanisms and pharmacokinetic interactions can indeed result in an enhanced activity and in an improved therapeutic index of cytotoxic agents. Moreover, as already seen, agents protecting from CT side effects can afford dose optimization and even dose escalation. Numerous molecules have been found to increase the CT therapeutic index in preclinical studies. Even though few of them have shown clinical relevance, we can say without any doubt that noncytotoxic agents are part of the medical approach to the cancer patient.

Below we discuss some noncytotoxic drugs that have been found to increase the effectiveness of CT in clinical studies.

6.3.1   Folinic Acid

5-Formyl-5,6,7,8-tetrahydrofolic acid (folinic acid or leucovorin; FA) has important physiological functions in the folate pool as donor of methyl, methylen, or other 1-carbon groups, and is used in anticancer therapy as a modulating agent for the drug 5-fluorouracil (5FU). In combination, FA has been shown to improve the therapeutic activity of 5FU in gastrointestinal malignancies and especially in colorectal cancer. FA is metabolized at the cellular level into the folate derivative 5,10-methylene-tetrahydrofolate (CH_2-THF), which stabilizes the complex formed by the 5FU anabolite, FdUMP, and its target enzyme, thymidilate synthetase (TS). TS plays a central role in cellular metabolic pathways and in DNA synthesis, being responsible for the catalytic conversion of deoxyuridilate to thymidilate through a methylation reaction requiring the folate cosubstrate CH_2-THF. One important action of 5FU is mediated by its conversion to FdUMP by the enzyme thymidine kinase or alternatively via ribonucleotide reductase: FdUMP is a potent competitive inhibitor of TS. Inhibition of TS can be mediated by the formation of an unstable binary complex between FdUMP and TS. However, a covalent ternary complex among FdUMP, TS, and CH_2-THF is thought to be the main determinant in the inhibition of TS. Although little is known about the final damage that leads the tumor cell to death, inhibition of TS has been postulated to be responsible for the antitumor effect. Based on experimental studies, several mechanisms of resistance to 5FU have been identified;[26] consequently, at least some of them are related with a reduced degree of TS inhibition (low intracellular levels of TS, altered enzyme kinetics, isoenzymes, low levels of 5FU activating enzymes). The antitumoral activity of 5FU is regulated not only by the level of metabolizing enzymes but also depends on the level of competing normal substrates. A high amount of TS in the tumor cell implies that it is difficult to inhibit the enzyme completely for a long period. A low level of reduced folates, such as CH_2-THF, decreases the stability of the inhibitory ternary complex. In some tumor models, resistance to 5FU has been associated with low preexisting levels of the reduced cofactor pools. FA can be metabolized intracellularly to the methylene form, and the ability to expand methylene pools by exogenous FA parallels the increase of cytotoxicity of 5FU.[27]

FA modulation of 5FU has been exploited in the treatment of all the tumors in which 5FU alone has shown some degree of activity: gastrointestinal adenocarcinomas, head and neck squamous cell carcinoma,[28] and breast cancer.[29] The main application of this combination, however, has been the treatment of advanced colorectal cancer, and, secondly, the adjuvant treatment of the same tumor. Many phase II studies have been performed suggesting an increased response rate, compared with 5FU alone, but also an increased mucosal toxicity (especially stomatitis and diarrhea, depending mainly on the schedule of 5FU administration). Phase III randomized studies designed to compare the activity of the 5FU-FA combination with 5FU alone (considered the standard treatment of advanced colorectal cancer) confirmed previous findings: in the metaanalysis published in 1992, which pooled the results of nine randomized studies,[30] the response rate obtained with the combination was significantly superior to that obtained with 5FU (23% vs. 11%), although this difference did not result in a significant improvement of long-term survival. On the contrary, an important phase III study conducted with the same aim but not included in the metaanalysis[31] has shown a significantly improved survival for patients receiving 5FU+FA. This trial also suggested that low FA doses (20 mg/m^2/d for 5 days) are as effective as higher doses (200 mg/m^2/d) in modulating 5FU activity.

In conclusion we can say that at least in advanced colorectal cancer, FA can effectively improve the therapeutic activity of 5FU: it is possible, however, that the higher cytotoxicity may also affect normal tissues (higher degree of mucositis), thus not resulting in any improvement in terms of therapeutic index.

6.3.2   Interferon

"Interferons" (IFNs) are a family of glycoproteins with immunomodulatory, antiviral, and antiproliferative activity extensively studied both as single agents and in association with cytotoxic drugs for their antitumoral properties. IFNs were originally classified by the cell type from which they were isolated (α, β, and γ). Further investigation has shown that each group consists of several antigenically distinct subtypes, with possible heterogeneous biological targets. Their action is mediated by specific cell surface receptors, which cause the activation of cytoplasmic IFN-stimulated gene-factors that are translocated into the nucleus and bind to specific promoter regions, thus modulating gene expression.[32] Shortly after introduction into clinical trials, IFNs were found to induce responses in tumors generally refractory to CT agents, such as melanoma and renal cell carcinoma, and to induce sustained complete remissions in a rare variety of chronic leukemia (HCL). The antiproliferative activity of IFNs as single agents and the different pattern of toxicity have been exploited for empirical combinations with antineoplastic drugs. Further preclinical studies showed IFNs to be capable of enhancing the cytotoxic activity of many antineoplastic drugs.[33] Laboratory studies have highlighted several possible biological mechanisms of the interaction between IFNs and CT agents that include both direct and indirect effects. These are briefly discussed below:

1. IFNs were introduced into antineoplastic CT for their activity on the immune system, the rationale being that activation of immune effectors contributed to the elimination of the minimal residual tumor cell burden that is not affected from CT. Thus, IFNs are used at the end of CT treatments as long-lasting maintenance therapy. Immunological effects include activation of NK activity, antibody-dependent cytotoxic T cells, macrophages, and T cells. In addition, IFNs can act directly on tumor cells by increasing expression of tumor-associated antigens, major histocompatibility antigens, and Fc receptors, which may make the tumor cell more immunogenic.

2. In human colon cancer cells IFNs have been found to reduce thymidylate synthetase levels and activity, thus blocking the compensatory rise in TS following treatment with 5FU.[34] This "biochemical modulation" appears to represent the basis for the greater-than-additive interaction of 5FU and IFN. In addition other mechanisms have been found to elucidate their synergistic activity: IFN increases the rate of 5FU transformation in fluorodeoxiuridine and in FdUMP, inducing a greater degree of TS inhibition. In some studies, IFNs have been found to decrease the uptake of thymidine and the activity of thymidylate kinase, thus reducing the utilization of exogenous thymidine. IFNα can protect mice from the toxic effects of 5FU. Clinical trials failed to uphold experimental findings: although some studies report a high response rate for the association of 5FU and IFN, in randomized trials the association did not result in any improvement of the cure rate when compared with 5FU alone.[35-37]

3. Complex cytokinetic interactions may take place between IFNs and cytotoxic drugs. IFN has been found to interfere with the cell cycle by blocking the transition from the G_0/G_1 to the S phase, and by slowing the passage through the S phase itself. The exposition to phase-specific drugs is thus prolonged and cells are more susceptible to the cytotoxic effect.
4. IFNs may alter the catabolism or clearance of specific cytotoxic drugs. IFNs have well-described inhibitory effects on cytochrome P-450 microsomal enzyme activity, which is involved in the metabolism of drugs such as doxorubicin and cyclophosphamide.[38]

The spectrum of antineoplastic agents modulated by IFNs is broad and includes agents with different mechanisms of action (alkylating agents, antimetabolites, anthracyclines, cisplatin, antibiotics). This suggests not only multiple mechanisms,[33] but also multiple levels of interaction, depending on the drug tested and the tumor type.

6.3.3   Modulation of Cisplatin Toxicity by Hydration and Diuretics

Water can be considered the oldest and simplest drug: its extensive use in oncology is justified by its potential in decreasing renal damage induced by nephrotoxic drugs that are excreted totally or in part through the kidneys. Furthermore, water has entered in the routine clinical practice due to its protection from toxicity when cisplatin is administered. Cisplatin (cis-dichlorodiamine platinum-II) is the most important member of a family of antineoplastic agents that play a central role in the treatment of several human solid tumors, including ovary, testicular, lung, head and neck, and bladder cancers. Its widespread clinical use has been limited by severe renal toxicity consisting of acute renal failure and in less serious cases of renal dysfunction manifested as a reduction in the glomerular filtration rate and a rise in creatinine serum levels, electrolyte imbalances, reduced production of erythropoietin, and related anemia.[39] Cisplatin-induced acute renal damage has been attributed to a direct interaction between the aquated form of the drug and tubular cell surface proteins, leading to a dysfunction of specific membrane carriers and enzymes and in some cases to tubular necrosis. The chronic damage seems to be related to the mechanism of cisplatin's cytotoxic activity, i.e., the formation of DNA adducts in renal tissue. Laboratory and clinical evidence has accumulated to suggest that the nephrotoxic potential of this drug can be ameliorated by the induction of an abundant diuresis during drug administration: a urine output of at least 100 ml/h has been shown to decrease nephrotoxicity in animals and humans.[40,41] The protective mechanism of hydration remains undefined: decrease in urinary platinum concentration or higher plasma clearance are possible, simple explanations. The role of electrolytes, however, must also be taken into account: animal and clinical studies show a marked reduction in renal failure when cisplatin is administered in isotonic or hypertonic saline compared to water,[42] without affecting its overall therapeutic activity. These beneficial effects of saline solutions may be related to the stabilization of the cisplatin molecule and to the minor amount of the aquated, toxic platinum species in the kidney.[43] The same mechanisms may account for the reduced renal damage found when diuretics (furosemide or mannitol) are administered along with cisplatin infusion: higher urine output and/or stabilization of the cisplatin molecule.[44,45]

6.3.4   Agents Revertant P-170–Mediated Multidrug Resistance

In some experimental tumor systems, p-170–mediated chemoresistance has been shown to be the major acquired obstacle to the action of drugs such as anthracyclines, vinca alkaloids, epipodophyllotoxins, and actinomycin D. These are natural product drugs with different mechanisms of action, whose intracellular accumulation is decreased in cells showing overexpression of the MDR-1 gene and of p-170. The degree of accumulation is inversely related to the presence of the 170-kd plasma membrane-associated glycoprotein (p-170) and to the expression of its encoding gene, named MDR-1. In general, p-170 is highly expressed in tumors intrinsically resistant to CT and in chemosensitive tumor cells after long lasting exposure to the drugs mentioned.

TABLE 6.2   Drugs Used in Preclinical and Clinical Trials as Revertant of Multidrug Resistance

Calcium channel blockers	Verapamil Diltiazem Nifedipine	Nimodipine Bepidril Nicardipine
Other cardiovascular medications	Dipyridamole Quinidine Amiodarone Reserpine Calmodulin inhibitors	Trifluoperazine Chlorpromazine Clomipramine Lidocaine
Hormonal compounds	Tamoxifen Toremifen	Megestrole acetate
Antibiotics and antineoplastic agents	Cefoperazone Ceftriaxone Erythromycin	Vincristine Vinblastine
Cyclosporins	Cyclosporin FK 506	SDZ PSC-833
Antimalarial compounds	Quimicrine	Quinine
Other drugs	Tumor necrosis factor Retinoids	SDZ 280–446

Many nonantineoplastic agents have been shown to revert p-170–mediated resistance (Table 6.2). These drugs belong to different pharmacologic classes, and, although their site of action is the cell layer, the mechanism of inhibition of the p-170 transport system is probably heterogeneous. The mechanism of action, however, is not well understood. Some studies have demonstrated an enhancement of p-170 phosphorylation at serine residues induced by verapamil and trifluoperazine. Other studies have shown that verapamil and diltiazem directly inhibit the binding of CT agents to p-170. It must also be noted that a single agent may have multiple types of interaction with p-170, and that each of the agents shown in Table 6.2 probably reverses CT resistance by other mechanisms as well: alteration of membrane fluidity, increase in tumor blood flow, modification of intracellular calcium concentrations, modulation of DNA repair and DNA damage (calmoduline inhibitors), pharmacokinetic effects (cyclosporin, cefoperazone, verapamil). In the preclinical setting, a common finding is the increased intracellular concentration of antineoplastic agents that is usually associated with an enhancement of cytotoxic activity. Moreover, in most cases the increased activity is more evident in cell lines made resistant to the cytotoxic agent tested.[46]

Until now clinical studies have failed to show a significant advantage in associating p-170 modulators to standard CT regimens. Phase I and phase II studies have been carried out combining cytotoxic drugs with verapamil, diltiazem, chlorpromazine, trifluoperazine, cyclosporins, quinidine sulfate, and tamoxifen. These studies, although frequently demonstrating the possibility of obtaining a percentage of responses in tumors refractory to CT, are often not conclusive. Schedules, drug doses, p-170–mediated resistance, and pharmacokinetic interactions are some of the issues often addressed empirically. Moreover, tissue concentrations of the modulator-enhancing cytotoxicity *in vitro* are often not reached because of the severe toxicities associated with the use of some of these agents (e.g., cardiovascular for verapamil, neurological for calmoduline inhibitors). Five phase III randomized trials[47-51] have been conducted in order to estabilish the effectiveness of verapamil when added to CT (Table 6.3). In only one study,[48] the use of verapamil was associated with an improved outcome, but the number of patients recruited for the study was small, thus ruling out any conclusive assertions. In the other cases, the combination of verapamil and CT did not result in a therapeutic advantage. The reasons of such findings are hypothesized by Dalton and co-workers:[49]

TABLE 6.3   Randomized Clinical Studies Assessing the Use of Verapamil in Combination with Antineoplastic Agents

Main Author	No. of Patients	Tumor	Chemotherapy	mg/die of Verapamil	Resp	Surv
Millward, M.J., 1993	72	NSCLC	VDS+IFO	480 d. 1–3	+	+
Milroy, R., 1993	226	SCLC	CTX+ADM+VCR	240 d 1–5	–	–
Dalton, S., 1994	127	Myeloma	VCR+ADM+DEX	240 d. 1–3 and 480 d. 4–12	–	–
Tsushima, T., 1995	96	Bladder*	ADM	25 d. 1–3	–	–
Wheeler, H., 1988	60	SCLC	ADM+VDS+VP16	480 d. 1–5	–	–

NSCLC: Non-small-cell lung cancer; SCLC: Small cell lung cancer; VDS: Vindesine; CTX: Cyclophosphamide; ADM: Adriamycin; VCR: Vincristine; DEX: Dexamethasone; Resp.: Advantage in response rate for Verapamil arm; Surv.: Advantage in survival for Verapamil arm.

*Vesical instillation with loco-regional CT in superficial bladder cancer.

1. Inadequate verapamil serum levels to significantly reverse p-170 function are attained in patients, due to the important cardiovascular side effects when it is administered in effective doses.
2. Verapamil is not the most effective chemosensitizer and other agents are needed to reverse p-170–mediated resistance.
3. p-170 is not a major contributor to drug resistance in the tumors studied and other mechanisms play a more important role.

The clinical relevance of the entire pattern of chemoresistance needs to be established in individual disease types. Experimental evidence shows that every single neoplasm develops, during its natural history, a specific pattern of resistance. Moreover, tumors such as non-small-cell lung cancer and ovary carcinoma do not usually exhibit p-170–mediated resistance: in these tumors, p-170 modulators probably do not affect the therapeutic index of cytotoxic drugs. Other membrane proteins have been described that are able to confer resistance independently of p-170 expression. The mechanisms of resistance and the roles of these membrane proteins are still unknown.

In conclusion, although the use of p-170 modulators cannot be considered a standard approach in cancer treatment, further research will most likely estabilish optimal scheduling and dosing of modulating agents, and will provide agents with higher therapeutic index and ad hoc combinations of multiple chemosensitizers.

References

1. Tsutsui, T., Suzuki, N., Maizumi, H., Barret, J. C., Vincristine sulfate induced cell transformation, mitotic inhibition, and aneuploidy in cultured Syrian hamster embryo cells, *Carcinogenesis*, 7, 131, 1986.
2. Jaxel, C., Taudou, G., Portemer, C., Mirambeau, G., Panijel, J., Duguet, M., Topoisomerase inhibitors induce irreversible fragmentation of DNA in concanavalin A stimulated splenocytes, *Biochemistry*, 27, 95, 1988.
3. Kaufmann, S., Induction of endonucleolytic DNA cleavage in human myelogenous acute leukemia cells by etoposide, camptothecin, and other cytotoxic anticancer drugs: a cautionary note, *Cancer Res* 49, 5870, 1989.
4. Epstein, R. J., Drug-induced DNA damage and tumor chemosensitivity, *J Clin Oncol*, 8 (12), 2062, 1990.
5. Tannok, I. F., Principles of cell proliferation: cell kinetics. In DeVita V. T., Hellman S., Rosenberg S. A., Eds., *Cancer: Principles and Practice of Oncology*, J.B. Lippincott, Philadelphia, 3–13, 1989.
6. Burchenal, J. H., Robinson, E., Johnston, S. F., The induction of resistance to N-10-methyl-pteroyl-glutamic acid in a strain of transmitted mouse leukemia, *Science*, 111, 116, 1950.
7. Goldie, J. H., Coldman, A. J., A mathematic model for relating the drug sensitivity of tumors to their spontaneous mutation rate, *Cancer Treat Rep*, 63, 1727, 1979.
8. Canellos, G. P., The dose dilemma, *J Clin Oncol*, 6(9), 1363, 1988.
9. Hryniuk, W. M., Dose-response is alive and well, *J Clin Oncol*, 4(8), 1157, 1988.
10. Coldman, A. J., Goldie, J. H., Impact of dose-intense chemotherapy on the development of permanent drug resistance, *Semin Oncol*, 14(4), Suppl 4, 29, 1987.

11. Von Hoff, D. D., Clark, G. M., Weiss, G. R., Marshall, M. H., Buchok, J. B., Knight, III, W. A., LeMaistre, C. F., Use of in vitro dose response effects to select antineoplastics for high-dose or regional administration regimens, *J Clin Oncol*, 4(12), 1827, 1986.

12. Gurney, H., Dodwell, D., Thatcher, N., Tattersall, M. H. N., Escalating drug delivery in cancer chemotherapy: a review of concepts and practice — part 2, *Ann Oncol*, 4, 103, 1993.

13. Carde, P., MacKintosh, R., Rosenberg, S. A., A dose and time response analysis of the treatment of Hodgkin's disease with MOPP therapy, *J Clin Oncol*, 1, 143, 1983.

14. Hryniuk, W. M., More is better, *J Clin Oncol*, 6, 1365, 1988.

15. De Vathaire, F., Arriagada, R., de The, H., Tarayre, M., Ruffie, P., Chomy, P., de Cremoux, H., Sancho-Garnier, H., Le Chevalier, T., Dose intensity of initial chemotherapy may have an impact on survival in limited small cell lung carcinoma, *Lung Cancer*, 8, 301, 1993.

16. Ozols, R. F., Ostchega, Y., Myers, C. E., Young, R. C., High-dose cisplatin in hypertonic saline in refractory ovarian cancer, *J Clin Oncol*, 3, 1246, 1985.

17. Ozols, R. F., Ostchega, Y., Curt, G., Joung, R. C., High-dose carboplatin in refractory ovarian cancer patients, *J Clin Oncol*, 5, 197, 1987.

18. Linch, D. C., Dose optimization and dose intensification in malignant lymphoma, *Eur J Cancer*, 30A(1), 122, 1994.

19. American Society of Clinical Oncology recommendations for the use of hematopoietic colony-stimulating factors: evidence-based, clinical practice guidelines, *J Clin Oncol*, 12, 2471, 1994.

20. Philip, T., Guglielmi, C., Chauvin, F., Hagenbeek, A., Van Der Lely, J., Bron, D., Sonneveld, P., Gisselbrecht, C., Cahn, J. Y., Harousseau, J. L., Coiffier, B., Biron, P., Somers, R., Autologous bone marrow transplantation (ABMT) vs. conventional chemotherapy (DHAP) in relapsed non-Hodgkin lymphoma (NHL): final analysis of the PARMA randomized study (216 patients), *Proc Am Assoc Clin Oncol*, 14, 390, abstr 1220, 1995.

21. Bezwoda, W. R., Seymour, L., Dansey, R. D., High-dose chemotherapy with hematopoietic rescue as primary treatment for metastatic breast cancer: a randomized trial, *J Clin Oncol*, 13, 2483, 1995.

22. Lieschke, G. J., Burgess, A. W., Granulocyte colony-stimulating factor and granulocyte-macrophage colony-stimulating factor. II, *N Engl J Med*, 327(2), 99, 1992.

23. Scanlon, K. J., Newman, E. M., Lu, Y., Priest, D. G., Biochemical basis for cisplatin and 5-fluorouracil synergism in human ovarian carcinoma cells, *Proc. Natl. Acad. Sci. U.S.A.*, 83, 8923, 1986.

24. Shirasaka, T., Shimamoto, Y., Ohshimo, H., Saito, H., Fukushima, M., Metabolic basis of the synergistic antitumor activities of 5-fluorouracil and cisplatin in rodent tumor models in vivo, *Cancer Chemother Pharmacol*, 32, 167, 1993.

25. Hainsworth, J. D., Greco, F. A., Etoposide: twenty years later, *Ann Oncol*, 6, 325, 1995.

26. Zhang, Z. G., Harstrick, A., Rustum, Y. M., Mechanisms of resistance to fluoropyrimidines, *Semin Oncol*, 19(2), Suppl 3, 4, 1992.

27. van der Wilt, C. L., Pinedo, H. M., Smid, K., Cloos, J., Noordhuis, P., Peters, G. J., Effect of folinic acid on fluorouracil activity and expression of thymidylate synthase, *Semin Oncol*, 19(2), Suppl 3, 16, 1992.

28. Vokes, E. E., Schilsky, R. L., Weichselbaum, R. R., Guaspari, A., Guarnieri, C. M., Whaling, S. M., Panje, W. R., Cisplatin, 5-fluorouracil, and high-dose oral leucovorin for advanced head and neck cancer, *Cancer*, 63, 1048, 1989.

29. Loprinzi, C. L., 5-fluorouracil with leucovorin in breast cancer, *Cancer*, 63, 1045, 1989.

30. The Advanced Colorectal Cancer Meta-analysis Project.: Modulation of fluorouracil by leucovorin in patients with advanced colorectal cancer: evidence in terms of response rate, *J Clin Oncol*, 10, 896, 1992.

31. Poon, M. A., O'Connell, M. J., Wieand, H. S., Krook, J. E., Gerstner, J. B., Tschetter, L. K., Levitt, R., Kardinal, C. G., Mailliard, J. A., Biochemical modulation of fluorouracil with leucovorin: confirmatory evidence of improved therapeutic efficacy in advanced colorectal cancer, *J Clin Oncol*, 9, 1967, 1991.

32. Caplen, H. S., Gupta, S. L., Differential regulation of a cellular gene by human interferon-gamma and interferon-alfa, *J Biol Chem*, 263, 332, 1988.

33. Wadler, S., Schwartz, E., Antineoplastic activity of the combination of interferon and cytotoxic agents against experimental and human malignancies: a review, *Cancer Res*, 50, 3473, 1990.

34. Chu, E., Zinn, S., Boarman, D., Allegra, C. J., Interaction of (gamma) interferon and 5-fluorouracil in the H630 human colon carcinoma cell line, *Cancer Res*, 50, 5834, 1990.

35. Kocha, W., 5-fluorouracil plus interferon alfa-2a vs. 5-fluorouracil plus leucovorin in metastatic colorectal cancer. Results of a multicenter multinational phase III study, *Proc Am Soc Clin Oncol* 12, 193, abstr 562, 1993.

36. York, M., Greco, F. A., Figlin, R. A., Einhorn, L., A randomized phase III trial comparing 5-FU with or without interferon alfa-2a for advanced colorectal cancer, *Proc Am Soc Clin Oncol* 12, 200, abstr 590, 1993.

37. Hill, M., Norman, A., Cunningham, D., Findlay, M., Nicolson, V., Hill, A., Joffe, J., Nicolson, M., Hickish, T., Royal Marsden phase III trial of fluorouracil with or without interferon alfa-2b in advanced colorectal cancer, *J Clin Oncol*, 13, 1297, 1995.

38. Renton, K. W., Mannering, G. J., Depression of hepatic cytochrome P-450 monooxygenase systems with administered interferon inducing agents, *Biochem Biophys Res Commun*, 73, 343, 1976.

39. Blachley, J. D., Hill, J. B., Renal and electrolyte disturbances associated with cisplatin, *Ann Intern Med*, 95, 628, 1981.

40. Cvitkovic, E., Spaulding, J., Bethune, V., Martin, J., Whitmore, W. F., Improvement of cis-dichlorodiammineplatinum (NSC 119875) therapeutic index in an animal model, *Cancer* 39, 1357, 1977.

41. De Simone, P. A., Yancey, R. S., Coupal, J. J., Effect of a forced diuresis on the distribution and excretion (via urine and bile) of ^{195}platinum when given as ^{195}platinum cis-dichlorodiammine platinum II, *Cancer Treat Rep*, 63, 951, 1979.

42. Ozols, R. F., Corden, B. F., Jacob, J., Wesley, M. N., Ostchega, Y., Young, R. C., High-dose cisplatin in hypertonic saline, *Ann Intern Med*, 19, 100, 1984

43. Litterst, C. L., Alterations in the toxicity of cis-dichlorodiammine platinum II and in tissue localization of platinum as a function of NaCl concentration in the vehicle of administration, *Toxicol Appl Pharmacol*, 61, 99, 1981.

44. Hayes, D. M., Cvitkovic, E., Golbey, R. B., Scheiner, E., Helson, L., Krakoff, I. H., High-dose cisplatinum-diamminedichloride, amelioration of renal toxicity by mannitol diuresis, *Cancer*, 39, 1372, 1977.

45. Ward, J. M., Grabin, M. E., Berlin, E., Young, D. M., Prevention of renal failure in rats receiving cis-diamminedichloroplatinum (II) by administration of furosemide, *Cancer Res*, 37, 1238, 1977.

46. Stewart, D. J., Evans, W. K., Non-chemotherapeutic agents that potentiate chemotherapy efficacy, *Cancer Treat Rev*, 16, 1, 1989.

47. Milroy, R., A randomized clinical study of verapamil in addition to combination chemotherapy in small cell lung cancer, *Br J Cancer*, 68, 813, 1993.

48. Millward, M. J., Cantwell, B. M. J., Munro, N. C., Robinson, A., Corris, P.A., Harris, A. L., Oral verapamil with chemotherapy for advanced non-small cell lung cancer: a randomized study, *Br J Cancer*, 67, 1031, 1995.

49. Dalton, W. S., Crowley, J. J., Salmon, S. S., Grogan, T. M., Laufman, L. R., Weiss, G. R., Bonnet, J. D., A phase III randomized study of oral verapamil as a chemosensitizer to reverse drug resistance in patients with refractory myeloma, *Cancer*, 75, 815, 1995.

50. Wheeler, H., Bell, D., Levi, J., A randomized trial of doxorubicin, etoposide and vindesine with or without verapamil in small cell lung cancer, *Proc Am Soc Clin Oncol* 7, 208, abstr 805, 1988.

51. Tsushima, T., Ohomori, H., Ohi, Y., Shirahama, T., Kawahara, M., Matsumura, Y., Ohashi, Y. I., Intravesical instillation chemotherapy of adriamycin with or without verapamil for the treatment of superficial bladder cancer: the final results of a collaborative randomized trial, *Cancer Chemother Pharmacol*, 35, S69, 1994.

52. Skipper, H. E., Schabel, F. M., Wilcox, W. S., Experimental evaluation of potential anticancer agents. XIII. On the criteria and kinetics associated with "curability" of experimental leukemias. *Cancer Chemotherapy Reports* 65, 1, 1964.

7

Antihypertensive Agents

Bodo Schwartzkopff
Heinrich-Heine-University,
Düsseldorf

Bodo E. Strauer
Heinrich-Heine-University,
Düsseldorf

7.1 Introduction

Arterial hypertension is a major risk factor for morbidity and mortality in the industrialized countries. The prevalence of high blood pressure increases with age, is greater for blacks than for whites, and in both races is greater in less-educated people. Despite a broad variety in the prevalence between the Western countries and even between regions of one country, a mean prevalence of high blood pressure of 15% to 20% can be assumed.[1]

According to national and international conventions, a blood pressure of more than 140 mmHg (systolic, SBP) and more than 90 mmHg (diastolic, DBP) on three different days is regarded as hypertensive. Nonfatal and fatal cardiovascular diseases (CVD), including coronary heart disease and stroke as well as renal disease, increase progressively with higher levels of both systolic and diastolic blood pressure. The relationships are strong, continuous, graded, consistent, independent, and predictive.[1,2]

High blood pressure leads to target-organ disease, including cardiac, cerebrovascular, peripheral vascular, renal and retinal manifestations that are associated with a several-fold increased risk of cardiovascular disease. Furthermore, organ damage is the most important factor for morbidity and disability. Main risk factors for CVD besides high blood pressure are dyslipidemia, cigarette use, diabetes mellitus, physical inactivity, and obesity.

The classification of high blood pressure respects these epidemiological finding by staging the systolic (every 20 mmHg) and diastolic (every 10 mmHg) values above 140 mmHg and above 90 mmHg, respectively.[1] The higher the stage of blood pressure the greater the risk of fatal and nonfatal cardiovascular events and renal disease. The mild stage of high blood pressure (SBP 140–159, DBP 90–99 mmHg) is the most common type of high blood pressure in the adult population. An increased risk for major coronary events was found at diastolic blood pressure levels of 88 to 95 mmHg (risk = 1.66) and even more above 95 mmHg (risk = 2.17) for white men.[3] Therefore, mild hypertension is responsible for a large proportion of the excess morbidity, disability, and mortality attributable to hypertension in the population.

Recent data indicate that even if 73% of patients with mild hypertension (blood pressure 140–160/90–99 mmHg) are aware of elevated blood pressure, 55% are treated, and a total of 29% of all patients have controlled blood pressure.[4]

Within the last 25 years several studies gave evidence that the treatment of elevated blood pressure is beneficial with respect to reduced cardiovascular mortality.[5] The incidence and prevalence of stroke, myocardial infarction, heart failure, and renal failure due to hypertension decreased according to population samples.[2,6] Thus, there is clear evidence that the treatment of high blood pressure is beneficial with respect to reduced morbidity and mortality rates in the population.

7.2   Cardiac Organ Manifestation in Arterial Hypertension

The heart is an important target organ in arterial hypertension leading to hypertensive heart disease. The coronary arteries, the myocytes, the interstitium as well as the microvasculature are affected. Arterial hypertension induces (a) hypertrophy of myocytes leading to left ventricular hypertrophy, (b) direct and indirect stimulation of interstitial fibroblasts causing fibrosis and, (c) structural and functional alteration of the coronary microcirculation.[7]

7.2.1   Myocardial Hypertrophy

Left ventricular hypertrophy is the structural mechanism of adaptation to chronic increased pressure or volume overload. Wall thickening normalizes the increased wall stress due to increased systolic pressure. Thus, a normal energetic demand per unit myocardial weight is achieved that guarantees a normal stroke volume to the periphery. Epidemiological studies revealed that 20% to 40% of hypertensive patients have left ventricular hypertrophy.[8] Right from the beginning, myocardial hypertrophy represents a condition of potential danger for malignant arrhythmias as well as for progressive heart failure.[8-10]

Epidemiological studies revealed that myocardial hypertrophy is an independent cardiovascular risk factor of ischemic, nonischemic, and proarrhythmic cardiovascular complications.[11] Besides pressure overload, molecular, hormonal, and tissue specific factors are involved in the development of hypertrophy. Experimental studies revealed the importance of the renin-angiotensin-aldosterone-system (RAAS) in hypertensive heart disease, especially the local intramyocardial angiotensin converting enzyme system (ACE).[7,12] Angiotensin II has been disclosed to be a growth stimulating factor by inducing protein synthesis and protooncogens that have trophic effects on smooth vascular muscle cells, fibroblasts, and myocytes. Angiotensin II is generated from angiotensin I by the local ACE-activity, mainly localized in the capillary network and to a lesser extent by the enzyme chymase that is independently generated from ACE-activity in mast cells.[13,14] Besides the renin-angiotensin system, an adrenergic stimulation promotes hypertrophy of myocytes. Noradrenalin-induced hypertrophy of myocytes in subhypertensive doses was found to be related to a stimulation of α-receptors in cell culture.[15,16] From the subcellular level noradrenalin, adrenalin, and isoproterenol activate protein kinases that induce protooncogenes.[17,18] The molecular mechanisms include the generation of fetal proteins that are not expressed in the nonhypertrophied myocardium, indicating an abnormal regulation of protein synthesis.[10,13,18]

7.2.2   Alterations of the Interstitium and Coronary Microvasculature
in Hypertensive Heart Disease

Besides the hypertrophy of the myocytes, the nonmyocytic cells of the interstitium are involved in the process of myocardial remodeling.[19] Interstitial collagen content is excessively generated by fibroblasts, leading to endomysial, perimysial, and perivascular fibrosis. Beside this reactive fibrosis, reparative fibrosis follows the loss of myocytes. Reparative fibrosis restitutes the integrity of tissue after cellular damage. Myocardial fibrosis increases myocardial stiffness, leading to systolic and diastolic dysfunction of the left ventricle.[20] Furthermore, coronary microcirculation is disturbed in arterial hypertension documented by a reduced maximal coronary blood flow and an increased minimal coronary resistance leading to a diminished coronary reserve.[21]

The functional consequences of a reduced coronary reserve are an inadequate supply with oxygen and substrates under increased metabolic demands. Morphologic investigations of transvenous endomyocardial biopsies revealed a thickening of the tunica media of arterioles.[22] Wall thickening and increased media/lumen ratio of arterioles reduce the maximal vasodilator capacity, according to Folkow et al.[23] Furthermore, every contraction of the vascular smooth muscle cells induces an even more marked lumen reduction than in normal media/lumen ratio.

Besides structural alterations, there are functional disturbances of the endothelium that regulate the tone of epicardial and intramyocardial coronary arteries and arterioles.[24-26] Endothelial cells produce and release vasodilator substances, including endothelium-derived relaxing factor (EDRF) identified as nitric oxide (NO), prostacyclin, and an endothelium-derived hyperpolarizing factor. Moreover, endothelium can produce vasoconstrictor substances such as cyclooxygenase-dependant endothelium-derived contracting factors (prostaglandin H_2, thromboxane) and endothelin.[26] Up to now the interactions of structural alterations and endothelial dysfunction, in terms of an alteration of release or production of vasodilators or by receptor-mediated release of vasoconstrictor products, are still undefined. Nevertheless, endothelial dysfunction and structural alterations of arterioles are main causes of microvascular ischemia.[22,24,25]

7.3 Goals of Antihypertensive Treatment

Blood pressure values above 140 mmHg systolic, and/or above 90 mmHg diastolic should be lowered according to national and international recommendations, respecting the criteria of evaluation of high blood pressure.[1] Appropriate diagnostic procedures are mandatory to exclude secondary causes of hypertension from the beginning that afford specific therapy. In essential hypertension, life-style modification, which includes weight reduction, increased physical activity, and moderation of dietary sodium and alcohol intake is recommended as definitive or adjunctive therapy.[1,2,27] In mild hypertension, lowered blood pressure (SBP = −9.1, DBP = −8.6 mmHg) and regressed left-ventricular hypertrophy were reported by life-style modification.[27] Nevertheless, drug treatment in combination with nutritional–hygienic intervention was more effective in preventing cardiovascular and other clinical events than nutritional–hygienic treatment alone.[27] Pharmacologic treatment is recommended by the majority of specialists, if blood pressure cannot be reduced below SBP 160 and/or DBP 95 mmHg by nutritional–hygienic treatment in most patients. In those patients with preexisting cardiovascular diseases and/or risk factors, therapy may be needed to lower blood pressure even below 140/90 mmHg.[1,2,28]

National and international conventions recommend diuretics, β-adrenoceptor antagonists, ACE inhibitors, calcium channel blockers, and α-adrenoceptor antagonists as the first line of antihypertensive medication.[1,27,28] Their use in the treatment of hypertension differs and has been changed over the years.[2]

Nevertheless, antihypertensive agents may also bear potential risks for the individual patient by drug-related side effects, negative influences on concomitant diseases, inadequate repair of target organ damages and less than expected survival rates for the population of hypertensive patients, thus missing the goal of full therapeutic success. Therapeutic goals of antihypertensive treatment should respect these aspects (Table 7.1).

TABLE 7.1 Main Goals in the Long-Term Treatment of Hypertension

1. Blood pressure reduction
2. Protection and reparation of target-organ disease
3. No adverse effects, improvement in quality of life
4. No negative influence on concomitant disease
5. Decrease in morbidity and mortality in the population

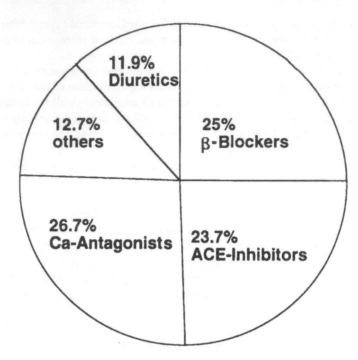

FIGURE 7.1   Percentage of prescriptions of antihypertensive drugs (mono- and combination-therapy) in 1995 in Germany (according to Institut für medizinische Statistik (IMS), Frankfurt).

7.3.1   Antihypertensive Drugs and Blood Pressure Reduction

All classes of antihypertensive agents (Figure 7.1) have been found to reduce blood pressure, even if there are large differences between different studies as two metaanalyses have shown.[29,30] With each antihypertensive drug the degree of fall in blood pressure is proportional to the initial blood pressure, with larger absolute falls occurring in those with high initial levels, as much as 100/60 mmHg in the individual patient.[29,30] It can be concluded that monotherapy with a first-line antihypertensive agent leads to a reduction in mean arterial blood pressure by 13% to 16% or in absolute values of 12 to 22 mmHg. Higher doses of monotherapy or a combination of two or more drugs achieve a better blood pressure reduction. Increased doses of an antihypertensive drug lead to a better blood pressure reduction, but side effects may increase.

Therefore, the combination of low-dose drugs may be beneficial regarding blood pressure reduction, avoiding those side effects that are dose-related. The reduction in absolute values also depends on the stage of hypertension. Collins et al.[5] reported the highest absolute reduction in diastolic blood pressure (−20 mmHg) in trials in which some or all patients had entry diastolic blood pressure >115 mmHg. In mild hypertension (Figure 7.2) the mean systolic and diastolic blood pressure (at the beginning 140/90 mmHg) were reduced by drug treatment on an average 12.3 mmHg in diastolic and 15.9 mmHg in systolic values. In the TOMHS study the change did not significantly differ between the various drugs.[27] Also in elderly patients blood pressure can be effectively lowered.[31] In elderly patients with diastolic hypertension, systolic blood pressure was reported to be reduced under antihypertensive drugs (mainly diuretics or β-adrenoceptor antagonists) by 15 to 20 mmHg, diastolic blood pressure by 5 to 9 mmHg in both sexes.[32,33] In isolated systolic hypertension in elderly patients the SHEP study reported a reduction of SBP by 12 mmHg and DBP by 4 mmHg.[34]

It was postulated that excessive reduction of diastolic blood pressure in patients with preexisting coronary disease actually causes coronary events, the so-called J-relationship (Figure 7.3).[35] It was supposed that a diastolic blood pressure below or above 85 to 90 mmHg is associated with an increased risk of cardiovascular mortality. The concept is supported by the finding of Polese et al.[36] who found the autoregulatory plateau of the coronary pressure-flow relationship shifted to higher perfusion pressures

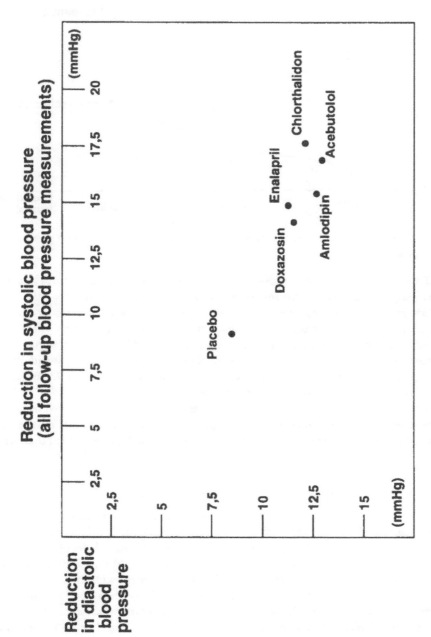

FIGURE 7.2 Blood pressure reduction in mild hypertension (TOMHS-study[27]).

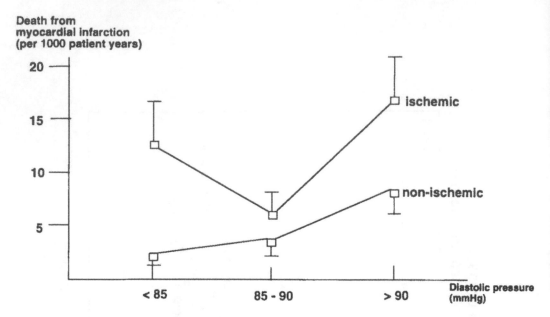

Death from myocardial infarction (per 1000 patient years)

FIGURE 7.3   J-curve/Hypothesis of increased myocardial infarction in critically lowered blood pressure (according to Cruickshank et al.[35]).

in hypertensive hypertrophy.[36] Nevertheless, blood pressure reduction is also accompanied by a reduction in myocardial oxygen consumption that improves tolerance for myocardial ischemia. Also the highly significant 25% reduction in the number of coronary events in the SHEP study argues against the J-relationship of diastolic blood pressure and coronary events.[34] The elderly patients in SHEP had an average diastolic blood pressure before treatment of only 77 mmHg, 61% had an abnormal electrocardiogram at base line, and 5% had a history of myocardial infarction. Under treatment with thiazides in first-line and β-blockers in second-line, diastolic blood pressure was reduced to 73 mmHg and 1.1 cardiovascular events per 100 patients each year could be prevented.

From these data it can be concluded that blood pressure reduction can be achieved by all first-line antihypertensive drugs and that combination therapy provides a greater absolute and relative reduction in systolic and diastolic blood pressure. In elderly patients systolic and diastolic blood pressure can be safely reduced.

7.3.2   Therapeutic Consequences of Cardio-Protection and Cardio-Reparation

Therapeutic interventions aim at cardio-protection before the development of myocardial and coronary complications that may lead to myocardial ischemia, and to diastolic and later on also to systolic dysfunction. Cardio-reparation can be regarded as the causal therapy. This includes regression of cardiac hypertrophy, normalization of myocardial structure including reversing of myocardial fibrosis, and restitution of the functional and structural integrity of the coronary microcirculation.[7]

Left ventricular hypertrophy regresses in most hypertensive patients when blood pressure is lowered by an antihypertensive drug (Figure 7.4). Nevertheless, in some patients normalization of left ventricular mass is not achieved.[8,9] With direct vasodilators (e.g., hydralazine) minimal or no regression of left-ventricular hypertrophy was reported, indicating that blood pressure lowering is not necessarily paralleled by a comparable reduction in left ventricular hypertrophy and might persist even if blood pressure is normalized (Figure 7.5). Experimental and clinical data indicate that therapeutic strategies leading to a lowered level of the growth factors noradrenalin and angiotensin II, for instance, by antiadrenergic drugs and ACE-inhibitors, most effectively improve myocardial hypertrophy.[37] Data from Koren et al.[40] support the hypothesis that a normalization of left ventricular mass might improve prognosis.

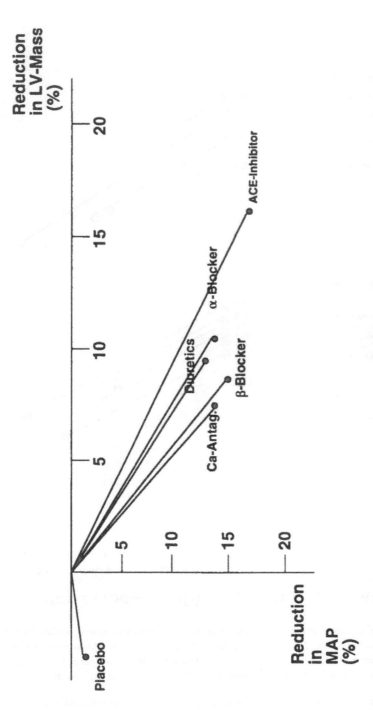

FIGURE 7.4   Blood pressure reduction (MAP = mean arterial pressure) and regression of left ventricular mass (LV-mass) (according to Dahlöf et al.[29]).

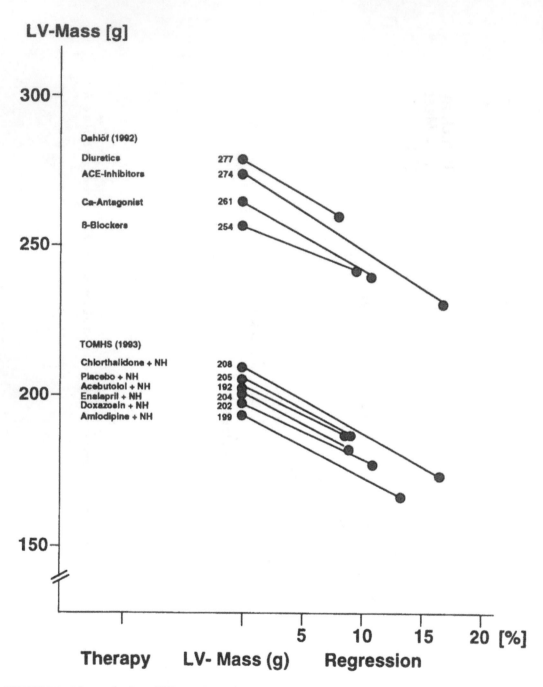

FIGURE 7.5 Mean reduction of LV-mass in moderate to severe hypertension with marked LV-hypertrophy[29] and in mild hypertension[27] with no or mild LV-hypertrophy. Patients with more pronounced LV-hypertrophy showed comparable percentage of regression in LV-mass, but absolute values are still increased at the end of the observation time compared with those patients with mild hypertrophy (NH = nutritional-hygienic measurements).

Besides the regression of left ventricular mass, reparation of left ventricular and vascular structure is desirable. Antihypertensive therapy with ACE inhibitors was associated with a reduction of the increased collagen content and an improved coronary reserve even exceeding the degree of regression in left ventricular hypertrophy.[38,39] In hypertensive patients receiving ACE-inhibitors morphologic investigations

TABLE 7.2   Cardioprotective and Cardioreparative Effects in Patients with Hypertensive Heart Disease

	Diuretics	β-Blockers	Calcium-Antagonists	α-Blockers	ACE-Inhibitors
Regression of LVH	Effective	Effective	Effective	Effective	Effective
Reverse of fibrosis	Not known	Not known	Not known	Not known	Effective
Repair of resistance vessels	Not known	Not known	Not known	Not known	Not known
Influence of endothelial function	Not known	Not known	Positive	Not known	Positive

of subcutaneous arteries from the gluteal region revealed an increase in the lumen diameter with reduced wall thickness. This was not observed with patients treated with β-blockers.[41] Therefore, improved coronary reserve under ACE-inhibitors might also be due to structural reparation of coronary microcirculation in hypertensive patients.[39,42] Calcium antagonists were also found to reduce left ventricular hypertrophy and to improve coronary reserve in hypertension.[42,43] β-adrenoceptor antagonist therapy resulted in a regression of myocardial hypertrophy accompanied by an improved coronary reserve. Coronary resistance, however, was only slightly lowered.[44] β-blockers improve the oxygen imbalance by reducing wall-stress and lowered oxygen demand, as well as by regression of left ventricular hypertrophy, but there is no evidence for reparation of structural components of the arteriolar vessel wall. Accordingly, investigations of subcutaneous arteries showed no reparation of resistance arteries under β-blockers.[41] An influence on collagen metabolism is only to be expected in the frame of regression of left ventricular hypertrophy.

Diuretics are reported to reduce left ventricular end-diastolic diameter and mass, especially in dilated excentric hypertrophy.[8,9] Excessive up-regulation of neurohumoral factors, e.g., renin under diuretics, may have some disadvantageous effects on coronary microcirculation as well as on myocardial fibrosis.[7,45] With diuretics cardio-reparation is not to be expected beyond the effects closely linked to the decline in blood pressure (Table 7.2).

7.3.3   Side Effects and Quality of Life

Antihypertensive agents share a number of potentially adverse reactions that are partly linked to their mode of action (Table 7.3) and that may necessitate termination of therapy. Jones et al.[46] reported about 37,643 hypertensive patients receiving a relevant drug in the time period identified. In 10,222 patients aged >40 years, treatment was changed. They now received at least one first-line antihypertensive agent (β-adrenoceptor antagonist, calcium channel blocker, angiotensin converting enzyme inhibitor, or diuretic) not prescribed before. After 6 months, only 40% to 50% of the patients were still on previously prescribed medication with any class of drug. It was pointed out that suboptimal compliance with treatment for hypertension can lead to further complications and higher costs by an increase in the number of prescriptions per patient. The authors concluded that patients terminate treatment early because of side effects rather than because of poor efficacy. Therefore, knowledge and adequate therapeutic handling of side effects, a cautious start with a new drug regimen respecting additional medication and diseases, and information to the patient about the necessity of the treatment, possible side effects, and confirmed controll dates may increase the acceptance of treatment.[1] The most important side effects of diuretics are reported to be light-headedness, hypokalemia, acute gout, and impotence.[47] There is a clear dose-dependency of these unwanted drug reactions. The most common adverse reactions with β-adrenoceptors are fatigue, lethargy, and cold peripheries. Other CNS side effects, bronchoconstriction, shortness of breath, gastrointestinal problems, sexual dysfunction, blurred vision, and dry eyes, occur only rarely.[48]

Calcium channel blockers may have disadvantages from activation of the sympathetic nervous system (dihydropyridine), associated with the development of palpitations, headache, sweating, tremor, and flushing in about 20% to 30% of patients. Edema is a frequent problem, especially with short-acting dihydropyridines. Adverse vasodilator effects are rarely seen with verapamil and diltiazem, which, however, may

TABLE 7.3 Antihypertensive Drugs: Mechanisms, Adverse Effects, and Drug Interactions

Type of Drug	Mechanisms	Major Adverse Effects	Relative or Absolute Contraindications	Drug Interactions
Diuretics, thiazides, and related agents				
Bendroflumethiazide, benzthiazide, chlorothiazide, chlorthalidone, cyclothiazide, hydrochlorothiazide, hydroflumethiazide, indapamide, metolazone, polythiazide, quinethazone, trichlormethiazide	Decreased plasma volume and decreased extracellular fluid volume; decreased cardiac output initially, followed by decreased total peripheral resistance with normalization of cardiac output; long-term effects include slight decrease in extracellular fluid volume	Hypokalemia, hypomagnesemia, hyponatremia, hyperuricemia, hyperglycemia, hypercholesterolemia, hypertriglyceridemia, sexual dysfunction, weakness; rare severe reactions: pancreatitis, bone marrow suppression, anaphylaxis	Hypokalemia, hypercalcemia, gout; ineffective in renal failure (serum creatinine ≥2.5 mg/dl) except for indapamide and metolazone	Cholestyramine and colestipol decrease absorption, non-steroidal anti-inflammatory drugs may antagonize diuretic effectiveness. Possible situations for increased antihypertensive effects: combinations of thiazides (especially metolazone) with furosemide can produce profound diuresis, natriuresis, and kaliuresis in renal impairment. Diuretics can raise serum lithium levels and increase toxic effects by enhancing proximal tubular reabsorption of lithium. Diuretics may make it more difficult to control dyslipidemia and diabetes
Loop diuretics				
Bumetanide, ethacrynic acid furosemide	See thiazides	Same as for thiazides, except hypercalcemia	Hypokalemia	
Potassium-sparing agents		Hyperkalemia	Renal failure	Combination with ACE-inhibitors may exaggerate hyperkalemia, danger in patients with nonsteroidal antiinflammatory drugs, or renal failure
Amiloride	Increased potassium reabsorption		Pregnancy	
Spironolactone	Aldosterone antagonist	Gynecomastia, mastodynia, menstrual irregularities, diminished libido in males		
Triamterene		Danger of renal calculi		

Drug	Effects	Side effects	Contraindications	Interactions
β-Blockers, β₁-selective, no ISA; Atenolol, betaxolol, bisoprolol, metoprolol	In the beginning, decreased cardiac output and increased total peripheral resistance that normalizes later on; decreased plasma renin activity	Bronchospasm (β₂-blockade), may aggravate peripheral arterial insufficiency, fatigue, insomnia, exacerbation of heart failure, masking of symptoms of hypoglycemia, hypertriglyceridemia, decreased high-density lipoprotein cholesterol (except for drugs with ISA), β₂-stimulation and α-β-blockers). Bradycardia (β₁-blockade), decreased glucose tolerance, reduces exercise tolerance, sexual dysfunction	Asthma, COPD, heart-block, sick sinus syndrome. Insulin-treated diabetes, patients with peripheral vascular disease.	NSAIDs may decrease effects of β-blockers. Rifampicin, phenobarbital, and smoking decrease serum levels of agents primarily metabolized by liver due to enzyme induction. Cimetidine may increase serum levels of β-blockers primarily due to enzyme inhibition in the liver. Quinidine may increase risk of hypotension. Combinations of diltiazem or verapamil with β-blockers have additive sinoatrial and atrioventricular node depressant effects and may also promote negative inotropic effects on failing myocardium. Combination of β-blockers and reserpine may cause marked bradycardia and syncope. β-Blockers may increase serum levels of lidocaine, theophylline, and chlorpromazine due to reduced hepatic clearance. Nonselective β-blockers prolong insulin-induced hypoglycemia and promote rebound hypertension due to unopposed α stimulation. All β-blockers mask adrenergically mediated symptoms of hypoglycemia and have potential to aggravate diabetes. β-Blockers may make it more difficult to control dyslipidemia. Phenylpropanolamine, pseudoephedrine, ephedrine, and epinephrine can cause elevations in blood pressure due to unopposed α-receptor-induced vasoconstriction.
β₁-selective, with ISA Acebutolol **Non-selective, no ISA** Propanolol, nadolol, timolol **Non-selective, with ISA** Alprenolol, penbutolol, carteolol, pindolol				
Non-selective with β₂ stimulation Celiprolol	Decreased total peripheral resistance			
α-β-blockers Labetalol, carvedilol	Same as β-blockers, plus α₁-blockade, decreased total peripheral resistance			
α₁-Blockers Doxazosin, prazosin, terazosin, bunazosin	Block postsynaptic α₁-receptors and cause vasodilation	Orthostatic hypotension, syncope, weakness, palpitation, headache		Concomitant antihypertensive drug therapy (especially diuretics), may increase chance of postural hypotension

TABLE 7.3   Antihypertensive Drugs: Mechanisms, Adverse Effects, and Drug Interactions

Type of Drug	Mechanisms	Major Adverse Effects	Relative or Absolute Contraindications	Drug Interactions
ACE-Inhibitors				
Benazepril, captopril, cilazapril, enalapril, fosinopril, lisinopril, perindopril, quinapril, ramipril, spirapril	Block formation of angiotension II, promoting vasodilatation, and decreased aldosterone; also increased bradykinin and vasodilatory prostaglandins	Cough, rash, angioneurotic edema, hyperkalemia; hypotension, especially in patients with high plasma renin activity or receiving diuretic therapy. Neutropenia, proteinuria (rarely).	Hyperkalemia, renal insufficiency; absolutely contraindicated in 2nd and 3rd trimesters of pregnancy, severe aortic stenosis, pericardial effusion, pericardial constriction, angioneurotic edema	Non-steroidal antiphlogistics (including aspirin and ibuprofen) may decrease blood pressure control. Antacids may decrease the bioavailability of ACE-inhibitors. Diuretics may lead to excessive hypotensive effects (hypovolemia). Hyperkalemia may occur with potassium supplements, potassium-sparing agents, and NSAIDs. ACE inhibitors may increase serum lithium levels. Potassium-sparing diuretics can cause reversible acute renal failure in patients with bilateral renal arterial stenosis or unilateral stenosis in solitary kidney and in patients with cardiac failure and with-volume depletion.
AT₁-receptor blockers of angiotensin II Losartan, iberatan, canderatan, valsartan	Block of AT₁-receptor of angiotensin II	Cough, anemia, immunologic system disease	Still undefined; principally the same as with ACE-inhibitor	Still undefined; principally the same as with ACE-inhibitor
Calcium antagonists Dihydropyridines: Amlodipine Felodipine Isradipine Nicardipine Nifedipine	Block inward movement of calcium ion across cell membranes and cause smooth-muscle relaxation	Headache, dizziness, peripheral edema, tachycardia, gingival hyperplasia, flushing	Use with caution in patients with heart failure. Dihydropyridine should not be used in acute myocardial infarction or unstable angina	Serum levels and antihypertensive effects of calcium antagonists may be diminished by interactions: rifampicin-verapamil; carbamazepine–diltiazem and verapamil; phenobarbital and phenytoin-verapamil. Cimetidine may increase pharmacologic effects of all calcium antagonists due to inhibition of hepatic metabolizing enzymes resulting in increased serum levels. Digoxin and carbamazepine serum levels and toxic effects may be increased by verapamil and possibly by

Drug	Mechanism	Side effects	Contraindications/Precautions	Drug interactions
Diltiazem Verapamil		Headache, dizziness, peripheral edema (less common than with dihydropyridines), gingival hyperplasia, constipation (especially verapamil), atrioventricular block, bradycardia	Contraindicated in patients with 2nd- or 3rd-degree heart block, or sick sinus syndrome	diltiazem. Serum levels of prazosin, quinidine, and theophylline may be increased by verapamil. Serum levels of cyclosporine may be increased by diltiazem, nicardipine, and verapamil; cyclosporine dose may need to be decreased.
Centrally acting α_2-agonists				
Clonidine, moxonidin, guanabenz, guanfacine, urapidil	Stimulate central α_2-receptors that inhibit efferent sympathetic activity	Drowsiness, sedation, dry mouth, fatigue, orthostatic dizziness, sexual dysfunction. Abrupt discontinuation may include rebound hypertension		Tricyclic antidepressants may decrease effects of centrally acting and peripheral norepinephrine depleters. Sympathomimetics, amphetamines, phenothiazines, and cocaine, may interfere with antihypertensive effects of guanethidine and guanadrel. Severity of clonidine withdrawal reaction can be increased by β-blockers. Monoamine oxidase inhibitors may prevent degradation and metabolism of norepinephrine released by tyramine-containing food and may cause hypertension; they may also cause hypertensive reactions when combined with reserpine or guanethidine. Methyldopa may increase serum lithium levels.
Methyldopa			May cause liver damage, fever, and Coombs-positive hemolytic anemia	
Peripheral-acting adrenergic antagonists	Inhibits catecholamine release from neuronal storage sites	Diarrhea, orthostatic and exercise hypotension. Lethargy, nasal congestion, depression.	Contraindicated in patients with history of mental depression or with active peptic ulcer.	
Guanadrel sulfate Guanethidine monosulfate Rauwolfia alkaloids Reserpine	Depletion of tissue stores of catecholamines			
Direct vasodilators				
Hydralazine	Direct vasodilation (primarily arteriolar)	Headache, tachycardia, fluid retention. Positive antinuclear antibody test	May precipitate angina pectoris in patients with coronary artery disease; diuretics should be added in fluid retention; β-blockers in reflex tachycardia, especially in patients with coronary artery disease. Lupus syndrome may occur (rare at recommended doses).	
Minoxidil		Hypertrichosis	May cause or aggravate pleural and pericardial effusions. Hypertrichosis.	

Adapted to Joint National Committee.[1] ISH = intrinsic sympathomimetic activity.

be associated with prolonged atrial arrest and atrioventricular conduction disturbances. These effects are even worsened in combination with β-adrenoceptor antagonists.[49] ACE-inhibitors may induce hypotension, headache, and frequently cough (1% to 30%). Angioneurotic edema and azotemia are rare side effects. Hypotension is most often observed in patients with heart failure, previous volume depletion by aggressive diuretic treatment, and in secondary hypertension with secondary hyperaldosteronism. To avoid hypotensive effects, small doses should be given at the beginning and supporting diuretic therapy should be carefully used.[50]

α-Adrenoceptor blockers may lead to postural hypotension, which is the most important and most frequent side effect. Orthostatic dizziness has been reported in up to 20% of patients. Other common adverse effects associated with these drugs are fatigue, headache, palpitations, and nausea, which occur in about 5% of patients. Most of these symptoms are mild and tend to diminish in severity with continuation of the drug (Table 7.3).[51]

In the TOMHS-study, the overall test for several quality of life indices indicated that there was a significant improvement in quality of life under antihypertensive treatment.[27] A greater improvement was reported for participants receiving acebutolol (p < 0.001) or chlorthalidone (p < 0.008) than for the given placebo group. Less pronounced improvements were seen with other treatment schedules not different significantly from the placebo group (Figure 7.6). Nevertheless, with these drugs a positive effect also has to be assumed because of an improvement compared to the quality of life before the treatment.

7.3.4   Antihypertensive Treatment and Concomitant Disease

In many hypertensive patients additional risk factors are present that significantly influence the clinical picture (Table 7.4). Hyperuricemia, smoking, diabetes mellitus, dyslipoproteinemia, adipositas, and insulin resistance may prevail in arterial hypertension. Adipositas, insulin resistance, and arterial hypertension are regarded as "metabolic syndrome" and represent an important risk factor for coronary artery disease and cardiovascular complications.[52] Therefore, additional treatment of the concomitant diseases is mandatory to reduce cardiovascular mortality. This will be more successful by the choice of a compatible antihypertensive drug.

Diuretics were found to have some disadvantageous effects on the plasma levels of lipids and glucose and also on glucose tolerance.[53,54] Therefore, in diabetic patients and in patients with dyslipoproteinemia diuretics should be given with caution (Tables 7.5 and 7.6). β-Adrenoceptor antagonists have no consistent effect on plasma lipids, but in the majority of patients they reduce HDL cholesterol while total cholesterol and LDL cholesterol remain unchanged. β-Adrenoceptor antagonists with intrinsic activity, α-, β-blocking or $β_2$-stimulating effects might improve dyslipoproteinemia.[54-56] Their influence on glucose and insulin metabolism requires careful attention in the follow-up, especially in elderly patients. In diabetic patients on insulin, hypoglycemia may occur more often under β-adrenoceptor blockers as the return of the blood sugar is delayed because of epinephrine antagonism. $β_1$-Selective antagonists are preferable for those susceptible to hypoglycemia, but all β-blockers may delay recovery. Diabetic patients not on insulin may more often become hyperglycemic under β-adrenoceptor antagonists. These drugs increased the risk for the development of diabetes mellitus four- to sixfold in 9- to 12-year follow-up studies.[54,58]

Calcium channel blockers are reported not to alter plasma lipids and glucose tolerance (Table 7.5 and 7.6).[54,56] ACE inhibitors do not influence lipids, but even improve glucose tolerance and decrease hyperinsulinemia. The favorable effects of the $α_1$-adrenoceptor antagonists both on serum insulin and on cholesterol and triglyceride levels led to suggestions that these drugs may have an outstanding role in the treatment of hypertensive diabetic patients.[54-56]

Renal failure may result from arterial hypertension and may be the cause of hypertension. In renal failure, blood pressure normalization is often associated with a slower progression of functional loss of the kidney, especially in diabetic patients. ACE-inhibitors are beneficial in patients with proteinuria and diabetic nephropathia.[59] This appears to result from an inhibition of angiogenesis and glomerulosclerosis by blocking angiotensin II-generation as well as increased kinin levels. Calcium channel blockers, especially diltiazem, were also reported to have nephroprotective effects.[60,62]

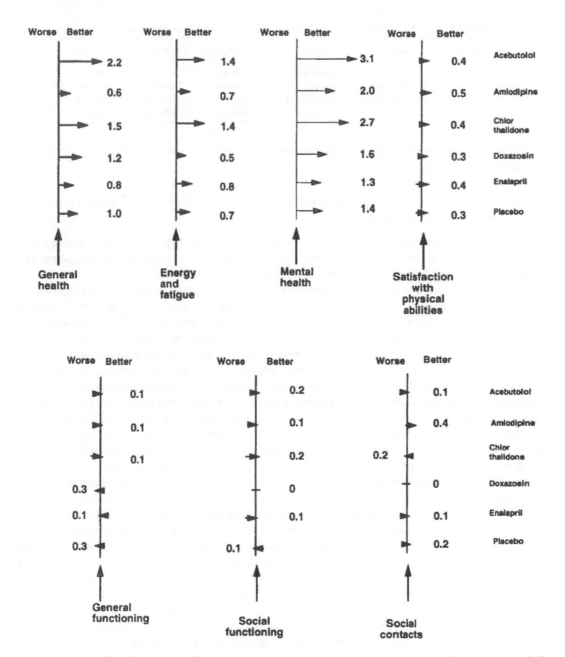

FIGURE 7.6 Quality of life under antihypertensive medication according to the score system of the TOMHS study.[27]

TABLE 7.4 Additional Risk Factors in Hypertensive Patients

	Percent of Patients
Physical inactivity	>50%
Hyperinsulinemia	50%
Dyslipidemia	>40%
Hypercholesterolemia > 240 mg/dl	25%
Decreased HDL cholesterol < 40 mg/dl	25%
Obesity	40%
Smoking	35%

TABLE 7.5   Possible Effect of Antihypertensive Drugs on Dyslipoproteinemia and Insulin

Risk factor	Diuretics	β-Blockers	ACE-Inhibitors	Ca-Antag.	α-Blockers
Cholesterol	Increase	Unchanged	(Reduce)	Unchanged	Reduce
HDL-Cholesterol	Reduce	(Reduce)	(Increase)	Unchanged	Increase
Glucose tolerance	Deteriorate	(Deteriorate)	Improve	Unchanged	Improve
Insulin	Increase	(Increase)	(Reduce)	Unchanged	Decrease

Facing the beneficial effects of newer antihypertensive drugs on concomitant diseases morbidity and mortality will hopefully further improve.

7.3.5   Effects of Antihypertensive Drugs on Morbidity and Mortality in Hypertensive Patients

Diuretics and β-adrenoceptor antagonists have been found to reduce hypertensive complications such as stroke, heart failure, renal failure, and events from coronary artery disease in the population. The benefit of antihypertensive treatment parallels the elevation of blood pressure.[5,31]

Collins et al.[5] reported a metaanalysis of fourteen randomized trials of antihypertensive drugs (mainly diuretics or β-adrenoceptor blockers) with a total of 37,000 individuals. A mean treatment duration of 5 years induced a decline of mean diastolic blood pressure by 5 to 6 mmHg. In prospective observational studies, a long-term difference of 5 to 6 mmHg in DBP is associated with a decline rate of 35% to 40% in stroke and 20% to 25% in coronary heart disease. In the cited treatment trials, stroke was reduced by 42% (95% confidence interval 33% to 50%, 289 vs. 484 events, 2p < 0,0001) suggesting that the epidemiologically expected stroke reduction appears rapidly. Coronary heart disease was reduced by 14% (95% confidence interval 4% to 22%; 671 vs. 771 events, p < 0.01). *Obviously, the expected reduction in coronary heart disease mortality was not completely achieved in this metaanalysis of mostly middle-aged patients.* Although newer studies in elderly patients promise a more pronounced reduction of coronary heart disease via blood pressure reduction, this is not the only factor in hypertensive patients that has to be taken into account for improving the prognosis (Figure 7.7).[32,34,63] Several factors might have influenced the results. The discrepancy between expected and observed efficacy in coronary morbidity and mortality could be caused by inadequate blood pressure reduction in 20% to 25% of the patients. Besides this, the interval of observation might have been too short to achieve a positive effect on atherosclerotic complications. In many studies patients received diuretics and/or β-adrenoceptor antagonists that may have adverse effects on lipids and glucose and may partially counteract the benefical effects of blood pressure lowering on atherosclerosis. With thiazides a loss of magnesium and potassium may occur that potentially causes proarrhythmogenic effects and may increase the risk of sudden death. Siscovick et al.[64] reported a population-based case study evaluating the influence of thiazide (hydrochlorothiazide and chlorthalidone) and potassium-sparing diuretic therapy on the risk of primary cardiac arrest compared with a β-adrenoceptor blocker therapy.[64] Compared with low-dose hydrochlorothiazide therapy (25 mg daily), medium doses (50 mg daily) were associated with a moderate increase in risk (odds ratio, 1.7) and high-dose therapy (100 mg daily) even induced a larger increase in risk (odds ratio, 3.6). The addition of a potassium-sparing drug to low-dose hydrochlorothiazide therapy was associated with reduced risk of cardiac arrest (odds ratio, 0.4 Figure 7.8). There was a significant trend from a high-dose to a low-dose regimen of thiazides. Corresponding to this finding, Hoes et al.[65] reported in a case-control study that the use of non-potassium sparing diuretics was associated with an increased risk of sudden cardiac death compared with the treatment with potassium-sparing diuretics, ACE-inhibitors, or calcium channel blockers.

Randomized studies evaluated the efficacy of β-blockers and diuretics. In the IPPSH study mortality rates from stroke and coronary heart disease did not differ following oxprenolol (a nonselective β-adrenoceptor blocker with intrinsic sympathomimetic activity) and an essentially diuretic antihypertensive treatment.[66] β-adrenoceptor blocker therapy was associated with significantly lower average blood

TABLE 7.6 Antihypertensive Drugs in Concomitant Diseases

	ACE-Inhibitor	β-Blocker	Calcium-Antagonist	Diuretics	α₁-Blocker	Other Agents
Elderly patient (>65 years)	X			X (Potassium-sparing)		X Clonidine
LV-Hypertrophy		X	X			
Coronary macroangiopathy	X		X			
Coronary microangiopathy	X		X			
After myocardial infarction	X	X				
Heart failure	X			X		
Renal failure	X caution: prolonged elimination		X	Creatinin >2 mg% loop diuretics		
Obstructive pulmonary disease	X	Contraindication	X		X	
Diabetes mellitus	X	Caution: nonselective β-blocker	X	Caution	X	
Gout				Contraindication		
Dyslipoproteinemia	Neutral	Caution	Neutral	Caution	X	
Prostatahyperplasia					X	
Gravidity	Contraindication	X	Contraindication	Contraindication		α-methyldopa, hydralazin

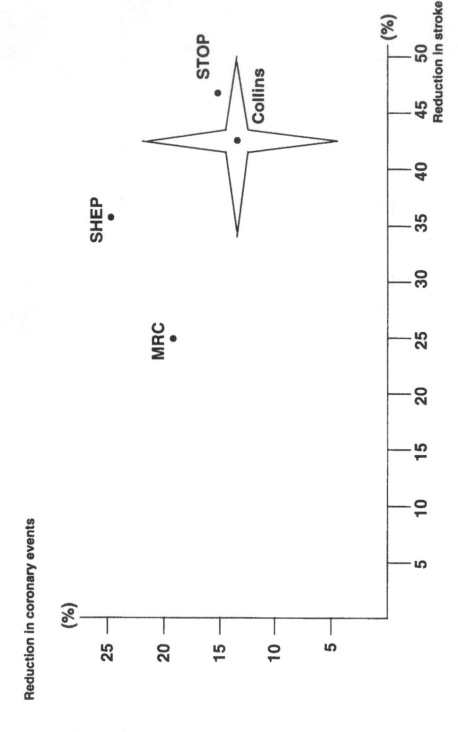

FIGURE 7.7 Antihypertensive drugs and reduction in coronary events and stroke in the MRC-trial (1992),[32] the Systolic Hypertension in the Elderly Program, (SHEP)(1991),[34] STOP-Hypertension (1990),[63] and in a review of 14 studies (mainly diuretics or β-blockers) reported from Collins et al. (1990).[5] The cross-bar reflects standard deviation. The observed reduction in coronary events is lower than the expected risk reduction in the majority of studies.

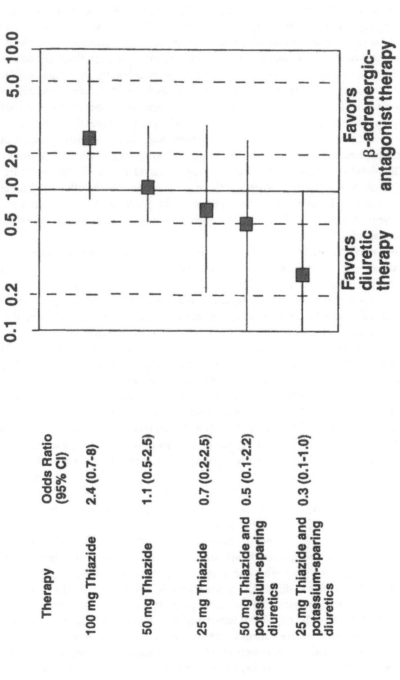

FIGURE 7.8   Risk of Primary Cardiac Arrest with thiazide therapy with and without potassium-sparing diuretic therapy, as compared with β-adrenergic-antagonist drug therapy, among patients treated with single antihypertensive drugs according to Siscovick et al.[64]

pressure and earlier electrocardiogram (ECG) normalization. Subgroup analysis revealed a significant benefit in cardiovascular mortality for nonsmoking men treated with β-blockers.[66] In the MRC trial of middle-aged hypertensives the coronary event rate was not significantly reduced by the diuretic bendrofluazide irrespective of smoking habits, nor was it reduced in smokers taking propranolol, but it was reduced in nonsmokers taking propranolol. In the HAPPHY trial there was a trend toward a difference in the incidence of CHD and stroke comparing β-blockers and diuretics in middle-aged men.[68] In the MAPHY study representing a subgroup analysis of the HAPPHY trial, metoprolol treatment was associated with a significantly lower mortality rate from coronary heart disease and stroke.[69] Total mortality was significantly lower in male, middle-aged smokers obtaining metoprolol.[69] In contrast to this, in the MRC-trial in elderly patients atenolol turned out to be significantly less effective in reducing stroke, coronary and all cardiovascular events than the combination of hydrochlorothiazide and amiloride.[32] Facing these data it is tempting to speculate that overall prognosis is improved by antihypertensive therapy but that some aspects might not to be influenced, thus limiting therapeutic success. In a case-controlled study Hoes et al. observed an increased risk of sudden death in hypertensive patients under β-adrenoceptor blockers compared with those receiving potassium-sparing diuretics, calcium channel blockers and/or ACE-inhibitors. Besides other factors, progression of atherosclerosis and incomplete cardioremodeling might account for the incomplete therapeutic success. On the other hand these drugs improve prognosis after myocardial infarction.[70] Facing these results, β-adrenoceptor blockers seem to be favorable in the treatment of middle-aged hypertensive men, smokers and in coronary artery disease, especially after myocardial infarction.

These data reveal the necessity to define the effects of the other antihypertensive drugs on cardiovascular mortality. Until now it could not be shown that calcium channel blockers reduce mortality more effectively in hypertensive patients.[70a] Psaty et al.[71] reported a case-control study comparing calcium channel blockers with β-adrenoceptor antagonists or diuretics in hypertensive patients. An increased risk for myocardial infarction was detected in hypertensive patients receiving high doses of calcium channel blockers.[71] This may result from negative inotropic effects, proarrhythmic effects, prohemorrhagic effects, proischemic effects from a coronary steal phenomenon and, for short-acting dihydropyridines of the first generation, a reflex increase in sympathetic activity possibly inducing plaque rupture. Negative inotropic effects, however, cannot sufficiently explain the results, since with β-adrenoceptor blockers, which also exhibit negative inotropic effects, the risk ratio even declined with increasing doses.[71] Moreover, in heart failure the long-acting dihydropyridine amlodipine was reported to improve the prognosis.[72] Messerli et al.[9] reported that under antihypertensive treatment with verapamil but not with diuretics the number of ventricular arrhythmias declined in parallel to regression of left ventricular hypertrophy. They concluded that the regression of left ventricular hypertrophy (representing the sum of structural abnormalities) is accompanied by a reduction of potentially hazardous arrhythmias. Also mortality rates tended to be improved following myocardial infarction.[73,74] Although a conclusive explanation for the increased risk of myocardial infarction in hypertensive patients on calcium channel blockers is lacking, high doses of short-acting dihydropyridine are not to be recommended.[28]

In hypertensive patients there are no data about mortality rates for ACE-inhibitors and α-adrenoceptor blockers. ACE-inhibitors, however, are beneficial in reducing mortality and morbidity rates in patients with heart failure and thus they are promising for hypertensive patients too.[75,76] In heart failure the combination of ACE-inhibitor, diuretic, digitalis glycoside, and the α-, β-adrenergic blocker carvedilol improved symptoms and prognosis.[77] From these promising data reported in patients with other than hypertensive diseases, it is to be hoped that newer antihypertensive drug regimes may improve morbidity and mortality also in hypertensive patients. Furthermore, a combination of antihypertensive drugs may facilitate blood pressure reduction, regression of left ventricular hypertrophy, and avoid some adverse effects by low-dose regimens in those hypertensive patients who do not respond adequately to monotherapy.[78,79]

References

1. Joint National Committee on Detection, Evaluation and Treatment of High Blood Pressure: The Fifth Report of the Joint National Committee on Detection, Evaluation and Treatment of High Blood Pressure. *Arch Intern Med*; 153: 186, 1993.
2. Kaplan, M.: *Clinical Hypertension*, 6th ed., Williams & Wilkins, Baltimore, MD, 1994.
3. Pooling Project Research Group: Relationship of blood pressure, serum cholesterol, smoking habit, relative weight and ECG abnormalities to incidence of major coronary events: final report of the pooling project. *J Chronic Dis*; 31: 201, 1978.
4. Burt, LV, Cutler, JA, Higgins, M, Horan, MJ, Labarthe, D, Whelton, P, Brown, C, Roccella, EJ: Trends in the prevalence, awareness, treatment, and control of hypertension in the adult U.S. population. *Hypertension*; 26: 60, 1995.
5. Collins, R, Peto, R, MacMakon, S, Herbert, P, Fiebach, NH, Eberlein, KA, Godwing, J, Qizilbash, N, Taylor, JO, Hennekens, CH: Blood pressure, stroke and coronary heart disease; Part 2, short-term reductions in blood pressure: overview of randomized drug trials in their epidemiological context. *Lancet*; 335: 827, 1990.
6. Bourassa, MG, Gurne, O, Bangdiwala, SI, Ghali, JK, Young, JB, Rousseau, M, Johnstone, DE, Yusuf, S: Natural history and patterns of current practise in heart failure. *J Am Coll Cardiol*; 22: 14A, 1993.
7. Weber, KT, Anversa, P, Armstrong, PW, Brilla, CG, Burnett, JC, Cruickshank, JM, Deveraux, RB, Giles, TD, Korsgaard, N, Leier, CV, Mendelsohn, FAO, Motz, W, Mulvany, MJ, Strauer, BE: Remodelling and reparation of the cardiovascular system. *J Am Coll Cardiol*; 20: 3, 1992.
8. Savage, DD, Garrison, RJ, Kannel, WB, Levy, D, Andersons, S, Stokes, J, III, Feinleib, M, Castelli, WP: The spectrum of left ventricular hypertrophy in general population; Sample: The Framingham study. *Circulation*; 75: Suppl. I, 26, 1987.
9. Messerli, FH, Neunez, BG, Nunez, MM, Garavaglia, GE, Schmieder, RE, Ventura, HO: Hypertension and sudden death. Disparate effects of calcium entry blocker and diuretic therapy on cardiac dysrhythmias. *Arch Intern Med*; 149: 1263, 1989.
10. Katz, A: Cardiomyopathy of overload: A major determinant of prognosis in congestive heart failure. *N Engl J Med*; 322: 100, 1990.
11. Koren, MJ, Deveraux, RB, Casale, PN, Savage, DD, Laragh, JH: Relation of left ventricular mass and geometry to morbidity and mortality in men and women with essential hypertension. *Ann Intern Med*; 114: 345, 1991.
12. Schunkert, H, Dzau, VJ, Tang, SS, Hirsch, TT, Apstein, CS, Lorell, BH: Increased rat cardiac angiotensin converting enzyme activity and m-RNA expression in pressure overload left ventricular hypertrophy. *J Clin Invest*; 86: 1913, 1990.
13. Dzau, VJ, Gibbons, GH, Cooke, JP, Omoigni, N: Vascular biology and medicine in the 1990s: Scope, concepts, potentials and perspectives. *Circulation*; 87: 705, 1993.
14. Urata, H, Healey, B, Stewart, RW, Bumpus, FM, Husain, A: Angiotensin II formatting pathways in normal and failing human hearts. *Circ Res*; 22: 883, 1990.
15. Laks, MN: Norepinephrine — the myocardial hypertrophy hormone? *Am Heart J*; 91: 674, 1976.
16. Simpson, PC: Norepinephrine-stimulated hypertrophy of cultured rat myocardial cell is an alpha$_1$-adrenergic response. *J Clin Invest*; 72: 732, 1983.
17. Morgan, HE, Baker, KM: Cardiac hypertrophy: Mechanical, neural and endocrine dependence. *Circulation*; 83:13, 1991.
18. Simpson, PC, Long, CS, Waspe, CE, Henrich, CJ, Ordahl, CP: Transcription of early developmental isogenes in cardiac myocyte hypertrophy. *J Mol Cell Cardiol*; 21: (Suppl 5) 77, 1989.
19. Schwartzkopff, B, Motz, W, Vogt, M, Strauer, BE: Heart failure on the basis of hypertension. *Circulation* (Suppl IV); 87: 66, 1993.
20. Weber, KT: Cardiac interstitium in health and disease: Remodelling of the fibrillar collagen matrix. *J Am Coll Cardiol*; 13: 1637, 1989.

21. Strauer, BE: Ventricular function and coronary hemodynamics in hypertensive heart disease. *Am J Cardiol*; 44: 999, 1979.

22. Schwartzkopff, B, Motz, W, Frenzel, H, Vogt, M, Knauer, S, Strauer, BE: Structural amd functional alterations of the intramyocardial coronary arterioles in patients with arterial hypertension. *Circulation*; 88: 993, 1993.

23. Folkow, B, Hallbäck, M, Lundgren, Y, Weiss, L: Structurally based increase of flow resistance in spontaneously hypertensive rats. *Acta Physiol Scand*; 79: 373, 1970.

24. Motz, W, Vogt, M, Rabenau, O, Scheler, S, Lückhoff, A, Strauer, BE: Evidence of endothelial dysfunction in coronary resistance vessels in patients with angina pectoris and normal coronary angiograms. *Am J Cardiol*; 68: 996, 1991.

25. Treasure, CB, Klein, JC, Vila, JA, Manoukian, SV, Renwick, GH, Selwyn, AP, Ganz, P, Alexander, RW: Hypertension and left ventricular hypertrophy are associated with impaired endothelium-mediated relaxation in human coronary resistance vessels. *Circulation*; 87: 86, 1993.

26. Lüscher, TF, Vanhoutte, PM: *Endothelium-Derived Vasoactive Factors*. CRC Press, Boca Raton; 1990.

27. Neaton, JO, Grimm, RH, Prineas, RJ et al. for TOMHS research group: Treatment of mild hypertension study. *J Am Med Assoc*; 270: 713, 1993.

28. Faulhaber, D: Hochdruckbehandlung in der Praxis. *Dtsch Ärzteblatt*; 92: 2455, 1995.

29. Dahlöf, B, Pennet, K, Hausson, L: Reversal of left ventricular hypertrophy in hypertensive patients. *Am J Hypertension*; 5: 95, 1992.

30. Cruickshank, JM, Lewis, J, Moore V, Dodel, C: Reversibility of left ventricular hypertrophy by differing types of antihypertensive therapy. *J Human Hypertension*; 6: 85, 1992.

31. Lever, AF, Ramsay, LE: Treatment of hypertension in the elderly. *J Hypertension*; 13: 571, 1995.

32. MRC working party; Medical research council trial of treatment of hypertension in older adults; principal results. *Br Med J*; 304: 405, 1992.

33. Amery, A, Birkenhager, W, Brixko, P: Mortality and morbidity results from the European Working party on High Blood Pressure in the Elderly Trial. *Lancet*; 1:1349, 1985.

34. SHEP, Cooperative Research Group: Prevention of stroke by antihypertensive drug treatment in older persons with isolated systolic hypertension: final results of the systolic hypertension in the elderly program. *J Am Med Assoc*; 265: 3255, 1991.

35. Cruickshank, JM, Thorp, JM, Zacharias, FJ: Benefits and potential harm of lowering high blood pressure. *Lancet*; 1: 581, 1987.

36. Polese, A, De Cesare, ND, Montorsi, P, Fabbiocchi, F, Guazz, M, Loaldi, A, Guazzi, MD: Upward shift of the lower range of coronary flow autoregulation in hypertensive patients with hypertrophy of the left ventricle. *Circulation*; 83: 845, 1991.

37. Motz, W, Vogt, M, Scheler, S, Strauer, BE: Pharmacotherapeutic effects of antihypertensive agents on myocardium and coronary arteries in hypertension. *Eur Heart J*, Suppl. D.; 13: 100, 1992.

38. Schwartzkopff, B, Motz, W, Strauer, BE: Repair of human myocardial structure by chronic treatment with ACE-inhibitors in hypertensive heart disease. *Circulation*; 90 Suppl. I: 343, 1994.

39. Motz, W, Strauer, BE: Improvement of coronary flow reserve after long-term therapy with enalapril. *Hypertension*; 27: 1031, 1996.

40. Koren, MJ, Deveraux, RB, Casale, PN, Savage, DD, Laragh, JH: Relation of left ventricular mass and geometry to morbidity and mortality in uncomplicated essential hypertension. *Ann Int Med*; 114: 345, 1991.

41. Schiffrin, EC, Deny, LY: Comparison of effects of angiotensin I converting enzyme inhibition and beta-blockade for 2 years on function of small arteries from hypertensive patients. *Hypertension*; 25(part 2): 699, 1995.

42. Vogt, M, Motz, W, Strauer, BE: Antihypertensive Langzeittherapie mit Isradipin. Verbesserung der koronaren Reserve bei Patienten mit arterieller Hypertonie und mikrovaskulärer Angina. *Arzneim.-Forsch/Drug Res.*; 44(4); 12: 1321, 1994.

43. Motz, W, Vogt, M, Scheler, S, Schwartzkopff, B, Kelm, M, Strauer, BE: Prophylaxe mit gefäßaktiven Substanzen. *Z. Kardiol*; 81: Suppl 4, 199, 1992.

44. Motz, W, Vogt, M, Scheler, S, Schwartzkopff, B, Strauer, BE: Verbesserung der Koronarreserve nach Hypertrophieregression durch blutdrucksenkende Therapie mit einem beta-Rezeptorenblocker. *Dtsch Med Wochenschr*; 118: 535, 1993.

45. Magrini, F, Regiani, P, Roberts, N, Meazza, R, Kulla, M, Zanchetti, A: Effects of angiotensin and angiotensin blockade on coronary circulation and coronary reserve. *Am J Med*; 84: Suppl. 3A:55, 1988.

46. Jones, JK, Gorkin, L, Lian, JF, Staffa, JA, Fletcher, AP: Discontinuation of and changes in treatment after start of new courses of antihypertensive drugs: a study of a United Kingdom population. *Br Med J*; 311: 293, 1995.

47. Ramsay, LE, Yeo, WW, Chadwick, IG, Jackson, PR: Diuretics, in Textbook of Hypertension. 1st Edition, Editor: Swales, JD: *Blackwell Scientific Publications*: Oxford, 1994, 1046.

48. Cruickshank, JM, Prickard, BNC: Beta-blockers, in Textbook of Hypertension. 1st Edition, Editor: Swales JD: *Blackwell Scientific Publications*: Oxford, 1994, 1059.

49. Swales, JD: Calcium Antagonists, in Textbook of Hypertension. 1st Edition, Editor: Swales, JD: *Blackwell Scientific Publications*: Oxford, 1994, 1111.

50. Gohlke, P, Unger, T: Angiotensin-converting Enzyme Inhibitors in Textbook of Hypertension. 1st Edition, Editor: Swales, JD: *Blackwell Scientific Publications*: Oxford, 1994, 1115.

51. Panfilov, VV, Reid, JL: Alpha-adrenoceptor antagonists, in Textbook of Hypertension. 1st Edition, Editor: Swales, JD: *Blackwell Scientific Publications*: Oxford, 1994, 1089.

52. Depres, JP, Lamarche, B, Mauriege, P, Cautin, B, Dagenais, GR, Moorjani, S, Lupien, PY: Hyperinsulinemia as an independent risk factor for ischemic heart disease. *N Engl J Med*; 334: 952, 1996.

53. Swales, JD: Antihypertensive drugs and plasma lipids. *Br Heart J*; 66: 409, 1991.

54. Kasiske, BL, Kalil, RS, Louis, TA: Effects of antihypertensive therapy on serum lipids. *Ann Int Med*; 122: 133, 1995.

55. Vyssoulin, GR, Karpassou, EA, Pitsavos, CE, Skoumas, YN, Paleologos, AA, Toutonzas, PK: Differentiation of β-blocker effects on serum lipids and apolipoproteins in hypertensive patients with normolipidaemie or dyslipidemia profiles. *Eur Heart*; 13: 1506, 1992.

56. Fogari, R, Zoppi, A, Malamani, GD, Marasi, G, Vanasia, A, Villa, G: Effects of different antihypertensive drugs on plasma fibrinogen in hypertensive patients. *Br J Clin Pharmacol*; 39: 471, 1995.

57. Skarfors, ET, Lithell, HO, Selincus, S, Aberg, H: Do antihypertensive drugs precipitate diabetes in predisposed men? *Br Med J*; 298:1147, 1989.

58. Bengtsson, C, Blohme, G, Lapidus, L, Lissner, L, Lundgren, H: Diabetes incidence in users and nonusers of antihypertensive drugs in relation to serum insulin, glucose tolerance and degree of adiposity: a 12-year prospective population study of women in Gothenburg, Sweden. *J Int Med*; 231: 583, 1992.

59. Maschio, G, Alberti, D, Janin, J, Locatelli, F: Effect of the angiotensin-converting-enzyme inhibitor benazepril on the progression of chronic renal insufficiency. *N Engl J Med*; 334: 939, 1996.

60. Bohlen, L, De Coutren, M, Weidmann, P: Comparative study of the effect of ACE-inhibitors and other antihypertensive agents on proteinuria in diabetic patients. *Am J Hypertension*; 7: 845, 1994.

61. Huchinson, FN, Webster, SK: Effects of ANG II receptor antagonist on albuminuria and renal function in passive Keymann nephritis. *Am J Physiol*; 32: F311, 1992.

62. Tolins, P, Raij, L: Comparison of converting enzyme inhibitor and calcium channel blocker in hypertensive glomerular injury. *Hypertension*; 16: 452, 1990.

63. Dahlof, B, Lindholm, LH, Hansson, L, Schersten, B, Ekbom, T, Wester, PO: Morbidity and mortality in the Swedish Trial in old patients with hypertension (STOP-Hypertension). *Lancet*; 338: 1281, 1991.

64. Siscovick, DS, Raghunathan, TE, Psaty, BM, Koepsell, TD, Wicklund, KG, Lin, X, Cobb, L, Rautaharju, PM, Copan, MK, Wagner, E: Diuretic therapy for hypertension and the risk of primary cardiac arrest. *N Engl J Med*; 330: 1852, 1994.

65. Hoes, AW, Grobbee, DE, Lubson, J, Man in't Veld, AJ, vd Does, E, Hofman, A: Diuretics, beta-blockers and the risk for sudden cardiac death in hypertensive patients. *Ann Int Med*; 123: 481, 1995.

66. IPPPSH Collaborative Group: Cardiovascular risk and risk factors in a randomized trial of treatment based on the beta-blocker oxprenolol: the international prospective primary prevention study in hypertension (IPPPSH). *J Hypertension*; 3: 379, 1985.

67. Medical Research Council Working Party, MRC trial of treatment of mild hypertension: principal results. *Br Med J*; 291: 97, 1985.

68. Wilhelmsen, L, Berglund, C, Elmfeldt, D, Fitzsimons, T, Holzgreve, H, Hosie, J et al.; Beta-blockers vs. diuretics in hypertensive men: Main results from the HAPPHY trial. *J Hypertension*; 5: 561, 1987.

69. Wikstrand, J, Warnold, I, Olsson, G, Tuomilehto, J, Elmfeldt, D, Berglund, G: Primary prevention with metroprolol in patients with hypertension. Mortality results from the MAPHY study. *J Am Med Assoc*; 259: 1976, 1988.

70. Yusuf, S, Peto, R, Lewis, J: Beta-blockade during and after myocardial infarction: an overview of the randomized trials. *Progr Cardiovasc Dis*; 27: 335, 1985.

71. Psaty, BM, Heckbert, SR, Koepsell, TD, Siscovick, DS, Raghunathan, TE, Weiss NS, Rosendaal, FR, Lemaitre, RN, Smith, NL, Wahl, PW, Wagner, EH, Furberg, CD: The risk of myocardial infarction associated with antihypertensive drug therapies. *J Am Med Assoc*; 274: 620, 1995.

72. Erdmann, E: Herzinsuffizienztherapie mit Calciumantagonisten? *Herz*; 20: Suppl. II: 1, 1995.

73. Yusuf, S, Held, P, Furberg, C: Update of effects of calcium antagonists in myocardial infarction or angina in light of the second Danish Verapamil Infarction Trial (Davit-II) and other recent studies. *Am J Cardiol*; 67: 1295, 1991.

74. Danish Study Group on Verapamil in myocardial infarction, effect on verapamil on mortality and major events after acute myocardial infarction (DA VIT II). *Am J Cardiol*; 318: 385, 1990.

75. SOLVD Investigators: Effect of enalapril on survival in patients with reduced left ventricular ejection fractions and congestive heart failure. *N Engl J Med*; 325: 293, 1991.

76. SOLVD Investigators: Effect of enalapril on mortality and the development of heart failure in asymptomatic patients with reduced left ventricular ejection fractions. *N Engl J Med*; 327: 685, 1992.

77. Packer, M, Bristow, MR, Colin, JN, Colucci, WS, Fowler, MB, Gilbert, EM: Effect of Carvedilol on the survival of patients with chronic heart failure. *Circulation*; 92: Suppl.8:673, 1995.

78. Osswald, H, Mühlbauer, B: The pharmacological basis for the combination of calcium channel antagonists and angiotensin converting enzyme inhibitors in the treatment of hypertension. *J Hypertension*; 13: 521, 1995.

79. Prisant, LM, Weir, MR, Papademetrion, V, Weber, MA, Adegbile, JA, Alemayehu, D, Lefkowitz, MP, Carr, MA: Low-dose drug combination therapy: an alternative first-line approach to hypertension treatment. *Am Heart J*; 130: 359, 1995.

8

Diuretics

Jianguo Zhi
Hoffmann-La Roche, Inc.

8.1 Introduction

Diuretics have been widely used for longer than half a century, frequently in the clinical management of edema and hypertension. Their major pharmacological effect is to increase urine flow, thereby promoting the net loss of water and electrolytes, particularly sodium ions (Na^+) from the body.

In comparison with angiotensin converting enzyme (ACE) inhibitors and calcium channel blockers, diuretics are cost-effective and, most important, they have been proven to reduce the risk of coronary heart disease and stroke in patients with hypertension. Three large-scale, long-term clinical trials in elderly patients with hypertension showed that low-dose diuretic therapy decreased the incidence of cardiovascular diseases, including morbidity and mortality from coronary heart disease.[1-3] Thus, diuretics, along with β-blockers, will remain first-line drugs for treatment of hypertension,[4] particularly in elderly patients,[5] and as a recommended pharmacological treatment of heart failure.[6]

However, the debate continues whether the long-term use of diuretics is safe to warrant their status as the preferred drugs of choice. The following is a summary of the analysis of the benefit/risk ratio for diuretics.

8.2 Mechanisms of Diuretic (Renal) Action

At present, diuretics can be classified by structure and, more adequately, by mechanism of action into four major groups: carbonic anhydrase inhibitors, thiazides and related diuretics, loop (high-ceiling) diuretics, and potassium-sparing diuretics. There are also three fixed-dose combination diuretics of a potassium-sparing diuretic with hydrochlorothiazide, and numerous fixed-dose combinations of other antihypertensive drugs almost exclusively with hydrochlorothiazide. Diuretics available in the U.S. (from

TABLE 8.1   Diuretics for Hypertension

Class	Drug	Daily Adult Dosage (mg)[d]	Onset of Action (h)	Duration of Action (h)	$t_{1/2}$ (h)[f]
Thiazide diuretics	Chlorothiazide[a]	125–500	2	6–12	1.5
(usually once daily)	Hydrochlorothiazide	12.5–25	2	6–12	2.5
	Hydroflumethiazide	12.5–50		6–12	12–27
	Methyclothiazide	2.5–5	2	24	
	Chlorthalidone	12.5–50	2.6	24–72	44
	Indapamide	2.5–5		24–36	10–22
	Metolazone[b]	1.25–5	1	12–24	4–5
	Quinethazone	50–100[e]	2	18–24	
Loop diuretics[a,b]	Bumetanide	0.5–5 in 2 or 3 doses	0.5–1	4	0.3–1.5
	Furosemide	20–320 in 2 doses	1	6–8	0.3–3.4
	Torsemide	5–20 in 1 or 2 doses	1	6–8	0.8–6.0
K+-sparing diuretics	Amiloride[c]	5–10 in 1 or 2 doses	2	24	21
	Triamterene	50–100 in 1 or 2 doses	2–4	7–9	4.2
	Spironolactone	12.5–100 in 1 or 2 doses	1	6–12	1.6

[a] Parenteral formulations are available when rapid onset of action (10–20 min) is needed.
[b] Can be used for renal failure subjects.
[c] Should be administered with food.
[d] From *Med Lett*, 37, 45, 1995. With permission.
[e] From PDR.[7]
[f] From Jackson.[10]
Note: The information on onset and duration of action is extracted from package inserts and literature sources.

1995 PDR[7]) and their basic pharmacodynamic and pharmacokinetic characteristics are listed in Table 8.1, with fixed-dose combinations involving diuretics in Table 8.2. The mechanisms of diuretic (renal) actions for these drugs are well established.[8-10]

TABLE 8.2   Fixed-Dose Combinations Involving Diuretics

Diuretics	Antihypertensive Class	Antihypertensive Drug
Hydrochlorothiazide	Vasodilator	Methyldopa
		Hydralazine
		Guanethidine
	Beta-blocker	Propranolol
		Bisoprolol
		Metoprolol
		Timolol
	ACE inhibitor	Benazepril
		Captopril
		Lisinopril
		Enalapril maleate
	Others	Reserpine–Hydralazine
		Reserpine
Chlorothiazide	Vasodilator	Methyldopa
		Reserpine

From PDR.[7]

8.2.1   Carbonic Anhydrase Inhibitors

These agents inhibit the carbonic anhydrase enzyme, predominantly at the proximal convoluted tubules, causing a reduction in hydrogen ions (H+) available for Na+/H+ exchange. Carbon dioxide (CO_2) reabsorption from the glomerular filtrate is suppressed, and bicarbonate (HCO_3^-) excretion is increased, which carries out sodium, water, and potassium. Increased urinary amounts of Na+, K+, and HCO_3^- result in an alkaline urine.

To maintain ionic balance, chloride (Cl⁻) is retained by the kidneys, resulting in hyperchloremic acidosis. The resulting metabolic acidosis eventually includes a refractory state or a decreased diuresis.

8.2.2 Thiazides and Related Diuretics

The most popular diuretic in this class is hydrochlorothiazide. The diuretic and saluretic effects of hydrochlorothiazide result from a drug-induced inhibition of NaCl reabsorption in the distal convoluted tubule (the primary site of action of thiazide diuretics, whereas the proximal tubule may represent a secondary site of action). The secretion of sodium and chloride and an accompanying volume of water is greatly enhanced; potassium excretion is also enhanced to a variable degree. Thiazide diuretics decrease both renal blood flow and the glomerular filtration rate. Although urinary excretion of bicarbonate is increased slightly, there is usually no significant change in urinary pH. Hydrochlorothiazide has a per mg natriuretic activity ~ 10 times (100 times for methyclothiazide) that of the prototype thiazide, chlorothiazide. However, at maximal therapeutic dosages, all thiazides are approximately equal in their diuretic/natriuretic effects.

All thiazides have antihypertensive properties, and may be used for this purpose either alone or to enhance the action of other antihypertensive drugs. Although the exact mechanism by which thiazides produce a reduction of elevated blood pressure is not known, sodium depletion and blood volume reduction appear to be involved. On the other hand, thiazides do not affect normal blood pressure. In addition to their diuretic effects, sulfonamide diuretics (chlorthalidone, indapamide, metolazone, and quinethazone) act on vascular function to reduce peripheral resistance, with little or no effect on cardiac output, rate, or rhythm.

These sulfonamide diuretics are pharmacologically similar to thiazides, although their duration of action is sufficiently long (24 hours) to allow once a day oral dosing. Chronic administration has little or no effect on glomerular filtration rate or renal plasma flow. They produce urinary excretion of sodium and chloride in approximately equivalent amounts as other thiazides, while potassium is excreted to a much lesser degree.

8.2.3 Loop (High-Ceiling) Diuretics

Furosemide, bumetanide, and torsemide act from within the lumen of the thick ascending limb of the loop of Henle, where they inhibit the Na⁺/K⁺/2Cl-carrier system. Furosemide has additional effects in the proximal tubule; the significance of these effects is unclear. Ethacrynic acid acts on the ascending limb of the loop of Henle and on the proximal and distal tubules. These diuretics increase renal blood flow without increasing the glomerular filtration rate.

8.2.4 Potassium-Sparing Diuretics

Spironolactone is a specific antagonist of the mineralocorticoid, aldosterone, acting primarily through competitive binding of receptors at the aldosterone-dependent Na⁺/K⁺ exchange site in the distal convoluted renal tubule. Spironolactone causes increased amounts of sodium and water to be excreted, while potassium is retained. It may be given alone and, most often, with other diuretic agents that act more proximally in the renal tubule.

Triamterene and amiloride inhibit the reabsorption of sodium ions in exchange for potassium and hydrogen ions at that segment of the distal tubule under the control of adrenal mineralocorticoids (especially aldosterone). This activity is not directly related to aldosterone secretion or antagonism; it is rather a result of a direct effect on the renal tubule. Since the fraction of filtered sodium reaching this distal tubular exchange site is relatively small, and the amount that is exchanged depends on the level of mineralocorticoid activity, the degree of natriuresis and diuresis produced by inhibition of the exchange mechanism is limited. Increasing the amount of available sodium and the level of mineralocorticoid activity by the use of more proximally active diuretics will increase the degree of diuretic action and potassium conservation.

TABLE 8.3 Therapeutic Uses of Diuretics

Thiazides	Loop Diuretics	K⁺-Sparing Diuretic Combinations	Other Combinations Involving Diuretics
Hypertension	Edematous disorders	Hypertension	Hypertension
Edematous disorders	Hyponatremia[a]	Edematous disorders	Edematous disorders
Osteoporosis	Acute renal failure	Hypokalemia	
Hypercalcemia	Fluid retention in cirrhosis	Hypomagnesemia	
Diabetes insipidus	Hypercalcemia		
	Renal tubular acidosis		
	Hypertension with renal insufficiency		

Extracted from package inserts and other literature sources.
[a]When combined with hypertonic saline.

Potassium-sparing diuretics occasionally cause increases in serum potassium, which can result in hyperkalemia. They do not produce alkalosis because they do not cause excretion of titratable acid and ammonium.

8.3   Therapeutic Uses

Diuretic and nondiuretic therapeutic uses of diuretics are summarized in Table 8.3. Key therapeutic uses for diuretics are two: hypertension and edema.

8.3.1   Hypertension

The link between renal disease and elevated blood pressure is the reason that diuretic therapy continues to be included in many treatment regimens. Diuretics are effective at lowering diastolic blood pressure by 5 to 6 mmHg on long-term (5 years) treatment in mild to moderate hypertension. In prospective observational studies, a long-term difference of 5 to 6 mmHg in usual diastolic blood pressure is associated with about 35% to 40% less stroke and 20% to 25% less coronary heart disease.[11]

Long-term therapy with diuretics decreases peripheral vascular resistance by depleting body sodium stores and reducing blood volume. Thiazides and related diuretics are the mainstay of antihypertensive diuretic therapy. They are indicated in the management of hypertension, either as the sole therapeutic agent to provide relief for mild and moderate hypertension or to enhance the effect of other antihypertensive drugs (e.g., sympatholytic or vasodilating agents) in the more severe forms of hypertension. Many thiazide-type diuretics are used to treat hypertension; hydrochlorothiazide and chlorthalidone are the most widely used. Many patients, particularly older ones, can be treated with small doses of diuretics equivalent to 12.5 mg to 25 mg of hydrochlorothiazide once daily. Doses as low as 6.25 mg are now used to enhance the effectiveness of other drugs while minimizing adverse effects. Metolazone and indapamide may be effective in patients with impaired renal function when thiazides are not.

Although thiazides are preferable to loop diuretics for the treatment of hypertension, loop diuretics might be needed to treat hypertension in patients with renal insufficiency (creatinine clearance less than 40 mL/min). In patients with normal renal function, loop diuretics are no more effective than thiazides for treatment of hypertension.

Potassium-sparing diuretics alone do not result in a great reduction in blood pressure and are not additive to the antihypertensive effects of thiazides. Consequently, the major therapeutic use of these drugs relates to their potassium-sparing properties.[12]

8.3.2   Edematous Disorders

While it is possible that diuretics lower blood pressure at doses that have little or no diuretic effect, daily adult dosage for edemas, usually two to three times higher than for hypertension, will produce pronounced diuretic actions.

By decreasing fluid and sodium retention and reducing ventricular preload by reducing filling pressure, thereby decreasing left ventricular volume and wall tension (low oxygen demand), diuretics relieve symptoms in patients with chronic heart failure. The success of diuretic treatment of edema in congestive heart failure is dependent on accompanying moderate restriction of sodium intake.[13] Diuretics that act on the loop of Henle, such as bumetanide or furosemide, as sole therapy, are more effective for treatment of heart failure than thiazide diuretics, such as hydrochlorothiazide, which act on the distal tubule.

Loop diuretics (bumetanide, furosemide, and torsemide) are indicated for the treatment of edema associated with congestive heart failure, chronic renal failure, and renal disease, including the nephrotic syndrome or hepatic disease. The intravenous injection is indicated when a rapid onset of diuresis is desired or when oral administration is impractical. The intensity of the diuretic effect of loop diuretics is primarily related to the urinary excretion rate, rather than to the blood concentration, of the diuretic.[14]

In some heart failure patients who have increased proximal tubular reabsorption of sodium, loop diuretics are limited in their efficacy because less sodium is delivered to the loop. Coadministration with a carbonic anhydrase inhibitor may cause a clinically significant diuresis because of the inhibition of sodium reabsorption in the proximal tubule by the carbonic anhydrase inhibitor, which was not possible with maximally tolerated doses of loop diuretics. Thus, theoretically, carbonic anhydrase inhibitors can be used as a diuretic in the few patients with severe edematous disorders who are poorly responsive or refractory to large doses of potent loop diuretics.[9]

Thiazides are indicated as adjunctive therapy for edema associated with congestive heart failure, hepatic cirrhosis, and corticosteroid and estrogen therapy. They are also employed for acute pulmonary edema (e.g., associated with acute left ventricular heart failure). In addition, they have been found useful for treatment of edema due to various forms of renal dysfunction such as the nephrotic syndrome, acute glomerulonephritis, and chronic renal failure.

Spironolactone is indicated for the management of edematous conditions for patients with congestive heart failure, cirrhosis of the liver accompanied by edema and/or ascites, and the nephrotic syndrome. Triamterene and amiloride are limited in their use for the clinical management of disorders involving abnormal fluid distribution, such as congestive heart failure, cirrhosis of the liver, and the nephrotic syndrome; also in steroid-induced edema, idiopathic edema, and edema due to secondary hyperaldosteronism.

8.3.3 Nondiuretic Uses

At present, carbonic anhydrase inhibitors are still used with moderate success for the following purposes: (1) as a systemic treatment of choice for glaucoma, reducing the rate of aqueous humor formation, (2) in refractory cases of petit mal epilepsy, where they act as an anticonvulsant and decrease the rate of spinal fluid formation (the mechanism of action of this therapeutic effect is unclear), and (3) to alkalinize the urine in the treatment of salicylate or barbiturate poisoning and in combination with bicarbonate to maintain electrolyte balance.

Thiazide diuretics are effective for prevention of renal calcium stones caused by hypercalciuria. Other potential indications include osteoporosis, e.g., corticosteroid-induced osteoporosis. There is some biological evidence to suggest that thiazide diuretics reduce bone loss by decreasing urinary calcium excretion, resulting in a net positive calcium balance. The increased bone density appears to be protective against hip fracture in elderly patients. Using the technique of metaanalysis, Jones et al.[15] concluded that current thiazide users have a 20% reduction in fracture risk and that long-term use may reduce fractures by a similar amount. Thus, they recommended that thiazide diuretics be considered as part of an approach to osteoporotic fracture prevention, particularly in hypertensive subjects.

Potassium-sparing diuretics are important drugs in treating primary aldosteronism and they are perhaps most commonly used with thiazide and loop diuretics to prevent or correct hypokalemia, especially when administered in the presence of cardiac glycosides. The most rational use of potassium-sparing diuretics is in patients who have actually become hypokalemic or hypomagnesemic rather than as prophylaxis for such electrolyte imbalances.[9] These drugs are more effective at attaining and maintaining homeostasis of these electrolytes than are exogenous supplements of the ions.[12]

TABLE 8.4   Frequent or Severe Adverse Effects of Diuretics

Thiazides	Loop Diuretics	K+-Sparing Diuretics
Hyperuricemia	Dehydration	Hyperkalemia
Hypokalemia	Circulatory collapse	Gastrointestinal disturbances
Hypomagnesemia	Hypokalemia	Rash[a, c]
Hyperglycemia	Hyponatremia	Headache[a]
Hyponatremia	Hypomagnesemia	Nephrolithiasis[b]
Hypercalcemia	Hyperglycemia	Hyponatremia[c]
Hypercholesterolemia	Metabolic alkalosis	Mastodynia[c]
Hypertriglyceridemia	Hyperuricemia	Gynecomastia[c]
Pancreatitis	Blood dyscrasias	Menstrual abnormalities[c]
Sexual dysfunction	Hypercholesterolemia	
Photosensitivity reactions	Hypertriglyceridemia	
May decrease excretion of lithium		
Rashes and other allergic reactions		

From *Med Lett*, 37, 45, 1995. With permission.
[a] Occurring to amiloride.
[b] Occurring to triamterene.
[c] Occurring to spironolactone.

8.4   Clinical Toxicities

Class adverse reactions of diuretics are summarized in Table 8.4, most of which are related to diuretic pharmacological effects, i.e., altered electrolyte balance and increased fluid loss. As with many drugs, idiosyncratic and allergic reactions may also occur during diuretic treatment.

8.4.1   Electrolyte Imbalance

Severe disturbances of potassium homeostasis (hypo- and hyperkalemia) can cause life-threatening cardiac complications, including cardiac arrhythmias and sudden death.[12] Diuretics remain the most important cause of clinically significant alteration in serum potassium levels.[16] Reduction in serum K+ levels (hypokalemia) is one of the most important adverse effects of many diuretics, particularly thiazide and loop diuretics. Those at risk include the elderly, women, patients with edematous states, and patients in whom higher doses and/or the more potent agents are used. In contrast, potassium-sparing diuretics may cause hyperkalemic metabolic acidosis, particularly in patients with renal insufficiency who already have a predisposition to hyperkalemia, or those who are taking K+ supplements, other drugs that decrease aldosterone secretion, or ACE inhibitors.

Large doses of thiazide and loop diuretics given more than once a day for long periods of time could induce negative magnesium balance and magnesium deficiency,[17] which predisposes the patient to arrhythmias and muscle weakness. It was recommended that if large doses of diuretics are used, magnesium levels should be checked and oral supplementation given when necessary.[6]

Severe hyponatremia is a disorder with a high mortality. Hyponatremia is a particularly dangerous side effect of thiazide use, most often seen in women, especially at high doses.[18]

8.4.2   Cardiac Arrest

Evidenced by a direct relation between the dose of thiazide and the odds-ratio of primary cardiac arrest, a case-control study found that use of diuretics for hypertension was associated with an increased risk of sudden death.[19] This increased incidence of sudden death has been postulated to be due to hypokalemia that could directly predispose the patient to ventricular arrhythmias and/or sudden death.[20] It is of interest to note that the use of non-potassium-sparing diuretics and β-blockers is equally associated with an increased risk, compared with a reference group treated primarily with potassium-sparing diuretics, for sudden death in hypertensive patients.[21] The risk for sudden cardiac death among recipients of non-potassium-sparing diuretics was more pronounced in those who had been receiving the diuretic for less than 1 year and in those aged 75 years or younger. A reduced risk, however, was identified among patients taking low-dose thiazide plus a potassium-sparing diuretic.[19]

8.4.3 Ototoxicity

High doses of loop diuretics may cause transient deafness in patients with renal failure, particularly if they are concomitantly treated with other ototoxic drugs such as aminoglycoside antibiotics.[9] Because of the ototoxic potential, the use of loop diuretics should also be avoided as much as possible in premature infants and neonates.

8.4.4 Allergic Reactions

Structurally, carbonic anhydrase inhibitors and thiazide diuretics are sulfonamides, which can cause blood dyscrasias and allergic skin reactions. Photoallergy appears to be the mechanism for thiazide-induced photosensitivity.[22] Noncardiogenic pulmonary edema appears to be an idiosyncratic reaction that occurs with some specificity with the thiazide diuretics.[23]

8.4.5 Metabolic Complications

Thiazide diuretics can result in insulin resistance, and insulin secretion may be inhibited, possibly associated with an induction of hypokalemia. Thiazides induce a short-term increase in serum cholesterol, low-density lipoprotein, and triacylglycerols, and cause a slight reduction in high-density lipoprotein. Long-term treatment with thiazides shows that adverse pharmacologic effects on lipid metabolism persist for many years.[24] However, in low doses, diuretic therapy has no long-term adverse effects on cholesterol and glucose level or quality of life.[25] Furthermore, it is of interest to note that although diuretics can worsen these coronary risk factors, which could theoretically offset the benefit of lowering blood pressure, they have significantly reduced the frequency of clinical myocardial infarction and stroke in the elderly patient with isolated systolic hypertension.[1-3] Nevertheless, if drug therapy for hypertension is required in obese and diabetic patients, diuretics must be used carefully, e.g., only low doses should be allowed, because of their potential metabolic effects.

8.5 Benefit/Risk Ratio

The goal of drug therapy is clinical efficacy achieved with relative safety. Diuretic therapy is justified only if the possible benefits outweigh the possible risks after considering the qualitative and quantitative impact of using a diuretic and the likely outcome if the diuretic is withheld. This decision depends on adequate clinical knowledge of the patient, of the disease and its natural history, and of the drug and its efficacy and potential adverse effects. As described above, for diuretics, clinical benefits and risks are well understood. The benefit/risk ratio varies with the type of diuretic, the indication and patient population, and it is affected by cost-effectiveness and availability of other treatments. To best use diuretics, several plausible approaches to maximizing their benefit/risk ratio are in place.

8.5.1 Dose Optimization

The therapeutic index, reflected by the ratio of the intensities of therapeutic to adverse effects, may not be a constant but rather be dependent on drug dose. For example, a drug may produce several different types of effect (either therapeutic or toxic), each with its own characteristic dose-intensity of effect relationship. Such a case is simulated[26] in Figure 8.1 (left panel); the three different relative intensities of effect vs. dose curves can be described by the Hill or sigmoid E_{max} equation:

$$E = \frac{E_{max} \cdot D^S}{ED_{50}^S + D^S}$$

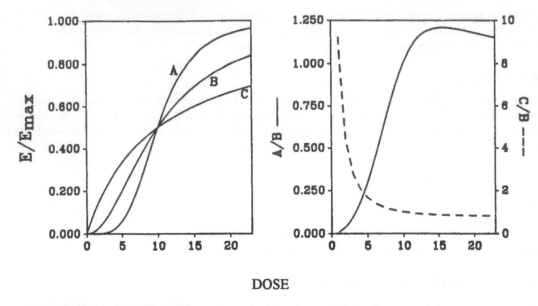

DOSE

FIGURE 8.1   Effect of the Hill equation exponent S on the relationship between dose and the ratio of the relative intensities of different pharmacologic effects of a hypothetical drug. *Left:* Simulated relative effect intensity vs. dose profiles of a drug having three different types of effect (A, B, and C). The Hill equations describing these relationships have the same ED_{50} (10 dose units) but different S values: four (for A), two (for B), and one (for C). *Right:* Relationship between the drug dose and the ratio A/B and C/B, respectively. From *Pharm Res*, 7, 697, 1990. With permission.

with the same ED_{50} but different values for the exponent S. In the equation, E is the intensity of the effect, E_{max} is the maximum attainable effect, D is the drug dose, ED_{50} is the drug dose that elicits one-half the intensity of the maximum effect, and S is a constant that defines the sigmoidicity of the relationship. The right panel in Figure 8.1 shows the ratio of the intensities of effects A and B and also the ratio of the intensities of effects C and B as a function of drug dose. Were one to define B as the therapeutic effect, and A and C as adverse effects, then it is evident that (1) the therapeutic index is drug dose dependent and (2) this dose dependence relates to differences in the relevant dose-intensity of effect relationship.

The left panel of Figure 8.2 depicts simulations of three other relative intensities of effect vs. drug dose curves, differing only in the value of ED_{50}. The drug dose dependency ratios A/B and C/B are illustrated in the right panel of Figure 8.2. Obviously, there can also be cases in which ED_{50} and S change concurrently and the special case where the ratio of the intensity of pharmacologic effect does not change with the drug dose (i.e., the impact of a change in E_{max} on the intensity of effect vs. drug dose relationship with ED_{50} and S being constant).

The simulation indicates that an acceptable benefit/risk ratio simply requires a proper selection of a drug dose range in which an optimized therapeutic index can be achieved. Effective utilization of this capability requires a thorough characterization and understanding of the relationship between drug dose and the intensity of therapeutic as well as adverse effects. Diuretics can be administered in single or divided doses, daily at high rates over short periods of time, or daily at lower rates over longer periods of time as long as the total diuretic effect per day is adequate. Thus, there is an opportunity to optimize the therapeutic index of diuretics (achieving a high ratio of Na^+ and urine excretion to K^+ excretion) by administering these drugs singly or in combinations at an appropriate rate.[26]

8.5.2   Combination Therapy

Judicious use of diuretic combinations may increase benefits and reduce risks, e.g., the well-known use of antikaliuretics with a thiazide. Thiazide diuretics have a steep dose–response curve that plateaus at

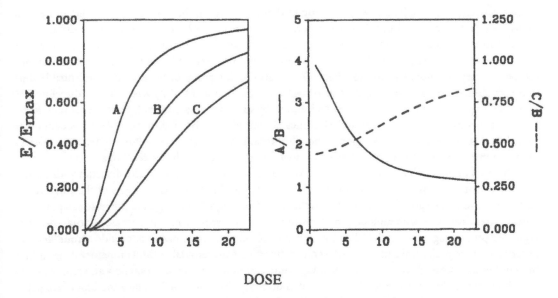

DOSE

FIGURE 8.2 Effect of ED_{50} value of the Hill equation on the relationship between dose and the ratio of the relative intensities of different pharmacologic effects of a hypothetical drug. See the caption to Figure 8.1 for additional information. The S value is 2 in all cases and ED_{50} is 5 (for A), 10 (for B), and 15 (for C) dose units. From *Pharm Res*, 7, 697, 1990. With permission.

low doses. At high doses, there is little additional antihypertensive benefit to be gained, but the occurrence of adverse effects increases markedly; thus, the benefit/risk ratio of increasing thiazide dosage in the higher ranges becomes increasingly unfavorable.[9] Similarly, adverse effects of potassium-sparing diuretics diminish the benefit/risk ratio of the monotherapy, limiting indications or narrowing the spectrum of activity for their use in rare disorders or in exceptional circumstances.

However, the combination of two diuretics with different but complementary mechanisms and sites of action, e.g., spironolactone/hydrochlorothiazide, provides additive diuretic and antihypertensive effects. At the same time, the spironolactone component helps to minimize hypokalemia and alkalosis characteristically induced by the thiazide component.

Treatment with hydrochlorothiazide or chlorothiazide with ACE inhibitors or other antihypertensive drugs (Table 8.2) also carries no risk of hypokalemia, most likely due to the very low dose of diuretics required for (antihypertensive) action and the associated potassium-sparing effect of ACE inhibitors.[27] The combinations of two diuretics (e.g., a thiazide plus a loop diuretic) are used to obtain additive or even synergistic natriuretic effects in patients who respond poorly to single agent therapy.[9,28]

Thiazide diuretics are available in fixed-dose combinations with potassium-sparing diuretics and with other antihypertensive drugs. Fixed-dose combination products can be advantageous,[27] but the dosage of each drug should be titrated separately. When the optimal maintenance doses correspond to the ratio in a combination product, taking fewer tablets may improve convenience and subsequent compliance.[9,29]

8.5.3 Individualization

Since drug therapy is more effective when individualized according to patient response, diuretics should be titrated to gain maximal response with the minimal dose possible to maintain that therapeutic response. Likewise, individual risk factors should be considered when deciding what ideal diuretics[30] and best dosing regimens should be used for the patient.[9] For example, in what ways might an adjustment of diuretic dosage regimen be accomplished in patients with renal failure, in which half-lives of loop diuretics are prolonged? Since delivery of loop diuretics to the lumen of the nephron, the site of action, is diminished in severe renal insufficiency, renal failure may reduce diuretic responsiveness. An increase in loop diuretic dose would deliver more drug to the site of action; this dose increase could also elevate ototoxicity and/or other nonrenal toxicities because of increased systemic exposure to the drug.

Certain populations (e.g., blacks) are more likely to respond to diuretics, making diuretics the first-choice agents in that population.[9] The elderly have shown the greatest benefit from diuretic antihypertensive therapy because of a favorable benefit/risk ratio in this group of hypertensive patients;[1-3,5] thus, diuretics should be the first drug class to consider for the treatment of hypertension in the elderly.[5,31] However, in young subjects, coronary risks induced by diuretics may outweigh the benefits to a degree sufficient to discourage diuretic usage. Diuretics should also be avoided, if possible, in patients with preexisting glucose intolerance, lipoprotein abnormalities, allergy to a sulfonamide, hypovolemia, hyponatremia, or gout due to potential aggravated diuretic-induced adverse reactions.

Individual patients who are concurrently taking other medications should take diuretics with caution because of the potential for a drug interaction. The K^+- and Mg^{2+}-depleting effects of the thiazide-like and loop diuretics can potentiate arrhythmias that arise from digitalis toxicity, and they also can potentiate ototoxicities associated with aminoglycoside antibiotics. Nonsteroidal antiinflammatory drugs, β-adrenergic receptor antagonists, and ACE inhibitors reduce plasma concentrations of aldosterone and can potentiate the hyperkalemic effects of K^+-sparing diuretics. Nonsteroidal antiinflammatory drugs inhibit the synthesis of prostaglandins or block prostaglandin secretion, limit renal blood flow and renal dilation, and thus reduce the antihypertensive effects of diuretics. Triamterene will reduce the bioavailability of concomitant thiazide diuretics. Thiazide diuretics should not be given concomitantly with lithium since they reduce its renal clearance by about 25%.

Many diuretics cross the placental barrier and appear in cord blood. The use of diuretics in pregnant women requires that the anticipated maternal benefits be weighed against the possible hazards to the fetus. Thus, the routine use of diuretics in an otherwise healthy woman is inappropriate. Diuretics also appear in breast milk; if use of a diuretic is deemed essential, the patient should stop nursing.

Finally, in some hypertensive patients, drugs other than diuretics may be given as primary therapy. For example, β-blockers should be given in patients with prior myocardial infarction and ACE inhibitors should be given in hypertensive patients with diabetes, especially diabetic nephropathy, and in patients with congestive heart failure.[32]

8.6   Concluding Remarks

It has been conclusively demonstrated that the risk of stroke, cardiovascular disease, and total mortality was significantly reduced in the diuretic-treated group compared to placebo. Diuretics are easy to take, cost-effective, and relatively safe antihypertensive drugs, when the dose is kept low. The choice of a diuretic depends on knowledge of its efficacy in different disease states as well as the complications associated with its use. Additionally, means for achieving greater selectivity in pharmacologic action and thereby for increasing the efficiency and safety of diuretic drugs are available: (1) selection of an appropriate diuretic drug dose, (2) use of drug combinations, and (3) individualization of therapy.

Acknowledgment

The author thanks Dr. Grzegorz S. Sedek and Ms. Angela T. Melia for their critical comments.

References

1. Dahlöf, B., Lindholm, L. H., Hansson, L., Scherstén, B., Ekbom, T., Wester, P-O., Morbidity and mortality in the Swedish Trial in Old Patients with Hypertension (STOP-Hypertension), *Lancet*, 338, 1281, 1991.
2. SHEP Cooperative Research Group, Prevention of stroke by antihypertensive drug treatment in older persons with isolated systolic hypertension: final results of the Systolic Hypertension in the Elderly Program (SHEP), *JAMA*, 265, 3255, 1991.

3. MRC Working Party, Medical Research Council trial of treatment of hypertension in older adults: principal results, *Br Med J*, 304, 405, 1992.

4. Psaty, B. M., Heckbert, S. R., Koepsell, T.D., Siscovick, D. S., Raghunathan, T. E., Weiss, N. S., Rosendaal, F. R., Lemaitre, R. N., Smith, N. L., Wahl, P. W., Wagner, E. H., Furberg, C. D., The risk of myocardial infarction associated with antihypertensive drug therapies, *JAMA*, 274, 620, 1995.

5. Lever, A. F., Ramsay, L. E., Treatment of hypertension in the elderly, *J Hypertension*, 13, 571, 1995.

6. Baker, D. W., Konstam, M. A., Bottorft, M., Pitt, B., Management of heart failure: I. Pharmacologic treatment, *JAMA*, 272, 1361, 1994.

7. *Physicians' Desk Reference*, 49th Edition, Medical Economics, Montvale, NJ, 1995, Section 3.

8. Rose, B. D., Diuretics, *Kid Int*, 39, 336, 1991.

9. Brater, D. C., Treatment of renal disorders and the influence of renal function on drug disposition, in *Clinical Pharmacology: Basic Principles in Therapeutics*, Third Edition, Melmon, K. L., Morrelli, H. F., Hoffman, B. B., Nierenberg, D. W., eds., McGraw-Hill, New York, 1992, Chap. 11.

10. Jackson, E. K., Diuretics, in Goodman & Gilman's *The Pharmacological Basis of Therapeutics*, Ninth Edition, Hardman, J. G., Limbird, L. E., Molinoff, P. B., Ruddon, R. W., Gilman, A. G., eds., McGraw-Hill, New York, 1996, chap. 29.

11. Collins, R., Reto, R., McMahon, S., Hebert, P., Fiebach, N. H., Eberlein, K. A., Godwin, J., Qizilbash, N., Taylor, J. O., Hennekens, C. H., Blood pressure, stroke, and coronary heart disease. Part 2, Short-term reductions in blood pressure; overview of randomized drug trials in their epidemiological context, *Lancet*, 335, 827, 1990.

12. Saggar-Malik, A. K., Cappuccio, F. P., Potassium supplements and potassium-sparing diuretics: a review and guide to appropriate use, *Drugs*, 46, 986, 1993.

13. Cody, R. J., Kubo, S. H., Pickworth, K. K., Diuretic treatment for the sodium retention on congestive heart failure. *Arch Int Med*, 154, 1905, 1994.

14. Brater, D. C., Pharmacodynamic considerations in the use of diuretics, *Annu Rev Pharmacol Toxicol*, 23, 45, 1983.

15. Jones, G., Nguyen, T., Sambrook, P. N., Eisman, J. A., Thiazide diuretics and fractures: can meta-analysis help?, *J Bone Mineral Res*, 10, 106, 1995.

16. Howes, L. G., Which drugs affect potassium?, *Drug Safety*, 12, 240, 1995.

17. Davies, D. L., Fraser, R., Do diuretics cause magnesium deficiency?, *Br J Clin Pharmacol*, 36, 1, 1993.

18. Sonnenblick, M., Friedlander, Y., Rosin, A. J., Diuretic-induced severe hyponatremia: review and analysis of 129 reported patients, *Chest*, 103, 601, 1993.

19. Siscovick, D. S., Raghunathan, T. E., Psaty, B. M., Koepsell, T. D., Wicklund, K. G., Lin, X., Cobb, L., Rautaharju, P.M., Copass, M. K., Wangner, E. H., Diuretic therapy for hypertension and the risk of primary cardiac arrest, *N Engl J Med*, 330, 1852, 1994.

20. Hoes, A. W., Grobbee, D. E., Peet, T. M., Lubsen, J., Do non-potassium-sparing diuretics increase the risk of sudden cardiac death in hypertensive patients? Recent evidence, *Drugs*, 47, 711, 1994.

21. Hoes, A. W., Grobbee, D. E., Lubsen, J., Man in 't Veld, A. J., van der Does, E., Diuretics, β-blockers, and the risk for sudden cardiac death in hypertensive patients, *Ann Intern Med*, 123, 418, 1995.

22. Diffy, B. L., Langtry, J., Phototoxic potential of thiazide diuretics in normal subjects, *Arch Dermatol*, 125, 1355, 1989.

23. Fine, S. R., Lodha, A., Zoneraich, S., Mollura, J. L., Hydrochlorothiazide-induced acute pulmonary edema, *Ann Pharmacother*, 29, 701, 1995.

24. Lind, L., Pollare, T., Berne, C., Lithell, H., Long-term metabolic effects of antihypertensive drugs, *Am Heart J*, 128, 1177, 1994.

25. Neaton, J. D., Grimm, R. H., Jr., Prineas, R. J., Stamler, J., Grandits, G. A., Elmer, P. J., Cutler, J. A., Flack, J. M., Shoenberger, J. A., McDonald, R., Lewis, C. E., Liebson, P. R., Treatment of mild hypertension study: final results, *JAMA*, 270, 713, 1993.

26. Zhi, J., Levy, G., Optimization of the therapeutic index by adjustment of the rate of drug administration or use of drug combinations: exploratory studies of diuretics, *Pharm Res*, 7, 697, 1990.

27. Opie, L. H., Fundamental role of angiotensin-converting enzyme inhibitors in the treatment of congestive heart failure, *Am J Card*, 75, F3, 1995.
28. Mouallem, M., Brif, I., Mayan, H., Farfel, Z., Prolonged therapy by the combination of furosemide and thiazides in refractory heart failure and other fluid retaining conditions, *Int J Card*, 50, 89, 1995.
29. Sica, D. A., Fixed dose combination antihypertensive drugs, *Drugs*, 48, 16, 1994.
30. Swanepoel, C. R., Which diuretic to use?, *Cardiovasc Drugs Ther*, 8, 123, 1994.
31. Cushman, W. C., Optimising diuretic therapy in elderly patients with hypertension, *Drugs Aging*, 7, 88, 1995.
32. Freis, E. D., The efficacy and safety of diuretics in treating hypertension, *Ann Intern Med*, 122, 223, 1995.

9

Antidiabetic Drugs

Pietro Compagnucci
Hospital of Camerino

Fausto Santeusanio
University of Perugia

9.1   Introduction

Diabetes mellitus (DM), a syndrome characterized by reduced insulin secretion and/or action, results in glucose intolerance, a subtle derangement of carbohydrate metabolism, and eventually in fasting hyperglycemia. Impairments in protein and fat metabolism are also present, and after a prolonged period of clinically overt or asymptomatic hyperglycemia chronic complications may become evident. Including microangiopathy (retinopathy, nephropathy), neuropathy, and macrovascular involvement, they are responsible for most of the morbidity and mortality of the disease. In developed countries an estimated 3% to 4% of the population is affected by DM.

Idiopathic DM is classified as type 1, insulin-dependent DM (IDDM), and non-insulin-dependent DM (NIDDM, 80% to 90% of all diabetic individuals) type 2[1,2] (Table 9.1). Secondary diabetes may be associated with pancreatic or endocrine diseases, or may be induced by drugs or chemicals (Table 9.2).

In IDDM insulin production by the β-cells of the islets of Langerhans is absent or severely deficient. Marked hyperglycemia is associated with glycosuria, ketonuria, polyuria, excessive thirst, fatigue, and weight loss. Severe decompensation that may lead to ketoacidotic coma may ensue, unless exogenous insulin is administered.

NIDDM is characterized by "insulin resistance," which prompts the pancreas to secrete more insulin, thus hyperinsulinemia is present. Insulin secretion is preserved except in advanced stages of the disease and insulin administration is not essential to maintain life. Patients may be asymptomatic; diagnosis is often made following routine urine or blood glucose tests revealing glycosuria and/or hyperglycemia.

TABLE 9.1   Characteristic Features of Type 1 (Insulin-Dependent) and Type 2
(Non-Insulin-Dependent) Diabetes Mellitus

	IDDM	NIDDM
Age of onset	Youth	Usually over age 35
Type of onset	Usually acute	Insidious
Body weight	Weight loss at onset	Usually obese
Symptoms	Thirst, polyuria, polydipsia	Absent or mild
Ketosis	Present	Absent
Endogenous insulin	Absent or minimal	Frequent hyperinsulinemia
Insulin treatment	Essential	Optional
Sulfonylureas	Not efficacious	Efficacious
Diet	Important	Essential (may suffice alone)
Vascular and neurologic complications	Usual after 5 or more yrs of disease	Frequent, may be present at onset
Islet cell antibodies	Present	Absent
Family history	Positive in 10%	Positive in 30%
Concordance in identical twins	50% concordance	Nearly 100% concordance

TABLE 9.2   Some of the Secondary Forms of Diabetes

1. Pancreatic disorders
   * Pancreatectomy
   * Pancreatitis
   * Hemochromatosis
2. Endocrinopathies
   * Glucagonoma
   * Cushing's syndrome
   * Acromegaly
   * Pheochromocytoma
   * Hyperthyroidism
3. Drugs or chemicals
   * Thiazide diuretics
   * Glucocorticoids
   * Oral contraceptives
   * Phenytoin
4. Genetic syndromes
   * Cystic fibrosis
   * Leprechaunism
   * Prader-Willi syndrome
   * Laurence-Moon-Biedle syndrome

Atherosclerotic heart disease, peripheral or cerebrovascular disease, or symptoms related to microangiopathy may draw attention to a previously undiagnosed type 2 DM.

Frequently associated with NIDDM, obesity contributes to insulin resistance. Obesity and NIDDM are both characterized by hyperinsulinemia. However, after glucose ingestion, insulin levels, though elevated, are lower in the diabetic obese than in the nondiabetic obese. It has recently been suggested that hyperglycemia and hyperinsulinemia are causally related to hypertension and dyslipidemia, thus explaining the high prevalence of atherosclerotic heart disease in NIDDM patients.[3]

Therapy for DM is based on diet, physical exercise, and hypoglycemic agents: insulin in IDDM and oral hypoglycemic drugs or insulin in NIDDM.

9.2   Diet and Physical Exercise in Diabetes Management

The diet of IDDM patients requires the ingestion of carbohydrate-containing food at specific time intervals matching with blood levels of injected insulin, in order to avoid wide fluctuations in blood glucose concentration and hypoglycemia. The absorption of ingested food must be synchronized with the time-activity curve of administered insulin. NIDDM patients are often obese and require a low-calorie diet to control weight and blood glucose.[4,5]

Physical activity should be encouraged. In IDDM subjects exercise programs improve physical fitness, psychological well-being, and social interactions. In NIDDM individuals physical exercise, along with diet and/or drug therapy, may improve glycemic control, reduce cardiovascular risk factors, and improve psychological well-being. Both IDDM and NIDDM patients who are about to start exercising should be carefully screened for retinopathy, neuropathy, nephropathy, hypertension, or silent myocardial ischemia, all of which may be exacerbated by physical activity. The exercise program should then be adjusted accordingly, and the patient should monitor glycemic response to exercise[6] to avoid hypoglycemic episodes.

In NIDDM patients diet and physical exercise, when correctly applied, may delay the use of oral hypoglycemic drugs or insulin for many years after diagnosis.

9.3   Insulin

Insulin is essential therapy for type 1 (insulin-dependent) DM and it may be required in type 2 (non-insulin-dependent) DM if blood glucose levels are not adequately controlled by diet and/or oral hypoglycemic agents. Insulin is also used in gestational diabetes, in the management of diabetic ketoacidosis and hyperosmolar nonketotic coma, in the perioperative period for both IDDM and NIDDM patients undergoing surgery, and in postpancreatectomy diabetes.

Insulin must be injected parenterally. Oral insulin administration, possibly providing more physiological portal concentration, is at present not feasible. In fact, the insulin peptide molecule is digested by the intestinal proteases. Encapsulation or incorporation of insulin into lipoproteins to form liposomes have proved unsatisfactory. Intranasal and rectal insulin administration are also unsuitable at present, although the plasma levels following the intranasal route may come close to the intravenous injection kinetic.

Insulin may be injected subcutaneously with syringes, pen injectors, pumps, subcutaneous injection ports, or jet injectors. Pen injectors offer clear advantages in multiple daily injection regimens because they are convenient and easily carried. Both short- and intermediate-acting insulin preparations are available in pen-injector cartridges, which are often more reliable than syringes. Continuous subcutaneous insulin infusion (CSII) by means of insulin infusion devices (pumps) has the main advantage of providing a constant basal plasma insulin level, which, like basal secretion of the normal pancreatic β-cells, suppresses hepatic glucose output, thus normalizing interprandial blood glucose levels. Boluses of insulin are delivered before meals to match the ingested nutrients. CSII uses short-acting insulin and is characterized by no subcutaneous insulin depot. Ketoacidosis may develop if therapy is accidentally interrupted, e.g., by pump failure, dislodgement of needle, catheter leakage, or any other reason. Selection of patients is fundamental for successful CSII therapy. Although blood glucose control is not clearly better than with a multiple injection regimen,[7] the incidence of severe hypoglycemia during long-term CSII therapy is reported to be significantly reduced compared with multiple daily insulin injections.[8]

The absorption kinetics of subcutaneously injected insulin is influenced by several factors, including type of insulin, site of injection, regional blood flow at the site of injection, volume and concentration of the injected dose. Inserting an insulin needle perpendicularly through the skin of a lean individual probably results in intramuscular injection and abnormally rapid absorption. Massage over the injection site accelerates the absorption of regular insulin more by dissociation of insulin hexamers than by increasing local blood flow. Absorption is most rapid when insulin is injected in the subcutaneous tissue of the abdominal wall. When short-acting insulin is mixed with intermediate-acting insulin containing zinc ions, some of the rapidly acting insulin is modified by the excess zinc, but this is not the case with protamine-containing intermediate-acting insulin. It must be remembered, however, that there is great intra- and inter-individual variability in insulin absorption kinetics. Variability is inversely related to the absorption rate of the insulin preparation, being maximal with the long-acting (ultralente). Self blood glucose monitoring is therefore essential to achieve good control of blood glucose concentration.

TABLE 9.3   Characteristics of Insulin Preparations

Type*	Additive**	Action (hours)***		
		Onset	Peak	Duration
Rapid				
Regular	None	0.5	2–4	5–8
Semilente	Zinc	0.5–1.0	2–8	12–16
Intermediate				
NPH (Isophane)	Protamine	1–2	4–12	18–24
Lente	Zinc	1–2	4–12	18–24
Slow				
Ultralente	Zinc	4–6	8–30	24–36
Protamine-zinc	Protamine and zinc	4–6	14–20	24–36

*Premixed insulin preparations are also available, such as regular/isophane at different relative concentrations (10%, 20%, 30%, 40%, 50% regular and, respectively, 90%, 80%, 70%, 60%, 50% isophane). Semilente and protamine-zinc insulin are seldom used.

**Insulin preparations have a cloudy appearance, except for regular insulin, which is a clear solution.

*** There is considerable inter- and occasionally intrapatient variability in the duration of action.

Regular insulin may also be injected intravenously or intramuscularly to treat acute metabolic derangements in DM patients. Intraperitoneal insulin delivery by pumps has been used for relatively long periods of time, with the advantage that the physiological portal route of insulin absorption allows the liver to extract the hormone and peripheral insulin levels are lower. In uremic patients insulin may be administered intraperitoneally along with dialysis fluid.

Healthy subjects produce 20 to 40 units of insulin per day, about half as basal secretion and about half in response to meals.[9] Normal pancreatic β-cells respond to meal ingestion with an immediate *first phase* of insulin secretion, which is followed by a more prolonged *second phase*, which persists as long as nutrient absorption lasts. Insulin is slowly absorbed after subcutaneous injection, and there is no peak plasma level, which is typical of the first phase of β-cell secretion. It must be injected about half an hour before meals to overcome the time lag.

9.3.1   Insulin Preparations, Human Insulin, and Insulin Analogs

Insulin preparations can be classified as short-, intermediate-, or long-acting (Table 9.3). Insulin in solution is short-acting; adding either the basic protein protamine or a small amount of zinc ions slows absorption and prolongs the action of subcutaneously injected insulin.

Until recently insulin was extracted from bovine and porcine pancreas. Pork and beef insulin differ from human insulin by, respectively, one and three amino acids. In the late 1970s progress in genetic engineering led to biosynthesis of human insulin by means of recombinant DNA (deoxyribonucleic acid) techniques. Biosynthetic human insulin was first administered to human volunteers in 1982 and is now in widespread use. At position B30, the extra hydroxyl group of threonine, present in place of alanine in human insulin, increases solubility, thus explaining the more rapid onset of action and the earlier peak action compared with porcine insulin. Beneficial when administering regular human insulin before meals, this effect may be a problem when a more prolonged action is desired, e.g., to counteract morning hyperglycemia after presupper administration of intermediate-acting human insulin. Whether human insulin attenuates autonomic symptoms of hypoglycemia (hypoglycemia unawareness) remains controversial.[10]

Insulin molecules are associated as hexamers in currently available insulin preparations. Compared with monomers, hexamers unfortunately show delayed absorption and lower insulin peak levels, which has stimulated the development of a number of insulin analogs with an earlier hypoglycemic action, thus the need to inject insulin 30 to 45 minutes before meals is avoided. The most promising preparation seems to be Lys(B28), Pro(B29) human insulin, in which amino acid residues in B28 and B29 occupy a reverse position with no association of insulin molecules. When injected with a meal, Lys-Pro insulin has the same effect as regular insulin given 30 minutes before eating; furthermore it may minimize the risk of postprandial hypoglycemia.[11]

Because of the importance of the basal rate of insulin secretion in healthy subjects, interest is focused on developing an ideal long-acting insulin analog. Without any significant peak of action, the compound should be able to maintain plasma insulin concentration at a constant level, thus making it unnecessary to use a continuous infusion device to suppress interprandial hepatic glucose output.

9.3.2 Intensive Insulin Therapy and the Prevention of Long-Term Complications of Diabetes Mellitus

Before the discovery of insulin by Banting, Best, Collip, and McLeod and its first clinical application in January 1922, IDDM was invariably lethal. At present, seventy-six years later, the main problem with diabetes is not survival, but control and prevention of chronic, long-term complications.

It has long been hypothesized that high blood glucose levels are responsible for the development of tissue damage in diabetes.[12] Hyperglycemia could act primarily through enzymatic and nonenzymatic protein glycosylation, and enhancement of the polyol pathway (Figure 9.1). In the *polyol pathway* the enzyme aldose-reductase reduces glucose to sorbitol, which tends to accumulate within the cells unless it is converted to freely diffusible fructose. Sorbitol accumulation may explain the damage to nerves, lens, retina, and vascular endothelium typical of DM.[13]

Protein glycation refers to the *nonenzymatic* linkage between glucose and the amino groups of proteins. The amount of the ketoamine is directly related to the mean concentration of glucose to which the protein is exposed during its lifespan. Glycated hemoglobin is an example. The reaction may progress to the so-called advanced glycosylation end-products (AGEs) resulting in damage of the affected proteins and tissues.[14,15] *Protein glycosylation* results from *enzymatic* linkage between carbohydrates and proteins. It is essential for the normal function of circulating proteins (fibrinogen, immunoglobulins, hormones such as FSH, LH, and TSH), of collagen, and of constituents of basement membrane. It is assumed that excessive protein glycosylation in the hyperglycemic diabetic patient may have deleterious consequences at various levels, including retinal capillaries and glomerular basement membrane.[16]

In 1993 the Diabetes Control and Complication Trial (DCCT) Research Group definitively demonstrated that achieving blood glucose values as close as possible to normal reduces the development (primary prevention) and slows the progression (secondary intervention) of diabetic complications (retinopathy, nephropathy, and neuropathy). In this study 1441 patients with IDDM were followed up for a mean of 6.5 years. At baseline no retinopathy was present in 726 subjects (primary prevention cohort) and mild retinopathy was observed in 715 subjects (secondary intervention cohort). Patients in both groups were randomly assigned to intensive or conventional therapy, and the onset and progression of retinopathy and other complications were assessed regularly. Intensive insulin therapy (three or more daily insulin injections or the use of an insulin infusion pump, along with home blood glucose monitoring) provided better blood glucose control than conventional therapy (one or two daily insulin injections).

Intensive insulin therapy reduced the risk of onset of diabetic retinopathy by 76% in the primary prevention cohort, and slowed progression by 54% in the secondary intervention cohort. In the two cohorts combined intensive therapy led to a 60% reduction in the onset of clinical neuropathy and to a 39% reduction in microalbuminuria. The study clearly concluded that intensive therapy, by producing better blood glucose control, delays the onset and slows the progression of diabetic retinopathy, nephropathy, and neuropathy in patients with IDDM.[17]

9.3.3 Intensive Insulin Therapy and the Risk of Hypoglycemia

Compared with conventional insulin treatment, intensive insulin therapy in DCCT reduced both mean glucose (231 mg/dl vs. 155 mg/dl) and HbA_{1c} levels (9.0% vs. 7.1%). However, this was associated with a marked increase in the incidence of severe hypoglycemia (62 vs. 19 episodes per 100 patient-years).

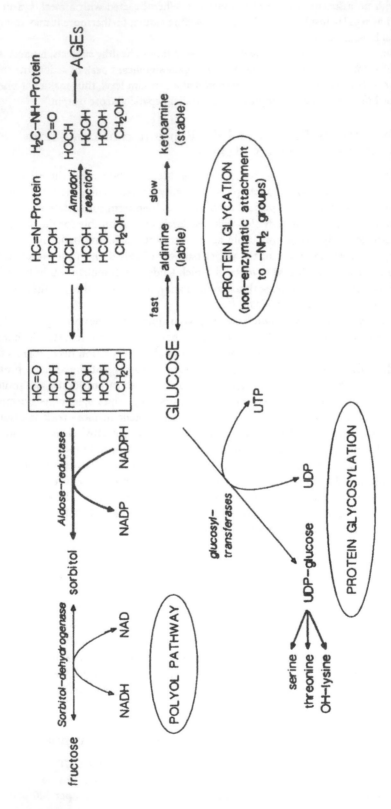

FIGURE 9.1   Glucose, at higher than normal concentration (as in DM), may enhance some metabolic pathways leading to deterioration of the proteins or cells affected: enzymatic and nonenzymatic protein glycosylation and sorbitol accumulation.

HbA$_{1c}$ values were directly correlated with the progression of retinopathy and inversely correlated with the risk of hypoglycemia. Undoubtedly, hypoglycemia was the major treatment-related side effect and its incidence was three times higher in the intensive therapy group than in the conventional treatment group.[17] When applying the conclusions of the study group in clinical practice, the risks of severe hypoglycemia may be overcome only by frequent blood glucose monitoring and by educational and dietary counseling[18,19] in order to achieve near-normoglycemia, with HbA$_{1c}$ values between 6.5% to 7.5%.

The DCCT trial was performed in IDDM patients aged 25 to 30 years.[17] In elderly NIDDM patients intensive insulin therapy–related hypoglycemia could, besides the neuroglycopenic and adrenergic responses, induce more serious side effects, such as acute myocardial infarction, acute cerebrovascular accidents, and even death. However, intensive therapy is rarely required for NIDDM patients. Dietary advice, weight control, and a physical exercise program[20] are usually sufficient to control blood glucose values,[21] although the sulfonylurea drugs, which are frequently prescribed, may induce hypoglycemia.

9.3.4 The Response to Hypoglycemia and "Hypoglycemia Unawareness"

In patients with long-lasting IDDM several studies have demonstrated delayed recovery from insulin-induced hypoglycemia,[22] due to impaired counterregulatory (mainly glucagon and epinephrine) response to hypoglycemia. Antibodies to insulin may also delay recovery from hypoglycemia by prolonging the circulating half-life of injected insulin.[23]

Diabetic patients may be unaware of hypoglycemia and thus not take early corrective action, such as food ingestion, to avoid the consequences of severe, prolonged neuroglycopenia. In young IDDM patients hypoglycemia unawareness may be associated with very tight glycemic control, and symptoms of hypoglycemia may be restored when glycemic control is slightly loosened,[24] with patient counseling and/or a closer monitoring of blood glucose values.[19] In patients with autonomic neuropathy the adrenergic dysfunction may be responsible for the lack of symptoms of hypoglycemia.

9.3.5 The Somogyi Effect and the Dawn Phenomenon

In healthy subjects overnight blood glucose levels are kept constant by the endocrine pancreas continuously delivering small amounts of insulin to the portal system. The intermediate-acting insulin that is usually injected subcutaneously before supper reaches a peak plasma insulin level 4 to 6 hours after injection, and circulating insulin levels decrease progressively 8 to 10 hours after injection. Waning insulin action, the possible occurrence of the so-called "*Somogyi effect*" and the physiological importance of the "*dawn phenomenon*" are mainly responsible for fasting hyperglycemia.

As clearly delineated in the original report,[25] the *Somogyi effect* consists of nocturnal hypoglycemia followed by rebound morning hyperglycemia produced by the release of counterregulatory hormones (catecholamines, cortisol, growth hormone) induced by hypoglycemia and by waning insulin action.[26] Although the Somogyi effect may be less frequent than originally thought,[27] its occurrence must be considered when dealing with a patient with fasting hyperglycemia. In this case blood glucose sampling in the early morning hours (about 3 a.m.) is recommended.

The *dawn phenomenon* consists of increased insulin requirements between 05.00 and 08.00 hours due to the counterregulatory effect of cortisol and growth hormone. In nondiabetic subjects a physiological early morning increase in cortisol levels follows sleep-related nocturnal growth hormone peaks, but a slight increase in the basal rate of insulin secretion counteracts this physiological insulin-resistant state. In diabetic patients, on the contrary, a morning decrease in the availability of insulin injected subcutaneously some 9 to 12 hours before leads to hyperglycemia. Furthermore, in IDDM patients an exaggerated nocturnal secretion of growth hormone may also make a contribution to the dawn phenomenon.[28]

9.3.6 Insulin Therapy and Hyperinsulinemia

Although still subject to debate,[29,30] hyperinsulinemia is generally accepted as a risk factor for macrovascular disease in both diabetic and nondiabetic subjects.[3] Several prospective studies have correlated hyperinsulinemia with the development of coronary heart disease.[31-33] Insulin stimulates proliferation of arterial smooth muscle cells and lipid synthesis in the arterial wall, suggesting a direct involvement in atherogenesis.[34,35]

Hyperinsulinemia may occur in all diabetic patients. In NIDDM endogenous hyperinsulinemia is a direct consequence of the insulin resistance, which is typical of the disease, particularly in obese patients.

In IDDM hyperinsulinemia derives from peripheral insulin injection. In nondiabetic subjects the liver extracts about 40% of the insulin coming from the pancreas through the portal vein,[36] and insulin concentration is lower in the peripheral than in the portal circulation. The liver, moreover, as the primary target of insulin action needs these high levels. After subcutaneous, intramuscular, or intravenous injection, insulin is delivered to the peripheral circulation, where it reaches the same concentration as in the liver, thus exposing peripheral tissues to abnormally high insulin levels. Hyperinsulinemia related to insulin therapy probably also explains the weight gain in diabetic patients on intensive therapy, which was well documented in the DCCT report. In this study, patients on intensive therapy gained a mean of 4.6 kg more than patients on conventional therapy after 5 years follow up.[17]

9.3.7 Antibodies to Insulin, Insulin Allergy, and Insulin Lipodystrophy

Subcutaneous administration of insulin results in antiinsulin antibody formation. The amino acid sequences of beef and pork insulin differ from the human, but even human insulin is immunogenic, probably because of structurally modified insulin molecules, or of the adjuvant properties of the subcutaneous injection, which facilitate aggregate formation. Monomeric insulin analogs should avoid this inconvenience.

Purified insulin preparations, particularly the recombinant human insulin used at present, are less immunogenic than the older impure animal preparations, and the clinical manifestations of antiinsulin antibodies actually occur very rarely. The presence of antiinsulin antibodies may induce local or general allergic reactions. They may delay the kinetics of insulin action because the insulin molecule binds to the antiinsulin antibody, increasing postprandial hyperglycemia, and with late hypoglycemia.[23]

Insulin lipodystrophy may result in insulin lipoatrophy and insulin lipohypertrophy. Insulin lipoatrophy is characterized by fat atrophy in subcutaneous tissue at the site of injection and may depend on the presence of insulin antibodies. Deposits of both insulin and antiinsulin antibodies have been observed by means of immunohistochemical methods in biopsies of lipoatrophic areas.[37]

Lipoatrophy may reverse with the injection of purified or human insulin in the affected area. If, however, insulin is repeatedly injected in the same area, lipohypertrophy may ensue owing to the lipogenic action of high insulin concentrations. These problems are rare with the highly purified human insulin used at present and can be avoided by rotating insulin injection sites.

9.4   NIDDM, Insulin, and Oral Hypoglycemic Agents

Table 9.4 shows the different sites of intervention in the therapy of type 2 DM. NIDDM develops because of insulin resistance, which is significantly influenced by obesity,[38] and which varies in degree in individual patients.[39] A relative, sometimes absolute, deficiency in insulin secretion is also present.[40] Some NIDDM patients may need insulin therapy; sometimes hyperosmolar nonketotic coma may develop unless exogenous insulin is given, especially in elderly subjects.

Both insulin resistance (impaired insulin action) and reduced insulin secretion (whether a relative, or absolute insulin deficiency) are responsible for hyperglycemia and other metabolic derangements, such as increased free fatty acids and triglyceride levels. Physiologically low portal insulin levels (e.g., basal secretion rate) are needed to suppress hepatic glucose production by glycogenolysis and gluconeogenesis. Therefore, a deficient basal insulin secretion is responsible for fasting hyperglycemia. On the other hand postprandial blood glucose level depends on diet and insulin-mediated peripheral glucose disposal at muscle and adipose tissue level, and so hyperglycemia may be a consequence of both impaired insulin action and/or secretion in response to meals.[41]

TABLE 9.4 Pathogenetic Approach to the
Management of NIDDM

1. Alteration in nutrient absorption
   • Dietary management (slow carbohydrates)
   • Guar gum, fibers
   • α-Glucosidase inhibitors
2. Stimulation of insulin secretion
   • Sulfonylureas
   • $α_2$-Adrenergic receptor antagonists
   • Gastrointestinal hormones (GIP, GLP-1)
   • Nonsulfonylurea secretagogues
3. Reduction of insulin resistance
   • Weight loss, physical exercise
   • Biguanides
   • Thiazolidinedione derivatives
4. Decrease of hepatic glucose production
   • Insulin and insulin analogs
   • Inhibitors of fatty acid release and oxidation

GIP: Gastric inhibitory peptide, or Glucose-dependent
insulin-secretory peptide
GLP-1: Glucagon-like peptide-1

The fasting blood glucose level is a reliable indicator of overall glycemic control in NIDDM patients.[42] Administering oral hypoglycemic agents and/or insulin aims at relieving basal hyperglycemia, but postprandial hyperglycemia is less easily controlled, even though blood glucose levels may return to basal level before the next meal. Dietary modification or slowing nutrient absorption may be beneficial.

9.4.1 NIDDM Therapy, Micro- and Macrovascular Complications

Therapy in NIDDM patients aims at control of hyperglycemia, thus alleviating symptoms, avoiding acute complications, and possibly reducing the burden of long-term complications. Although the DCCT trial[17] was conducted with IDDM subjects, it is logical to suppose that NIDDM patients also benefit from better blood glucose control. The University Group Diabetes Program (UGDP) controversy has hampered the validation of this concept, which hopefully will be demonstrated by the ongoing United Kingdom Prospective Diabetes Study (UKPDS).

In the multicentric UGDP study, 1027 newly diagnosed, adult-onset diabetic patients were randomized from 1961 to 1966 in 12 United States centers to one of five treatment regimens: placebo, tolbutamide 1.5 g/day, a standard daily dose of insulin based on the patient's body surface area, a variable dose of insulin based on glucose control, and phenformin 100 mg/day. After about 9 years the study was stopped because of an apparent excess mortality in the tolbutamide group.[43] The conclusions have been extensively analyzed and criticized,[44] and the initial aim of evaluating the efficacy of different therapeutic regimens in preventing the vascular complications of NIDDM patients has not been attained. In fact, the tolbutamide-treated patients presented an excess of cardiovascular risk factors, i.e., higher cholesterol, and blood glucose levels, arterial calcification and digitalis use, which may well explain the excess mortality.[45]

In the ongoing UKPDS multicentric randomized study, 5102 newly diagnosed type 2 diabetic subjects were recruited between 1977 and 1991 in 23 centers in the United Kingdom.[46] The study will be concluded in 1998 when the mean time since randomization will be 11 years and a response will be available as to whether improved blood glucose control in type 2 diabetes prevents long-term complications, and whether any specific therapy is particularly advantageous. After an initial 3-month follow up period, 4209 subjects were asymptomatic with fasting plasma glucose between 6.0 and 15.0 mmol/l (108 to 270 mg/dl) and were randomized to conventional or intensive therapy. Patients in the conventional therapy arm have been managed primarily with diet alone, whereas intensive therapy is based on sulfonylurea or insulin and, in obese subjects, metformin.

Results at 6 years show better blood glucose control in the intensive therapy group, with a progressive deterioration of glucose control over time in both groups (conventional therapy 1 and 6 years after randomization: median fasting plasma glucose 8.2 and 9.5 mmol/l, median HbA$_{1c}$ 6.8% and 8.0%; intensive therapy 1 and 6 years after randomization: median fasting plasma glucose 6.8 and 7.8 mmol/l, median HbA$_{1c}$ 6.1% and 7.1%). The study demonstrates that intensive therapy maintains better blood glucose control than conventional therapy (median HbA$_{1c}$ level over 6 years 6.6% and 7.4%, respectively), and hopefully it will also be possible to assess the cost-benefit ratio of different therapeutic options in NIDDM.

At present, recommendations for therapy in NIDDM are based on extrapolation of the DCCT results. This may be valid for microvascular complications (retinopathy, nephropathy, and neuropathy),[21] but therapeutic intervention in NIDDM patients may negatively influence already present vascular risk factors, such as hyperinsulinemia.[35] Undoubtedly, theoretical considerations suggest that treatment strategies aimed at reducing risk factors such as the formation of glucose-dependent glycation products, obesity, cholesterol levels, and hypertension will be beneficial in NIDDM subjects.[47]

Dietary management is the mainstay of therapy for NIDDM patients. Oral hypoglycemic agents and/or insulin are added if diet alone does not achieve an adequate blood glucose control. The oral hypoglycemic agents that are at present available for clinical use are sulfonylureas, biguanides, and acarbose.

9.4.2   Insulin in NIDDM

When insulin treatment is required in type 2 diabetic subjects, multiple daily injections should be administered as in type 1 diabetics. However, the therapeutic objectives may differ. In type 2 diabetics under 60 years of age, without serious micro- and macroangiopathic complications, the aim of treatment should be similar to that recommended by the DCCT (HbA$_{1c}$ between 6.5% and 7.5%, preprandial plasma glucose below 120 to 140 mg/dl, and postprandial plasma glucose below 150 to 160 mg/dl). In older patients, with chronic diabetic complications, HbA$_{1c}$ should be maintained below 8.5%. This may avoid the risk of hypoglycemia and the use of high doses of exogenous insulin.

9.4.3   Sulfonylureas

Sulfonylurea drugs are chemically related to antibacterial sulfonamides whose hypoglycemic action was accidentally detected in the early 1940s. The first sulfonylurea agents, carbutamide, tolbutamide, chlorpropamide, acetohexamide, and tolazamide, were introduced to clinical use in the early 1950s. The later development of more potent compounds such as glibenclamide, glipizide, gliquidone, and gliclazide led to "second generation" sulfonylureas. Table 9.5 shows some pharmacokinetic and therapeutic characteristics of the commonly used sulfonylurea drugs.

TABLE 9.5   Pharmacokinetic and Therapeutic Characteristics of Common Sulfonylurea Drugs

Generic Name	Used Since	Daily Dose	Duration of Action
First generation			
Tolbutamide	1956	1.0–3.0 g	6–10 h
Chlorpropamide	1957	100–500 mg	60 h
Acetohexamide	1962	0.25–1.5 g	12–18 h
Tolazamide	1962	100–750 mg	16–24 h
Second generation			
Glibenclamide (Glyburide)	1969	2.5–20 mg	24 h
Glipizide	1971	2.5–20 mg	15–24 h
Gliquidone	1975	60–180 mg	5 h
Gliclazide	1979	80–320 mg	16–24 h

Sulfonylureas stimulate insulin secretion without affecting insulin synthesis.[48] They depolarize the plasma membrane of pancreatic β-cells by closure of the ATP-dependent K^+ channels. Depolarization opens voltage-dependent Ca^{2+} channels, which activates exocytosis of insulin. Indeed glibenclamide is routinely used to assess the ATP-dependent K^+ channel in electrophysiological experiments. Both in myocardial and in vascular smooth muscle ATP-dependent K^+ channels are known to open as a protective mechanism during ischemia. In animal models the outcome of experimental myocardial infarction was significantly worsened by sulfonylurea-derivative closure of ATP-dependent K^+ channels. On the other hand, in ATP-dependent K^+ channels closure reduces the incidence of ventricular fibrillation.[49] New insulin stimulatory compounds, benzoic acid derivatives, such as glymepiride and repaglinide, may be pancreas-specific, without any action on myocardial and vascular receptors.[50] The question of the safety of sulfonylureas in NIDDM patients with cardiovascular disease has recently been reviewed.[51]

The stimulatory effect of sulfonylureas on pancreatic β-cells does not restore the first phase of insulin secretion, but the second phase is enhanced. Sulfonylureas and plasma glucose have similar stimulating effects on the second phase of insulin secretion in both healthy and NIDDM subjects. A direct relationship exists between plasma glucose levels and second-phase insulin secretion, and sulfonylureas shift the dose–response curve to the left: more insulin is secreted for any given glucose concentration. This effect persists during long-term treatment, but the sulfonylurea-induced drop in blood glucose may reduce insulin secretion in absolute terms. Extrapancreatic effects of sulfonylureas, such as improved insulin action, have been postulated but not clearly demonstrated. The reduction in blood glucose levels may itself explain the effects on insulin resistance. This is further confirmed by the lack of effect of sulfonylurea drugs in C-peptide deficient IDDM patients.[52]

During sulfonylurea therapy improvements in the lipid profile,[53] platelet aggregation, and the microcirculation[54] have been reported and are probably due to better blood glucose control.

9.4.3.1 Primary and Secondary Sulfonylurea Failure

The term *sulfonylurea failure* indicates that satisfactory blood glucose control has not been achieved, even at highest doses.

Primary failure occurring within the first few months of treatment usually depends on the incorrect use of sulfonylureas in insulin-dependent diabetic patients, but may be observed in highly hyperglycemic nonobese NIDDM individuals. In these cases β-cell function has markedly deteriorated and insulin therapy is mandatory. About 15% to 20% of NIDDM patients at diagnosis are said to have primary failure to sulfonylureas.[55] In obese NIDDM patients a hypocaloric diet has a key role in restoring responsiveness to sulfonylureas.

Secondary failure refers to the patients who, after several months or years of successful sulfonylurea treatment, are no longer satisfactorily controlled. The frequency of secondary failure increases with the duration of diabetes and has been estimated to affect about 3% to 5% of patients taking sulfonylureas each year.[56] Poor compliance with diet is often responsible, along with the increased insulin resistance that is mainly caused by weight gain in diabetic patients under therapy. True secondary failure is related to a deterioration in β-cell function over time. Hyperglycemia itself, due to its toxic effect upon β-cell function, could contribute to sulfonylurea failure. It has been suggested that even prolonged treatment facilitates sulfonylurea resistance. A summary of factors responsible for failure of treatment with sulfonylureas is listed in Table 9.6.

When secondary failure is suspected, all contributing factors should be considered and eliminated when possible. If no positive result is achieved, conversion to insulin therapy must not be postponed in nonobese subjects, but adding metformin can be tried initially in obese NIDDM patients.

9.4.3.2 Hypoglycemia and Other Side Effects of Sulfonylurea Therapy

Hypoglycemic episodes can occur during sulfonylurea therapy and tend to be more severe with long-acting compounds such as chlorpropamide and glibenclamide. Hypoglycemia is more frequent in the late afternoon, when insulin sensitivity is physiologically enhanced or increased by physical activity, and it is often overlooked since it may present with aspecific symptoms, such as irritability and weakness.

TABLE 9.6   Factors Responsible for Failure of
Sulfonylurea Treatment

Primary failure
  • IDDM (late onset)
  • Marked hyperglycemia
Secondary failure
  • Progressive decrease in β-cell function (true failure)
  • Increased body weight
  • Poor compliance with diet
  • Sedentary life-style
  • Stress, intercurrent illness
  • Diabetogenic drugs (steroids...)
  • Inadequate sulfonylurea dose
  • Hyperglycemia impairs sulfonylurea absorption

TABLE 9.7   Clinically Relevant Interactions Between Sulfonylureas and Other Drugs

Increase in hypoglycemic action
  • By displacing sulfonylureas from protein binding sites: nonsteroidal antiinflammatory drugs, sulphonamides, warfarin, fibrates
  • By inhibiting sulfonylurea metabolism: warfarin, alcohol, cimetidine
  • By increasing insulin secretion: nonsteroidal antiinflammatory drugs
  • By reducing renal excretion of sulfonylureas: probenecid, allopurinol
Decrease in hypoglycemic action
  • By increasing hepatic metabolism of sulfonylureas: rifampin, barbiturates
  • By reducing insulin secretion: amphetamines, β-adrenergic drugs, phenytoin
  • By inducing peripheral insulin resistance: thiazide diuretics, corticosteroids, estrogens, catecholamines
Reduction of counterregulatory response to hypoglycemia*
  • β-blockers and other sympatholytic drugs
Concurrent use of hypoglycemic drugs
  • Insulin, alcohol, aspirin

*Both sulfonylurea- and insulin-induced

In elderly NIDDM patients with coronary or carotid artery disease sulfonylurea-induced hypoglycemia may be responsible for vascular accidents and even death. Therefore, long-acting sulfonylureas should be avoided and low doses are recommended[57] in elderly individuals.

As NIDDM subjects are generally middle-aged or elderly, and are often taking other drugs besides sulfonylureas, interactions with other drugs are of particular concern. One example is the potentiating effect exerted by commonly used nonsteroidal antiinflammatory drugs, which act by stimulating insulin secretion and by displacing protein-bound sulfonylurea molecules. The latter mechanism also explains the enhancement of hypoglycemic action induced by oral anticoagulants. Hypoglycemia may be exacerbated by alcohol, which inhibits gluconeogenesis. Several classes of drugs may affect the hypoglycemic action of sulfonylureas by a number of mechanisms (see Table 9.7 for summary).

Other side effects of sulfonylurea therapy include sensitivity reactions such as skin rashes and cholestatic jaundice, and, more rarely, blood dyscrasias such as agranulocytosis, thrombocytopenia, and hemolytic anemia.

The use of chlorpropamide may be associated with hyponatremia and water retention due to inappropriate sensitivity of the renal tubules to antidiuretic hormone.

9.4.3.3   Guidelines for Sulfonylurea Therapy

Sulfonylurea therapy should be started in obese or normal-weight diabetic subjects, usually diagnosed after age 30, when blood glucose levels are not adequately controlled by dietary therapy and an exercise program. Underweight diabetic patients are more likely to be insulin-dependent and not responsive to sulfonylureas. Maturity-Onset Diabetes of Youth (MODY) is usually diagnosed before age 30 and does respond to sulfonylureas.

Sulfonylureas are contraindicated in IDDM patients during pregnancy and in gestational diabetes mellitus; during acute stress, such as major surgery or trauma and severe infections; in patients with kidney or liver failure with a history of hypersensitivity reactions to sulfonylurea; and in those prone to develop severe hypoglycemia.

Available data do not demonstrate any significant differences in the efficacy and side effects of the different sulfonylurea compounds. Compliance may be better with long-acting sulfonylureas since they are administered once or twice daily, but they increase the risk of hypoglycemia, particularly in the elderly.

Blood glucose control during sulfonylurea therapy can best be assessed in the morning before breakfast and in the evening before supper. The morning measurement is a reliable measure of overall glucose control[42] and is closely related to HbA_{1c}. The evening measurement is particularly useful for detecting hypoglycemic episodes.

9.4.4 Combined Sulfonylurea and Insulin Therapy

Sulfonylureas have been administered concurrently with insulin, but the results have been extensively debated,[58,59] mainly because they vary from patient to patient, and no large trial on a homogeneous group of subject has yet been performed.

Combined therapy may be started when sulfonylurea and metformin do not achieve satisfactory fasting blood glucose control. An intermediate or long-acting insulin, such as ultralente or isophane, is injected at bedtime, whereas sulfonylurea is given during the day. The dose of insulin is usually less than 15 to 25 U. Metformin is better stopped in order to avoid potentially harmful side effects. Theoretically associating insulin and sulfonylurea administration should improve basal insulin levels, thus reducing hepatic glucose output during the night and therefore fasting blood glucose. Consequently less exogenous insulin may be needed. The intermittent administration of insulin in sulfonylurea-treated patients has also been advocated in order to improve pancreatic β-cell insulin secretion.[60]

Combined therapy is more complex, since it requires the use of two drugs, and its clinical usefulness is temporary. Therefore, it should be reserved for only a limited number of patients.[61] In cases of true secondary failure to sulfonylurea, whether alone or in combination with metformin, we recommend switching to multiple daily insulin treatment.

9.4.5 Metformin

The use of biguanides dates back to the Middle Ages when *Galega officinalis* was a known antidiabetic remedy. In 1920 guanidine was recognized as the active principle of *Galega*, and in the late 1950s the guanidine derivatives metformin and phenformin were introduced into clinical practice as oral hypoglycemic drugs. In the late 1970s phenformin was withdrawn from clinical use in many European and North American countries, as it is associated with an increased risk of lactic acidosis. Metformin is now widely used for the treatment of diabetes in monotherapy or in combination with sulfonylureas.

Metformin does not stimulate insulin secretion; it enhances peripheral and hepatic insulin sensitivity. Reduction of both gluconeogenesis and hepatic glucose output is its main metabolic effect, thus fasting blood glucose concentrations are lowered. This is important, because the fasting plasma glucose level influences glucose control throughout the day. In contrast to sulfonylureas, which decrease blood glucose levels by increasing insulin secretion, high levels of both glucose and insulin are reduced during metformin therapy. Metformin also has a beneficial effect on plasma lipid concentrations.[62,63]

Metformin reduces blood glucose levels to the same degree as sulfonylureas with no gain, but rather often a loss in body weight.[64] Reduction of hyperinsulinemia is observed in metformin therapy, so this drug is preferentially used in overweight NIDDM subjects. Furthermore, malabsorption, or a mild anorectic action possibly related to gastrointestinal side effects particularly at high doses, may contribute to weight loss, which is due to a reduction in adipose tissue mass but not in lean mass.[62]

Because metformin does not stimulate insulin secretion, it is considered an antihyperglycemic, rather than an hypoglycemic, oral drug. Unlike the sulfonylureas, there is no hypoglycemia even after ingestion of very large doses.[57] Elevated blood glucose levels are reduced by metformin only in the presence of insulin and no blood glucose reducing effect is usually observed in normal subjects.

9.4.5.1 Side Effects of Metformin

Gastrointestinal side effects are relatively common in patients taking metformin, and include anorexia, nausea, metallic taste, abdominal discomfort and cramps, flatulence, and diarrhea. Consequently some nutrients, vitamin B12, and folate may be malabsorbed. These adverse effects can be minimized by starting with low doses of the drug (500 or 850 mg daily).

The more severe side effect of biguanides is lactic acidosis, which is often lethal, but is extremely rare with metformin. In fact, the dimethyl residue of metformin is less lipophilic than the phenylethyl residue of phenformin and reaches lower intramitochondrial concentrations. By determining a more marked inhibition of the respiratory process in the mitochondria, phenformin significantly decreases lactic acid oxidation, thus increasing the risk of lactic acidosis.[65]

As mentioned above, renal failure, by impairing biguanide excretion, predisposes to lactic acidosis, especially in the elderly. Both hepatic failure and alcohol abuse may favor the accumulation of lactate, thus potentiating the effect of biguanides. Severe infection, dehydration, heart failure, and shock also facilitate the development of lactic acidosis.

9.4.5.2 Guidelines for Metformin Therapy

NIDDM subjects, who remain hyperglycemic despite a correct dietary and exercise therapy, are candidates for sulfonylurea or metformin therapy. The mechanisms of action of sulfonylurea and metformin differ significantly: the former stimulates insulin secretion, the latter increases insulin sensitivity. Obese NIDDM subjects are often hyperinsulinemic and sulfonylureas may aggravate hyperinsulinemia and its adverse effects.[35,36] The weight loss associated with metformin therapy may be beneficial in obese NIDDM patients. However, we must not forget that reduction of food intake is all that is required in the vast majority of overweight NIDDM patients.

The different mechanisms by which sulfonylureas and metformin reduce elevated blood glucose levels may favor combination therapy, which has a beneficial effect on blood glucose and plasma lipid levels.[63]

Special caution is required when metformin is prescribed, either alone or in combination, in elderly subjects. Renal, hepatic, cardiovascular, and respiratory function should be carefully monitored, and the drug stopped if some impairment is detected. Metformin should be suspended in acute medical and surgical emergencies, in order to offset the risk of lactic acidosis. When NIDDM patients require insulin therapy due to persistent marked hyperglycemia and weight loss or ketosis, metformin should be suspended because of the risk of side effects, and because combining insulin and metformin has not been definitively established as providing any clear improvement in blood glucose control.

9.4.6 Acarbose

Acarbose is an α-glucosidase inhibitor that acts by competitively inhibiting the α-glucosidases in the intestinal brush border. These enzymes are able to convert nonabsorbable dietary starch and sucrose into absorbable monosaccharides, e.g., glucose (Figure 9.2). Enzyme inhibitors delay this conversion, slowing the formation and consequently the absorption of monosaccharides, thus reducing postprandial blood glucose concentration. Both starch and sucrose are influenced, whereas lactose and glucose itself are not.[66]

Many studies in experimental animals and in healthy and non-insulin-dependent diabetic (NIDDM) subjects have shown that acarbose decreases postprandial blood glucose, with a lesser reduction in fasting blood glucose, plasma triglycerides, and postprandial insulin levels.[67,68] The main effect of acarbose in subjects fed complex carbohydrates seems to be a delayed absorption of glucose with reduced postprandial glycemic peaks. In long-term studies on NIDDM patients, acarbose, as monotherapy,[69,70] and in combination with metformin, sulfonylureas, or insulin,[71] significantly reduced glycosylated hemoglobin levels.[69] Acarbose and sulfonylureas have a similar effect on blood glucose control in NIDDM patients when diet alone fails.[71] The use of acarbose in IDDM patients is at present less well defined, although it should theoretically improve postprandial glucose control and reduce the risk of late postmeal hypoglycemia.

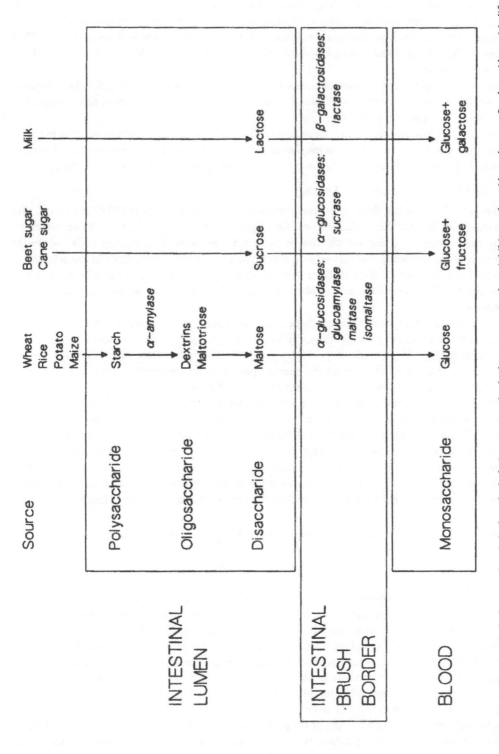

FIGURE 9.2 Main dietary carbohydrates and carbohydrate hydrolysing enzymes in the human gut. Acarbose inhibits α-glucosidases, but not β-galactosidases. Modified from Santeusanio and Compagnucci,[73] with permission.

9.4.6.1 Tolerability and Side Effect of Acarbose

Acarbose is only minimally absorbed from the gut and no systemic adverse effects have been demonstrated after long-term administration.[72,73] The drug allows undigested carbohydrates to pass into the large bowel, where they are fermented, causing flatulence, bloating, and diarrhea. These symptoms, which occur in approximately 30% to 60% of patients, tend to decrease with time and seem to be dose-dependent. They are minimized by starting therapy with low doses (such as 50 mg thrice daily), which may already be effective in many patients.

An increase in serum hepatic transaminases observed in earlier studies in the U.S., where doses of acarbose up to 900 mg daily were used, has not been reported with the lower doses recommended at present (150–300 mg daily, up to 600 mg daily).

Because of its mechanism of action, acarbose does not induce hypoglycemia, but may enhance hypoglycemic reactions when associated with sulfonylureas or insulin. As the drug slows starch and sucrose digestion and absorption, glucose must be used to correct hypoglycemia during acarbose therapy.

9.4.6.2 Guidelines for Acarbose Therapy

Sulfonylureas and injected insulin may increase plasma insulin concentrations in NIDDM subjects, who are often insulin-resistant and hyperinsulinemic. Consequently, although blood glucose levels may revert to near normal values, hyperinsulinemia may be increased, thus facilitating atherogenesis and weight gain; the risks of hypoglycemia and serious side effects rise considerably, particularly in the elderly.

Metformin may be used in obese diabetic subjects with hyperinsulinemia and insulin resistance. It should be avoided, however, particularly in older subjects with even mild cardiovascular, respiratory, or kidney disease.

Acarbose delays the absorption of carbohydrates without inducing malabsorption. It does reduce blood glucose concentrations, but not dramatically, particularly postprandially, and produces a consistent fall of about 0.5% to 1.0% in glycosylated hemoglobin. The benefit is evident within the first weeks of treatment and is maintained throughout the therapy period, as documented in clinical trials usually lasting for more than 6 to 12 months. This effect may delay the need for other traditional antidiabetic drugs, and acarbose may effectively be used as a first-line drug, after diet and exercise therapy have failed.[74] On the other hand, even when diabetic patients are not well controlled with sulfonylureas or biguanides, acarbose may lower blood glucose levels, thus postponing insulin therapy. Studies in IDDM patients have also shown that acarbose decreases insulin requirements and the risk of nocturnal hypoglycemia.[75]

In summary, acarbose may be used as monotherapy, or as a valid alternative to metformin when biguanides are contraindicated, or to sulfonylureas in the elderly, for whom hypoglycemia is particularly dangerous. Furthermore, acarbose may be usefully combined with metformin or sulfonylureas.

9.5 New Drugs in the Therapy of Diabetes Mellitus

Table 9.4 summarizes several alternative approaches to sulfonylureas, metformin, and acarbose in the treatment of NIDDM.

Stimulation of insulin secretion is only possible as long as pancreatic β-cell function is preserved, and secondary failure to sulfonylureas mainly depends on progressive β-cell dysfunction. Nevertheless, several new nonsulfonylurea insulin secretagogues are currently under investigation. Glimepiride and repaglinide are benzoic acid derivatives that induce insulin secretion by means of a sulfonylurea-like mechanism. Whether they offer any clinical advantages over the sulfonylureas remains to be definitively confirmed. Such α_2-adrenergic receptor antagonists as idazixan and midaglizol may act by reducing the excessive α_2-adrenergic tone that is responsible for impaired insulin secretion. However, these drugs are not clinically useful, since they may increase blood pressure.[76]

Thiazolidinedione derivatives are a class of drugs that reduce blood glucose levels in experimental type 2 diabetes in animals by enhancing insulin sensitivity. Ciglitazone, proglitazone, englitazone, and pioglitazone are some of the compounds that have already been studied.[77] However, because of the increased sensitivity of some organs to the growth-promoting effect of insulin, caution must be exercised when using these drugs in human beings. Cardiomegaly, and a mild anemia that may be related to hypertrophy of bone marrow adipocytes, have been observed in animal models. However, recent clinical trials in NIDDM patients have provided some promising results.[78]

Fatty acid metabolism is altered in NIDDM patients, because adipose tissue releases excessive amounts of nonesterified fatty acids (NEFA), which may in turn impair peripheral glucose disposal (the "glucose-fatty acid" cycle of Randle).[79,80] Excess adipose tissue in NIDDM subjects may be a source of NEFA, thus explaining the insulin resistance associated with obesity. Unfortunately, reducing NEFA concentration pharmacologically has proved disappointing.[81]

Fatty acid oxidation in the liver stimulates gluconeogenesis, so inhibition of NEFA oxidation has been advocated as a useful therapeutic approach in NIDDM subjects.[82] The clinical use of agents such as etomoxir and clomoxir has, however, been hampered by side effects, principally cardiac hypertrophy, in experimental animals.

Glucagon-like peptide-1 (GLP-1) exists in two forms, GLP-1(7-37) and GLP-1(7-37) amide, the latter being the natural form in humans. GLP-1 is secreted from enteroglucagon-producing cells in the small intestine and derives from the preproglucagon molecule. Along with the gastric inhibitory polypeptide (or glucose-dependent insulinotropic polypeptide, GIP) it constitutes the "incretin" system that potentiates glucose-induced insulin secretion. In type 2 diabetic patients with fasting hyperglycemia GLP-1 infusion increased plasma insulin concentrations, decreased glucacon levels, and normalized plasma glucose.[83] Subcutaneous injection of GLP-1 reduced postprandial hyperglycemia in non-insulin-dependent diabetic patients.[84] The disadvantage of GLP-1 is its brief duration of action, and clinical use is limited by lack of results in long-term treatment.

At present, only the α-glucosidase inhibitor acarbose is currently used along with insulin, sulfonylureas, and metformin in the therapy of diabetes mellitus.[81]

Acknowledgments

The excellent linguistic assistance of Ms. G.A. Boyd, B.A., D.LIT., in the manuscript preparation is gratefully acknowledged.

Glossary of Abbreviations

AGEs	Advanced glycosylation end-products
ATP	Adenosine triphosphate
CSII	Continuous subcutaneous insulin infusion
DCCT	Diabetes control and complication trial
DM	Diabetes mellitus
DNA	Deoxyribonucleic acid
FSH	Follicle-stimulating hormone
GIP	Gastric inhibitory polypeptide (or glucose-dependent insulinotropic polypeptide)
GLP-1	Glucagon-like peptide-1
IDDM	Insulin-dependent diabetes mellitus
LH	Luteinizing hormone
Lys	Lysine
MODY	Maturity-onset diabetes of youth
NEFA	Non-esterified fatty acids
NIDDM	Non insulin-dependent diabetes mellitus
Pro	Proline
TSH	Thyroid-stimulating hormone (thyrotropin)
UGDP	University group diabetes program
UKPDS	United Kingdom prospective diabetes study

References

1. National Diabetes Data Group, Classification and diagnosis of diabetes mellitus and other categories of glucose intolerance, *Diabetes*, 28, 1039, 1979.
2. American Diabetes Association, Position statement: guide to diagnosis and classification of diabetes mellitus and other categories of glucose intolerance, *Diabetes Care*, 20 (Suppl. 1), S21, 1997.
3. DeFronzo, R.A., Ferrannini, E., Insulin resistance: a multifaceted syndrome responsible for NIDDM, obesity, hypertension, dyslipidemia, and atherosclerotic heart disease, *Diabetes Care*, 14, 173, 1991.
4. American Diabetes Association, Position statement: nutrition recommendations and principles for people with diabetes mellitus, *Diabetes Care*, 19 (Suppl. 1), S16, 1996.
5. Franz, M.J., Horton, E.S., Bantle, J.P., Beebe, C.A., Brunzell, J.D., Coulston, A.M., Henry, R.R., Hoogwerf, B.J., Stackpoole, P.W., Nutrition principle for the management of diabetes and related complications, *Diabetes Care*, 17, 490, 1996.
6. American Diabetes Association, Position statement: diabetes mellitus and exercise, *Diabetes Care*, 19 (Suppl. 1), S30, 1996.
7. Calabrese, G., Bueti, A., Santeusanio, F., Giombolini, A., Zega, G., Angeletti, A., Cartechini, M.G., Brunetti, P., Continuous subcutaneous insulin infusion treatment in insulin-dependent diabetic patients: a comparison with conventional optimized treatment in a long-term study, *Diabetes Care*, 5, 457, 1982.
8. Bode, B.W., Steed, R.D., Davidson, P.C., Reduction in severe hypoglycemia with long-term continuous subcutaneous insulin infusion in type I diabetes, *Diabetes Care*, 19, 324, 1996.
9. Polonsky, K.S., Rubenstein, A.H., Current approaches to measurement of insulin secretion, *Diabetes Metab Rev*, 2, 315, 1986.
10. Pickup, J., Human insulin, *Br Med J*, 299, 991, 1989.
11. Pampanelli, S., Torlone, E., Lalli, C., Del Sindaco, P., Ciofetta, M., Lepore, M., Bartocci, L., Brunetti, P., Bolli, G.B., Improved post-prandial metabolic control after subcutaneous injection of a short-acting insulin analog in IDDM of short duration with residual pancreatic β-cell function, *Diabetes Care*, 18, 1452, 1995.
12. Pirart, J., Diabetes mellitus and its degenerative complications: prospective study of 4,400 patients observed between 1947 and 1973, *Diabetes Care*, 1, 168, 252, 1978.
13. Gabbay, K.H., The sorbitol pathway and the complications of diabetes, *N Engl J Med*, 295, 443, 1976.
14. Brownlee, M., Cerami, A., Vlassara, H., Advanced glycosylation end products in tissues and the biochemical basis of diabetic complications, *N Engl J Med*, 318, 1315, 1988.
15. Bucala, R., Cerami, A., Vlassara, H., Advanced glycosylation end products in diabetic complications, *Diabetes Reviews*, 3, 258, 1995.
16. Spiro, R.G., Search for a biochemical basis of diabetic microangiopathy, *Diabetologia*, 12, 1, 1976.
17. The Diabetes Control and Complications Trial Research Group, The effect of intensive treatment of diabetes on the development and progression of long-term complications in insulin-dependent diabetes mellitus, *N Engl J Med*, 329, 977, 1993.
18. Diabetes Control and Complications Trial Research Group, Implementation of treatment protocols in the diabetic control and complication trial, *Diabetes Care*, 18, 361, 1995.
19. Bolli, G.B., Fanelli, C.G., Unawareness of hypoglycemia, *N Engl J Med*, 333, 1771, 1995.
20. Pollet, R.J., El-Kebbi, I.M., The applicability and implications of the DCCT to NIDDM, *Diabetes Reviews*, 2, 413, 1994.
21. Nathan, D.M., Inferences and implications. Do results from the Diabetes Control and Complications Trial apply in NIDDM? *Diabetes Care*, 18, 251, 1995.
22. Kleinbaum, J., Shamoon, H., Impaired counterregulation of hypoglycemia in insulin-dependent diabetes mellitus, *Diabetes*, 32, 493, 1983.

23. Bolli, G., De Feo, P., Compagnucci, P., Cartechini, M.G., Angeletti, G., Santeusanio, F., Brunetti, P., Gerich, E.J., Abnormal glucose counterregulation in insulin-dependent diabetes mellitus. Interaction of anti-insulin antibodies and impaired glucagon and epinephrine secretion, *Diabetes*, 32, 134, 1983.

24. Fanelli, C., Pampanelli, S., Calderone, S., Lepore, M., Annibale, B., Compagnucci, P., Brunetti, P., Bolli, G.B., Effects of recent, short-term hyperglycemia on responses to hypoglycemia in humans. Relevance to the pathogenesis of hypoglycemia unawareness and hyperglycemia-induced insulin resistance, *Diabetes*, 44, 513, 1995.

25. Somogyi, M., Kirstein, M., Insulin as a cause of extreme hyperglycemia and instability, *Week Bull St Louis Med Soc*, 32, 498, 1938.

26. Bolli, G.B., Gottesman, I.S., Campbell, P.J., Haymond, M.W., Cryer, P.E., Gerich, J.E., Glucose counter-regulation and waning of insulin in the Somogyi phenomenon (post-hypoglycemic hyperglycemia), *N Engl J Med*, 311, 1214, 1984.

27. Tordjman, K.M., Havlin, C.E., Levandoski, L.A., White, N.H., Santiago, J.V., Cryer, P.E., Failure of nocturnal hypoglycemia to cause fasting hyperglycemia in patients with insulin-dependent diabetes mellitus, *N Engl J Med*, 317, 1552, 1987.

28. Campbell, P.J., Bolli, G.B., Cryer, P.E., Gerich, J.E., Pathogenesis of the dawn phenomenon in patients with insulin-dependent diabetes mellitus: accelerated glucose production and impaired glucose utilization due to nocturnal surges in growth hormone, *N Engl J Med*, 312, 1473, 1985.

29. Jarret, R.J., In defence of insulin: a critique of syndrome X, *Lancet*, 340, 469, 1992.

30. Reaven, G.M., Laws, A., Insulin resistance, compensatory hyperinsulinemia, and coronary heart disease, *Diabetologia*, 37, 948, 1994.

31. Welborn, T.A., Wearne, K., Coronary heart disease and cardiovascular mortality in Busselton with reference to glucose and insulin concentrations, *Diabetes Care*, 2, 154, 1979.

32. Pyörälä, K., Savolainen, E., Kaukola, S., Haapakoski, J., Plasma insulin as coronary heart disease risk factor: relationship to other risk factors and predictive value over 9.5 year follow up of the Helsinki Policemen Study population, *Acta Med Scand*, 701 (Suppl), 38, 1985.

33. Fontbonne, A., Charles, M.A., Thibult, N., Hyperinsulinaemia as a predictor of coronary heart disease mortality in a healthy population: The Paris prospective Study, 15-year follow-up, *Diabetologia*, 34, 356, 1991.

34. Stout, R.W., Insulin and atheroma — an update, *Lancet*, 1, 1077, 1987.

35. Jarret, R.J., Is insulin atherogenic? *Diabetologia*, 31, 71, 1988.

36. Chap, Z., Ishida, T., Chou, J., Hartley, C.J., Entman, M.L., Brandenburg, D., Jones, R.H., Field, J.B., First-pass hepatic extraction and metabolic effects of insulin and insulin analogs, *Am J Physiol*, 252, 209, 1987.

37. Reeves, W.G., Allen, B.R., Tattersall, R.B., Insulin-induced lipoatrophy: evidence for an immune pathogenesis, *Br Med J*, 280, 1500, 1980.

38. Arner, P., Pollare, T., Lithell, H., Different aetiologies of type 2 (non insulin-dependent) diabetes mellitus in obese and non-obese subjects, *Diabetologia*, 34, 483, 1991.

39. Banerji, M.A., Lebovitz, H.E., Insulin-sensitive and insulin-resistant variants of NIDDM, *Diabetes*, 38, 784, 1989.

40. Ward, W.K., Beard, J.C., Halter, J.B., Pfeifer, M.A., Porte, D., Jr., Pathophysiology of insulin secretion in non insulin-dependent diabetes mellitus, *Diabetes Care*, 7, 491, 1984.

41. De Fronzo, R.A., The triumvirate: beta cell, muscle, liver. A collusion responsible for NIDDM, *Diabetes*, 37, 667, 1988.

42. Holman, R.R., Turner, R.C., The basal plasma glucose: a simple relevant index of diabetes, *Clin Endocrinol*, 14, 279, 1981.

43. University Group Diabetes Program, A study of the effects of hypoglycemic agents on vascular complications in patients with adult-onset diabetes. II. Mortality results, *Diabetes*, 19 (Suppl 2), 789, 1971.

44. Kolata, G.B., Controversy over study of diabetes drugs continues for nearly a decade, *Science*, 203, 986, 1979.
45. Kilo, C., Miller, J.P., Williamson, J.R., The crux of the UGDP: spurious results and biologically inappropriate data analysis, *Diabetologia*, 18, 179, 1980.
46. United Kingdom Prospective Diabetes Study Group, U.K. Prospective Diabetes Study 16. Overview of 6 years' therapy of type II diabetes: a progressive disease, *Diabetes*, 44, 1249, 1995.
47. Wolffenbuttel, B.H.R., van Haeften, T.W., Prevention of complications in non insulin-dependent diabetes mellitus (NIDDM), *Drugs*, 50, 263, 1995.
48. Gerich, J.E., Oral hypoglycemic agents, *N Engl J Med*, 321, 1231, 1989.
49. Billman, G., Avendano, C.E., Halliwill, J.R., Burroughs, J.M., The effects of the ATP-dependent potassium channel antagonist glyburide on coronary blood flow and susceptibility to ventricular fibrillation in anesthetized dogs, *J Cardiovasc Pharmacol*, 21, 197, 1993.
50. Smits, P., Thien, T. Cardiovascular effects of sulphonylurea derivatives. Implications for the treatment of NIDDM? *Diabetologia*, 38, 116, 1995.
51. Leibowitz, G., Cerasi, E., Sulphonylurea treatment of NIDDM patients with cardiovascular disease: a mixed blessing? *Diabetologia*, 39, 503, 1996.
52. Groop, L.C., Sulfonylureas in NIDDM, *Diabetes Care*, 15, 737, 1992.
53. Taskinen, M.R., Beltz, W.F., Harper, I., Fields, R.M., Schonfeld, G., Grundy, S.M., Howard, B.V., Effects of NIDDM on very-low-density lipoprotein triglyceride and apolipoprotein metabolism: studies before and after sulfonylurea therapy, *Diabetes*, 35, 1268, 1986.
54. Paton, R.C., Kernoff, P.B.A., Wales, J.K., Effects of diet and gliclazide on the hemostatic system of noninsulin-dependent diabetics, *Br Med J*, 283, 1018, 1981.
55. Balodimos, M.C., Camerini-Dávalos, R.A., Marble, A., Nine years experience with tolbutamide in the treatment of diabetes, *Metabolism*, 15, 269, 1966.
56. Groop, L.C., Pelkonen, R., Koskimies, S., Bottazzo, G.F., Doniach, D., Secondary failure to treatment with oral antidiabetic agents in noninsulin-dependent diabetes, *Diabetes Care*, 9, 129, 1986.
57. Seltzer, H.S., Drug-induced hypoglycemia. A review of 1418 cases, *Endocrinol Metabol Clin North Am*, 18, 163, 1989.
58. Lebovitz, H.E., Pasmantier, R.M., Combination insulin-sulfonylurea therapy, *Diabetes Care*, 13, 667, 1990.
59. Riddle, M.C., Evening insulin strategy, *Diabetes Care*, 13, 676, 1990.
60. Yki-Järvinen, H., Esko, N., Eero, H., Marja-Ritta, T., Clinical benefits and mechanisms of a sustained response to intermittent insulin therapy in type 2 diabetic patients with secondary failure, *Am J Med*, 84, 185, 1988.
61. Ward, E.A., Ward, G.M., Turner, R.C., Effect of sulphonylurea therapy on insulin secretion and glucose control of insulin-treated diabetics, *Br Med J*, 283, 278, 1981.
62. Stumvoll, M., Nurjhan, N., Perriello, G., Dailey, G., Gerich, J.E., Metabolic effects of metformin in non insulin-dependent diabetes mellitus, *N Engl J Med*, 333, 550, 1995.
63. De Fronzo, R.A., Goodman, A.M., and the Multicenter Metformin Study Group, Efficacy of metformin in patients with non insulin-dependent diabetes mellitus, *N Engl J Med*, 333, 541, 1995.
64. U.K. Prospective Study of Therapies of Maturity-onset Diabetes, I. Effect of diet, sulphonylurea, insulin or biguanide therapy on fasting plasma glucose and body weight over one year. Multicentre study, *Diabetologia*, 24, 404, 1983.
65. Crofford, O.B., Metformin, *N Engl J Med*, 333, 588, 1995.
66. Jenkins, D.J.A., Taylor, R.H., Goff, D.V., Fielden, H., Misiewicz, J.J., Sarson, D.L., Bloom, S.R., Alberti, K.G., Scope and specificity of Acarbose in slowing carbohydrate absorption in man, *Diabetes*, 30, 951, 1981.
67. Balfour, J.A., McTavish, D., Acarbose. An update of its pharmacology and therapeutic use in Diabetes Mellitus, *Drugs*, 46, 1025, 1993.
68. Lebovitz, H.E., Oral antidiabetic agents. The emergence of α-glucosidase inhibitors, *Drugs*, 44 (Suppl 3), 21, 1992.

69. Santeusanio, F., Ventura, M.M., Contadini, S., Compagnucci, P., Moriconi, V., Zaccarini, P., Marra, G., Amigoni, S., Bianchi, W., Brunetti, P., Efficacy and safety of two different dosages of Acarbose in noninsulin-dependent diabetic patients treated by diet alone, *Diab Nutr Metab*, 6, 147, 1993.

70. Hoffmann, J., Spengler, M., Efficacy of 24-week monotherapy with acarbose, glibenclamide, or placebo in NIDDM patients, *Diabetes Care*, 17, 561, 1994.

71. Chiasson, J.L., Josse, R.G., Hunt, J.A., Palmason, C., Rodger, N.W., Ross, S.A., Ryan, E.A., Tan, M.H., Wolever, T.M.S., The efficacy of acarbose in the treatment of patients with noninsulin-dependent diabetes mellitus. A multicenter controlled clinical trial, *Ann Intern Med*, 121, 928, 1994.

72. Hollander, P., Safety profile of Acarbose, an α-glucosidase inhibitor, *Drugs*, 44 (Suppl 2), 47, 1992.

73. Santeusanio, F., Compagnucci, P., A risk–benefit appraisal of acarbose in the management of noninsulin-dependent diabetes mellitus, *Drug Safety*, 11, 432, 1994.

74. American Diabetes Association, Consensus statement. The pharmacological treatment of hyperglycemia in NIDDM, *Diabetes Care*, 18, 1510, 1995.

75. McCulloch, D.K., Kurtz, A.B., Tattersall, R.B., A new approach to the treatment of nocturnal hypoglycemia using alpha-glucosidase inhibition, *Diabetes Care*, 6, 483, 1983.

76. Elliott, H.L., Jones, C.R., Vincent, J., Lawrie, C.B., Reid, J.L., The alpha adrenoceptor antagonist properties of idazoxan in normal subjects, *Clin Pharmacol Ther*, 36, 190, 1984.

77. Hofmann, C.A., Colca, J.R., New oral thiazolidinedione antidiabetic agents act as insulin sensitizer, *Diabetes Care*, 15, 1075, 1992.

78. Kumar, S., Boulton, A.J.M., Beck-Nielsen, H., Berthezene, F., Muggeo, M., Persson, B., Spinas, G.A., Donoghue, S., Lettis, S., Stewart-Long, P., for the Troglitazone study group, Troglitazone, an insulin action enhancer, improves metabolic control in NIDDM patients, *Diabetologia*, 39, 701, 1996.

79. Randle, P.J., Hales, C.N., Garland, P.B., Newsholme, E.A., The glucose fatty-acid cycle. Its role in insulin sensitivity and the metabolic disturbances of diabetes mellitus, *Lancet*, 1, 785, 1963.

80. Ferrannini, E., Barrett, E.J., Bevilacqua, S., DeFronzo, R.A., Effect of fatty acids on glucose production and utilization in man, *J Clin Invest*, 72, 1737, 1988.

81. Rachman, J., Turner, R.C., Drugs on the horizon for the treatment of type 2 diabetes, *Diabetic Med*, 12, 467, 1995.

82. Foley, J.E., Rationale and application of fatty acid oxidation inhibitors in treatment of diabetes mellitus, *Diabetes Care*, 15, 773, 1992.

83. Nauck, M.A., Kleine, M., Örskov, C., Holst, J.J., Willms, B., Creutzfeldt, W., Normalization of fasting hyperglycemia by exogenous glucagon-like peptide 1(7-36 amide) in type 2 (non insulin-dependent) diabetic patients, *Diabetologia*, 36, 741, 1993.

84. Gutniak, M.K., Linde, B., Holst, J.J., Efendic, S., Subcutaneous injection of the incretin hormone glucagon-like peptide 1 abolishes post-prandial glycemia in NIDDM, *Diabetes Care*, 17, 1039, 1994.

10

Anticonvulsant Drugs and Drugs for Parkinson's Disease

Mervyn J. Eadie
University of Queensland

Epilepsy and Parkinson's disease are two of the main neurological disorders for which reasonably satisfactory drug therapies are available. The considerations regarding the advantages and disadvantages of drug treatment of epilepsy and Parkinson's disease are so different that the two disorders are more easily discussed separately. In epilepsy the manifestations of the disorder appear intermittently and, at best, the available treatments may lead to what is in effect cure; in Parkinson's disease the symptoms are continuously present once they have appeared, and while treatment may relieve them for a time the benefits of therapy ultimately diminish as the disorder progresses despite the treatment.

10.1   Epilepsy and Anticonvulsant Therapy

The drugs currently available for use as anticonvulsant therapy do not act by curing the processes underlying epilepsy, but interfere with the development and spread of epileptic seizure activity. Anticonvulsant therapy is therefore symptomatic, yet full suppression of seizure mechanisms sometimes seems to result in their becoming permanently inactive.

10.1.1   Benefits

The benefits to be derived from anticonvulsant therapy come from controlling epileptic seizures and thus reducing their consequences for the sufferer's pattern of life. In this connection the word "control" is usually taken to mean reducing the frequency and/or the severity of the patient's individual epileptic seizures. However, it is useful to draw a distinction between the benefits that are opened up by rendering the patient totally free from all seizures and those that come from merely decreasing a patient's seizures.

10.1.1.1   Complete Seizure Control

Two main lines of benefit accrue from rendering the patient completely free from all clinical manifestations of epilepsy. First, while the successful treatment is continued the patient is in effect not epileptic, is not at risk of the dangers to life and limb that exist during seizures, and need not submit to the limitations in patterns of living that society imposes on the person with active epilepsy for his or her protection and that of the local community. Second, if a patient's epileptic seizures are fully controlled for a long enough period, often a matter of several years, there is a prospect that the underlying epileptic process may have gone into remission, so that the therapy can ultimately be ceased without seizures returning, i.e., the patient's epilepsy is in effect cured.

10.1.1.1.1 Health Advantages from Full Seizure Control

Patients may be injured or die during epileptic seizures, and persons with active epilepsy are at a statistically increased risk of sudden unexpected death.[1] Some such deaths may occur during unwitnessed seizure but others probably do not and remain unexplained. Studies have shown that sudden unexpected death in epileptics is often associated with unusually low or unusually high plasma antiepileptic drug concentrations.[2] This suggests that the unexplained deaths are related to the seizure disorder or its treatment. Death in seizures themselves may be due to neurogenic inhibition of cardiorespiratory function, to aspiration, to the consequences of sudden arterial hypertension (particularly if there is preexisting cardiovascular disease), or to injury from failing or drowning during an attack. If a seizure evolves into convulsive status epilepticus the risk of death is still higher, and there is then also the hazard of tissue injury from metabolic acidosis. In the presence of full seizure control, fatalities during seizures should not occur, though there is no real evidence that the hazard of sudden inexplicable death in those with epilepsy will be avoided if the seizures are fully controlled, though it is usually assumed that this will be the case.

Nonfatal bodily injury during seizures can occur as part of the seizures themselves, particularly if they involve violent motor activity, as in bilateral tonic-clonic fits. Injury may also occur as a result of impaired consciousness or falling during seizures. The muscle contractions of the seizure can cause soft tissue injury such as muscle damage, a bitten tongue, a joint injury, e.g., a shoulder dislocation, or vertebral fractures, whereas falling or impaired consciousness may lead to head injury, bruises, abrasions, or burns, depending on the circumstances in which the seizure has occurred. Full seizure control avoids these contingencies.

There are also psychological consequences from having seizures, and living in anxiety of suffering further seizures, which will nearly always occur quite unpredictably. Various continuing emotional reactions may develop, ranging from depression through reasonable or disproportionate anxieties, to bravado or hysteria. Growing acceptance of the knowledge that a patient's seizures are fully controlled can progressively dispel such emotional disturbances and allow his or her personality or revert to its more usual characteristics.

10.1.1.1.2 Social Advantages from Full Seizure Control

The social advantages that derive from full seizure control vary considerably from person to person, and depend heavily on the individual's life situation. Toward one extreme, to continue to experience seizures may constitute relatively little additional social handicap for an intellectually retarded individual who is essentially unemployable in today's technologically complex society. Such a person may have had to forgo certain recreational activities in which a seizure could be dangerous, e.g., swimming alone, bicycle riding. Otherwise, the seizures may have made virtually no difference in the patient's life-style. Full seizure control might then offer little more advantage than a widening of the spectrum of recreational activities that is possible. The benefits of full seizure control would be proportionately much less than in, for example, the self-employed small businessman who may have found that loss of his vehicle driving license and perhaps the occurrence of a seizure during some business dealing may have made it difficult or impractical for him to remain in his occupation.

10.1.1.1.3 "Cure" of Epilepsy

When epilepsy is not due to progressive brain pathology, the available data suggest that there are about three chances in four that the disorder will go into remission after one or a small number of seizures.[3] However, this prognosis applies only for persons treated with anticonvulsants. No data are available for untreated patients, so that it is uncertain whether the remission occurs because of, or merely in the presence of, the therapy. Nonetheless, there is evidence that early and effective treatment is associated with an increased chance of remission,[4] suggesting that the anticonvulsant therapy produces benefit. Thus achieving full seizure control by means of anticonvulsant therapy probably increases the chance of an enduring cure of epilepsy, which completely removes the disadvantages of having suffered from the disorder and allows the patient the opportunity to attain his or her epilepsy-free potential.

10.1.1.2 Improved Seizure Control

10.1.1.2.1 Benefits for Health
The presence of fewer seizures simply reduces the disadvantages for health that are due to seizures but does not remove the disadvantages in the way that full seizure control does. While partial control of seizures decreases most of the physical risks associated with attacks, there is no evidence that it lessens the danger of sudden unexplained death. Partial control of seizures may not decrease the psychological disturbances that are experienced in response to having seizures to the extent that the seizures decrease in frequency or severity. Uncertainty as to when a seizure may occur appears to be a major determinant of the psychological response to having seizures, and this uncertainty continues until it becomes clear to the patient that all his or her seizures have ceased.

10.1.1.2.2 Social Benefits from Improved Seizure Control
Except probably for holding a vehicle driving license, which is usually regarded as an all or nothing matter, decreasing the frequency of a patient's seizures tends to offer proportionate advantages in diminishing the social limitations imposed by having epilepsy. In relation to holding a driving license, the effect of anticonvulsant therapy in decreasing seizures only becomes significant when there has been a long enough seizure-free period to permit a license to be held again. This period appears to vary from one state or country to another, and may also vary with the pattern of the individual's seizure disorder.

10.1.1.2.3 Remission of Epilepsy
Improved but incomplete seizure contol is unlikely to enhance the possibility of achieving remission and cure of epilepsy. Complete seizure control is required for this.

Imperfect but improved seizure control that is achieved by anticonvulsant therapy offers tangible benefits for the epileptic patient, but these benefits yield substantially less advantage for the patient than those that apply if anticonvulsant therapy can produce sustained full control of all seizures.

10.1.2 Risks of Anticonvulsant Therapy

The use of anticonvulsants carries risks and disadvantages as well as benefits. The overall degree of risk, and the severity of each individual risk, can be quantitated up to a point, but the disadvantage each risk holds for the patient varies with the life situation of the individual. The adverse events of treatment, which comprise these risks, can be subdivided pragmatically into those that are:

1. Idiosyncratic but catastrophic
2. Idiosyncratic and inconvenient, at least at their onset
3. Dose dependent, when they may be regarded as (a) manifestation of overdosage, which should normally be recognized and reversed, or (b) insidious and perhaps delayed, and possibly not readily recognized because of their deferred appearance
4. Teratogenic

10.1.2.1 Idiosyncratic Catastrophic Events

Such events are rare during anticonvulsant therapy and also unpredictable. If they were not rare, the drugs that cause them probably would never have come into use, or remained in use once safer agents became available. Despite its antiseizure efficacy and otherwise reasonably desirable adverse effect profile, an anticonvulsant such as methoin (Mesantoin), whose use was associated with a 1% or 2% risk of aplastic anemia[5] could not survive successfully in the marketplace. Similarly, the recently introduced felbamate, the first drug of proven efficacy in the Lennox-Gastaut syndrome,[6] had to be withdrawn from use when cases of aplastic anemia and hepatotoxicity began to appear in its wake.[7]

Aplastic anemia, and hypoplastic development of various hematological cell lines, very rarely complicate treatment with the anticonvulsants in common contemporary use. If such a complication is suspected, a causal relationship is rarely proven by rechallenge with the suspect drug — the therapy being taken is assumed to be culpable, is withdrawn, and the blame is attributed to it in the absence of any more probable cause.

In rare instances, use of phenytoin, carbamazepine, the barbiturate anticonvulsants, and ethosuximide has been associated with extremely severe skin reactions that threaten life, e.g., exfoliative dermatitis.[8] The newer anticonvulsant agents (vigabatrin, lamotrigine, gabapentin) have not yet been widely enough used to know if they carry the same hazard of catastrophic skin reactions.

Valproate use can be complicated by two very uncommon but potentially lethal adverse effects. The first is a hepatitic reaction with liver failure, which, up to 1994, was responsible for at least 132 fatalities worldwide.[9] The second, probably an order of magnitude less common, is pancreatitis.[10] Both adverse effects can occur in the same individual simultaneously.[11] The overall risk of fatal hepatoxicity is calculated to be of the order of 1 in 10,000 patients, but is higher in young infants, particularly those taking multiple anticonvulsants.[12] The hepatotoxicity appears to be related to impaired branched-chain fatty acid β-oxidation,[13] but no predictive test is so far available. It is difficult to obtain data for the proportion of nonfatal cases of valproate-associated hepatotoxicity, but some adults do recover, though others do not. The existence of this rare but potentially lethal reaction calls into question the wisdom of using valproate for varieties of epilepsy in which other agents have a reasonable record of efficacy, e.g., absence epilepsy, juvenile myoclonic epilepsy.

10.1.2.2   Idiosyncratic Initially Inconvenient Events

Minor, initially annoying dermatological reactions to anticonvulsants are much more common than disastrous ones. A morbilliform rash, beginning on the chest, appears in the second week of therapy in 5% to 10% of persons who begin to take phenytoin,[14] and there is probably a similar incidence of skin reaction in the early stages of therapy with carbamazepine.[15] These rashes normally disappear quickly if therapy is ceased, but if treatment with the causative agent is continued the rash can become more extensive and ultimately develop into a generalized dermatitis with evidences of systemic involvement, e.g., eosinophilia, hepatitis, lymphadenopathy, and fever. These dermatological reactions to phenytoin and carbamazepine are generally considered to contraindicate further use of these drugs in patients who have reacted to them.

A skin rash may also occur early in the use of lamotrigine.[16] It is reversible on cessation of therapy, and is said not to recur if the drug is subsequently reintroduced into treatment more gradually. Other anticonvulsants only rarely seem responsible for various patterns of skin rash, nearly all of which are reversible on cessation of therapy.

Other uncommon, and presumably idiosyncratic, reversible, adverse effects that involve other organs are known, e.g., hepatitis (phenytoin, carbamazepine),[17] nephritis (phenytoin),[18] pneumonitis (phenytoin),[19] lymphadenopathy with a histology reminiscent of Hodgkin's disease (phenytoin),[20] heart block (carbamazepine),[21] porphyria (all microsomal mixed-oxidase inducing anticonvulsants),[22] systemic lupus erythematosus (all established anticonvulsants except, perhaps, the benzodiazepines).

Such idiosyncratic unwanted effects of anticonvulsants are mainly of nuisance value during therapy. They inconvenience patients and may cause them anxiety and short-term discomfort, but so long as their nature is recognized promptly and the offending agent is withdrawn, they rarely do lasting harm. However, they limit the range of agents available to treat a given patient's epilepsy.

10.1.2.3   Dose-Dependent Unwanted Effects

Unlike the idiosyncratic adverse effects, this category of unwanted effect will occur in all patients who receive the drug once its dose is high enough. The advent of such adverse effects sets the upper limit of dosage for the drug in question in a given patient. Such adverse effects are generally well known and should not cause more than temporary problems for the intelligent patient if the treating clinician is alert and takes prompt action to modify drug dosage once the presence of the effect is known. There is a greater problem when such adverse effects develop insidiously or have a delayed onset, and do not produce very obvious symptoms so that they may not be recognized readily if patients are hesitant to complain.

These dose-dependent risks are discussed below in relation to the organ system involved.

10.1.2.3.1 Central Nervous System
10.1.2.3.1.1 Higher-level Function
Nearly all anticonvulsants produce a dose- and concentration-dependent depression of higher-level cerebral functions. This begins with a slowing of intellectual processes and a diminution in alertness. In children this can produce an insidious decline in educational performance and in adults a slowing and lessening of efficiency in the carrying out of the patient's occupation. It may take time for the existence of these disadvantageous effects to be recognized, and in the meantime they may lead to secondary anxieties and other emotional disturbances. Various neuropsychological tests can be used to detect the minor degrees of such intellectual impairment before it becomes clinically obvious. However, it is important that such tests measure only the aspects of brain function that are of concern. It has been suggested that the demonstration of the apparently greater effect of phenytoin than carbamazepine on intellectual function[23] depended on the use of neuropsychological tests whose results were determined in part by motor coordination, which is more affected by phenytoin than carbamazepine.

If anticonvulsant doses are taken higher, these effects on intellectual function and alertness will become increasingly obvious to patient and treating physician. Only an imperceptive physician, and a remarkably acquiescent patient, will allow more severe adverse effects to continue for more than a few days, unless it is clear that satisfactory seizure control can be obtained only at these doses, that the advantages of the treatment still outweigh its disadvantages, and that no alternative therapy is available. It is the minor insidious effects on cognitive function that may pass unnoticed and be allowed to persist for months or years. The real question is the amount of harm they do relative to the benefits that the anticonvulsant therapy offers. These cognitive effects appear to be reversible on dosage reduction, but if they have persisted for a long time before they are reversed a younger patient's education may have lagged to an extent that cannot easily be remedied.

Certain anticonvulsants seem more prone to produce adverse cognitive effects than others. Phenobarbitone and its congeners enjoy a bad reputation in this regard, as do benzodiazepine anticonvulsants, e.g., clonazepam. In several studies phenytoin has seemed more undesirable than carbamazepine, with valproate distinctly less likely to cause problems in equieffective anticonvulsant dosage. There has not yet been time for the recently introduced agents vigabatrin, lamotrigine, and gabapentin to find their place in this hierarchy of anticonvulsants that affect cognitive function. It is also clear that there are exceptions to the above rankings in some individuals. For instance, ordinarily therapeutic doses of valproate can very infrequently cause stupor in patients.[24]

Any of the anticonvulsants, in what are ordinarily therapeutic doses, may cause psychological disturbances in occasional subjects. Rarely, seizure control produced by an anticonvulsant, e.g., valproate, vigabatrin, can be associated with the development of an acute psychotic reaction — so-called forced normalization. Clonazepam use may bring out hitherto unnoticed irritability and aggressiveness in occasional patients. Such psychological disturbances should be quite temporary unless their relation to the anticonvulsant therapy is not recognized and the therapy is not revised once they appear.

10.1.2.3.1.2 Brain Stem Function
Phenytoin and carbamazepine can cause dose- and concentration-related brain stem and midline cerebellar disturbances that produce ataxia of gait and later double vision, nausea, and vomiting.[25,26] These effects ordinarily do not occur at drug concentrations sufficient to control epilepsy, though they can be troublesome at therapeutic concentrations in persons who make unusual demands on their balance mechanisms, e.g., ballet dancers, yachtsmen. Such effects should be readily recognizable by the alert physician, are reversible, and should not be allowed to persist for more than a few days.

10.1.2.3.2　Peripheral Nervous System

Long-term phenytoin and carbamazepine intake can cause a subclinical peripheral neuropathy,[27] recognizable on clinical neurophysiological studies. Such a neuropathy will rarely come to clinical attention, but there is also very little evidence that it ever causes clinical problems.

10.1.2.3.3　Gums

Phenytoin, but not other anticonvulsants, causes softening of the gums and gum hypertrophy,[28] which may be made worse by poor dental hygiene. The consequences are partly cosmetic, but the periodontal changes are taken seriously by contemporary dentists and are often regarded by them as an indication for substituting another anticonvulsant in place of phenytoin. The problem is particularly relevant in the child and adolescent. By the time it becomes noticeable the patient may be well established on phenytoin therapy, which is in all other aspects entirely satisfactory, so that there may be an understandable reluctance to change treatment.

10.1.2.3.4　Body Hair

Phenytoin intake can cause overgrowth of hair on the limbs and trunk.[29] Particularly if the hair is dark, this can constitute an appreciable cosmetic disadvantages. Particularly in young girls, the possibility of its future appearance is a relative contraindication to initiating phenytoin therapy.

10.1.2.3.5　Bone

Continual use of phenytoin, phenobarbitone, and perhaps carbamazepine can lead to hypocalcemia and to osteomalacia.[30] Such bony changes develop only after prolonged drug intake. The mechanism seems related to drug-induced alterations in vitamin D metabolism. It is not clear whether changes have ever been recorded, let alone become symptomatic in situations where nutritional standards are high, and where there is sufficient exposure to sunlight for vitamin D to be activated in the skin. Nonetheless, there are concerns that anticonvulsant use may increase the risk of postmenopausal osteoporosis, though whether these concerns represent more than theoretical speculation is another matter.

In the age group in which epilepsy is most often treated it should nearly always be possible to prevent the bony complications of anticonvulsant therapy by appropriate dietary management.

10.1.2.3.6　Folates

Chronic phenytoin or phenobarbitone intake leads to a fall in plasma and red cell folate concentration[31] and, rarely, to a macrocytic anemia.[32] In the great majority of instances the biochemical folate deficiency remains subclinical. There were earlier suggestions that it might be responsible for the cognitive deficiencies associated with anticonvulsant intake,[33] but this possibility no longer seems to be taken seriously. The folate depletion can be remedied by oral folate supplementation, but too high a folate dose can cause a fall in plasma phenytoin and phenobarbitone concentration,[34] thus possibly impairing seizure control.

10.1.2.3.7　Hyponatremia

Carbamazepine intake, and more so use of the carbamazepine congener oxcarbazepine,[35] can cause dose-dependent hyponatremia, which can not only produce symptoms but occasionally cause further seizures.[36] Dosage reduction, or substitution of another anticonvulsant, may become necessary.

10.1.2.3.8　Plasma Biochemical Parameters

Anticonvulsant intake may be associated with various alterations in plasma biochemical parameters, e.g., γ-glutamyl transpeptidase activity[37] and immunoglobulin concentrations.[38] These alternations are not associated with symptoms and constitute no disadvantage for the patient unless the treating practitioner fails to recognize their significance and embarks on unnecessary investigations.

With a sensible and informed patient and an alert treating practitioner, the dose-related adverse effects of anticonvulsant therapy should provide little disadvantage for the patient unless failure of all appropriate alternative therapy forces the attempt to push drug dosage into a range where benefits must be traded off against unwanted effects of therapy. The main exception to this generalization lies in the failure of patient and practitioner to recognize minor cognitive and personality alterations produced by conventional doses of anticonvulsant drugs.

10.1.2.4 Teratogenic Effects

If a pregnant woman takes anticonvulsants, her risk of giving birth to a malformed child is roughly doubled or trebled, though there is still between 18 and 19 chances in 20 that the offspring will be quite normal. Further, there is no completely unambiguous evidence that it is taking the anticonvulsants rather than having epilepsy, which requires anticonvulsant therapy, that is mainly responsible for the association between the therapy and the malformations. The risks are higher if multiple anticonvulsants are taken and the drugs are used in higher dosage and for a longer time,[39] but these particular risk factors are also all measures of the severity of a given patient's epilepsy. Anticonvulsant therapy is associated with a variety of congenital malformations, but there appear to be two association-specific, though uncommon, malformations: the occurrence of spina bifida in some 1% to 2% of pregnancies in which valproate is taken[40] and in some 0.5% of those in which carbamazepine is used.[41]

There is no unambiguous evidence available that cessation of anticonvulsant therapy during pregnancy reduces the risk of having a malformed offspring. The risk is lower in epileptic women who are untreated during pregnancy,[39] but the reason why these women are untreated never appears to be explained in the literature. Possibly their epilepsy was already inactive, or milder than that in the average patient, before they became pregnant. Withdrawal of anticonvulsants during pregnancy is likely to enhance the risk of seizures during pregnancy and there is some evidence that seizures can have deleterious consequences for the products of conception.[42]

Clearly anticonvulsant therapy in pregnancy is associated with risks, not so much for the patient as for the patient's offspring. It has been suggested that these risks can be reduced by (1) ensuring that the lowest adequate anticonvulsant dose is used during pregnancy[39] (though such a dosage should have been in place prior to pregnancy if therapy had been managed optimally) (2) giving folate supplements[39] to reduce the risk of spina bifida, and (3) giving vitamin K during labor,[43] to prevent hemorrhage in the neonate. Certainly there are sensible precautions in the current state of knowledge, but there seems no data available as to the extent to which they actually reduce the risk to the fetus. Of course, prenatal screening, e.g., measuring amniotic fluid α-fetoprotein content at 16 weeks of gestation or carrying out ultrasound examination at 18 to 20 weeks may detect spina bifida in pregnancies in which valproate or carbamazepine is taken and thus offer a possibility for avoiding the long-term difficulties associated with an offspring suffering that particular malformtion.

Assuming that the fetal malformations are due to the maternal anticonvulsant intake, and are not consequences of the disorder for which the anticonvulsant must be used, it is necessary not only to consider the risk rate (mentioned above), but the degree of disadvantage for mother and offspring that the risk carries. Disregarding spina bifida as being essentially avoidable or preventable, though at a cost, some of the malformations associated with exposure to anticonvulsants in pregnancy are major,[44] e.g., facial and palatal cleft, diaphragmatic hernias, various congenital heart malformations, but they are surgically correctable. However, many malformations are minor[44] and are largely cosmetic, e.g., slightly distorted fingertips, various degrees of hypertelorism, and scarcely warrant remedial action.

10.1.3 Reconciliation of Benefits and Risks of Anticonvulsant Therapy

The decision to use anticonvulsant drugs in the patient with epilepsy should always involve an attempt to balance the maximum benefit that the therapy can reasonably be expected to provide against the likely risk the treatment will hold over the probable duration of therapy. In this attempted reconciliation the patient's own aspirations and views need to be taken into consideration. Further, the reconciling of benefits and risks must take into account not only the likely situation in the immediate future, but that which may well apply in several years' time. There is a school of thought that is more concerned with not doing harm than with doing good. Those who subscribe to it and place undue weight on the early, and largely reversible, unwanted effects of anticonvulsant therapy may be reluctant to accept the potential benefit from a commitment to long-term anticonvulsant intake. Yet the main dividends from anticonvulsant therapy are medium and long-term ones, and appropriate therapeutic decisions are not possible until this is appreciated. In the treated individual, the balance between risk and benefit does not necessarily remain static as time passes. Life circumstances change and with time some risks recede and others increase, whereas the potential benefits increase the longer the patient remains seizure free. The guiding principle remains — the risks must never be allowed to outrun the benefits that are realistically attainable, without an attempt being made to redress the situation.

10.2   Parkinson's Disease and its Drug Treatment

The functional deficit in Parkinson's disease seems to depend mainly on deficient dopaminergic neurotransmission in the pathway linking the pars compacta of the substantia nigra to the striatum. The agents that are most effective in restoring neurological function have dopaminergic actions, though these are achieved through different biochemical mechanisms. Thus levodopa is simply the immediate metabolic precursor of dopamine and increases presynaptic dopamine concentrations by a mass action (the concurrent use of levodopa with L-aromatic aminoacid decarboxylase inhibitors that cannot pass through the blood-brain barrier simply avoids wasteful and therapeutically undesirable extraneural synthesis of dopamine). Selegiline (a monoamine oxidase type B inhibitor) blocks one metabolic pathway involved in dopamine degradation and thus helps prolong and increase the postsynaptic actions of the neurotransmitter. Whether it also has a more basic action on the biochemical pathogenesis of Parkinsonism is currently argued.[45] Bromocriptine and pergolide are direct agonists at postsynaptic dopamine receptors. While a number of centrally acting antimuscarinic type anticholinergic agents have anti-Parkinsonian actions, at least one such substance, benztropine, is also a dopamine receptor agonist.[46] The mechanism of action of amantadine has remained elusive. It was originally thought to have dopaminergic actions, perhaps achieved by facilitating the release of preformed dopamine from axon terminals. It is now known to be a glutamine (N-methyl-D-aspartate type) receptor antagonist that may rather selectively block the effects of a rapidly arriving series of glutamate molecules.[47] This action is thought to decrease the excitatory input from the subthalamo-pallidal pathway on the neural circuits whose excessive output onto the neocortex is responsible for the manifestations of Parkinsonism.[48]

In the light of these mechanisms of anti-Parkinsonian drug action it is possible to discuss the benefits and risks of anti-Parkinsonian drug therapy largely in terms of altered striatal dopaminic function.

10.2.1   Benefits of Anti-Parkinsonian Therapy

Unlike the situation in epilepsy, the disturbance of function in Parkinson's disease is present continuously. For the patient to receive the maximum of possible benefit, continuous control of Parkinsonian manifestations is desirable. The main functional deficit in the syndrome is usually analyzed in terms of the consequences of rigidity, bradykinesis, and tremor. Dopaminergic therapy is reasonably successful in correcting Parkinsonian rigidity and bradykinesis so long as these disturbances are not too severe, but the response of tremor is less satisfactory, and sometimes is not noticeable until the therapy has been in use for some weeks or months. Occasionally, relief of rigidity can remove an influence that has been dampening the tremor, so that institution of therapy may seem to make the tremor worse for a time. The anticholinergic agents have some reputation for efficacy against Parkinsonian tremor, but the effect usually is relatively modest.

Dopaminergic therapy is purely symptomatic in Parkinsonism. The underlying disease process, of unknown nature in the idiopathic disease, progresses relentlessly while treatment simply eases the symptoms. There is no prospect of cure, and no evidence that dopamine replacement therapy alters the natural history of the underlying disease process. In these respects the treatments available for Parkinson's disease differ from those available for epilepsy, where many of the underlying pathological mechanisms can become permanently inactive while being treated. Anti-Parkinsonian therapy therefore offers only symptomatic relief that is not necessarily complete and tends to diminish with the passage of time. The duration of benefit varies from patient to patient, the period probably depending on the rate of degeneration of dopaminergic neurons in the pars compacta of the substantia nigra.

While the effects of dopaminergic therapy in Parkinson's disease decrease with time, the treatment may prolong the sufferer's life expectancy quite significantly, as compared with that which applied before levodopa became available. While some of this prolongation may be due to concurrent improvements in medical care, part is likely to be due to postponement of the complications of Parkinsonian immobility, e.g., venous thrombosis, pulmonary embolism, chest infection.

The benefits to be derived from anti-Parkinsonian drug therapy depend on the stage the disease has reached and the dominant clinical manifestations when treatment began.

10.2.1.1 Early and Mild Disease

If the patient has mild, usually early, Parkinson's disease and the exclusive or dominant complaint is of tremor, treatment may have relatively little to offer if there is no evidence of any limitation of mobility. It is unlikely that the available drugs will significantly decrease a true Parkinsonian resting tremor to an extent that will satisfy the patient, and the apparent failure of treatment, and any adverse effects that happen to occur, are likely to disappoint the patient. The intelligent patient may be better managed by explanation and no drug therapy at this stage.

However, should there be some restriction of mobility at this stage and the patient be aware that this is limiting his or her motor performance, the response to dopaminergic therapy can be extremely gratifying, resulting in full or nearly full restoration of mobility, perhaps leaving no readily detectable trace of the disorder while the treatment effect is present, unless Parkinsonian tremor is also present.

10.2.1.2 Moderate Disease

In moderately severe Parkinson's disease there is nearly always restriction of mobility, which troubles the patient, and tremor (if present) is relatively less prominent. Dopaminergic therapy, probably in higher dose than in milder disease, can still provide substantial amelioration of disability in most instances.

10.2.1.3 Severe Disease

It is unlikely that the contemporary patient will reach the stage of severe Parkinson's disease without having received one or more forms of dopaminergic therapy. Some patients may have reached this stage relatively quickly because their disorder is poorly responsive to, or unresponsive to, levodopa, or because the drug cannot be tolerated. They may then be receiving treatment with a postsynaptically acting dopaminergic agent, e.g., bromocriptine, pergolide, but on the whole these are not quite as effective as levodopa itself. If the disorder is also refractory to these agents, or they are otherwise unsatisfactory, a centrally acting anticholinergic or amantadine, or both, might be in use, but in tolerable dosages these agents are not potent enough to produce more than relatively minor (though still useful) benefit.

Considerably more often, the patient with severe disease will be receiving levodopa, often supplemented by selegiline and perhaps a postsynaptic dopamine agent. The treatment will produce quite useful benefit over part of each dosage interval, but there will be one or more of a variety of problems at other stages of the interval. These problems do not seem to be so much adverse effects of the therapy as consequences of the interplay between the therapy and the increasing neuronal devastation in the substantia nigra and striatum. The pathogenesis of the individual disturbances is inadequately understood, though some seem related to altered levodopa pharmacokinetics and others possibly to altered properties of the increasingly depleted population of nigro-striatal neurons.[49] The therapeutic problems include:

1. An increasing shortening of the duration of effect of each levodopa dose
2. Periods of choreiform dyskinesia at the expected time of maximum dopaminergic effect, and of Parkinsonian rigidity and bradykinesis before the next levodopa intake is due
3. Periods of sudden loss of levodopa effect, with freezing of the capacity for movement — the so-called "on-off" effect
4. Dystonic leg and foot movements at times of expected minimum levodopa effects
5. Hallucinations and paranoid ideation (which are also to some extent adverse effects of the therapy). By this stage a degree of dementia may be present as the pathology of the disorder unfolds.

Overall, dopaminergic therapies can provide significant symptomatic benefits in mild and moderate Parkinsonism, and considerably improve the health-related quality of life. In later stages and more advanced disease there are still benefits, but the capacity of the therapy to produce continuous control of disease manifestations has usually become increasingly exhausted.

It is worth mentioning the incidental benefit that amantadine therapy offers in Parkinson's disease in that it can lead to fewer chest infections in patients, because of the drug's anti-influenza A_2 actions.

10.2.2   Risks of Anti-Parkinsonian Therapy

There are a few idiopathic adverse events of anti-Parkinsonian therapy, e.g., pulmonary fibrosis from the ergot–derived postsynaptic dopamine agonists, but the great majority of the risks of anti-Parkinsonian treatment are dose-related expressions of the pharmacodynamics of the agents used.

10.2.2.1   Dopaminergic Agent Effects

10.2.2.1.1   Extraneural Effects

If extraneural levodopa metabolism is not completely blocked by a sufficient dose of an aromatic L-aminoacid decarboxylase inhibitor, some dopamine and other catecholamines are likely to form outside the nervous system. The resultant increased circulating dopamine may cause nausea and perhaps vomiting, but there is very little risk of cardiac arrhythmia, though fear of this was a source of concern when levodopa first became available.

10.2.2.1.2   Nigrostriatal Dopaminic Effects

If Parkinsonism is overfully corrected by dopaminergic agents, choreiform dyskinesias occur, particularly at the expected times of the maximum postdosage dopaminergic effect. When a mild degree of such dyskinesis is present, the patient often experiences an optimal degree of freedom of movement. He or she may therefore be willing to accept a degree of such intermittent involuntary movement as a price for an enhanced overall motor performance. Although there is no general agreement that this is necessarily undesirable, the patient's appearance when the dyskinesia is present may tend to distress his or her relatives. Also some patients who experience this enhanced freedom of movement insist on taking progressively increasing drug doses to continue to boost their motor performance. One has the impression that such patients develop late stage treatment failure problems earlier than would have otherwise been expected.

10.2.2.1.3   Meso-Limbic Dopaminic Effects

Dopaminergic therapy, particularly that provided by the ergot derivatives, tends to cause, or worsen, hallucination and paranoid ideation in Parkinsonian patients. The hallucinations usually distress the patient and the relatives and care givers. Sometimes they may lead to abnormalities of behavior that produce social and occasionally legal difficulties. The dopamine D_4 receptor antagonist clozapine is reported to relieve the hallucinations without compromising control of the Parkinsonism.[50] If this is not available, the dose of the dopaminergic agent must be reduced, or a D_1 and/or D_2 antagonist added to the patient's therapy to relieve the hallucinations. However, either of these courses of action is likely to impair the control of the patient's Parkinsonism.

10.2.2.1.4   Other Effects

Dopaminergic agents and Parkinson's disease itself both tend to cause a fall in blood pressure. The combination of the two can increase the possibility of symptoms due to postural hypotension.

10.2.2.2   Amantadine Effects

Amantadine does not often produce serious adverse effects, but it may sometimes cause hallucinations, postural hypotension, dependent edema, and the peculiar mottled skin appearance of livedo reticularis. Handicap from the latter is purely cosmetic. The hallucinations and the hypotension may be little more than a nuisance, but can occasionally be severe enough to necessitate a change of therapy. If the cause of the edema is not recognized, the patient may be subjected to unnecessary investigations to determine its etiology.

If amantadine is the only anti-Parkinsonian therapy used, and its dose is ceased abruptly, the patient may become very severely Parkinsonian after some hours, and perhaps unable to swallow. This can cause considerable management difficulties.

10.2.2.3 Centrally Acting Anticholinergics

Use of centrally acting anticholinergics carries an appreciable risk of dose-dependent adverse effects. These effects include confusion, drowsiness, hallucinations, blurred vision, dry mouth, constipation, and a tendency to urinary retention. These adverse effects are usually only a mild annoyance to the patient unless the anticholinergic is used in higher dosage as the main source of the patient's anti-Parkinsonian treatment. The adverse effects may then appreciably limit the patient's quality of life.

10.3   Reconciliation of Benefits and Risks

In epilepsy, the benefits to be derived from treatment are often relatively remote, and have to be taken on trust until enough time has passed for the response to become clear, though the risks of treatment are often immediately apparent. In contrast, the benefits from treating Parkinson's disease and many of the risks of the treatment are often clear from the outset of therapy. Therefore, at an early stage in treatment patient and treating practitioner usually find it relatively easy to offset the disadvantages of anti-Parkinsonian therapy against its advantages and to decide whether the therapy should be continued or modified. The problems come later, when the treatment appears to become less effective, or relatively ineffective, and the patient tends to perceive it as producing adverse effects, though the apparent limitations of the therapy are really at least partly due to the evolving pathology of the disease. Unfortunately, the best therapeutic endeavors are not likely to produce a great deal of improvement in this situation though attempts to withdraw anti-Parkinsonian therapy, if the patient insists on it, are likely to show that the existing benefit to risk ratio is not too far from the optimum achievable.

References

1. Klenerman, P., Sander, J. W. A. S. and Shorvon, S. D., Mortality in patients with epilepsy: a study of patients in long term residential care. *J. Neurol. Neurosurg. Psychiatr.* 56, 149, 1993.
2. Lund, A. and Gormsen, H., The role of antiepileptics in sudden death. *Acta Neurol. Scand.* 72, 444, 1985.
3. Shorvon, S. D., The temporal aspects of the prognosis of epilepsy. *J. Neurol. Neurosurg. Psychiatr.* 47, 1157, 1984.
4. Oller-Daurella, L. and Oller, L. F.-V., Influence of 'lost time' on the outcome of epilepsy. *Eur. Neurol.* 31, 175, 1991.
5. Robins, M. M., Aplastic anaemia secondary to anticonvulsants. *Am. J. Dis. Child.* 104, 614, 1962.
6. Felbamate Study Group in Lennox-Gastaut Syndrome, Efficacy of felbamate in childhood epileptic encephalopathy (Lennox-Gastaut syndrome). *N. Engl. J. Med.* 328, 29, 1993.
7. Dichter, M. A., Integrated use of old and new antiepileptic. *Curr. Opin. Neurol.* 8, 95, 1995.
8. Schmidt, D. and Kluge, W., Fatal toxic epidermal necrosis following re-exposure to phenytoin: a case report. *Epilepsia.* 24, 440, 1983.
9. Konig, St. A., Siemes, H., Blaker, F., Boenigk, E., Grob-Selbeck, G., Hanefeld, F., Haas, N., Kohler, B., Koelfen, W., Korinthenberg, R., Kurek, E., Lenard, H.-G., Penin, H., Penzien, J. M., Schunke, W., Schultze, C., Stephani, U., Stute, M., Traus, M., Weinmann, H.-M. and Scheffner, D., Severe hepatotoxicity during valproate therapy: an update and report of eight new fatalities. *Epilepsia,* 35, 1005, 1994.
10. Williams, L. H. P., Reynolds, R. P. and Emery, J. L., Pancreatitis during sodium valproate treatment. *Arch. Dis. Childh.* 58, 543, 1983.
11. Dickinson, R. G., Bassett, M. L., Searle, J., Tyrer, J. H. and Eadie, M. J., Valproate hepatotoxicity: a review and report of two instances in adults. *Clin. Exptl. Neurol.* 21, 79, 1985.

12. Dreifuss, F. E., Santilli, N., Langer, D. H., Sweeny, K. P., Moline, K. A. and Melander, K. B., Valproic acid hepatic fatalities: a retrospective review. *Neurology* 36, 379, 1987.
13. Eadie, M. J., McKinnon, G. E., Dunstan, P. R., MacLaughlin, D. B. and Dickinson, R. G., Valproate metabolism during hepatotoxicity and associated with the drug. *Quart. J. Med.* (184) 77, 1229, 1990.
14. Leppik, I. E., Lapora, J. and Lowerson, R., Seasonal incidence of phenytoin allergy unrelated to plasma levels, *Arch. Neurol.* 42, 120, 1985.
15. Chadwick, D., Shaw, M. D. M., Foy, P., Rawlins, M. D. and Turnbull, D. M., Serum anticonvulsant concentrations and the risk of drug induced skin eruptions. *J. Neurol. Neurosurg. Psychiatr.* 47, 642, 1984.
16. Betts, T., Goodwin, G., Withers, R. M. and Yuen, A. W. C., Human safety of lamotrigine. *Epilepsia* 32 (Suppl. 1), S17, 1991.
17. Gram, L. and Bentsen, K. D., Hepatic toxicity of antiepileptic drugs, a review. *Acta Neurol. Scand.* 68 (Suppl. 97), 81, 1985.
18. Hyman, L. R., Ballow, M. and Kniester, M. R., Diphenylhydantoin interstitial nephritis. Roles of cellular and humoral immunogenic injury. *J. Pediatr.* 92, 915, 1978.
19. Chamberlain, D. W., Hyland, R. H. and Ross, D. J., Diphenylhydantoin-induced lymphocytic interstitial pneumonia. *Chest* 90, 458, 1986.
20. Saltzstein, S. L. and Ackermann, L. V., Lymphadenopathy induced by anticonvulsant drugs and mimicking clinically and pathologically malignant lymphoma. *Cancer* 12, 164, 1959.
21. Hamilton, D. V., Carbamazepine and heart block. *Lancet* 1, 1365, 1987.
22. Larson, A. W., Wasserstrom, W. R., Felsher, B. and Shih, J. C., Post traumatic epilepsy and acute intermittent porphyria, effect of phenytoin, carbamazepine and clonazepam. *Neurology* 28, 824, 1978.
23. Dodrill, C. B. and Troupin, A., Psychotropic effects of carbamazepine in epilepsy: a double-blind comparison with phenytoin. *Neurology* 27, 1023, 1977.
24. Marescaux, C., Warter, J. M., Micheletti, G., Rumbach, L., Coquillat, G. and Kurtz, G., Stuporous episodes during treatment with sodium valproate: report of seven cases. *Epilepsia* 23, 297, 1982.
25. Kutt, H. and McDowell, F., Management of epilepsy with Diphenylhydantoin sodium. *J. Am. Med. Assoc.*, 203, 969, 1968.
26. Livingstone, S., Villamater, C., Sakate, Y. and Pauli, L. L., Use of carbamazepine in epilepsy. Results in 87 patients, *J. Am. Med. Assoc.*, 200, 116, 1967.
27. Danner, R., Electrophysiological effects of diphenylhydantoin and carbamazepine on the peripheral and central nervous system. *Acta Neurol. Scand.* 65, 668, 1982.
28. Angelopoulos, A. P. and Goaz, P. W., Incidence of Diphenylhydantoin gingival hyperplasia. *Oral Surg.* 34, 898, 1972.
29. Livingstone, S., *Comprehensive Management of Epilepsy in Infancy, Childhood and Adolescence.* Springfield, IL, Charles C Thomas, 1972.
30. Richens, A. and Rowe, D. J. F., Disturbance of calcium metabolism by anticonvulsant drugs. *Brit. Med. J.*, 4, 73, 1970.
31. Klipstein, F., Subnormal serum folate and macrocytosis associated with anticonvulsant drug therapy. *Blood* 23, 68, 1964.
32. Stokes, J. B. and Fortune, C., Megaloblastic anaemia associated with anticonvulsant drug therapy. *Aust. Ann. Med.* 7, 118, 1958.
33. Reynolds, E. H., Mental effects of anticonvulsants, and folic acid metabolism. *Brain* 91, 197, 1968.
34. Olesen, O. V. and Jensen, O. N., The influence of folic acid on phenytoin (DPH) metabolism and the 24-hours fluctuation in urinary output of 5-(p-hydroxyphenyl)-phenylhydantoin (HPPH). *Acta Pharmacol. Toxicol.* 28, 265, 1970.
35. Ballardie, F. and Mucklow, J. C., Partial reversal of carbamazepine-induced water intolerance by demeclocycline. *Brit. Med. J.*, 17, 763, 1984.
36. Pendelbury, S. C., Moses, D. K. and Eadie, M. J., Hyponatraemia during oxcarbazepine therapy. *Human Toxicol.* 8, 337, 1989.

37. Abe, Y., Tamagawa, K. and Eguchi, M., Liver function during antiepileptic drug therapy: serum γ-glutamyl transpeptidase activity. *Brain Develop.* 5, 274, 1973.

38. Aarli, J. A. and Tonder, O., Effect of antiepileptic drugs on serum and salivary IgA. *Scand. J. Immunol.* 4, 391, 1975.

39. Kaneko, S. and Kondo, T., Antiepileptic agents and birth defects: Incidence, mechanism and prevention. *CNS Drugs* 3, 41, 1995.

40. Robert, E. and Guibaud, P., Maternal valproic acid and congenital neural tube defects. *Lancet* 2, 937, 1982.

41. Rosa, F. W., Spina bifida in infants of women treated with carbamazepine during pregnancy. *N. Engl. J. Med.* 324, 674, 1991.

42. Minkoff, H., Schaffer, R. M., Delke, I. and Grunebaum, A. N., Diagnosis of intracranial hemorrhage in utero after a maternal seizure. *Obstet. Gynecol.* 65 (Suppl.), 22S, 1985.

43. Solomon, G. E., Hilgartner, M. W. and Kutt, H., Coagulation defect caused by diphenylhydantoin. *Neurology* 22, 1165, 1972.

44. Janz, D., On major malformations and minor anomalies in the offspring of parents with epilepsy: review of the literature. In Janz, D., Bossi, L., Helge, H., Richens, A. and Schmidt, D., eds., *Epilepsy, Pregnancy and the Child.* New York, Raven Press, 1985, 211–222.

45. Parkinson Study Group, DATATOP: a multicentre controlled clinical trial in early Parkinson's disease. *Arch. Neurol.* 46, 1052, 1989.

46. McKillop, D. and Bradford, H. F., Comparative effects of benztropine and nomifensine on dopamine uptake and release from striatal synaptosomes. *Biochem. Pharmacol.* 30, 2753, 1981.

47. Greenberg, D. A., Glutamate and Parkinson's disease. *Ann. Neurol.* 35, 639, 1994.

48. Kopin, U., The pharmacology of Parkinson's disease therapy: an update. *Annu. Rev. Pharmacol. Toxicol.* 32, 281, 1993.

49. Wooten, G. F., Progress in understanding the pathophysiology of treatment-related fluctuations in Parkinson's disease. *Ann. Neurol.* 24, 363, 1988.

50. Friedman, J. H. and Lannon, M. C., Clozapine in the treatment of psychosis in Parkinson's disease. *Neurology* 39, 1219, 1989.

11

Antihistamines and Antiserotonins

Giovanni Passalacqua
University of Genoa

Giorgio Walter Canonica
University of Genoa

11.1   Antihistamines: Introduction

Since their discovery, antihistamines (the term usually refers to H_1 antagonists) have appeared to be highly effective drugs for the symptomatic treatment of allergic diseases, even if marked by troublesome side effects. In recent years pharmacological research has produced more potent molecules devoid of severe side effects and possessing favorable additional properties: the so-called new or nonsedating antihistamines. Recently questions about the safety of these drugs have arisen. The sedative effect, the possible arrhythmogenic effect, and the reported carcinogenic action justify a revision of the literature and a careful evaluation of the benefit/risk ratio.

After their discovery, H_2 antagonists reached central importance in clinical practice for the treatment of peptic diseases and are now widely used. Reports of adverse effects due to this class of drugs have not been so important and frequent as for H_1 antagonists. Nevertheless, because of their large use, often in long-term treatments and in some countries "over the counter," a benefit/risk ratio evaluation appears necessary.

11.2   Overview of the Histaminergic System

Histamine (β-aminoethylimidazole) was identified by Dale in 1910, and subsequently it was recognized as a key mediator of allergic disorders: hay fever, urticaria, and anaphylaxis. In humans histamine is stored in mast cells (and basophils), so it is particularly abundant in skin, gut, and bronchial submucosa. Several stimuli, including physical, chemical, and IgE-mediated, can provoke mast cell degranulation and histamine release. Histamine is rapidly metabolized by histamine-N-methyltransferase and diamine oxydase.

0-8493-2791-1/99/$0.00+$.50
© 1999 by CRC Press LLC

The mechanism of action of histamine was first clarified with the identification of H_1 receptors.[1] The H_1 receptor is largely distributed on smooth muscle cells of the bronchial tree, vessels, and gut. Its stimulation provokes peripheral vasodilation, increased permeability, and bronchoconstriction.

The histaminergic system was better characterized with the later discovery of the H_2 receptor.[2] This receptor is distributed on smooth muscle cells of peripheral vessels and in the gastric tissue. Its stimulation results in a dramatic increase of gastric acid secretion; the histamine-mediated secretagogue action appears to be quantitatively the most important, although interacting with acetylcholine and gastrin. The H_2 receptor also mediates, to a certain degree, the vasodilatory response to histamine, but the H_1 receptor has an affinity much more pronounced to its ligand than the H_2 ones. Nevertheless, the vasodilator effect induced in humans by histamine infusion can be completely inhibited only by the association of H_1 and H_2 blockers.

Finally, the H_3 receptor is apparently represented only in the central nervous system: its function has not yet been clearly elucidated, although a selective antagonist (thioperamide) is available for clinical research.[3]

11.3　First Generation or "Older" H_1 Antagonists

The first H_1 antagonists ("old" or "classic" antihistamines) became commercially available in the 1940s. Promethazine, diphenhydramine, chlorpheniramine, triprolidine, pyrilamine maleate, and hydroxyzine are highly effective in relieving allergic symptoms, but they are also marked by many troublesome side effects. The better known side effect, sedation, limits their clinical use, since it is often so important and severe as to impose the interruption of therapy. The sedative effect is consequent to the lipid solubility and to penetration of the blood-brain barrier.[4] Furthermore, older antihistamines exert additional side actions, commonly observed in clinical practice: constipation, cough, urine retention, xerostomia, nausea, vomiting. These effects are likely due to a certain degree of blockage to cholinergic, α-adrenergic, and serotoninergic receptors.[5] The extrahistaminergic activity probably also contributes to the sedative effect. Finally, older antihistamines are competitive antagonists with short half-lives: therefore, multiple daily doses are usually needed to maintain a satisfactory clinical effect. For these reasons old antihistamines are currently not used for the treatment of allergic disorders, such as perennial or seasonal allergic rhinitis; for these diseases the benefit/risk ratio is, in general, unfavorable (Table 11.1)

Nevertheless, the sedating and antipruritic effects may sometimes be useful, for example, in chronic urticaria and atopic dermatitis; in fact, the first generation antihistamine hydroxyzine is still employed

TABLE 11.1　H_1 Antihistamines: General Aspects

Old Antihistamines	New Antihistamines
Effects	Effects
H_1 antagonism	H_1 antagonism (high selectivity)
Cholinergic, adrenergic, serotoninergic antagonism	Antiallergic: inhibition of mediator release during allergic inflammation
Side effects	Side effects
Sedation	Sedation (weak or absent)
constipation, fatigue, urinary retention, seizures, headache, tachycardia	Arrhythmogenic (rare)
Clinical uses	Clinical uses
Antiallergic	Antiallergic
Antiemetic	(rhinitis, conjunctivitis, atopic dermatitis, chronic urticaria)
Antipsychotic	Asthma (?)
Anti-motion sickness	
Adjuncts in anaphylaxis treatment	

in European countries for the treatment of such disorders. On the other hand, physicians have learned how to take advantage of the side effects exerted by older antihistamines (phenothiazines in particular), on the central nervous system and successfully employ them as antipsychotic, sedative, and antiemetic drugs or sometimes to counter motion sickness.[6] Finally, since old antihistamines can be administered intravenously, they are also used as adjuncts, after adrenaline, in the emergency treatment of anaphylaxis.

11.4 The New "Nonsedating" H_1 Antagonists

11.4.1 General Pharmacological Aspects

The unfavorable ratio between clinical benefits and untoward effects of the old H_1 antagonists induced pharmacological research scientists to improve H_1 selectivity, pharmacokinetics, and tolerability of antihistamines, resulting in the production of new molecules. After the synthesis of ketotifen and oxatomide, which still showed a certain sedative effect, a large number of nonsedating antihistamines were produced: acrivastine, astemizole, azelastine, cetirizine, ebastine, levocabastine, loratadine, terfenadine. The efficacy and safety of these molecules are supported by many experimental data and clinical studies. In addition, at present, other new molecules (emedastine, epinastine, mizolastine, noberastine, setastine) are undergoing clinical trials.[7]

The newer antihistamines have a very high affinity for the H_1 receptor and negligible affinity for the other ones. Moreover, they are large molecules (see Figure 11.1) and have low lipid solubility; thus the blood-brain barrier penetration and the related sedative effect are reduced.

The binding of new antihistamines to H_1 receptors appears less reversible than for the older compounds and their effects do not extinguish rapidly. Indeed, the half-lives of the new antihistamines are largely variable, ranging from 2 hours for acrivastine to 9.5 days for demethyl-astemizole (an active metabolite of astemizole), but the pharmacodynamics are not predictable on the basis of the half-lives.[8,9] The generation of active metabolites and the low reversibility of the binding usually prolong the clinical effects independent of the blood concentration of the drug. Some characteristics of the available compounds are reported in Table 11.2.

Terfenadine (60–120 mg), loratadine (10 mg), cetirizine (10 mg), administered orally, promptly suppress the wheal and flare reaction; the suppression persists for 12 to 24 hours. Similar results are obtained with acrivastine, even if its half-life is shorter than that of other new antihistamines and therefore requires a more frequent administration. On the other hand, astemizole needs long periods of treatment to achieve a potent and long-lasting inhibition of wheal and flare reaction. Almost all new antihistamines are metabolized in the liver and produce active metabolites. Cetirizine and acrivastine differ from the other drugs because they are excreted in the urine without being modified.

Some interesting additional properties of the new antihistamines have recently been described. These drugs appear able, under specific experimental conditions, to inhibit the generation of mediators involved in the allergic inflammation, and to exert a mild bronchodilator effect; for these reasons the term "antiallergic drugs" has been suggested for the new antihistamines.[10,11]

New H_1 antagonists represent, in terms of H_1 receptor selectivity and lack of sedation, a homogeneous class of drugs, even if the pharmacokinetics profile of the various drugs differs. In the following section we will, for this reason, describe in general their side effects and safety.

11.4.2 Safety of the New H_1 Antagonists

11.4.2.1 Sedation

The term *sedation* describes a wide range of personal experiences: drowsiness, loss of alertness, likelihood of falling asleep, decreased concentration, and a global reduction of psychomotor performance. Actually,

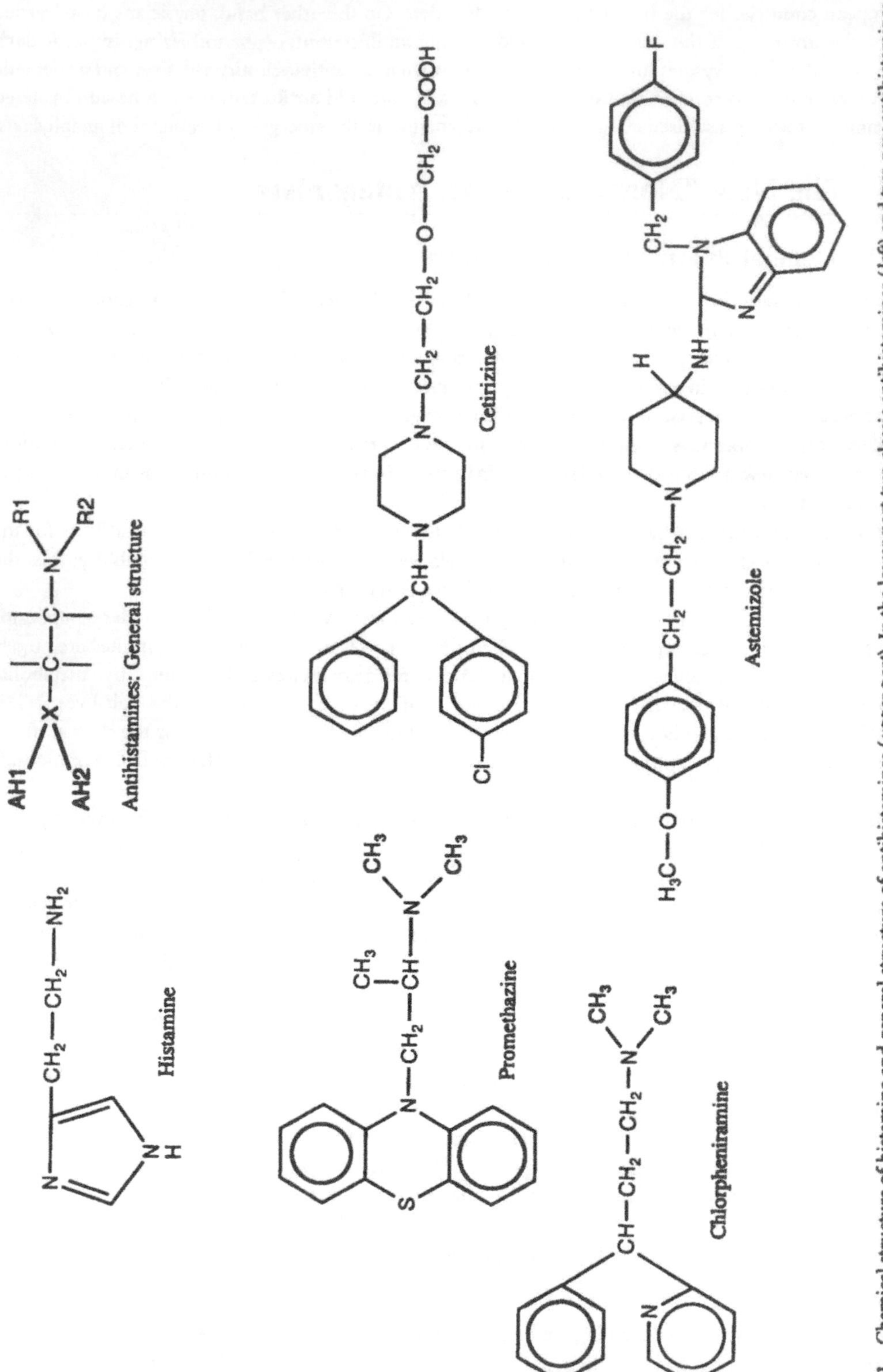

FIGURE 11.1   Chemical structure of histamine and general structure of antihistamines (*upper part*). In the lower part two classic antihistamines (*left*) and two new antihistamines (*right*) are shown for a comparison

TABLE 11.2 Pharmacological Properties of Some New Antihistamines

Drug/Active Metabolite	Dose mg	Half Life	PB (%)	C_{max} (µg/L)	T_{max} (h)	Metabolism
Astemizole	10–30	24 h	97	.8–1.2	.3–.9	Liver
Demethyl-astemizole		9.5 d	na		.4–1	Liver
Azelastine	2–16	25 h	78–	3–5.9	42	Liver
Demethylazelastine		43 h	88			Liver
Cetirizine	10–20	6–11 h	93	250–580	.5–1.5	>60% unchanged
Ebastine	10–50	10 h	na	12	2.5–4.7	Liver
Carebastine		10 h	na	na	na	Liver
Loratadine	20–40	7–11 h	99	4.7–21.6	.7–1.3	Liver
Decarboethoxyloratadine		11–23 h	23–70		.9–2.2	Unchanged
Terfenadine	60–240	16–23 h	97	1.3–4.5	.8–1.1	Liver
Carboxylic acid		17 h	70	2.2–3.6	3	Liver

PB = Protein Binding; C_{max} = max concentration; T_{max} = time to achieve C_{max}; na = not available

histamine is a mediator of the central nervous system and the histaminergic system is known to affect alertness, vigilance, and slow-wave activity during sleep.

The problem of sedation is of great importance for the safety of workers and drivers and for the school performances of children. Obviously, the large employment of H_1 antagonists requires a rigorous scientific measurement of the sedating effect, which is usually performed through clinical and instrumental tests.[12,13] Driving tests (actual or simulated), psychomotor tests, Stanford autoevaluation scale for sleepiness, electroencephalogram, and acoustic evoked potentials are the most frequently used evaluation methods. Driving tests are simple and safe (driving and flying simulators can be used), and they are particularly suitable for a global assessment of psychomotor performance. On the other hand, each of the available psychomotor tests (i.e., visual-motor coordination, reaction time, short-term memory, alertness, etc.) investigates mainly a specific performance. A valid study should be conducted on healthy volunteers, in double blind fashion, and comprising a comparison with ascertained sedative antihistamines (usually triprolidine or chlorpheniramine). The possible additive effect of alcohol or other sedative drugs has also to be evaluated.

A large number of trials satisfying the quoted characteristics is available in literature: all the studies provide concordant results about the lack of significant sedation of the new antihistamines, as summarized below.[13] Astemizole, up to 30 mg/day, appears devoid of any sedative effect when administered both in a single dose and multiple doses. Loratadine shows a measurable sedating effect only at a dose of 40 mg (4 times higher than the recommended one), but it is devoid of any effect on psychomotor performances at the usual dosages.[14] Terfenadine does not show any sedative effect at single doses of 60, 120, 180 mg/day, while a sedative effect appears with 240 mg, which overcomes the recommended dose.[15] On the other hand, the dose of 120 mg t.i.d. does not affect psychomotor performance. Cetirizine, at a dose of 10 mg, also appears devoid of any sedative effect even if administered for 7 days.[16] Finally, the nasal and conjunctival administration of levocabastine does not interfere with psychomotor performance and acrivastine at 4 mg, 8 mg, and 16 mg single doses does not affect the psychomotor tests, as well as ebastine in a 10 mg single dose.[13]

Presently, up to 20 well designed and controlled studies are reported in literature. The results from these studies agree in demonstrating a lack of significant sedative effect from the new antihistamines compared to the older compounds. These results are obtained with the dosages recommended by the manufacturer (Table 11.3), whereas a sedative effect appears for some compounds when such dosages are exceeded.

11.4.2.2 Arrhythmogenic Effect

The histaminergic system exerts little but not negligible actions on cardiac electric activity.[17] Histamine increases the sinus node rate, slows down the conduction of the atrioventricular node via H_1 receptors, and rarely increases the automaticity of the node, or induces abnormal automaticity in Purkinje fibers and working ventricular cells. Thus, an interference with these effects, can partially explain the arrhythmogenic effects described for some of the newer antihistamines.

TABLE 11.3   New H$_1$ Antihistamines: Recommended Doses

Drug	Formulation	Suggested Daily Dose
Azelastine	Tablet 2 mg	2 mg b.i.d.
Acrivastine	Tablet 8 mg	8 mg b.i.d., 8 mg t.i.d.
Astemizole	Suspension 2 mg/ml	10 mg b.i.d.
	Tablet 10 mg	
Cetirizine	Tablet 10 mg	10 mg o.d.
	Drops 10 mg/ml	
Ebastine	Tablet 10 mg	10 mg o.d.
Levocabastine	Eyedrops 0.5 mg/ml	1 drop/eye b.i.d.
	Nasal spray .5 mg/ml	2 spray/nostril b.i.d.
Loratadine	Tablet 10 mg	10 mg o.d.
Terfenadine	Suspension 6 mg/ml	60 mg b.i.d., 120 mg o.d.
	Tablet 60 mg, 120 mg	

o.d. = once daily; b.i.d. = twice a day; t.i.d. = three times a day

In 1975, Hollister reported electrocardiographic alterations (T-wave lowering and flattering, QT prolongation) in patients consuming hydroxyzine, but the problem received no further interest, until new reports about the arrhythmogenic effects of newer antihistamines appeared. In particular, several cases of QT interval prolonging, torsade de pointes, and ventricular dysrhythmias due to terfenadine or astemizole have been published since 1986: some cases were fatal and most of them were near fatal.[13]

Nevertheless, a careful revision of the literature shows that in many cases the assumed doses largely exceeded those suggested by the manufacturers, while in several other cases the patients concurrently consumed drugs interfering with cytochrome P450 metabolism (ketoconazole and cefaclor). Finally, a small number of patients presented previous arrhythmias or impaired liver function. Almost all of the newer antihistamines undergo hepatic metabolization via the cytochrome P450 system: any interference with this metabolic pathway may produce a parent drug accumulation and a quinidine-like effect. The possible macrolide interference with antihistamine metabolism has been depicted since 1983 by Pessayre, and several studies confirmed the impairing effect of erythromycin, ketoconazole, or itraconazole on the metabolism of terfenadine.[18,19] On the other hand, studies conducted in healthy subjects failed to demonstrate any cardiac rhythm or conduction disturbances. To our knowledge, no case of arrhythmogenic effect from acrivastine, cetirizine, levocabastine, or oxatomide has been published. The incidence of cardiotoxic adverse effects appears almost negligible compared to the great number of patients and treatments daily administered all over the world.[20]

11.4.2.3   Tumor Growth

A recent experimental study on mice by Brandes et al.[21] raised the problem of the possible carcinogenic effect of some antihistamines. A population of mice with induced melanoma or fibrosarcoma was intraperitoneally administered astemizole, loratadine, hydroxyzine, doxylamine, or cetirizine for 21 days. Then the weights of tumors were assessed and compared to those of control groups. The increase of the tumor growth appeared maximal for astemizole and loratadine, followed by hydroxyzine, while doxylamine and cetirizine appeared comparable to placebos.

These findings are interesting from a speculative point of view, but they are not consistent with clinical experience: no report of suspected carcinogenity is available in more than 50 years of clinical trials and current use of antihistamines. The results in rodents are not immediately transferable to humans because of the different experimental conditions and the different cellular metabolic systems. Thus, at present the benefit/risk ratio and the clinical use of the antihistamines have not to be modified at all, as stated by the FDA.[22]

11.4.3   Conclusions

An overall review of the available literature, in addition to everyday clinical experience, allows us to conclude for a very favorable benefit/risk ratio of the new antihistamines. These drugs are highly effective in the treatment of several allergic disorders: perennial and seasonal allergic rhinoconjunctivitis, chronic urticaria, and atopic dermatitis; these diseases are presently the ideal indications for the employment of the quoted drugs. The additional antiallergic properties, the mild bronchodilator effect, and the possible preventive effect on asthma onset are to be further evaluated, but they may suggest the future use of new "antiallergic" antihistamines in seasonal rhinoconjunctivitis complicated by mild or intermittent asthma. Furthermore, the quoted properties represent an interesting perspective for future pharmacological research. The well-known sedative effect of the older antihistamines is largely avoidable with the new compounds, which allow a safe long-term treatment. The possible cardiotoxic effect is extremely rare and it appears usually related to an overdose or an altered drug metabolism with abnormal accumulation and consequent effect on cardiac repolarization.

Some simple rules are to be observed to increase antihistamines safety: not exceeding the doses recommended (Table 11.3), avoiding the concurrent administration of drugs that are recognized to interfere with antihistamine metabolism (azolic antifungals or macrolides), taking particular care in subjects with significant liver function impairment (risk of an abnormal drug accumulation), or in subjects with previous rhythm disturbances (prolonged QT, AV block).

The newer antihistamines are useful and safe drugs that are easy to handle. Their employement requires, however, a careful evaluation of each clinical case, mainly in order to recognize possible contraindications.

11.5 H$_2$ Receptor Antagonists

11.5.1 General Pharmacological Aspects

The discovery of the histamine H$_2$ receptor and its involvement in stimulating gastric acid secretion induced the pharmacological research to the synthesis of selective antagonists to the H$_2$ receptor to be used as antisecretive drugs.[2] The first compound (burimamide) was available in the early 1970s, but the first commercially available and worldwide used H$_2$ antagonist was cimetidine. This drug has been followed, during subsequent years, by other powerful compounds, namely ranitidine, famotidine, nizatidine, and roxatidine. Recently, niperotidine (which also has anticholinergic effects) was commercialized in Italy.

H$_2$ antagonists are well and rapidly absorbed in the gut; bioavailability is limited to about 50% because of the hepatic first-pass effect. In this sense nizatidine differs from the other compounds, reaching a biovailability of about 90%. The half-lives range from 2 to 3 hours, but the effects of ranitidine and the subsequent compounds last for about 12 hours, because of the high receptor affinity. All these drugs are excreted largely unmetabolized in the urine. The commercially available H$_2$ antagonists (Table 11.4) have a high selectivity to the H$_2$ receptor and negligible affinity for the other ones. Despite their high affinity and potency, the physiological effects on extragastric H$_2$ (e.g., blood vessels) receptors are actually negligible at the therapeutic doses. Moreover, these drugs have low lipophilicity, so the blood-brain barrier penetration is of little importance, and usually does not cause sedation or similar adverse effects on the central nervous system.

TABLE 11.4　H$_2$ Antagonists

Drug	Formulation	Recommended Dose
Cimetidine	Tablet 200 mg, 400 mg	400 mg o.d./400 mg b.i.d.
Famotidine	Tablet 40 mg	40 mg o.d./40 mg b.i.d.
Niperotidine	Tablet 230 mg, 460 mg	230 mg b.i.d./460 mg o.d.
Nizatidine	Tablet 150 mg, 300 mg	150 mg b.i.d./300 mg o.d.
Ranitidine	Tablet 150 mg, 300 mg	150 mg b.i.d./300 mg o.d.
	Ampoule 50 mg	50–100 mg iv/24 hours
Roxatidine	Tablet 75 mg, 150 mg	75 mg b.i.d./150 mg o.d.

b.i.d. (bis in die) = twice a day; o.d. = once daily

H_2 antagonists dramatically reduce gastric acid secretion, also blocking acetylcholine- and gastrin-induced acid output. The pepsin release from chief cells is also severely inhibited by these agents, while the intrinsic factor production is minimally decreased and vitamin B_{12} absorption is not affected by H_2 antagonist therapy. The treatment of "peptic" diseases represents the elective and virtually unique indication of H_2 antagonists. They are usually employed in the treatment of gastric and duodenal ulcer, leading to ulcer healing in about 75% and 90% of the cases, respectively, after 8 weeks of treatment. Indeed relapses of ulcers are not infrequent after termination of the therapy: for this reason long-term treatments are usually required to prevent relapses. H_2 antagonists are also used in the treatment of erosive gastritis, reflux esophagitis, and in the prevention of anastomosis ulceration after surgery. The treatment of Zollinger-Ellison syndrome usually requires very high doses of H_2 antagonists to achieve a sufficient inhibition of acid output; for this reason the new ATPase pump inhibitors are more suitable for the treatment of this disorder. Finally, since the vasodilation provoked by massive histamine release is partially mediated by H_2 receptors, H_2 antagonists are sometimes administered intravenously as adjuncts in the treatment of anaphylaxis and acute, severe urticaria.

11.5.2   Safety of H_2 Antagonists

"The H_2 receptor antagonists is perhaps one of the safest classes of compounds that has ever been introduced. Side effects occur in less than 1% of the patients...".[23] This consideration is directly derived from the small number of reported severe adverse effects compared to the wide use of H_2 antagonists. Actually the low penetration through the blood-brain barrier and the negligible physiological importance of extragastric H_2 receptors may explain the safety and the excellent tolerability of these drugs.

11.5.2.1   Cimetidine

This imidazole derivative presents an overall incidence of side effects in about 5% of the treated patients. Nevertheless, the most common adverse effects (constipation, headache, dizziness, nausea) are generally of limited severity and do not require the interruption of therapy.[24] The cases of interstitial nephritis, agranulocytosis, and cardiac arrest related to cimetidine are indeed anecdotal. Actually, cimetidine inhibits to a certain degree the cytochrome P450 activity; thus it can affect the metabolism of other drugs. In this regard, theophylline, procainamide, and warfarin are compounds more susceptible to accumulation. On the other hand, this effect is not observed with the newer H_2 antagonists, which do not contain an imidazolic ring.

Some cases of gynecomastia, impotence, and loss of libido have been described in long-term treatment with cimetidine. These effects are likely due to the binding to androgen receptors and to an enhanced prolactin secretion: they are usually reversible, but preclude the use of the drug in children.[25,26] Finally, some cases of mental confusion have been described in elderly patients following cimetidine treatment, though no psychomotor impairment has been observed in healthy patients.[27,28] Consequently, special care has to be paid when administering cimetidine to elderly subjects or to patients with severely impaired liver and renal function: in those cases, the use of newer H_2 antagonists is preferable.[23]

11.5.2.2   Ranitidine

Ranitidine is a basic substitute furan: in general its safety profile appears more favorable than that of cimetidine.[29] Headache and constipation/diarrhea occur in about 2% to 3% of treated patients and rarely require the termination of treatment; similar data have been obtained in long-term treatments.[30,31] No significant interference with the hypothalamic-pituitary-gonadal axis has been observed, and only three cases of gynecomastia, which disappeared after the termination of the treatment, have so far been described.[29] Nine cases of possible ranitidine-related liver damage have been reported, plus a single case of fatal acute hepatitis.[31] A few cases of bradycardia or atrioventricular block have also been described, following rapid intravenous ranitidine administration and high blood concentrations. The possible, although rare, adverse effects of ranitidine on the central nervous system include (in addition to headache) confusion, disorientation, hallucinations, and delirium. Indeed, these effects have often been described in patients with underlying severe diseases.[32] Headache occurs in less than 3% of the cases and it is promptly resolved by the termination of treatment.[33]

11.5.2.3 Famotidine and Allied New H_2 Antagonists

Famotidine, a guanidinothiazole group member, is one of the most potent H_2 antagonists currently available for the treatment of peptic ulcer. Famotidine is about forty times more potent than cimetidine and eight times more potent than ranitidine and is capable of producing ulcer healing in a high percentage of patients.[34] It does not interact with the cytochrome P450 metabolic system, nor with androgen receptors; thus it appears not to affect the pharmacokinetics of drugs that are metabolized in the liver.[35] The neurological effects (headache, confusion, insomnia) appear in percentage comparable to other H_2 antagonists, and the overall incidence of side effects is less than 1%.[36,37] Nizatidine and roxatidine have pharmacokinetic/pharmacodynamic characteristics and safety profiles very similar to famotidine and a comparable rate of side effects.[38,39] Niperotidine, has recently been commercialized: its efficacy appears at least comparable to the other compounds, but there is still no data on its clinical safety.

11.5.3 Conclusions

Since in a large number of cases long-term treatment with H_2 antagonists are required to prevent ulcer recurrence, the benefit/risk ratio represents an extremely important factor for these drugs.[40]

H_2 antagonists are of course among the safest drugs currently available. Cimetidine, the first commercially available one, seems to cause the highest number of side effects. The number of adverse effects reported in the literature is, however, negligible compared to the number of doses administered. The recent H_2 antagonists (ranitidine, nizatidine, and famotidine) appear to be even safer and, for this reason, preferable to cimetidine, expecially for long-term treatment. The literature suggests that particular care has to be taken when H_2 antagonists are administered intravenously: a slow infusion has to be performed to avoid the rare but possible adverse cardiac effects. Furthermore, elderly subjects or patients suffering from severe liver disease appear to be more susceptible to developing adverse psychiatric effects; thus, in these cases a reduction of the doses is suggested. Finally, as an indirect effect, H_2 antagonists may modify the absorption rate of other drugs by altering the intragastric pH, but this action can be partially avoided by a proper administration schedule.

An overall review of the literature suggests a favorable benefit/risk ratio of H_2 antagonists, since the severe adverse effects are extremely rare, fatalities are only anecdotal, and the most common untoward effects are of little severity and are often self-resolving.

11.6 Antiserotonins

11.6.1 Introduction

The pharmacology of the serotoninergic system is complex, since serotonin is largely distributed in the human body and its effects are mediated by a conspicuous number of specific receptors with different and sometimes overlapping actions. Although serotonin receptor antagonists are available in the clinical practice, their indications are not so well defined as for H_1 and H_2 antihistamines, and their clinical use is of course less widespread: a small number of serotonin antagonists (ketanserin, cyproheptadine, 5-HT_3 antagonists) are currently used. These considerations represent a valid reason for a careful evaluation of the benefit/risk ratio of antiserotonins.

11.6.2 Serotonin: Physiological Functions and Receptors

Serotonin (5-hydroxytryptamine, 5-HT) was chemically defined in 1948, although its existence had been suspected since 1930.[41] The largest amount of serotonin (80%) in humans is stored in enterochromaffin cells of the gut, and the remaining amount is distributed in platelets and the central nervous system.

Serotonin is synthesized from tryptophan; it is stored in secretory granules and, after secretion, is rapidly degradated by monoamine-oxydases.

In humans serotonin exerts a surprisingly wide range of physiological effects: it acts as a mediator in the central nervous system; it is involved in the hemostatic process; it mediates both relaxation and contraction of smooth muscles; and it stimulates or inhibits nerve ending activity. Such a variety of actions implies the existence of several receptors with differentiated distribution and variable affinity. Today three main classes of serotonin receptors are well known, namely 5-HT$_1$, 5-HT$_2$, 5-HT$_3$ plus the recently described 5-HT$_4$.[42] Furthermore, 5-HT$_1$ receptors have been divided in five subtypes, on the basis of pharmacologic *in vitro* responses. Briefly the 5-HT$_1$ receptor stimulation inhibits the release of acetylcholine and noradrenaline in the nervous system, while in the cardiovascular system it provokes vasodilation or vasoconstriction in different districts. The 5-HT$_2$ receptor is responsible for smooth muscle cell contraction in vessels, bronchi, and gut and for an increased vascular permeability. Finally, the 5-HT$_3$ receptor is largely distributed in the peripheral nervous system and induces the release of different neurotransmitters.

The experimental classification of 5-HT receptors and their actions are not always sufficient to explain the effects of serotonin agonists and antagonists. In fact, the *in vivo* effects of antiserotonins often differ from those expected on the basis of the experimental models. Today, a small number of relatively selective 5-HT receptor antagonists are known (e.g., ketanserin, cyproheptadine, and "setrons"), and their clinical use has, therefore, to be carefully selected.

11.6.3 Ketanserin

11.6.3.1 General Characteristics and Clinical Effects

Ketanserin is a 5-HT$_2$ antagonist devoid of agonist properties and also exerting a weak α-$_1$ adrenoceptor antagonism. The rationale for the use of ketanserin as an antihypertensive agent is derived from the observation that stimulation of 5-HT$_2$ receptors results in blood pressure elevation.

Actually, ketanserin administered orally or intravenously reduces blood pressure in hypertensive patients, but not in healthy subjects. At a dose of 40 mg b.i.d. ketanserin has an efficacy comparable to metoprolol 100 mg b.i.d., α-methyldopa 500 mg b.i.d., or captopril 50 mg t.i.d.[43] Indeed, the mechanism of action of ketanserin is not clear, since the selective 5-HT$_2$ antagonsist ritanserin does not have any hypotensive effects; a central effect or the α_1-adrenoceptor antagonism has been proposed to explain the effect of ketanserin. Another possibly advantageous property of ketanserin is the ability of inhibiting serotonin-induced platelet aggregation; this effect is more pronounced in elderly subjects.[44] Ketanserin is well absorbed in the gut; its bioavailability is about 50% and it circulates almost completely bound to plasma proteins. It is metabolized in the liver to ketanserinol and its half-life is about 14 hours after a single dose (29 hours at steady state).

11.6.3.2 Safety

The safety of ketanserin has been assessed in a large number of placebo-controlled studies and also in comparison to other antihypertensive agents. In a study performed on a large population (up to 4000 subjects) the adverse effects reported were: dizziness (9.7%), tiredness (9.4%), edema (4.7%), and dry mouth (3.5%).[45] Similar percentages have been found in the remaining studies. None of the quoted effects was severe. A possibly important adverse effect of ketanserin is cardiac rhythm disturbance (QT$_c$ interval prolongation and T-wave flattening); this problem has been variously investigated. A significant QT prolongation appears to occur at doses of 40 mg twice a day or higher, while intravenous administration of ketanserin 10 mg does not alter the cardiac rhythm.[46] Today, 13 cases of torsade de pointes have been reported, but in most cases there were concurrent aggravating factors such as previous prolonged QT or hypokalemia. The quinidine-like effect appears more likely to occur when ketanserin is administered in association with antiarrhythmic drugs or diuretics inducing a potassium loss.[47] In particular, hypokalemia appears to be a critical factor for the onset of adverse cardiac effects. Finally, no significant interference of ketanserin with prolactin secretion or renin activity has been demonstrated.

11.6.3.3 Conclusions

Ketanserin appears to be an effective antihypertensive agent, particularly suitable for elderly patients or in subjects suffering from peripheral arteriopathies. Nevertheless, the possible quinidine-like effect excludes its use in patients with electrolyte imbalance, or receiving potassium-excreting diuretics or antiarrhythmic drugs. Until now the claimed favorable effects of ketanserin on microcirculation are not well defined.

Since a large number of antihypertensive agents, with well-defined mechanisms of action and effectiveness are available, ketanserin should be considered a drug given priority only in selected patients.[43]

11.6.4   5-HT$_3$ Receptor Antagonists

11.6.4.1   General Characteristics

The development of 5-HT$_3$ antagonists is strictly related to cancer chemotherapy-induced emesis. The observation that metoclopramide at high concentrations inhibits the effects of serotonin on guinea pig guts and high doses of metoclopramide reduce cisplatin-induced emesis leads to hypotheses of a 5-HT$_3$ receptor involvement in chemotherapy-induced vomiting. The demonstration of the antiemetic activity of a 5-HT$_3$ antagonist, renzapride, rapidly leads to the synthesis of several powerful compounds: granisetron, ondansetron, and tropisetron are the ones currently available, but new compounds (e.g., MDL 72222 and MDL 743147EF) are undergoing clinical trials. Although at first 5-HT$_3$ antagonists were proposed for the treatment of migraine, their use in chemotherapy-induced emesis promptly appeared the most important and promising application.

Ondansetron, granisetron, and tropisetron are selective 5-HT$_3$ antagonists with powerful antiemetic action. The mechanism of action of these drugs is still a matter of debate: the first hypothesis concerned an action on central nervous system 5-HT$_3$ receptors, while recent experimental data suggest that the main site of action of 5-HT$_3$ antagonists is the gut innervation.[48]

All the three cited compounds are commercially available for both oral and intravenous administration (Table 11.5). Ondansetron, a carbazole compound, shows a bioavailability of about 60%, plasma peak levels are attained about 1 hour after an 8 mg oral dose and the half-life ranges between 3.5 and 4.5 hours (but up to 6 hours in elderly patients). Tropisetron, an azabicyclo derivative, has a high bioavailability (about 90%) and it is poorly metabolized during the first pass; its half life ranges between 7.3 and 8.6 hours, but in "poor metabolizing" subjects it can be up to 40 hours. Granisetron has similar characteristics, but the half life after an intravenous dose of 40 to 80 μg/kg varies from 3.5 hours in healthy subjects to 11 hours in cancer patients. All three drugs are metabolized in the liver via the cytochrome P450 system.[49]

TABLE 11.5   5-HT$_3$ Antagonists

Compound	Therapeutic Regimen		
Ondansetron			
tablet 4, 8 mg	4–8 mg iv	+	1 mg/hr iv in the
ampoule 4, 8 mg	30 min before chemotherapy		subsequent 24 hrs
	OR		
	8 mg orally	+	8 mg orally t.i.d.
	2 hrs before chemotherapy		in the following 5 days
Granisetron			
tablet 1 mg	3 mg iv	+	3 mg iv
ampoule 3 mg	15 min before chemotherapy		in the following 24 hrs
Tropisetron			
tablet 5 mg	5 mg iv	+	5 mg orally/day
ampoule 5 mg	15 min before chemotherapy		for the subsequent 5 days

11.6.4.2   Efficacy and Safety

Granisetron, tropisetron, and ondansetron have been demonstrated as highly effective in preventing or treating chemotherapy- and radiation-induced emesis in a large number of studies, both in the classic ferret model and in humans. In particular, they show high efficacy in chemotherapy protocols employing the strongly emetogenic compound cisplatin (20–120 mg/m^2). The effectiveness of the three drugs is almost similar and generally superior to the antiemetic protocols employing multiple drugs (e.g., prochlorperazine, metoclopramide, dexamethasone). Furthermore, granisetron appears to be effective also in the treatment of anticipatory vomiting.[50] On the other hand, the cited 5-HT$_3$ antagonists are usually less effective in preventing chemotherapy-induced delayed emesis.[49] A summary of the protocols employing 5-HT$_3$ antagonists is given in Table 11.5.

The safety profile of 5-HT$_3$ antagonists compared with current antiemetic treatments is in general very favorable. Granisetron, at the dose of 40 to 160 μg/kg, provokes the following side effects with low incidence: headache (14%), constipation (4%), somnolence (2%), and diarrhea (1%). In contrast to an incidence of about 6% for metoclopramide, no extrapyramidal side effect has been reported.[51] Further-more, no toxicity has been described with doses up to 300 μg/kg. Similar results have been reported with ondansetron: headache occurs in 15% of the patients and diarrhea in 20%, while no extrapyramidal effect is present.[52] An increase of liver enzymes has been described in about 6% of the patients consuming ondansetron, although it is not easy to differentiate between the effects of ondansetron itself and cisplatin.[53] Today a small number of controlled trials is available for tropisetron, but its safety profile does not seem to differ from the other compounds in comparative studies and dose finding studies.

11.6.4.3   Conclusions

5-HT$_3$ receptor antagonists are highly effective in preventing the cancer chemotherapy-induced emesis when highly emetogenic drugs (e.g., cisplatin) are employed. The safety profiles of the commercially available compounds appear satisfactory, and the reported adverse effects (headache, constipation, fatigue) are largely accceptable compared to the therapeutic effect. Actually, since other antiemetic drugs are available, the use of 5-HT$_3$ antagonists should be limited to the severe emesis induced by cancer chemotherapy refractory to conventional treatments. In the latter case they appear to be safe and effective drugs and their use is justified even in view of the high costs.

11.6.5   Cyproheptadine

Cyproheptadine is a phenothiazine-like compound showing both H$_1$ and 5-HT$_2$ receptor antagonism and anticholinergic activity. Nevertheless, it is traditionally considered as an antiserotonin, since the 5-HT$_2$ receptor's antagonism is more pronounced. Cyproheptadine is available as a tablet (4 mg) and a syrup (0.4 mg/ml); it is well absorbed in the gut and metabolized via cytochrome P450, with a half-life of about 6 hours. The pharmacological actions of cyproheptadine are more complex, which can be derived from the simple basis of its affinity to the single receptors mentioned above. In the past years it has been proposed as a therapeutic agent for Cushing's syndrome, migraine, carcinoid syndrome, or hyperin-sulinemia on the basis of clinical trials.[54-56] Moreover, it has been reported to interfere at various degrees with growth hormone secretion, pituitary-gonadal axis, and thyroid hormone metabolism.

Like other phenothiazines, cyproheptadine penetrates the blood brain barrier and exerts a by no means negligible anticholinergic activity; thus it has sedative, antiemetic, and antipsychotic effects. Furthermore, cyproheptadine exerts a particular and specific effect: it increases appetite, leading to significant weight gain. This fact, probably due to a complex and not fully understood action on neurotransmitters, was first considered a troublesome side effect, and subsequently used to treat emaciation, anorexia, and loss of appetite in children, and tumoral and anorexic patients.

In the past, one of the most common clinical indications of cyproheptadine was the treatment of skin disorders (e.g., chronic urticaria, mastocytosis, cold urticaria) or allergic rhinitis. At present, because of the availability of highly effective and selective H$_1$ antagonists, the use of cyproheptadine as an antihist-amine has been almost completely abandoned. Cyproheptadine still represents a possible therapeutic option in the management of cachexia, anorexia, and appetite disorders, but its employment nevertheless requires a careful case-by-case evaluation.[57] In conclusion, the complex and multiple pharmacological actions of cyproheptadine make the differentiation between advantageous and untoward effects more difficult; in any case, its use as an antihistamine or antimigraine drug should be avoided, since compounds possessing well-known efficacy and selectivity are available.

Glossary of Abbreviations

b.i.d. (bis in die): twice a day
o.d.: once daily
t.i.d (tris in die): three times a day

References

1. Ash, A. S. F., Schild H. O., Receptors mediating some actions of histamine, *Br J Pharmacol*, 27, 427, 1966.
2. Black W., Dunian W. A. M., Durant C. J., Ganellin C. R., Parsons E. M., Definition and antagonism of histamine H_2 receptors, *Nature*, 236, 385, 1972.
3. Arrang J. M., Garbarg M., Lancelot J. C., Lecomte J. M., Schwartz J. C., Highly potent and selective ligands for histamine H_3 receptors, *Nature*, 327, 117, 1987.
4. Goldberg M. J., Spector R., Chiang C. K., Transport of diphenhydramine in the central nervous system, *J Pharmacol Exp Ther*, 240, 717, 1987.
5. Timmermann H., Factors involved in the incidence of central nervous system effects of H_1 blockers, In *Therapeutic Index of Antihistamines*, Church, M. K. and Rhioux, J. P., eds, Hogrefe & Huber, New York, 1992.
6. Mitchelson F., Pharmacological agents affecting emesis. A review, *Drugs*, 43(3), 295, 1992.
7. Janssens, M. M. L., Howarth, P., The antihistamines of the nineties, *Clin Rev Allerg*, 11, 111, 1993.
8. Simons, F. E. R., Simons, K. J. Pharmacokinetic optimization of histamine H_1 receptors antagonists, *Clin Pharmacokinet*, 21, 372, 1991.
9. Simons, F. E. R., Simons, K. J., Second generation H_1 receptor antagonists, *Ann Allerg*, 66, 5, 1991.
10. Bousquet, J., Campbell, A., Michel. F., Antiallergic activities of antihistamines, In *Therapeutic Index of Antihistamines*, Church, M. K. and Rhioux, J. P., eds, Hogrefe & Huber, New York, 1992, 57.
11. Holgate, S. T., Finnerty, J. P., Antihistamines in asthma, *J Allerg Clin Immunol*, 83, 537, 1989.
12. Passalacqua, G., Scordamaglia, A., Ruffoni, S., Parodi, M. N., Canonica, G. W., Sedation from H_1 antagonists: evaluation methods and experimental results, *Allergol Immunopathol*, 21, 79, 1993.
13. Simons, F. E. R., H_1 receptor antagonists: comparative tolerability and safety, *Drug Safety*, 10, 350, 1994.
14. Bradley, C. M, Nicholson, A. M., Studies on the central effects of the H_1 antagonist loratadine, *Eur J Clin Pharmacol*, 32, 419, 1987.
15. Betts, T. A, Edson, A. E., Furlong, P. I., The effects of single doses of 120 and 240 mg of terfenadine on driving and other measures of psychomotor performance including visual evoked responses, *Ann Allerg*, 66, 98, 1991.
16. Levander, S., Stahle-Blackdahl, M., Hagemark, O., Peripheral antihistamines and central sedative effects of single and continuous oral doses of cetirizine and hydroxyzine, *Eur J Clin Pharmacol*, 41, 435, 1991.
17. Wolff, A. A., Levi, R., Histamine and cardiac arrhythmias, *Circ. Res*, 58, 1, 1986.
18. Honig, P. K., Woosley, R. L., Zamani, K., Conner, D. P., Cantilena, L. R., Changes in the pharmacokinetics and electrocardiographic pharmacodynamics of terfenadine with concomitant administration of erythromycin, *Clin Pharmacol Ther*, 52, 231, 1992.
19. Pohjola-Sintonen, S., Viitasalo, M., Toivonene, L., Neuvonen, P., Torsades de pointes after terfenadine-itraconazole interaction, *Br Med J*, 306, 186, 1993.

20. Hanrahan, J. P., Choo, P. W., Carlson, W., Greineder, D., Faich, G.A., Antihistamines associate sudden death, ventricular arrhythmias, syncope and QT interval prolongation: a comparison of terfenadine and other antihistamines, *Post Marketing Surv,* 6, 23, 1992.

21. Brandes, L. J., Warrington, C., Arron, R. J., Bogdanovic, P. R., Fang, W., Queen, G. M., Stein, D. A., Tong, J., Zaborniak, C. L. F., LaBella, F. S., Enhanced cancer growth in mice administered daily humans-equivalent doses of some H_1 antihistamines: predictive *in vitro* correlates, *J Nat Cancer Inst,* 86, 770, 1994.

22. FDA reviews antihistamine mouse study, *FDA Talk Paper,* May 17th, 1994.

23. Deakin, M., Williams, J.G., Histamine H_2 receptor antagonists in peptic ulcer disease: efficacy in healing peptic ulcer, *Drugs,* 44, 709, 1992.

24. Freston, J. W., Cimetidine II. Adverse reactions and patterns of use, *Ann Int Med,* 97, 728, 1992.

25. Funder, J. V., Mercer, J. E., Cimetidine, a histamine H_2 receptor antagonist occupies androgen receptors, *J Endocrinol Metab,* 48, 198, 1979.

26. Kelly, D.A., Do H_2 receptor antagonists have a therapeutical role in childhood ? *J Ped Gastroenterol Nutr,* 19, 270, 1994.

27. Sonnenblick, M., Rosin, A. J., Weissberg, N., Neurological and psychiatric side effects of cimetidine, *Postgrad J Med,* 58, 415, 1982.

28. Orr, W. C., Duke, J. C., Imes, N. K., Mellow, M. H., Comparative effects of H_2 receptor antagonists on subjective and objective assessment of sleep, *Aliment Pharmacol Ther,* 8, 203, 1994.

29. Wormsley, G. K., Safety profile of ranitidine, *Drugs,* 46, 975, 1993.

30. Dawson, J., Richards, D. A., Stables, R., Dixon, G. T., Cockel, R., Ranitidine: pharmacology and clinical use, *J Clin Hosp Pharm,* 8, 1, 1983.

31. Lewis, J. H., Safety profile of long-term H_2 antagonist treatment, *Aliment Pharmacol Therap,* 5, 49, 1991.

32. Cantu, T. G., Korek, J. S., Central nervous system reactions to histamine-2 receptor blockers, *Ann Int Med,* 144, 1027, 1991.

33. Epstein, C. M., Histamine H_2 antagonists and the nervous system. *Am Fam Physician,* 32, 109, 1985.

34. Schunack, W., What are the differences between the H_2 receptor antagonists? *Aliment Pharmacol Ther,* 1, 493S, 1987.

35. Famotidine. The ACG committee on FDA related matters with primary authorship by G.Friedman, American College of Gastroenterology, *Am J Gastroenterol,* 82, 504, 1987.

36. Chichmanian, R. M., Mignot, G., Spreux, A., Jean-Girard, C., Hofliger, P., Tolerance of famotidine. Study of network of sentinel physicians in pharmaco-vigilance, *Therapie,* 47, 239, 1992.

37. Saigenij, K., Fukutomi, H., Nakazawa, S., Famotidine: postmarketing clinical experience, *Scand J Gastroenterol,*134, 34, 1987.

38. Di Ciommo, V., Ferrario, F., Nizatidine a meta-analytical study, *Clin Ther,* 143, 201, 1993.

39. Indue, M., Clinical studies on the use of roxatidine for the treatment of peptic ulcer in Japan, *Drugs,* 35, 114, 1988.

40. Freston, J. W., H_2 receptor antagonists and duodenal ulcer recurrence: analysis of efficacy and commentary on safety, *Am J Gastroenterol,* 82, 1242, 1987.

41. Rapport, M. M., Green, A. A., Page, I. H., Serum vasoconstrictor (serotonin). IV. Isolation and characterization, *J Biol Chem,* 176, 1243, 1948.

42. Bradley, P. B., Engel, G., Feniuk, W., Fozard, J. R., Humphrey, P. P. A., Proposal for the classification and nomenclature of functional receptors for serotonin, *Neuropharmacol,* 25, 563, 1986.

43. Brogden, R. N., Sorkin, E. M., Ketanserin. A review, *Drugs,* 40, 907, 1990.

44. Amsten, R., Fetkovska, N., Terracin, F., Pletscher, A., Buhler, F. R., Serotonin metabolism and age-related effects of antihypertensive therapy with ketanserin, *Drugs,* 36, 61, 1988.

45. Janssens, M., Symoens, J., Prevention of arterosclerotic complications with ketanserin. Results of a large intervention trial, *J Drug Ther Res,* 14, 138, 1989.

46. Cameron, H. A., Waller, P. C., Ramsay, L. E., Prolongation of the QT interval by ketanserin, *Postgrad Med J,* 64, 112, 1988.

47. PACK claudication substudy investigators. Random placebo-controlled double blind trial of ket-anserin in claudicants. Changes in claudication distance and ankle systemic pressure, *Circulation*, 80, 1544, 1989.
48. Andrews, P. R. L., Davis, C. J., Bingham, S., Davidson, G. H. S., Hawthorn, J., The abdominal visceral innervation and the emetic reflex: pathways, pharmacology and plasticity, *Can J Physiol Pharmacol*, 68, 325, 1990.
49. Aapro, M. S., HT_3 receptor antagonists: an overview of their present status and future potential in cancer therapy-induced emesis, *Drugs*, 42, 551, 1990.
50. Aapro, M. S., Kirchner, V., Terrey, J. P., The incidence of anticipatory nausea and vomiting after repeat cycle chemotherapy: the effect of granisetron, *Br J Cancer*, 69, 957, 1994.
51. Tabona, M. V., An overview on the use of granisetron in the treatment of emesis associated with cytostatic chemotherapy, *Eur J Cancer*, 26, s37, 1990.
52. Bryson, J. C., Finn, A. L., Plagge, P. B., Twaddel, T. P. H., Brenckman, W. D., The safety profile of ondansetron from clinical trials, *Proc Am Soc Oncol*, 9, 328, 1990.
53. Smith, R. N., Safety of ondansetron, *Eur J Cancer Clin Oncol*, 25, s19, 1989.
54. Moertel, C. G., Kvols, L. K., Rubin, J., A study of cyproheptadine in the treatment of metastatic carcinoid tumor and the malignant carcinoid syndrome, *Cancer*, 67, 33, 1991.
55. Lance, J. W., Anthony, M., Sommerville, B., Comparative trial of serotonin antagonists in the management of migraine, *Br Med J*, 8, 327, 1970.
56. Mylecharane, E. J., $5\text{-}HT_2$ receptor antagonists and migraine therapy, *J Neurol*, 238, S45, 1991.
57. Tchekmedyan, N. S., Halpert, C., Ashley, J., Heber, D., Nutrition in advanced cancer: anorexia as an outcome variable and target of therapy, *J Parenter Enter Nutr*, 16, 88S, 1992.

12

Cytokines and Immunomodulators

David J. Peace
Loyola University Medical Center

Warren Wong
Loyola University Medical Center

Karen T. Ferrer
University of Illinois Medical Center

12.1 Introduction

Cytokines have emerged as an important new class of drugs for the treatment of malignancies, infectious diseases, hematological cytopenias, and other disorders.[1] Currently nine different recombinant and natural cytokines are approved for systemic administration in the United States, including erythropoietin, granulocyte colony stimulating factor, granulocyte-macrophage colony stimulating factor, interleukin-2, interferon αn3, interferon α2a, interferon α2b, interferon β1b, and interferon γ 1B. Several dozen other cytokines are now available as recombinant proteins suitable for *in vivo* evaluation. Many therapeutic applications have been explored at the preclinical level with encouraging results. However, a number of cytokines with therapeutic effects in experimental models have failed to enter the mainstream of medical practice, either because they have had limited clinical benefit or because they have been associated with prohibitive toxicities. The pleiotropic biological activity of most cytokines has complicated the use of cytokines as systemic drugs. The following review focuses on the nature of the biology of cytokines and its implications for the evaluation, application, and benefit/risk analysis of cytokines as pharmacological agents.

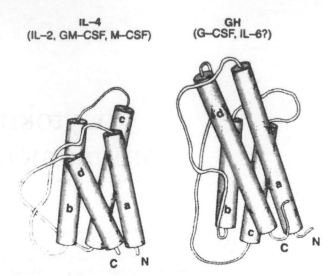

FIGURE 12.1   Schematic representation of "short" and "long" α-helical cytokines. The depicted molecules are indicated in bold lettering at the top of the figure; representative members of each class are listed in parentheses. See text for further details. Reprinted with permission.[3]

12.2   General Principles of Cytokine Biology

12.2.1   Definition and Classification

Cytokines are small, intercellular signaling molecules that regulate cell growth, differentiation, and function.[1] Many types of cells, notably lymphocytes and macrophages, produce and secrete cytokines in response to specific activating signals. Cytokines participate as paracrine and autocrine factors within complicated regulatory networks to elicit and coordinate various biological processes, including cellular and humoral immunity, inflammation, chemotaxis, wound healing, and hematopoiesis. Individual cytokines commonly are classified according to their principal biological action, e.g., hematopoietic, immunomodulatory, proinflammatory, antiinflammatory and chemotactic. However, functional classifications do not reflect the fact that the biological actions of most cytokines are pleiotropic and dependent on the context in which they are evaluated.[1,2]

Five major families of cytokines have been identified on the basis of distinguishing structural features.[3] Several cytokine families belong to larger superfamilies that encompass other growth and hormonal factors. The "α-helical" cytokines are structurally related to growth hormone and constitute the largest cytokine family. This family includes a preponderance of cytokines with primary hematopoietic and immunomodulatory activities. The α-helical cytokines are small glycoproteins that share a conserved core of four α-helices arranged with a unique up-up-down-down orientation. Two subfamilies of differing helical lengths have been delineated (Figure 12.1).[3] Short helices consisting of approximately 15 amino acids are characteristic of interleukin (IL)-2, IL-4, IL-5, granulocyte-macrophage colony stimulating factor (GM-CSF), and macrophage colony stimulating factor (M-CSF) and are predicted for IL-3, IL-7, IL-9, IL-13, and stem cell factor (SCF). Long helices consisting of approximately 25 amino acids are characteristic of growth hormone (GH), interferon (IFN)β and granulocyte colony stimulating factor (G-CSF) and are predicted for IFNα, IL-6, IL-10, IL-11, leukemia inhibitory factor (LIF), erythropoietin (Epo), and ciliary neurotrophic factor (CNTF). IFNγ is a structurally related, but unique, α-helical cytokine.[4]

The other major structural cytokine families are represented by cytokines with a bowl-like β-trefoil structure, e.g., IL-1α and IL-1β; homodimeric cytokines with a distinctive knot of interlinking cystine loops, e.g., transforming growth factor (TGF)-β$_1$, TGF-β$_2$ and TGF-β$_3$; homotrimeric cytokines with a jelly roll-like pyramidal structure, e.g., tumor necrosis factor (TNF)α and TNFβ; and monomeric cytokines with

TABLE 12.1   Abbreviations

Abbreviation	Term
α	alpha
β	beta
γ	gamma
CNS	central nervous system
CNTF	ciliary neurotrophic factor
Epo	erythropoietin
G-CSF	granulocyte colony stimulating factor
GH	growth hormone
GM-CSF	granulocyte-macrophage colony stimulating factor
HLA	human leukocyte antigen
ICAM	intracellular adhesion molecule
IFN	interferon
IGIF	interferon-gamma inducing factor
IL	interleukin
LAK	lymphokine-activated killer cell
LFA-1	leukocyte function-associated antigen
LGL	large granular lymphocyte
LIF	leukemia inhibitory factor
M-CSF	monocyte colony stimulating factor
MIF	migration inhibition factor
MIP	macrophage inflammatory protein
MPL	myeloproliferative leukemia factor
NK	natural killer
OSM	oncostatin M
PF	platelet factor
PRL	prolactin
R	receptor
SCF	stem cell factor
TGF	transforming growth factor
TNF	tumor necrosis factor
Tpo	thrombopoietin

a single α-helix and multiple β-strands, e.g., IL-8, platelet factor-4 (PF-4) and macrophage inflammatory protein-1β (MIP-1β).[3] Members of the last family are particularly small (approximately 8 to 10 kD) and act over short distances to mediate chemotaxis and other inflammatory effects, hence their designation as "chemokines."[5] Other cytokines recently have been identified, e.g., IFNγ-inducing factor (IGIF), that exhibit unique structural features and may represent novel, undefined cytokine families.[6]

12.2.2   Mechanisms of the Biological Activity of Cytokines

The biological effects of cytokines are mediated by specific receptors that are displayed on the surface of responding cell types. Remarkable progress has been made in recent years toward the elucidation of the structure and physiology of cytokine receptors.[7-12] Many different receptors have been cloned and sequenced. In general, cytokine receptors are complex structures with multiple functional domains. Several distinct receptor families have been identified that correspond to the aforementioned structural cytokine families.[10-12]

Receptors for the α-helical cytokines include one or more type I transmembrane proteins with an amino-terminal extracellular domain, a single membrane-spanning domain, and a variable cytoplasmic domain that lacks intrinsic kinase activity.[10,12] The ligand-binding extracellular receptor domain includes fibronectin type III-like subdomains with multiple conserved residues.[12] Two different subclasses of α-helical cytokine receptors have been identified on the basis of a distinguishing sequence motif within the fibronectin type III subdomain that is proximal to the cell membrane.[12] Class 1 receptors that are utilized by various lympho-hematopoietic growth factors contain the consensus sequence tryptophan-serine-x-tryptophan-serine (WSXWS), where "x" is a nonconserved residue (Figure 12.2).[10,12] Class 2 α-helical cytokine receptors that are utilized by the interferons and IL-10 do not contain the WSXWS motif.[12]

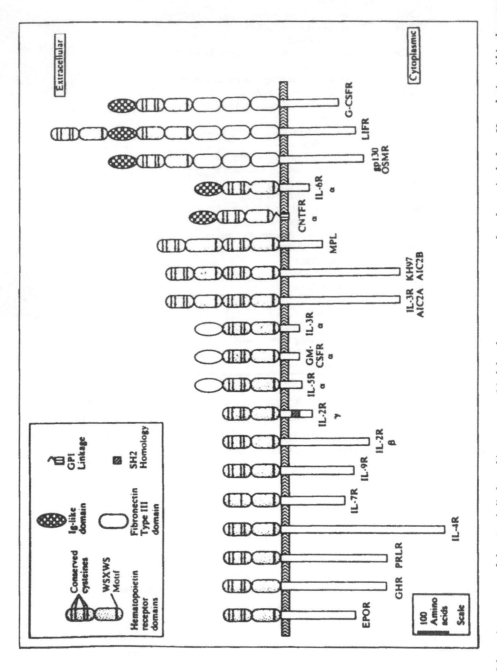

FIGURE 12.2 Schematic structures of class 1 α-helical cytokine receptors. Modular elements are represented as shown in the box. Homologies within the cytoplasmic domains are generally very limited and confined to short stretches of amino acid sequence. For those receptors that exist in more than one isoform with different length cytoplasmic domains the longer one is depicted. Reprinted with permission.[10]

Receptors for members of the IL-1, TNF, and TGF-β cytokine families also include type I transmembrane proteins. Two different receptors with multiple extracellular immunoglobulin-like domains have been identified that bind to both IL-1α and IL-1β. Similarly, two different receptors, TNF-RI (p55) and TNF-RII (p75), with reiterative cysteine-rich extracellular domains, have been identified that bind to both TNFα and TNFβ. There are three different receptors for TGF-β isoforms, including two type I transmembrane receptors with intracellular serine/threonine kinase subdomains.[13] Receptors for chemokines are structurally distinct and consist of proteins with seven transmembrane domains.[5]

Many cytokine receptors are composed of multiple subunits. Homodimeric, heterodimeric, and trimeric receptors have been described.[1,10,11] The multicomponent nature of cytokine receptors confers functional versatility and the potential for complex regulatory control. This is exemplified by the multiple forms of IL-2 receptor (IL-2R) that are variably displayed by quiescent and activated T lymphocytes. Normally, T lymphocytes exist in a dormant state, wherein they constitutively express basal levels of low-affinity IL-2R, composed of two different class 1 α-helical cytokine receptor chains, p75 β and p64 γ. T cell activation by specific antigenic stimulation results in the up-regulation of the p75 β-chain and *de novo* expression of a third, structurally distinct IL-2R chain, p55 α. The recruitment of the p55 α-chain to the IL-2R complex markedly increases the receptor's affinity for IL-2.[14] Activated T cells that express the high-affinity heterotrimer are preferentially induced to proliferate by ambient IL-2. Intriguingly, the IL2R p64 γ-chain also associates with ligand-binding receptors for a number of other α-helical cytokines that support T cell proliferation, including IL-4, IL-7, IL-9, and IL-15.[1,2,15] Other functionally related groups of cytokines have been identified that bind to receptors that share interchangeable component receptor chains. The phenomenon of "shared" receptor components may account, in part, for the functional redundancy of different cytokines.

The binding of cytokines to cognate cell surface receptors activates signal transduction pathways that control myriad cellular processes, including proliferation, differentiation, metabolism, motility, and migration. The nature of the transduced signal is determined by the precise composition and status of the receptor, while it is occupied with a cytokine ligand. In the example of the high-affinity trimeric IL-2R, signaling is mediated by the interaction of specialized regions of the cytoplasmic domains of both the p75 β-chain and the p64 γ-chain with multiple nonreceptor phosphotyrosine kinases. The p75 β-chain recruits several different phosphotyrosine kinases including p56[kk], Jak-1 and syk, whereas the p64 γ-chain recruits the Jak-3 phosphotyrosine kinase to the IL-2R complex.[16,17] The recruitment and activation of phosphotyrosine kinases by IL-2 stimulation causes the phosphorylation of multiple arrays of intracellular proteins and, ultimately, the expression of various genes, including c-Fos, c-Jun, c-Myc, and bcl-2.[8,16]

All evidence suggests that heterogeneous arrays of signal transduction also can be evoked by the stimulation of other cytokine receptors. In general, class 1 and class 2 α-helical cytokine receptors have the ability to recruit one or more types of nonreceptor phosphotyrosine kinases and to activate multiple intracellular signal pathways, including the ras-raf-map kinase pathway and various Jak-Stat pathways.[8,9,17] The signal transduction mechanisms of other types of cytokine receptors have not been as well defined, but generally rely on initial triggering of serine/threonine phosphokinases, rather than phosphotyrosine kinases. Nonreceptor serine/threonine kinases are recruited by receptors for members of the IL-1 and TNF cytokine families through a mechanism that involves the hydrolysis of sphingomyelin to ceramide. Receptors for chemokines also recruit and activate nonreceptor serine/threonine kinases.[5] In contrast, receptors for TGFβ utilize intrinsic serine/threonine kinase domains for signal transduction.

The multiplex and multistep nature of cytokine signal transduction provides many potential opportunities for signal modulation and differentiation. Variations in the level of individual components of the surface receptor complex or of the corresponding intracellular signal pathways can alter both the quantity and quality of signal throughput. Cytokines elicit protean and, at times, antithetical responses, depending on many different factors that affect the status of various aspects of signal transduction. In most circumstances, the specific properties of cytokine signaling that account for the unique functional effects of particular cytokines have not been defined.

The distribution of specific cognate receptors within a recipient is an important determinant of the constitutional effects of exogenously administered cytokines. Cytokines that bind to widely expressed receptors, e.g., IL-1, IL-6, IL-12, TNFα, and IFNγ, mediate highly pleiotropic effects when administered systemically. Cytokines that bind to tissue-restricted receptors, e.g., the hematopoietic growth factors, generally have more focused biological effects. However, a number of tissue-restricted cytokines mediate distal pleiotropic effects by inducing the synthesis and release of secondary cytokines. The assessment of the direct biological actions of individual cytokines often is complicated by the confounding influence of secondarily induced cytokines.

12.2.3   Biological Effects of Cytokines

The biological activities of selected cytokines are shown in Table 12.2. The variety and scope of cytokine function is striking. Well known for their ability to control the growth and function of lymphocytes and other hematopoietic cells, cytokines also regulate the growth, differentiation, and function of many other cell types. Virtually all somatic cells express receptors for multiple cytokines and, therefore, potentially can be affected by cytokine stimulation. Cytokines facilitate and modulate many normal cellular functions. They also mobilize and coordinate multifaceted host responses to many different types of insult. Cytokines play a central role in eliciting and driving host immune and inflammatory responses and in directing the repair and reconstruction of damaged tissues.

Individual cytokines can have heterogeneous effects on particular types of cells. IL-2, for example, has multiple effects on subpopulations of lymphocytes. IL-2 promotes the proliferation of activated B and T lymphocytes; the differentiation of cytotoxic lymphocytes, including lymphokine-activated killer (LAK) cells, natural killer (NK) cells, and cytotoxic T lymphocytes (CTL); the up-regulation of lymphocyte surface molecules, including LFA-1, ICAM-1, and HLA determinants; and the production and release of secondary cytokines, including TNFα and IFNγ.[1] Although IL-2 is a primary growth factor for T lymphocytes, it has been shown that IL-2 also programs T lymphocytes to undergo apoptosis.[18]

The unique biological effects of individual cytokines often have been difficult to differentiate from the overlapping functions of other cytokines.[1,19] Gene knockout techniques have become an important method for the determination of the specific *in vivo* functions of designated cytokines. Homologous deletions of the genes for IL-2, for example, have been used to examine the role of IL-2 during T lymphocyte ontogeny. Remarkably, IL-2 does not appear to be required for many aspects of T lymphocyte development.[20] Newborn mice lacking endogenous IL-2 exhibit a normal composition of thymic and peripheral T lymphocyte subsets. Ultimately, IL-2 knockout mice succumb to a hyperproliferative T cell disorder, demonstrating that IL-2, paradoxically, is required for the down-regulation of T lymphocyte activation and growth *in vivo*.[20,21] Similar knockout strategies have been used to assess the essential roles of many other cytokines and their corresponding receptors.

Cytokines have been implicated in the pathophysiology of many diseases. Either insufficient or excessive production of cytokines can have pathogenic consequences. Impaired production of endogenous erythropoietin, for example, is an important cause of the anemia associated with chronic renal insufficiency. Conversely, excessive or aberrant production of endogenous cytokines has many pathogenic effects, ranging from distorted metabolism, e.g., cachexia, to overt tissue injury and cell death. The overproduction of cytokines with pronounced proinflammatory effects, including IL-1, IL-6, TNFα and IFNγ, have been especially implicated in the pathogenesis of various diseases. TNFα, has been extensively studied in this regard.[22,23]

Elevated levels of TNFα have been detected in the serum or affected tissues of patients with many different diseases including bacterial sepsis, graft-versus-host disease, systemic vasculitis, rheumatoid arthritis, idiopathic pulmonary fibrosis, cerebral malaria, multiple sclerosis, and psoriasis.[1,22] In experimental models of many of these disorders, the administration of neutralizing antibodies with specificity for TNFα attenuates or abrogates the pathological manifestations of the disease.[24] Anti-TNFα antibodies have been used successfully to reduce or prevent the mortality associated with severe endotoxemia, graft-versus-host disease, and cerebral malaria in preclinical studies. Anti-TNFα antibodies also have been used to attenuate the course of acute graft-versus-host disease and cerebral malaria in clinical trials.[24,25] Efforts to reduce the morbidity and mortality of bacterial sepsis and endotoxic shock in humans by

TABLE 12.2 Cytokines and Their Biological Activity

Cytokine	Biological activity[a]
IL-1	Induction of acute phase response; activation of osteoclasts; increase in expression of IL-2 receptors; endogenous pyrogen; induction of endothelial cell secretion of TNFα; proliferation of fibroblasts, endothelial cells, thymocytes, and type 2 helper T cells; differentiation of pre-B cells; induction of antibody secretion by B cells; activation of macrophages; induction of macrophage production of IL-6, prostaglandin E, and G-CSF
IL-2	Proliferation of activated T cells, B cells, and LGL cells; induction of NK, LAK, CTL, and monocyte cytotoxicity; production of lymphokines by T cells
IL-3	Proliferation and differentiation of stem cells, mast cells, neutrophils, erythroid cells, and megakaryocytes
IL-4	Proliferation of activated B and T cells, NK cells, thymocytes, and mast cells; regulation of antibody isotypes, i.e., IgE and IgG4; induction of B cell proliferation; inhibition of IL-2-induced LAK activity
IL-5	Production of IgA and IgM isotypes; proliferation and differentiation of eosinophils
IL-6	Proliferation of normal B lymphocytes and plasmacytomas; production of antibodies; proliferation and differentiation of T cells; induction of acute phase response; expression of IL-2 and IL-2 receptors; induction of megakaryocyte differentiation; synergistic promotion of multipotential colony cell growth; activation of NK cells
IL-7	Proliferation of progenitor B cells, progenitor T cells, and mature T cells
IL-8	Chemoattraction of neutrophils, basophils, and T cells; augmented expression of adhesion molecules; augmented activation of neutrophils
IL-9	Proliferation of T cells, mast cells, and megakaryoblastic cell lines
IL-10	Inhibition of cytokine synthesis by type 1 helper T cells, e.g., IL-2 and IFNγ; proliferation of B cells, thymocytes, and mast cells; inhibition of inflammatory cytokine production by activated macrophages, e.g., IL-1, IL-6 and TNFα
IL-11	Proliferation of plasmacytomas, megakaryocytic, and macrophage progenitor cells
IL-12	Proliferation and differentiation of type 1 helper T cells; induction of CTL and LAK cells
IL-13	Proliferation of B cells; inhibition of inflammatory cytokine production, e.g., IL-1, IL-6, and TNFα; induction of IgE synthesis
IL-14	Proliferation of activated B cells; inhibition of immunoglobulin synthesis
IL-15	Proliferation of activated T cells and LAK cells
IFNα/β	Inhibition of tumor cell proliferation; suppression of viral activity; immunomodulation; induction of class I HLA antigens
IFNγ	Inhibition of tumor cell proliferation; suppression of viral activity; proliferation of B cells; activation of macrophages and B cells; induction of class I and II HLA antigens; augmentation of antibody secretion
Epo	Proliferation and differentiation of erythroid progenitors
G-CSF	Proliferation and differentiation of granulocytic myeloid progenitors
GM-CSF	Proliferation and differentiation of myeloid and lymphoid progenitors; activation of granulocytes and macrophages; chemotaxis of neutrophils and monocytes
MIF	Inhibition of macrophage migration; activation of macrophages
Tpo	Proliferation and differentiation of megakaryocytic progenitors
TGF-β	Induction of matrix protein synthesis by fibroblasts and endothelial cells; proliferation of fibroblasts; inhibition of the proliferation of lymphocytes, endothelial, and epithelial cells
TNFα	Proliferation of fibroblasts; induction of monocyte and endothelial cell cytokine production, e.g., IL-1, IL-6, and GM-CSF; induction of acute phase response; activation of neutrophils; induction of procoagulant cofactor production; increase in bone resorption; induction of collagen synthesis; increase in lipoprotein lipase activity; cytolysis and cytostasis of tumor cells

[a] Selected biological effects.[1,2,19,34]

treatment with anti-TNFα antibodies have not been as encouraging and, apparently, have been harmful to some patients.[26]

12.3 Evaluation and Application of Systemically Administered Cytokines

12.3.1 Preclinical and Clinical Evaluation of Cytokines

The initial identification and characterization of a cytokine is often prompted by the ability of the cytokine to mediate a specific biological effect *in vitro*. The *in vitro* activity may serve as a surrogate marker for a corresponding *in vivo* activity and often engenders expectations of therapeutic effects that stimulate further preclinical and clinical evaluation. However, *in vitro* assays rarely, if ever, provide a full estimation of the potential *in vivo* activity of a cytokine. *In vitro* bioassays are unable to assess the many distal biological effects that cytokines mediate through their ability to trigger cascades of secondary cytokines *in vivo*. Cytokines normally function as components of complex networks and are subject to many different facilitatory and inhibitory influences. Thus, *in vivo* models are essential to assess thoroughly the toxicity and therapeutic efficacy of promising candidate cytokines.

Initial safety and efficacy testing of natural and recombinant human cytokines is usually performed in rodents and rabbits. In general, the evaluation of a human protein in a xenogeneic model may be confounded by the predisposition toward immunogenic responses that potentially can attenuate or mask the biological activities of the protein. Apart from such masking phenomenon, human cytokines exhibit considerably reduced potencies in rodent models. Some human cytokines, e.g., IFNγ and IFNβ, have little or no activity in rodents. Although an absence of toxicity in xenogeneic species is valuable in allaying concerns about nonspecific toxic effects, the use of species in which candidate cytokines are active is obviously more informative. Nonhuman primates often are superior to rodents for safety evaluation, because they exhibit a much higher degree of interspecies homology and shared biological activity with humans.

Most homologous cytokines are extremely potent and exhibit steep dose–response curves when administered *in vivo*. Maximally tolerated doses and toxicity profiles of recombinant cytokines are promptly established during initial clinical trials. Importantly, encountered toxicities may be highly dependent upon the conditions and regimens examined. Subtle variations in the schedule of cytokine administration can markedly affect treatment outcome. A dramatic demonstration of the schedule-dependency of cytokine-mediated toxicities was recently provided by phase II clinical studies of recombinant IL-12. Unexpected lethal toxicities were observed in patients treated with high doses of IL-12, when a small "priming" dose of IL-12 was omitted prior to the high-dose therapy.[27] Schedule-dependent toxicities also have been described for other proinflammatory cytokines. The often unanticipated toxicities associated with the systemic administration of cytokines have necessitated special vigilance during both the initial and later phases of clinical evaluation of these drugs.

The therapeutic efficacy of systemically administered cytokines also can be substantially altered by subtle changes in treatment regimens. Thus, the determination of the most effective treatment scheme can be problematic. Preclinical models have limited predictive value for clinical efficacy and relevant surrogate markers that correlate with therapeutic outcome often are not available to guide treatment design and adjustments. Attempts to manipulate immunological responses for the treatment of various disorders, including malignancies and infectious diseases, have been especially challenging, because the targeted responses are so complex and poorly defined. Optimal therapeutic effects of immunomodulatory cytokines may be achieved at doses far below the maximally tolerated doses of these agents. Low-dose regimens of IL-2, for example, are substantially less toxic and apparently as effective as much higher dose regimens for the treatment of malignancies.[28] Dose–response curves can be nonlinear, with high doses of cytokine exerting diminished therapeutic efficacy compared to lower doses. A bell-shaped dose–response curve has been demonstrated for IFNγ in both preclinical and clinical trials.[29] These observations have theoretical implications for the rational design of cytokine treatments. However, in the absence of appropriate surrogate end points, intended therapeutic effects are the only practical guide for the development of the most efficacious treatment regimens. In most circumstances, empirical testing and adaptation over sequential clinical trials remains the most reliable, albeit cumbersome, approach.

12.3.2 Therapeutic Applications of Systemically Administered Cytokines

Cytokines currently approved for clinical use in the United States and their indications are listed in Table 12.3. To date, the most consistent clinical benefits have been obtained with cytokines that have limited pleiotropy and defined biological effects, i.e., hematopoietic colony stimulating growth factors. Other cytokines with more pleiotropic effects, including IL-2 and the interferons, are approved for the management of severe infectious, neoplastic, and immunological disorders, but have been associated with adverse effects and lower therapeutic indices.

TABLE 12.3 Clinically Approved Cytokines and Their Indications

Cytokines	Indications[a]
Epo	Anemia of chronic renal disease
	AIDS-related anemia
	Anemia secondary to chemotherapy
G-CSF	Myeloid reconstitution after myelosuppressive chemotherapy
	Myeloid reconstitution after bone marrow transplantation
	Mobilization of peripheral blood progenitor cells for collection
GM-CSF	Myeloid reconstitution after bone marrow transplantation
	Myeloid reconstitution after myelosuppressive chemotherapy in elderly patients with acute myelogenous leukemia
	Mobilization of peripheral blood progenitor cells for collection
IFNα	AIDS-related Kaposi's sarcoma
	Chronic viral hepatitis B
	Chronic viral hepatitis C
	Chronic viral hepatitis non-A, non-B
	Chronic myelogenous leukemia
	Condylomata acuminata
	Hairy cell leukemia
	Malignant melanoma (adjuvant)
IFNβ	Relapsing-remitting multiple sclerosis
IFNγ	Chronic granulomatous disease
IL-2	Renal cell carcinoma

[a] Clinical indications approved by the United States Food and Drug Administration.

Many innovative cytokine strategies are being pursued at the preclinical and the clinical levels. Several recently discovered hematopoietic cytokines hold great promise as remedies for various cytopenias that are not amenable to currently approved colony stimulating factors. Novel immunomodulatory cytokines, such as IL-12 and IL-15, are being evaluated for their ability to potentiate specific immune responses against tumors and infectious agents, either directly or as vaccine adjuvants.[30] Other cytokines have attracted attention for their ability to favorably modify the pattern of endogenous cytokine production associated with specific immune responses or to limit the production and pathogenic effects of "deleterious" endogenous cytokines associated with various diseases. As favorable synergistic activities of cytokines have come to light through preclinical studies, clinical investigations have begun to focus on the utility of combined cytokine strategies for many different diseases.[29] Cytokines also are being investigated for their ability to potentiate the therapeutic effects of other classes of pharmacological agents. The increasing complexity of cytokine-based therapeutic strategies has been associated with a corresponding increase in the complexity of the benefit/risk analysis of such regimens.[29,31]

12.4 Toxicities of Systemically Administered Cytokines

12.4.1 General Mechanisms of Cytokine-Mediated Toxicity

Exogenously administered cytokines are associated with many, often serious, toxicities. These insults primarily result from the exaggerated physiological effects of cytokines' interactions with their cognate receptors. Most cytokines are highly pleiotropic and, therefore, elicit protean adverse effects when they are administered in supraphysiological doses. The toxic manifestions of cytokine treatment generally depend on sustained receptor occupancy by the cytokine ligand and usually subside once the cytokine is withdrawn. Like all heterologous proteins, recombinant cytokines also can elicit more sustained nonspecific toxic effects and specific immunological responses. However, the toxicities induced by excessive receptor stimulation have been the greatest impediments to the therapeutic use of cytokines.

The distribution and physiological status of specific cognate receptors are important determinants of the toxic effects of cytokines. Many cytokines, notably the proinflammatory cytokines, e.g., IL-1, IL-6,

IL-12, TNFα, and IFNγ, directly bind to most tissues throughout the body and elicit heterogeneous toxicities when they are administered systemically. Cytokines that bind to tissue-restricted receptors, e.g., hematopoietic colony stimulating factors, usually are associated with more limited spectra of toxicities, unless they are able to induce the synthesis and release of other cytokines. Secondarily induced cytokines can extend the adverse impact of cytokine treatment to tissues that lack cognate receptors for the primary cytokine and can amplify the deleterious effects of an exogenously administered cytokine on tissues that mutually express receptors for both the primary and secondary cytokine.

The final outcome of a cytokine's interaction with its cognate receptors is influenced by many variables. Cytokines act within a milieu of positive and negative regulatory signals that determine the level of receptor expression and the status of intracellular signaling mechanisms within target tissues. Proximate regulatory mechanisms are intimately linked to overarching biological programs that influence cytokine activity and potential toxicities. Circadian cycles, for example, have a demonstrable effect upon the lethal toxicity of systemically administered TNFα. Ninefold differences in the probability of TNFα-induced mortality have been observed in experimental animals, as a function of the time of day that the animals are treated with TNFα.[32]

Cytokines normally are deployed in a coordinated, multiarrayed fashion in time and space. Small, otherwise innocuous, doses of exogenous cytokines can have serious toxic effects when they are administered at inopportune times in the course of various pathological processes. Cytokines also can mediate opposing effects, depending on the manner in which they are administered. Exogenous IL-2, for example, can either exacerbate or attenuate graft-versus-host reactions following bone marrow transplantation, based on the method of drug administration.[33] Unexpected and paradoxical toxicities are not uncommon for this class of drugs and present a considerable challenge for the design and implementation of treatment regimens.

12.4.2   Toxicity Profiles of Selected Cytokines

Considerable experience with cytokine-induced toxicities has been acquired through numerous preclinical and clinical trials. A complete compendium of the toxicities elicited by exogenous cytokines is beyond the scope of this review. However, the toxicity profiles of representative cytokines illustrates the diverse and overlapping nature of the adverse effects of systemically administered cytokines. The comparative toxicities of cytokines approved for clinical use in the United States are shown in Table 12.4. More extensive descriptions of the toxicity profiles of selected cytokines are provided below.

12.4.2.1   Tumor Necrosis Factor α

Despite encouraging antitumor and immunomodulatory effects in preclinical models, TNFα has not been useful as a systemic therapeutic agent in humans, because it is extremely toxic. Intravenous administration of maximally tolerated doses of TNFα (approximately 200 µg/m²) causes severe constitutional symptoms including fever, rigors, nausea, vomiting, and diarrhea.[34] Profound hypotension resulting from generalized vasodilatation and fluid extravasation is dose-limiting. The marked decline in peripheral vascular resistance caused by exogenous TNFα is most likely mediated by nitric oxide. TNFα directly induces the expression of nitric oxide synthetase in endothelial cells.[35] TNFα also induces the release of chemokines and the expression of adhesion molecules on the surface of endothelial cells, resulting in chemotaxis and accumulation of activated leukocytes. The interaction between activated leukocytes and the endothelium subjects the latter to high concentrations of prostaglandins, superoxides, proteases, and other toxic products of activated leukocytes and further contributes to endothelial injury and increased permeability.

Virtually all organs are vulnerable to injury by TNFα. Proinflammatory, hemostatic, metabolic, cytostatic, and cytocidal effects of TNFα cause various types of organ damage. Specific organ effects of high doses of systemic TNFα include myocardial inflammation and patchy necrosis, pulmonary congestion, diffuse pulmonary infiltrates, hepatic inflammation, hepatic necrosis, glomerulonephritis, and tubular necrosis.[22,34] Pharmacological doses of TNFα disrupt the pituitary-adrenal axis and other endocrine

TABLE 12.4 Toxicities of Clinically Approved Cytokines

Site	Epo[b]	G-CSF[b]	GM-CSF[b]	IFN α/β[c]	IFN γ[b]	IL-2[c]
Constitutional[d]	±	−	+	+++	++	+++
Cardiac	−	−	−	+	−	++
Hypotension	−	−	−	+	−	+++
Capillary leak	−	−	−	−	−	+++
Renal	−	−	−	+	−	+++
Hepatic	−	−	−	++	+	+++
Diarrhea	−	−	−	++	−	+++
CNS	−	−	−	++	−	++
Neuropathy	−	−	−	+	−	−
Arthralgia	−	+	+	+	±	++
Rash	−	−	−	+	+	+++
Anemia	−	−	−	+	−	++
Thrombocytopenia	−	−	−	++	−	++/+++
Leukopenia	−	−	−	++	−	++
Infections	−	−	−	±	−	+/++
Autoimmune	−	−	−	+	−	+

[a] Key: −: not a known toxicity; ±: possible toxicity; +: low toxicity (grade 1 or 2 toxicity in less than 20% of patients); ++: moderate toxicity; +++: severe toxicity (grade 1 or 2 toxicity in more than 80% of patients and/or grade 3 or 4 toxicity in more than 10% of patients)

[b] Based on randomized studies[48,49]

[c] Based on phase I and II studies[36,37,40,50]

[d] Fever, chills, nausea, fatigue, anorexia, myalgia

pathways.[22] TNFα also has indirect catabolic effects and promotes net energy expenditure. Chronically elevated serum levels of TNFα decrease erythropoietin production and directly suppress hematopoiesis. Systemically administered TNFα also has many immunomodulatory effects that can disrupt normal immune regulation and provoke autoimmune responses. Immune-mediated hemolytic anemias and thrombocytopenias have been observed during treatment with TNFα.[22]

12.4.2.2 Interleukin-2

IL-2 has been extensively studied as a systemic drug administered by various schedules and regimens. Dose-limiting toxicities of hypotension and severe vascular leak are observed in humans when IL-2 is administered at 6×10^5 IU/kg by intravenous boluses every 8 hours or at lower doses when administered by continuous intravenous infusion. High-dose IL-2 therapy is associated with many constitutional symptoms and organ-specific toxicities, including myocarditis, ventricular and supraventricular cardiac arrhythmias, pulmonary edema, pulmonary infiltrates, gastrointestinal ischemia, diarrhea, hyperbiliru-binemia, prerenal azotemia, oliguria, dermatitis, mental confusion, and somnolence.[36,37] IL-2 induces anemia, thrombocytopenia, and lymphopenia. IL-2 also has multiple direct effects on lymphocytes. Exogenously administered IL-2 impairs normal immunoregulatory mechanisms and has both immuno-suppressive and autoimmune effects.[36,37] Patients treated with high-dose IL-2 therapy have an increased incidence of catheter-related infections that has been attributed to an acquired impairment of neutrophil chemotaxis.[36] The spectra of toxicities elicited by high-dose IL-2 and TNFα overlap extensively, in part because IL-2 induces the production and release of endogenous TNFα.[38] Many of the toxic manifestations of high-dose IL-2 treatment can be attenuated in preclinical models by the administration of neutralizing antibodies specific for TNFα.[39]

12.4.2.3 Interferon α

The α-interferons are pleiotropic cytokines with direct effects on many different tissues. The adverse effects of IFNα generally are tolerable within therapeutically useful dose ranges. Daily doses of 1 to 9 million units/m^2 of IFNα cause constitutional symptoms including fever, fatigue, headaches, myalgias, arthralgias, and anorexia.[40] These symptoms are usually transient and can be controlled with corticosteroids or nonsteroidal antiinflammatory drugs. At higher doses, IFNα induces a profound fatigue syndrome with accompanying neurological and psychological manifestations that are dose-limiting.[40] High-dose IFNα induces encephalopathic changes and frontal lobe dysfunction characterized by a flattened affect and psychomotor retardation. IFNα is also myelosuppressive. Patients develop leukopenia soon after the initiation of IFNα therapy. Protracted therapy with IFNα causes a normochromic, normocytic anemia. Other organ-specific effects of IFNα therapy include vasomotor changes, nausea and vomiting, proteinuria, and various endocrine and metabolic abnormalities, including alterations in serum levels of progesterone, estrogen, high-density lipids, and various electrolytes.[40] IFNα has numerous immunomodulatory effects and elicits autoimmune reactions. Approximately 40% of patients treated with extended courses of IFNα develop autoantibodies to various tissue antigens.[41] Antinuclear, antismooth muscle and antithyroid antibodies have been detected most frequently. Immune-mediated thyroiditis, hemolytic anemias, and thrombocytopenias also have been observed. All forms of IFNα are immunogenic and can elicit neutralizing antibodies.[40]

12.4.2.4   Hematopoietic Cytokines

Pure hematopoietic growth factors, including erythropoietin and colony stimulating factors, bind to receptors that are expressed primarily by subpopulations of hematopoietic progenitor cells. Accordingly, their principal biological effects are limited to the induction of hematopoietic cell growth and differentiation. In general, systemically administered hematopoietic growth factors induce mild or negligible toxic effects. Fever, chills, and flulike symptoms, including bone and muscle aches, develop occasionally, but are usually transient. The most serious toxicities associated with the use of these factors arise from excessive growth of the targeted hematopoietic cell lineages, i.e., leukocytosis (leukostasis) and erythrocytosis (secondary polycythemia), when drug administration is not appropriately monitored.

12.5   Benefit/Risk Ratios of Systemically Administered Cytokines

12.5.1   General Considerations

The benefit/risk ratios of individual cytokines vary according to the intrinsic nature of the cytokine and the clinical application for which the cytokine is considered. Cytokine therapies have been associated with a wide spectrum of benefit/risk ratios, ranging from treatments with consistent therapeutic effects and minimal toxicities to treatments with modest therapeutic benefits and life-threatening toxicities. The benefit/risk ratios for clinically approved cytokine applications are shown in Table 12.5.

TABLE 12.5   Benefit/Risk Ratios of Clinically Approved Cytokines

Cytokine	Benefit/Risk Ratio
Epo	High benefit/low risk
G-CSF	Moderate benefit/low risk
GM-CSF	Moderate benefit/low risk
IFNα	Moderate–high benefit/moderate risk
IFNβ	Moderate benefit/moderate risk
IFNγ[a]	High benefit/low risk
IL-2	Low benefit/high risk

[a] Benefit/risk ratio of IFNγ as treatment for chronic granulomatous disease[49]

The complexity of the biological activities of cytokines has had important ramifications for the benefit/risk outcomes of this class of drugs. Most systemically administered cytokines used for therapy have ancillary effects that do not contribute to intended results and that, beyond a certain threshold, adversely affect recipients. Although toxic cytokine regimens have been used justifiably for the management of highly morbid and lethal conditions that are not amenable to alternative therapies, many potentially beneficial applications for less morbid conditions have been precluded by the unacceptable toxicities of systemic cytokine therapy.

The multifaceted nature of cytokine activity implies that in many circumstances the toxicities associated with cytokine treatment might result from mechanisms that diverge from the mechanisms that are responsible for therapeutic efficacy. This has stimulated various strategies to improve the benefit/risk ratio of cytokine therapy by selectively modifying toxic vs. therapeutic effects of treatment. Such efforts have become increasingly sophisticated as the elucidation of the mechanisms for the therapeutic and the toxic effects of cytokine treatment have revealed exploitable differences.

12.5.2   Determinants of Cytokine Benefit/Risk Ratios

Many variables have been recognized that affect the outcome of cytokine treatment for different diseases. The physiological condition of the recipient, specific characteristics of the disease, and pharmacological attributes of the cytokine regimen are especially important determinants of treatment outcome. Genetic differences between patients contribute to heterogeneous therapeutic and toxic effects of cytokine treatment. Other less conspicuous variables also can influence the relative benefit/risk ratio of specific treatment regimens.

Performance status is an important predictive factor for patients' ability to tolerate the adverse effects of cytokine treatment and to respond favorably to treatment. Elderly and debilitated patients do not tolerate cytokine-mediated toxicities as well as younger and less compromised patients. Furthermore, the therapeutic efficacy of cytokines may be reduced in physiologically compromised patients, as a consequence of reduced numbers or sensitivity of relevant target cell types. For example, patients who have been treated with multiple cycles of cytotoxic chemotherapy can develop impaired responses to colony stimulating factors, due to cumulative injury to their hematopoietic progenitor cells. Likewise, severely immunocompromised patients are intrinsically incapable of responding to various immunomodulatory cytokines.

The stage and character of the disease are major determinants of the efficacy of cytokine therapy. Infectious, autoimmune, and malignant diseases are more or less amenable to cytokine treatment depending on their extent at the time of therapy. Patients with chronic infections and inflammatory conditions, such as viral hepatitis and multiple sclerosis, derive the most benefit from specific cytokine treatment, if the therapy is administered prior to the development of substantial tissue injury. Similarly, patients with malignancies are most amenable to cytokine treatment during early stages of their disease, while tumor burdens are low. The most favorable benefit/risk ratios for the treatment of chronic myelogenous leukemia with IFNα, for example, are achieved during the early chronic phase of the disease.[42]

Pharmacological variables that influence the efficacy and toxicity of cytokine therapy include dosage, formulation, and the route and schedule of administration. Minor differences in drug formulation can result in significant differences in the relative benefit/risk ratio of treatment. It has been reported, for example, that recombinant GM-CSF produced in prokaryotic cells is more toxic than the homologous recombinant protein produced in eukaryotic cells.[43] The influential role of drug dose and schedule on treatment outcome already has been noted. Importantly, alterations in various pharmacological parameters sometimes can differentially affect the efficacy and toxicity of cytokine treatment, resulting in considerable shifts in the overall benefit/risk ratio of the treatment.

12.5.3   Strategies to Optimize the Benefit/Risk Ratios of Cytokine Therapy

Many proposed cytokine applications have not been introduced into clinical practice, because they are associated with unfavorable benefit/risk ratios. Furthermore, some cytokine applications that have been approved for clinical use are associated with marginal benefit/risk ratios. Several general strategies to improve the benefit/risk ratios of cytokine treatments have been considered, including improved selection of potential candidates for treatment, augmentation of the therapeutic effects of treatment, and selective attenuation of the toxicities associated with treatment.

The therapeutic efficacy of many cytokine regimens varies considerably from patient to patient. Heterogeneous outcomes are especially common for immunomodulatory regimens used to treat malignant, infectious, and inflammatory conditions. One obvious strategy to optimize the benefit/risk ratio of such treatments is to restrict the treatment to individuals most likely to respond favorably. Unfortunately, in many circumstances adequate predictive factors are not available to enable appropriate patient selection. Progress in this area has been contingent on more thorough elucidation of the mechanisms that underlie the therapeutic effects of empirical cytokine regimens. Along these lines, it has been observed that approximately one third of patients with hepatitis C are resistant to treatment with IFNα. DNA markers that correlate with the responder phenotype recently have been identified.[44] Such markers may lead to more selective application of IFNα therapy, thus sparing patients who have the nonresponder phenotype from the adverse effects and costs of ineffectual therapy.[44] Similar analyses across a broad spectrum of infections, malignancies, and other diseases that might be amenable to cytokine modulation should identify subpopulations of patients for whom various cytokine regimens are most appropriate.

Ideally, the elucidation of the mechanistic basis of the beneficial effects of cytokine treatment will engender novel methods to overcome nonresponder phenotypes and will enable consistent therapeutic results to be achieved for most potential recipients. These methods might involve specific pharmacological manipulations or the development of combination regimens that enlist the complementary and synergistic effects of other cytokines or other classes of drugs to achieve improved therapeutic results. One exciting approach to emerge recently has been the use of molecular techniques to generate novel hybrid cytokines, e.g., IL-3/GM-CSF fusion protein-PIXY321, that may have superior therapeutic effects compared to their native counterparts.[45]

Alternative approaches to improve the benefit/risk ratio of cytokine therapy include methods to selectively reduce the adverse effects of cytokine treatment. In circumstances where therapeutic responses result solely from localized cytokine effects, it might be feasible to circumvent the systemic effects of cytokine treatment by directed delivery of drug to relevant sites *in vivo*. Site-specific cytokine treatment has been accomplished by the locoregional infusion of cytokines and by the use of special carriers, e.g., specific monoclonal antibody conjugates that target attached cytokines to selected tissues. Cytokines also have been selectively expressed in targeted tissues by gene transfer techniques. Gene therapy strategies currently are being pursued most actively for the induction of antitumor immune responses. Autologous tumor cells have been transduced with genes encoding various immunomodulatory cytokines including TNFα, GM-CSF, and IL-2, then used as vaccines in order to potentiate specific antitumor immune responses.[46]

Many diseases are not amenable to locoregional therapy. In these circumstances, efforts to favorably alter the benefit/risk ratio of cytokine treatment are focused on methods to selectively attenuate cytokine-mediated toxicities without compromising therapeutic efficacy. Systemic recombinant cytokines often cause toxic effects through the induction of secondary cytokines. Various strategies have been adopted to reduce or block the harmful effects of these secondary cytokines. Pharmacological agents that inhibit the synthesis of endogenous proinflammatory cytokines or their distal mediators are commonly used in conjunction with IL-2 and IFNα therapy to mitigate the toxic effects of treatment. Methods to interrupt defined cytokine circuits with ligand-specific or receptor-specific monoclonal antibodies or with soluble recombinant receptor antagonists are currently being studied in various preclinical and clinical models.[24] The development of cytokine congeners that selectively activate particular subsets of cognate cytokine receptors represents another promising strategy to elicit intended therapeutic effects while avoiding the systemic toxicities elicited by the parent compound. As an example, congeners of TNFα recently have been created that preferentially stimulate TNF-RI or TNF-RII receptors. Congeners with specificity for TNF-RI exhibit lower proinflammatory effects than wild type TNFα, yet retain tumoricidal activity, suggesting that they might have superior benefit/risk ratios for tumor therapy.[47]

12.6 Benefit/Cost Ratios of Cytokine Therapy

Recombinant cytokines are proprietary drugs that have proven to be quite costly for patients. In 1995, the *per annum* cost of recombinant erythropoietin for the treatment of anemia due to chronic renal failure was $5,637 and the *per annum* cost of recombinant interferon α2a for the treatment of Kaposi's sarcoma was $56,116.[48] The high consumer costs of recombinant cytokines result, in part, from the considerable research and development expenditures that are required for the clinical introduction of these drugs, but, in addition, result from exceptionally strong clinical demand. Recombinant cytokines have become essential components of technically demanding and expensive treatment programs, such as chronic dialysis and hematopoietic stem cell transplantation, which has accentuated the underlying demand for these agents.

Mounting political and economic pressures to contain the costs and maximize the efficiency of health care delivery have led to increased scrutiny of the benefit/cost ratios of cytokine therapies. Cytokines with modest levels of therapeutic efficacy or low therapeutic indices are especially vulnerable to unfavorable benefit/cost assessments. However, agents with high therapeutic indices, but "excessive" costs, are not immune to rigorous benefit/cost analyses in a managed care environment that increasingly is focused on achieving maximal economic efficiency.

12.7 Summary and Perspective

Nearly 40 years have elapsed since Lindemann originally described the antiviral effects of interferon. During this interim more than 100 cytokines have been identified. In addition, numerous "man-made" cytokines and cytokine-conjugates have been created for possible therapeutic use. The induction of specific biological responses by the administration of exogenous cytokines has become an important therapeutic strategy and an area of expanding clinical investigation that is in a nascent phase. Cytokine-mediated effects of particular therapeutic interest include the induction of hematopoietic cell growth and differentiation; the regulation of the differentiation and function of various immune effector cells; the mediation of direct cytostatic and cytocidal effects against neoplastic and infected cells; and the inhibition of "deleterious" endogenous cytokines associated with various diseases.

Although many cytokines have potentially therapeutic properties, they also have protean adverse effects that have limited their introduction into clinical practice. In light of the complex and multifaceted nature of cytokines' biological activities, it is probable that in many circumstances the specific toxic effects of cytokine treatment may be separable from intended therapeutic effects at several different mechanistic levels. Distinct responder cell populations, receptor subtypes, or intracellular signaling pathways may mediate therapeutic, as opposed to toxic, cytokine effects. As the precise mechanisms of cytokine-mediated toxicity and therapeutic efficacy are elucidated, the ability to design rational strategies to limit the toxicity and to augment the efficacy of cytokine therapy will undoubtedly increase. The prospects for substantial improvements in the benefit/risk ratios of this unique and expanding class of pharmacological agents are extremely promising.

References

1. Aggarwal, B. B. and Pocsik, E., Cytokines: from clone to clinic, *Arch. Biochem. Biophys.*, 292, 335, 1992.
2. Paul, W. E., Pleiotropy and redundancy: T cell-derived lymphokines in the immune response, *Cell*, 57, 521, 1989.
3. Davies, D. R. and Wlodawer, A., Cytokines and their receptor complexes, *FASEB J.*, 9, 50, 1995.
4. Trotta, P. P. and Nagabhusham, T. L., Gamma interferon, protein structure and function, in *Interferon: principles and medical applications*, Baron, S., Coppenhauer, D. H., Dianzani, F., Fleischmann, W. R., Jr., Hughes T. K., Jr., Klimpel, G. R., Niesel, D. W., Stanton, G. J., and Tyring, S. K., eds., *Univ. Texas*, Galveston, 1992, 117.

5. Murphy, P. M., The molecular biology of leukocyte chemoattractant receptors, *Annu. Rev. Immunol.*, 12, 593, 1994.

6. Okamura, H., Tsutsui, H., Komatsu, T., Yutsudo, M., Hakura, A., Tanimoto, T., Torigoe, K., Okura, T., Nukada, Y., Hattori, K., Akita, K., Namba, M., Tanabe, F., Konishi, K., Fukuda, S., and Kurimoto, M., Cloning of a new cytokine that induces IFN-γ production by T cells, *Nature*, 378, 88, 1995.

7. Paul, W. E. and Seder, R. A., Lymphocyte responses and cytokines, *Cell*, 76, 241, 1994.

8. Taniguchi, T., Cytokine signaling through nonreceptor protein tyrosine kinases, *Science*, 268, 251, 1995.

9. Kishimoto, T., Taga, T., and Akira S., Cytokine signal transduction, *Cell*, 76, 253, 1994.

10. Cosman, D., The hematopoietin receptor superfamily, *Cytokine*, 5, 95, 1993.

11. Baird, P. N., Dandrea, R. J., and Goodall, G. J., Cytokine receptor genes: structure, chromosomal location, and involvement in human disease, *Leuk. Lymphoma*, 18, 373, 1995.

12. Bazan, J. F., Structural design and molecular evolution of a cytokine receptor superfamily, *Proc. Natl. Acad. Sci. U.S.A.*, 87, 6934, 1990.

13. Cheifetz, S., Weatherbee, J. A., Tsang, M. L., Anderson J. K., Mole J. E., Lucans, R., and Massague, J., The transforming growth factor-β system, a complex pattern of cross-reactive ligands and receptors, *Cell*, 48, 409, 1987.

14. Smith, K. A., Interleukin-2: Inception, impact, and implications, *Science*, 240, 1169, 1988.

15. Giri, J. G., Ahdieh, M., Eisenman, J., Shanebeck, K., Grabstein, K., Kumaki, S., Namen, A., Park, L. S., Cosman, D., and Anderson, D., Utilization of the beta and gamma chains of the IL-2 receptor by the novel cytokine IL-15, *EMBO J.*, 13, 2822, 1994.

16. Minami, Y. and Taniguchi, T., IL-2 signaling: recruitment and activation of multiple protein tyrosine kinases by the components of the IL-2 receptor, *Curr. Opin. Cell. Biol.*, 7, 156, 1995.

17. Miyazaki, T., Kawahara, A., Fujii, H., Nakagawa, Y., Minami, Y., Liu, Z. J., Oishi, I., Silvennoinen, O., Witthuhn, B. A., Ihle, J. A., and Taniguchi, T., Functional activation of Jak1 and Jak3 by selective association with IL-2 receptor subunits, *Science*, 266, 1045, 1994.

18. Lenardo, M. J., Interleukin-2 programs α/β T-lymphocytes for apoptosis, *Nature*, 353, 858, 1991.

19. Le, J. and Vilcek, J., Cytokines with multiple overlapping biological activities, *Lab. Invest.*, 56, 234, 1987.

20. Schorle, H., Holtschke, T., Hünig, T., Schimpl, A., and Horak, I., Development and function of T cells in mice rendered interleukin-2 deficient by gene targeting, *Nature*, 352, 621, 1991.

21. Kramer, S., Schimpl, A., and Hunig, T., Immunopathology of Interleukin (IL) 2-deficient mice: thymus dependence and suppression by thymus-dependent cells with an intact IL-2 gene, *J. Exp. Med.*, 182, 1769, 1995.

22. Vassalli, P., The pathophysiology of tumor necrosis factors, *Annu. Rev. Immunol.*, 10, 411, 1992.

23. Beutler, B. and Cerami, A., Tumor necrosis, cachexia, shock, and inflammation: a common mediator, *Annu. Rev. Biochem.*, 57, 505, 1988.

24. Piquet, P. F., Chang, H. R., Tiberghien, P., Wijdenes, J., and Herve, P., Use of antibodies to cytokines, in *Clinical applications of cytokines: role in pathogenesis, diagnosis and therapy*, Oppenheim, J. J., Rossio J. L., and Gearing, A. J. H., eds. *Oxford Univ. Press*, New York, 1993, 258.

25. Holler, E., Kolb, H. J., Mittermuller, J., Kaul, M., Ledderose, G., Duell, T., Seeber, B., Schleuning, M., Hintermeier-Knabe, R., Ertl, B., et al., Modulation of acute graft-versus-host disease after allogeneic bone marrow transplantation by tumor necrosis factor (TNF alpha) released in the course of pretransplant conditioning: role of conditioning regimens and prophylactic application of a monoclonal antibody neutralizing human TNF alpha (MAK195F), *Blood*, 86, 890, 1995.

26. van der Poll, T. and Lowry, S. F., Tumor necrosis factor in sepsis: mediator of multiple organ failure or essential part of host defense?, *Shock*, 3, 1, 1995.

27. Cohen, J., IL-12 deaths: explanation and a puzzle, *Science*, 270, 908, 1995.

28. Yang, J. C., Topalian, S. L., Parkinson, D., Schwartzentruber, D. J., Weber, J., Ettinghausen, S. E., White, D. E., Steinberg, S. M., Cole, D. J., Kim, H. I., Levin, R., Guleria, A., MacFarlane, M. P., White, R. L., Einhorn, J. H., Seipp, C. A., and Rosenberg, S. A., Randomized comparison of high-dose and low-dose intravenous Interleukin-2 for the therapy of metastatic renal cell carcinoma: an interim report, *J. Clin. Oncol.*, 12, 1572, 1994.

29. Talmadge, J. E., Synergy in the toxicity of cytokines: preclinical studies, *Int. J. Immunopharmacol.*, 14, 383, 1992.

30. Afonso, L., Scharton, T. M., Vieira, L. Q., Wysocka, M., Trinchieri, G., and Scott, P., The adjuvant effect of Interleukin-12 in a vaccine against *Leishmania major, Science*, 263, 235, 1994.

31. Stryckmans, P., Duff, G. W., and Fracchia, G. M., Efficacy and safety of cytokines for human therapy: report of an EC study group, *Cytokine*, 5, 180, 1993.

32. Hrushesky, W. J. M., Langevin, T., Kim, Y. J., and Wood, P. A., Circadian dynamics of tumor necrosis factor α (Cachectin) lethality, *J. Exp. Med.*, 180, 1959, 1994.

33. Sykes, M., Pearson, D. A., and Szot, G. L., IL-2 induced GVHD protection is not inhibited by cyclosporine and is maximal when IL-2 is given over a 25 h period beginning on the day following bone marrow transplantation, *Bone Marrow Transpl.*, 15, 395, 1995.

34. Sidhu, R. S. and Bollon, A. P., Tumor necrosis factor activities and cancer therapy: a perspective, *Pharmacol. Ther.*, 57, 79, 1993.

35. Hibbs, J., Westenfelder, C., Taintor, R., Vavrin, Z., Kablitz, C., Baranowski, R. L., Ward, J. H., Menlove, R. L., McMurry, M. P., Kushner, J. P., and Samlowski, W. E., Evidence for cytokine-inducible nitric oxide synthesis from L-arginine in patients receiving Interleukin-2 therapy, *J. Clin. Invest.*, 89, 867, 1992.

36. Bruton, J. K. and Koeller, J. M., Recombinant Interleukin-2, *Pharmacotherapy*, 14, 635, 1994.

37. Lotze, M. T., Matory, Y. L., Rayner, A. A., Ettinghause, S. E., Vetto, J. T., Seipp, C. A., and Rosenberg, S. A., Clinical effects and toxicity of Interleukin-2 in patients with cancer, *Cancer*, 58, 2764, 1986.

38. Mier, J. W., Vachino, G., van der Meer, J. W., Numerof, R. P., Adams S., Cannon, J. G., Bernheim, H. A., Atkins, M. B., Parkinson, D. R., and Dinarello, C. A., Induction of circulating tumor necrosis factor (TNF alpha) as the mechanism for the febrile response to Interleukin-2 (IL-2) in cancer patients, *J. Clin. Immunol.*, 8, 426, 1988.

39. Fraker, D., Langstein, H. N., and Norton, J. A., Passive immunization against tumor necrosis factor partially abrogates Interleukin 2 toxicity, *J. Exp. Med.*, 70, 1015, 1989.

40. Quesada, J. R., Toxicity and side effects of interferons, in *Interferon: principles and medical applications*, Baron, S., Coppenhauer, D. H., Dianzani, F., Fleischmann, W. R., Jr., Hughes T. K., Jr., Klimpel, G. R., Niesel, D. W., Stanton, G. J., and Tyring, S. K., eds., *Univ. Texas*, Galveston, 1992, 427.

41. Preziati, D., LaRosa, L., Covini, G., Marcelli, R., Rescalli, S., Persani, L., Del Ninno, L., Meroni, P. L., Colombo, M., and Beck-Peccoz, P., Autoimmunity and thyroid function in patients with chronic active hepatitis treated with recombinant interferon alpha-2a, *Eur. J. Endocrinol.*, 132, 587, 1995.

42. Giralt, S., Kantarjian, H., and Talpaz, M., The natural history of chronic myelogenous leukemia in the interferon era, *Semin. Hematol.*, 32, 152, 1995.

43. Dorr, R. T., Clinical properties of yeast-derived vs. *Escherichia coli*-derived granulocyte-macrophage colony stimulating factor, *Clin. Ther.*, 15, 19, 1993.

44. Enomoto, N., Sakuma, I., Asahina, Y., Kurosaki, M., Murakami, T., Yamamoto, C., Ogura, Y., Izumi, N., Marumo, F., and Sato, C., Mutations in the nonstructural protein 5A gene and response to interferon in patients with chronic hepatitis C virus 1b infection, *N. Engl. J. Med.*, 334, 77, 1996.

45. Curtis, B. M., Williams, D. E., Broxmeyer, H. E., Dunn, J., Farrah, T., Jeffery, E., Clevenger, E., DeRoos, P., Martin, U., Friend, D., Craig, V., Gayle, E., Price, V., Cosman, D., March, C. J., and Park, L. S., Enhanced hematopoietic activity of a human granulocyte/macrophage colony-stimulating factor-interleukin 3 fusion protein, *Proc. Natl. Acad. Sci., U.S.A.*, 88, 5809, 1991.

46. Pardoll, D., New strategies for active immunotherapy with genetically engineered tumor cells, *Curr. Opin. Immunol.*, 4, 619, 1992.

47. Barbara, J. A. J., Smith, W. B., Gamble, J. R., van Ostade, X., Vandenabeele, P. V., Tavernier, J., Fiers, W., Vadas, M. A., and Lopez, A. F., Dissociation of TNFα-cytotoxic and proinflammatory activities by p55 receptor- and p75 receptor-selective TNFα-mutants, *EMBO J.*, 13, 843, 1994.

48. *Physicians GenRx — 1995*, Denniston P. L., ed., Mosby-Year Book, St. Louis, 1995.

49. The International Chronic Granulomatous Disease Cooperative Study Group, A controlled trial of interferon gamma to prevent infection in chronic granulomatous disease, *N. Eng. J. Med.*, 324, 509, 1991.
50. Rosenberg, S. A., Yang, J. C., Topalian, S. L., Schwartzentruber, D. J., Weber, J. S., Parkinson, D. R., Seipp, C. A., Einhorn, J. H., and White, D. E., Treatment of 283 consecutive patients with metastatic melanoma or renal cell cancer using high-dose bolus interleukin 2, *JAMA*, 271, 912, 1994.

13

The Benefit/Risk Ratio of Analgesic–Antipyretics and Antiinflammatory Agents: A Review of Epidemiologic Evidence

Giuseppe Traversa
Istituto Superiore di Sanità

Nicola Magrini
University of Bologna

13.1   Introduction

A discussion of the benefit/risk profile of nonsteroidal antiinflammatory drugs (NSAIDs) should consider the fact that for most of their clinical indications NSAIDs are symptomatic drugs (i.e., they do not modify the natural history of the disease) and at the same time may cause a broad spectrum of adverse reactions. Unfortunately, there is no evidence that the safety of newer NSAIDs is improving with time. In the past few years several epidemiologic studies have consistently confirmed the gastrointestinal toxicity of these drugs and the presence of clinically relevant differences among individual NSAIDs.

As is commonly observed for other drugs, newly marketed NSAIDs tend to be presented as safer and more effective than prototypical compounds, whereas epidemiologic studies evaluating the actual risks have not confirmed the premarketing expectations. A recent study performed in three countries (United Kingdom, United States, and Spain) in the time period 1974 to 1993 shows that 29 newly marketed drugs (3% to 4% of all new drugs) were subsequently discontinued for safety reasons; among these 9 were NSAIDs.[1]

The assessment of efficacy and toxicity of NSAIDs is of particular importance since these drugs are among the most frequently used (considering both prescription and over-the-counter use). Drug utilization data from Scandinavian countries (where these statistics are regularly available since the 1970s)

0-8493-2791-1/99/$0.00+$.50
© 1999 by CRC Press LLC

show an overall level of about 30 Defined Daily Doses (DDD) per 1000 inhabitants per day in 1992 for all NSAIDs (excluding aspirin and paracetamol) with an increasing use in the last decade (about 20 DDD in 1980).[2,3] These figures from overall sales of NSAIDs could be translated into a 3% prevalence, assuming an average continuous use (with full compliance). The actual prevalence of use is much higher in a given year since only a minority of patients takes these drugs on a regular basis. In the United States, prescriptions for NSAIDs made during 1991 amounted to 3.8% of all prescriptions;[4] nearly 15% of the population was treated with NSAIDs during 1984.[5] Similar data were observed in other countries. In Italy in 1991, for example, the corresponding figures were 35 DDD per 1000 inhabitants per day; prescriptions for NSAIDs represented almost 5% of all prescriptions, and 20% of the general population received at least one prescription per year.

The benefit/risk profile of NSAIDs presents two main areas of interest. The first one concerns the evaluation of the benefit/risk profile of NSAIDs as a class; in particular, the benefit/risk profile varies according to clinical indications, characteristics of the users (mainly age and clinical characteristics) and patterns of use (mainly dose, duration, and concomitant use of other drugs). The second one concerns the differences among various NSAIDs.

In the present review we will try to examine and summarize the available evidence concerning the risks of NSAIDs as a class and for each single drug in order to evaluate the benefit/risk profile for their main clinical indications.

13.2 Risks Related to Nonsteroidal Antiinflammatory Drugs

13.2.1 Gastroduodenal Damages

The available evidence shows that NSAID use causes gastroduodenal erosions and ulcers, with their complications represented by bleedings, perforations, and deaths. What is a matter of debate is the magnitude of the risk of developing the adverse event and in particular a serious complication. Many nonexperimental (case-control and cohort studies) and experimental studies (randomized controlled trials) have been performed in the area of NSAID-related gastroduodenal damage, and this review is focused on the results of the more relevant ones.

13.2.1.1   The Increase in the Occurrence of Gastroduodenal Lesions

Two measures of risk will be used in the following: (1) the cumulative incidence, which expresses the probability of developing the event of interest within a defined period of time, and (2) the relative risk (or odds ratio, OR), which says how many times the incidence of the event of interest is more (or less) frequent among exposed subjects in comparison to nonexposed subjects (in our case, users of NSAIDs in comparison to nonusers).

The estimates of the risk measures vary primarily in relation to the end point considered and the severity of the lesions. Endoscopic studies carried out among osteoarthritis patients treated with NSAIDs have documented the presence of peptic ulcer in a considerable proportion of subjects. After 2 to 3 months of treatment 15% to 25% of patients free of lesions at the beginning of the study developed an ulcer;[6-8] roughly two thirds are gastric and the remaining ones duodenal. It has to be noted that most of the lesions have been recognized, because in these trials all patients underwent endoscopy as part of a study protocol regardless of the presence of symptoms. Given these incidence rates among users, the corresponding relative risk can be estimated to be even greater than 100, i.e., users of NSAIDs would develop a peptic ulcer (generally a silent one) with a frequency 100 times higher (or more) than nonusers.[9]

With the objective of guiding clinical practice, it is important to focus on clinically relevant damages. Overall, the metaanalyses performed show that current users of NSAIDs appear to experience a three- to fourfold increase in the frequency of serious gastroduodenal damages in comparison to nonusers[9-11] (Table 13.1). Though these metaanalyses show some differences in the estimates, no clear trend is observed by severity of complication: the relative risks are 2.4 for gastrointestinal bleedings, 7.7 for perforations,

TABLE 13.1   NSAID Use and Occurrence of Gastroduodenal (GD) Lesions

Estimates of the increase in the risk of GD lesions among NSAID users

	Odds Ratios
Overall	3–4
Age[10]	
≤60 years	3.2
>60 years	7.1
History of GD lesions[11]	
No	2.4
Yes	4.8
Dose[16]	
Low	2.5
Medium	4.5
High	8.6
Duration of use[11]	
<1 month	8.0
1–3 months	3.3
>3 months	1.9

Estimates of the incidence of GD lesions among NSAID users

	Incidence
GD ulcer (symptomatic or asymptomatic) detected through repeated endoscopy, within 3 months from starting NSAID use[6-8]	~20%
Ulcer complications (bleeding, perforation) among relatively old NSAID users (mean age: 68 years) with relatively long duration of use (up to 6 months)[12]	
Overall	1.0%
Patients <75 years without history of GD lesions	0.4%
Patients ≥75 years with history of GD lesions	5.3%
GD ulcer (symptomatic) among relatively young NSAID users (<65 years) with relatively short duration of use (mean duration: 2 months)[21]	0.1%

[10] Bollini, 1992; [11] Gabriel, 1991; [12] Silverstein, 1995; [16] Langman, 1994; [21] Lanza, 1955.

and 4.8 for deaths.[11] The difference in the magnitude of the relative risk with the above-mentioned endoscopic studies is striking and depends almost entirely on the different outcomes considered: severe symptomatic vs. asymptomatic injuries. Moreover, whereas endoscopic studies have shown a greater incidence of gastric rather than duodenal ulcer, almost no difference in the relative risks by site of lesion is observed in the metaanalyses of nonexperimental studies.[9,11]

Besides this average risk for an "average" patient receiving an "average" therapeutic regimen one should consider that age, comorbidity, past history of peptic ulcer together with doses of NSAID, and duration of use strongly influence the probability of developing a gastroduodenal ulcer or its complications.

The relative risk increases with age, but no appreciable difference is observed between males and females. For elderly patients (older than 60) the summary relative risk is 7.1, whereas for younger patients the corresponding risk is 3.2.[10] In a recent large randomized controlled trial (8843 patients) aimed at studying the gastroprotective effect of misoprostol in the prevention of serious NSAID-induced gastrointestinal complications, after controlling for 18 potential risk factors, patients aged 75 years or older experienced a 2.5 increase in the relative risk of developing complications as opposed to younger patients.[12] Age was an independent risk factor when considered as a continuous variable, as well as when dichotomized with a cut point at age 65. The same analysis showed that patients with a previous history of peptic ulcer, or gastrointestinal bleeding, experienced, respectively, a 2.3 or a 2.6 increase in the relative risk of serious gastrointestinal complications. These figures are fairly similar to those reported by a metaanalysis of nonexperimental studies: the relative risk is 4.8 for patients with a previous history and 2.4 for those without such a history.[11]

13.2.1.2   Effect of Dose and Duration of Therapy

A dose response relation in the occurrence of clinically relevant gastrointestinal lesions has been demonstrated in several experimental and nonexperimental studies. The investigation of gastrointestinal adverse events, which occurred in those experimental studies aimed at evaluating the effect of antithrombotic prophylaxis with aspirin, is of interest, since largely different doses for different durations have been considered. The U.K. Transient Ischaemic Attack Trial allowed a comparison between the 1200 mg and 300 mg daily dose of aspirin: the odds ratios for gastrointestinal bleeding were 2.8 and 1.6 for the 1200 mg and 300 mg daily dose, respectively.[13] A dose response relation was also found in the Dutch trial in which 30 mg and 283 mg of aspirin were compared.[14] It is worthwhile noting also that daily doses lower than 100 mg (75 mg and 30 mg) of aspirin are associated with bleeding complications.[15] In nonexperimental studies, the definition of dose is generally less precise. In a recent case-control study daily doses were defined on the basis of recommendations of the British National Formulary.[16] The relative risk of ulcer complications increased from 2.5 to 4.5, and to 8.6 among users of low, medium, and high doses, respectively (Table 13.1). In another study, a dose response relation affected both new and long-term users: almost a doubling in the ulcer hospitalization rates was shown among users of more than 1.75 daily doses in comparison to those receiving no more than 0.75 daily doses.[17]

NSAIDs cause an acute damage to the gastroduodenal mucosa and the risk drops quickly after cessation of use. Studies that have defined current use of NSAIDs by using different time windows have shown that the increase in the risk is greater in the 1 to 4 weeks immediately following the beginning of therapy. Gabriel estimated that the relative risk for less than 1 month of NSAID use was 8.0; for longer than 1 month but less than 3 months, 3.3; and for longer than 3 months, 1.9[11] (Table 13.1). Moreover, it has been observed that new (incident) users experience a greater increase in the relative risk than chronic users. These findings are coherent with two partly overlapping phenomena: the development of mucosa adaptation to continuous NSAID therapy, and the so-called "depletion of susceptibles" among users. Though a large proportion of subjects who receive NSAIDs develop some sort of gastroduodenal damage, in the majority of cases the lesions resolve even during continuation of drug use.[18] It is consequently not surprising that endoscopic studies conducted among asymptomatic subjects have shown a considerably larger increase in the occurrence of lesions than studies focused on severe clinical complications. The term *depletion of susceptibles* simply reflects the fact that chronic users of NSAIDs (as of any other drug) represent a selected population of survivors among the overall users.[19,20] The population of chronic users is "depleted" by those susceptible subjects who have already developed the gastroduodenal adverse event. This also explains why, in some studies, chronic users tend to show even a "protection" against gastroduodenal damages.

13.2.1.3   The Risk for Individual Patients

Most studies provide relatively similar estimates of the risk of developing symptomatic gastroduodenal events among NSAID users. The estimates are also consistent when the analyses are carried out by subgroup of risk factor (age, previous history, dosage, duration, etc.). Instead, a greater variability seems to affect the estimates of incidence of gastrointestinal events among users. Other than the already mentioned risk factors, the discrepancies mainly depend on the definition of outcomes (symptomatic ulcers, complications, deaths), the procedures for diagnostic confirmation, the setting where the diagnosis is made (hospital admissions, outpatient office visits), the baseline levels of the outcomes in the nonuser population. Nonetheless, it is certainly important to provide both physicians and patients with at least the order of magnitude of the individual risk (probability) of developing a defined event when using NSAIDs (Table 13.1). In a large cohort study conducted among 68,028 subjects younger than 65 years who received NSAIDs for a relatively short duration (2 months, on average), 1.1 per 1000 developed a confirmed symptomatic ulcer (with or without hemorrhage) during or in the 30 days following the end of treatment.[21] In the aforementioned misoprostol study of 8843 patients (mean age, 68 years) receiving NSAID therapy for a 6-month period, 1% of the patients (without concomitant misoprostol therapy) developed a gastrointestinal complication. The incidence of complications was estimated to range from a minimum of 0.4% among young patients without a history of peptic ulcer, to 5.3% among those with age greater than 75 and a history of peptic ulcer to a 9% among elderly patients with a history of both peptic ulcer and bleeding and concomitant cardiac disease.[12] Perforations, the most severe complication of peptic ulcer (with an estimated case fatality rate of 5%), were experienced by 7 users of NSAIDs (out of the 4439 patients without concomitant misoprostol therapy), corresponding to an incidence of 1.6 per 1000.

13.2.1.4 Prophylaxis of NSAID-Induced Lesions

The importance of NSAID-induced gastroduodenal damages bolstered research into the potentially preventive effect of concurrent therapy with gastroprotective agents. Preliminary studies indicate that omeprazole is effective in reducing the occurrence of gastroduodenal lesions, though further studies are needed to provide more precise estimates of the effect in the overall and by site of ulcer. H_2-receptor antagonists and misoprostol have been studied more thoroughly and thus provide sound evidence. A recent metaanalysis included 22 randomized controlled trials and 1955 patients. The outcome was the diagnosis of ulcer, either symptomatic or asymptomatic, detected through serial endoscopy. For misoprostol there is strong evidence of a reduction of both gastric (OR = 0.1) and duodenal (OR = 0.3) ulcer; H_2-receptor antagonists do not prevent gastric ulcer but do reduce (OR = 0.4) the frequency of duodenal ulcer.[22] In a subsequent endoscopic study of 285 patients with arthritis receiving long-term NSAID therapy, high doses of famotidine (40 mg twice daily) were effective in reducing the incidence of duodenal ulcer (OR = 0.2) and, to a lesser extent, of gastric ulcer (OR = 0.4).[7]

Recently, the results of a very large randomized controlled trial involving 8843 patients with rheumatoid arthritis who were receiving long-term NSAID therapy (6 months) demonstrated that misoprostol (200 µg four times daily), in comparison to placebo, reduces the occurrence of severe gastrointestinal complications by 40%.[12] Complications were experienced by 25 of 4404 (5.7 per 1000) patients in the misoprostol group compared with 42 of 4439 (9.5 per 1000) patients in the placebo group. When all symptomatic ulcers where considered, either complicated or uncomplicated, the reduction was as high as 57%. This difference is coherent with the even greater protective effect of misoprostol for uncomplicated ulcers that was shown in endoscopic studies.[22] The incidence of serious complications was strongly related to the presence of risk factors. The drop-out rate (39%), mainly for diarrhea and abdominal pain, was large in both groups but greater in patients receiving misoprostol (42%) than in those receiving placebo (36%).

How can all this information be incorporated in a process of rational indication of cotherapy with misoprostol? What are the elements to be discussed with the patients to achieve a presumably informed decision? First, the possibility to stop NSAID use should be considered: in a 6 month period 4 out of 10 treated patients, whether or not receiving misoprostol, did withdraw. Second, a reduced daily dose might be as effective as higher doses and with a decrease in the risk equivalent to cotherapy (the issue of drug of choice will be considered afterward). Third, the probability of developing a complication is a function of the patients' risk factors. For example, in the lower risk group misoprostol would reduce in 6 months the incidence of serious complications from about 4 per 1000 to 2.4 per 1000; in other words 625 patients have to be treated to avoid one complication. Conversely, among elderly patients with a history of ulcer and other risk factors, the incidence would decrease from around 9 per 100 to 5.4 per 100, corresponding to 1 saved complication every 28 patients. Expressed monetarily, the cost of preventing a complication would range from $276,900 in the lowest risk group to $12,500 in the highest risk group.[23]

13.2.1.5 The Differences in Risk of Individual NSAIDs

With the occurrence of well-documented gastroduodenal lesions, a particularly delicate issue relates to the comparative evaluation of benefits and risks of different NSAIDs with special focus on newly marketed ones. The efficacy of available NSAIDs is considered largely similar.[5,24] Clinical trials that demonstrate efficacy in general are not sufficiently powerful to evaluate the relative safety of individual substances. It is then necessary to rely on nonexperimental studies to compare the risk of adverse events of NSAIDs, though at least three potential sources of bias have to be considered. First, new NSAIDs may preferentially be given to high risk groups of patients, particularly those with a history of peptic ulcer. Second, users of new NSAIDs may receive greater diagnostic attention, leading to the discovery of a greater proportion of subjects with lesions. Third, new substances may prevalently be used for the treatment of acute conditions. To deal with these potential sources of bias, most studies have adjusted for risk factors (first of all, history of peptic ulcer and age), have focused on severe events (thus limiting the likelihood of differential diagnostic attention), and have considered the duration of use.

As already reported in the previous paragraphs, aspirin (acetylsalicylic acid) users show risks similar to those found among users of other NSAIDs (considered as a group). A recently performed case-control study on low-dose aspirin for the prevention of cardiovascular disease showed that enteric-coated aspirin is less gastrotoxic than standard preparations.[25] At the end of 1994, more than 15 studies compared the gastroduodenal toxicity of various NSAIDs.[16] Their findings provide a frame of reference to distinguish between lower risk substances and higher risk ones. Among the studies considered by Langman, ibuprofen use is generally associated with the lowest risk, and azapropazone presented a distinctly elevated risk (in the two studies in which it was included). Within the substances in the lower range of risk, there are diclofenac, naproxen, and indomethacin. Within those in the upper range of risk there is piroxicam. These findings were confirmed in a recent metaanalysis.[26] In the only study in which a new marketed agent (ketorolac) was compared with other NSAIDs it was shown that ketorolac users experienced the highest risk of developing a peptic ulcer.[27]

Different isoforms of the enzyme prostaglandin endoperoxide synthase (cyclo-oxygenase 1 and 2, Cox 1, and Cox 2) have been isolated in different tissues. The possibility that selective inhibitors of the isoform expressed in inflamed tissues (Cox 2) might cause less gastroduodenal damage has stimulated efforts in the development of selective Cox 2 inhibitors. To date, the evidence is too scanty to say whether this hypothesis will have any relevance for clinically important end points.

13.2.1.6   Gastroduodenal Lesions Attributable to NSAID Use

Many authors have attempted to quantify the proportion of gastroduodenal ulcers that occurs in the population as a result of NSAID use. On the basis of relative risk and of level of NSAID use in the population, one can calculate the proportion of total cases that would be theoretically preventable had the exposure been withdrawn. Using the relative risk it is possible to estimate the proportion of exposed cases that occurred because of the exposure (the so-called attributable risk among exposed: ARe = (RR-1)/RR). For instance, if the relative risk of developing a gastroduodenal ulcer among NSAID users is 3, the proportion of exposed cases attributable to the exposure itself is 66% (i.e., (3-1)/3). To calculate the attributable risk in the general population (of users and nonusers of NSAIDs), this figure will be multiplied by the proportion of cases who are exposed to NSAIDs. For instance, if 30% of cases is current user of NSAIDs, about 20% (i.e., 66% × 30%) of the total cases of gastroduodenal ulcer in the population would be theoretically preventable by abstaining from use.

There is a wide variability in the estimates of the total proportion of cases of gastroduodenal ulcer attributable to NSAID use. This variability depends on the estimates of relative risks and of NSAID use that are obtained from epidemiologic studies. With relative risks of 3 to 4, and estimates of NSAID use among cases ranging between 15% and 60%, the theoretical proportion of total cases of peptic ulcer attributable to NSAIDs would range between 10% and 45%. These figures do not represent a realistic estimate of the amount of peptic ulcer that could be preventable in the population, principally because we cannot afford a total elimination of NSAID. Nonetheless, the reduction of doses and the use of less gastrotoxic NSAIDs would probably prevent a substantial proportion of gastroduodenal ulcers and their complications.

13.2.2   Liver Injury

13.2.2.1   The Overall Risk of Liver Injury

The severity of drug-induced liver injury ranges from asymptomatic test abnormalities to fulminant hepatic failure, and no distinctive clinical or histological pattern can be recognized. Only a limited number of studies tried to quantify the risk of acute liver injury among NSAID users, and the relative safety of specific agents. The main reason lies in the rarity of severe adverse events. In the first large study in this area,[28] a population of 228,392 subjects who received 1.5 million prescriptions for NSAIDs during the period 1982 to 1986 was followed up for acute liver injury hospitalization. The crude cumulative incidence of hospitalization in the 60 days following a prescription was 7 per 100,000. The corresponding odds ratio for current users of NSAIDs, in comparison to nonusers, was 1.8. Despite the already low level of hospitalization for acute liver injury among NSAID users, it is reasonable to assume that the actual incidence of the adverse event was even lower. It is in fact likely that some patients with a "true" diagnosis of infectious hepatitis or cholelithiasis may have inflated the total number of cases (instead of being excluded because of alternative causal explanations). In particular, some patients did not undergo tests to exclude infectious hepatitis (A or B) and the diagnostic tests for hepatitis C were not available at the time of the study. In this regard, it has been estimated that 67% of Non A–Non B hepatitis resulted positive for anti-HCV when the test was performed.[29]

Despite the possible overestimation of the incidence, the study of Pérez-Gutthann provides additional findings on risk factors.[28] Subjects older than 65 years had a twofold increase in the risk (OR = 2.3) compared to younger ones, whereas no difference was observed between males and females. Since most users are elderly, it is of interest to consider the effect of concomitant use of other medications, in particular those associated with liver injury (e.g., antibiotics and anticonvulsives). Among patients with simultaneous use of NSAIDs and potentially hepatotoxic drugs the odds ratio of acute liver injury was as high as 11.6.

In a larger and more recent study, out of a population of about 4 million inhabitants, 625,307 patients received prescriptions of NSAIDs (a total of 2.13 million prescriptions), during the period October 1987 to August 1991.[30] Within 60 days following the prescription, 23 events of acute liver injury were identified (after having excluded those attributable to alternative etiology), which gives a cumulative incidence of 3.7 per 100,000 users. Also this study showed an increase in the risk of acute liver injury among concomitant users of other hepatotoxic drugs, whereas no increase with age was observed.

A recent review of quantitative studies of the risk of serious hepatic injury among NSAID users concludes that symptomatic adverse events are extremely rare and almost all of what occurs is mild disease.[31]

13.2.2.2   The Differences in Risk of Individual NSAIDs

Among different NSAIDs, sulindac appears to carry the greatest hepatotoxicity. The study conducted by Pérez-Gutthann, though insufficient to detect statistically significant differences among individual NSAIDs, showed an adjusted odds ratio ranging from 1.2 for ibuprofen to 5.0 for sulindac (for the other four NSAIDs considered in the study — diclofenac, indomethacin, naproxen, and piroxicam — the estimated odds ratios did not exceed 2.6).[28] In the study by García Rodríguez, the risk of developing an acute liver injury among sulindac users was at least twelve times as high as for any other NSAID.[30] Also a study that focused mainly on outpatient events of hepatic injury showed the greatest incidence following sulindac exposure, although the confidence intervals were wide.[32] In the analysis of 1100 cases of drug-induced hepatic injury reported between 1978 and 1987 to the Danish Commitee on Adverse Drug Reactions, 97 cases were induced by NSAIDs (18 were caused by sulindac and 17 by ibuprofen).[33] On the basis of consumption data, it was estimated that the incidence of hepatic injury was 18 times higher among sulindac users than among ibuprofen users. The difference is striking and points in the same direction as the other studies.

Among analgesics and antipyretics, the dose-dependent hepatotoxicity of paracetamol has been widely described, particularly in cases of overdose in suicide attempts. A single dose of 10 to 15 g may be hepatotoxic and doses exceeding 20 to 25 g are potentially fatal. The hepatotoxic interaction between paracetamol and alcohol drinking is also well recognized. In a study of patients admitted for severe liver damage following paracetamol overdose (an average single dose of 30 to 36 g), mortality was twice as high among patients whose alcohol consumption exceeded predefined guidelines.[34] There is also evidence of liver injury when paracetamol (acetaminophen) is given in therapeutic doses to alcoholics. Finally, many reports signaled that even nonalcoholics may develop hepatic injury after taking therapeutic doses of paracetamol. In the study by Friis, among the 18 cases of liver injury suspected to be caused by paracetamol, only two patients were reported to have taken more than 4 g/die (though in 17 of the 18 cases more than one drug was taken).[33]

13.2.3   Renal Disease and Blood Pressure

13.2.3.1   Renal Disease

Since the findings of a phenacetin-induced chronic renal injury, a general concern has been raised on the possibility that other analgesics and antiinflammatory drugs might also be nephrotoxic. Long-term use of phenacetin appeared to cause renal papillary necrosis. This observation led to the description of an "analgesic nephropathy," though there exist no distinctive clinical criteria for its definition or diagnosis. To date, there is evidence supporting an etiologic role also of paracetamol, the major metabolite of phenacetin, in the development of chronic nephropathy. Among long-term daily users of paracetamol, a two- to threefold increase in the occurrence of chronic renal disease has been observed in two case-control studies.[35,36] Nonetheless, these epidemiologic studies did not show an increase of renal disease among aspirin users. Given the hypothesized mechanism of NSAID nephrotoxicity, namely the inhibition of prostaglandin mediated renal function, one would expect similar findings for aspirin and other NSAIDs. In a case-control study (554 patients and 516 controls), an increased risk for chronic renal disease (with severity ranging from minor renal insufficiency to end stage renal disease) was observed only when the analysis was focused on men aged more than 65 years (OR = 10).[37] On the contrary, no increase was shown among younger men and all females (regardless of age).

Among short-term effects of NSAID use there is a reversible acute renal failure of hemodynamic origin (generally within 24 hours after administration). This effect is usually moderate or even asymptomatic, and the development of oliguric or anuric renal failure can only be observed in a very small proportion of NSAID users.[38] This acute renal failure is ischemic in origin and is due to the inhibition of renal prostaglandin synthesis, which can be critical in subjects with decreased renal perfusion (e.g., patients with congestive heart failure, preexisting renal disease, major surgery). On the other hand, short-term administration of NSAIDs to healthy individuals is considered to have little if any measurable consequence on renal function.[38]

Although several case-reports and clinical observations point at NSAIDs as a possible cause of acute renal failure, epidemiologic evidence quantifying this adverse effect is scanty. In a cohort of more than 50,000 NSAID users, no case of hospitalization for acute renal disease was reported.[39] In another cohort study of 114 elderly patients (mean age 87 years), the serum urea nitrogen level increased by 50% in 13% of the patients after initiation of short-term NSAID therapy; concurrent use of diuretics and high doses of NSAIDs were found as additional risk factors.[40] No statistically significant changes in serum creatinine or potassium levels were noted; serum urea nitrogen level returned to baseline 2 weeks after discontinuation of NSAIDs. In a large, randomized controlled trial, no difference in the levels of serum urea was observed during the follow up (1 to 7 years) of 2449 patients receiving placebo, or 300 and 1200 mg of aspirin.[13] It remains to be noted that these indicators of renal function are only partially sensitive predictors of clinically evident renal disease.

More studies are certainly needed in this area to provide estimates of both the occurrence of NSAIDs-induced nephrotoxicity and the concomitant role of other risk factors (mainly history of cardiovascular diseases and prior renal disorders). On the one hand, it is important to establish whether acute renal toxicity, as opposed to long-term effects, is of interest. On the other hand, given the expected very low incidence of NSAID-induced renal disease (in particular of acute renal failure),[41] very large cohorts of NSAID users need to be enrolled. Furthermore, two requirements have to be fulfilled: a great accuracy in the ascertainment of exposure histories to avoid misclassification of subjects and a clear definition of temporal relation to avoid a possible confounding by indication (i.e., illnesses seen among NSAID users may represent a long-term effect of disease processes that caused patients to take NSAIDs).

13.2.3.2 Effect on Blood Pressure

On the basis of the antiprostaglandin effect of NSAIDs and of the clinical and experimental observation that NSAIDs can cause fluid retention (usually mild in normal subjects), their role in provoking an increase of blood pressure has been studied. Although the findings are partially conflicting, most recommendations underline a cautious use of these drugs in the elderly (at increased risk of musculoskeletal disorders and hypertension) and the possible reduction of the effect of antihypertensive medications. Recently, a metaanalysis of 54 clinical trials relevant to 1324 subjects has been conducted to evaluate the effects of NSAIDs on blood pressure.[42] Only acute use of NSAIDs was evaluated, the mean duration of therapy being 15 days. The overall effect was modest. Among normotensive subjects (not using antihypertensive medications), the increase in mean arterial pressure (MAP = one third systolic + two thirds diastolic pressure) was 1.1 mmHg. Hypertensive subjects experienced a 3.3 mmHg increase in MAP, though the increase was almost halved when the results were adjusted by dietary salt intake. The restriction of the analysis to trials in which NSAIDs were compared to placebo showed indomethacin to be associated with a 3.9 mmHg increase in MAP. No difference with placebo was observed for the two other most frequently studied NSAIDs, sulindac and naproxen.

In another metaanalysis, 50 clinical trials relevant to 771 patients were considered.[43] Again, the duration of NSAID use was relatively short (<15 days in 70% of the trials). On the whole, NSAIDs elevated MAP by 5.0 mmHg: the elevation was 5.4 mmHg among controlled hypertensive subjects (in particular, users of beta-blockers and diuretics) and 1.1 mmHg among normotensive ones. The most marked elevation was produced by piroxicam and the least hypertensive effect by sulindac and aspirin.

The findings of the metaanalyses are consistent with regard to normotensive subjects: little if any effect on blood pressure is observed. The elevation in blood pressure among hypertensive subjects appears coherent with the findings reported in the renal effects paragraph. Nonetheless, discrepancies in the magnitude of blood pressure elevation are present. Larger studies are needed, in particular to evaluate the effects on blood pressure of long-term use of NSAIDs.

In this regard, long-term effects of both low and medium doses of aspirin on blood pressure were assessed in a large, randomized controlled trial aimed at evaluating antithrombotic prophylaxis with aspirin. The study population (2449 patients) was randomized to placebo, 300 mg, and 1200 mg of aspirin, and followed up for a maximum of 7 years.[13] A general fall in blood pressure was observed in the three groups during the study period, and considered a consequence of active surveillance. Blood pressure during follow up was almost identical in the three groups.

13.2.4 Hematological Reactions

Despite the many reports implicating NSAIDs in the development of hematological reactions (mainly agranulocytosis and aplastic anemia), only few epidemiologic studies have been carried out.[44,45] The estimated incidence of agranulocytosis and aplastic anemia is very low: about 6 and 2 per million inhabitants per year, respectively. Nonetheless, given the severity of these conditions it is of interest to describe both the comparative risk of different NSAIDs and the absolute risk of NSAID use.

Pyrazolon derivatives (e.g., phenylbutazone, oxyphenbutazone, dipyrone) are the NSAIDs most frequently associated with severe hematological reactions, and for this reason they have been withdrawn from the market in many countries. The International Agranulocytosis and Aplastic Anemia Study found a 5.2 increase in the odds ratio of agranulocytosis among dipyrone users, though the estimates of the odds ratio ranged widely across regions, from 0.9 to 33.3. No increase was observed among users of other pyrazolons (OR = 1.2). Among other non-aspirin NSAIDs no association was found (OR = 0.9) after the exclusion of indomethacin, which in turn showed a distinctly elevated risk (OR = 8.9).[44] These results are partly conflicting with those of the other case-control study, which was aimed at evaluating the association between NSAID use (in the 30 days before the event) and hospitalization for neutropenia.[45] The adjusted odds ratio for NSAIDs as a class was 4.2. The study population (75 patients hospitalized with neutropenia and 276 controls) was too limited to evaluate individual NSAIDs, but the risk was fairly similar for different subgroups of NSAIDs (e.g., indole derivatives, propionic acid derivatives). Moreover, the association was found only for mild disease: though about 20% of the patients with neutropenia had agranulocytosis (neutrophil count below 500/μL) none were exposed to NSAIDs.

In conclusion, the risk of hematological reactions is low, unpredictable, associated with short-term therapies, and, with the exception of a greater concern on pyrazolones, the evidence is insufficient to choose among single NSAIDs. Thus, is the available information useful for clinical decision making? Unfortunately, the most reasonable answer is no, at least until larger studies will be able to establish the comparative risk of individual NSAIDs.

13.3   The Benefit/Risk Profile of Nonsteroidal Antiinflammatory Drugs in their Main Indications

13.3.1   Fever

Considerable evidence suggests a potentially conflicting role of fever in both potentiating and inhibiting the resistance to infection.[46] Very high body temperature (above 40°C) is associated with injury to the brain and other organs, and treatment is absolutely necessary. On the contrary, low-grade fever (below 39°C) may represent an adaptive advantage for the subject, making questionable the rationale for anti-pyretic treatments unless the patient feels particularly uncomfortable.[47] Despite the potential for beneficial effect of low-grade fever, use of antipyretics is now an accepted routine in everyday practice. This is especially the case in pediatrics, where most of the clinical studies have been conducted. In children, lowering febrile temperature appears to have only modest influence on mood and activity; moreover, no efficacy is shown in the reduction of fever-associated convulsions.[48]

All NSAIDs have shown efficacy in reducing elevated body temperature, but paracetamol is almost universally considered, at least among children, the drug of choice. This recommendation is based on the adverse events associated with NSAIDs as a class, and in particular because of the gastroduodenal damages that occur even after short-course therapies. In addition, the use of aspirin among children and adolescents has been associated with the occurrence of a severe and very rare encephalopathy, Reye's syndrome. Though the incidence of the syndrome is very low (6 to 7 cases per million children per year)[48] and despite the controversies about the causal relation with aspirin, there is a wide consensus on the contraindication of aspirin in the treatment of children with chickenpox or influenza.

Other than paracetamol, ibuprofen is the drug most widely studied in the treatment of fever. In a very large randomized controlled trial, 84,192 children were assigned to receive paracetamol (12 mg/kg) or ibuprofen (5 or 10 mg/kg).[49] No case of either anaphylaxis or acute renal failure was observed. Four children (7.2 per 100,000) were hospitalized for acute gastrointestinal bleeding among those receiving ibuprofen (two in each dosage group), whereas none occurred in the paracetamol group. The difference was not statistically significant, though entirely consistent with available knowledge on gastrotoxicity of NSAIDs.

Though ibuprofen is considered as second choice drug (after paracetamol) for the treatment of fever in children, it remains to identify those conditions in which this second choice antipyretic would be fully justified. For adults, depending on the clinical anamnesis of the patient, aspirin and ibuprofen might as well be suggested.

13.3.2   Musculoskeletal Disorders

The most frequent use of NSAIDs is related to the symptomatic treatment of osteoarthritis, either with or without an inflammatory component. Available evidence suggests that analgesic and antiinflammatory doses of NSAIDs show similar efficacy on different outcomes (e.g., reducing pain and improving mobility).[50] It has also to be noted that the use of NSAIDs in the routine management of osteoarthritis has been questioned on the basis of a suspected adverse effect on the evolution of the underlying condition. The reduction of pain may in fact favor a progression of the articular damage through an increased joint loading.[51]

The efficacy of different NSAIDs is fairly similar. Given the variability in individual response, patients who do not respond to one NSAID may respond to another. Different trials have also compared the efficacy of paracetamol with NSAIDs given in analgesic doses.[50,52,53] Patients treated with paracetamol (4 g daily) and ibuprofen (1200 mg daily) obtained similar improvements in an overall health assessment score.[51] Nonetheless, pain on walking and at rest decreased to a greater extent in the ibuprofen group than in the paracetamol group. Two other studies compared paracetamol (2 g daily) with diclofenac (100 mg daily). In one study, significantly improved mobility scores and pain were observed following diclofenac in comparison with paracetamol.[53] In the other one, diclofenac scored better than paracetamol on each of the considered outcomes (pain, stiffness, escape analgesia), though the differences did not reach statistical significance because of the limited number of patients.[52] More studies are needed to compare full doses of paracetamol (4 g daily) with analgesic doses of NSAIDs. On the basis of the available findings, paracetamol may be considered a therapeutic alternative to NSAIDs essentially among patients at high risk of developing gastroduodenal damages. Nonetheless, as reported in the previous paragraphs, both the potential hepatotoxicity of paracetamol especially among alcohol drinkers and the possibility of renal impairment among long-term users certainly need to be considered.

For rational management of musculoskeletal disorders (mostly osteoarthritis), the following recommendations may be taken into account:

- There are alternative and concurrent measures to NSAID use: loss of weight, rest, avoidance of aggravating activities, structured physical exercise.
- Use the minimum effective analgesic doses (Table 13.2).[54]
- Use less gastrotoxic NSAIDs.
- Discuss with the patient the duration of treatment, which, in general, should be relatively short (1 to 2 weeks).
- On the basis of the anamnesis of the patient consider the option paracetamol/NSAIDs
- History of peptic ulcer represents a contraindication to NSAIDs. If an NSAID is to be used anyhow, consider the role of concomitant therapy with misoprostol or H_2-receptor antagonists especially in elderly patients.

13.3.3 Postoperative Pain

Several clinical studies have shown the efficacy of NSAIDs in relieving postoperative pain in both minor and major surgery. The available evidence suggests that NSAIDs may reduce by about a third the requirements for opioids in the postoperative days. In some studies, equivalence in the efficacy of the two analgesic regimens has been observed. Moreover, the combination of NSAIDs and opioids provides better analgesia than the one reached by opioids alone. The analgesia is related to doses with a plateau effect, and no improvement is achieved by starting NSAID use in the preoperative day. The most frequently studied NSAIDs are: diclofenac, ibuprofen, indomethacin, and ketorolac. Only few comparative studies are available and, as for other indications and despite the general tendency to adopt newer NSAIDs, there is no evidence of greater efficacy of any individual NSAID. Moreover, no study has been carried out to compare the safety of different NSAIDs. In this regard, the most important adverse events that have been reported relate to gastrointestinal bleedings and renal impairments (see previous paragraphs).

Hemorrhagic complications, other than gastrointestinal ones, have also been observed as a consequence of the antiplatelet activity of NSAIDs. The benefit/risk profile can be expected to be similar to the one observed for the antiplatelet prophylaxis among surgical patients (though both risks and benefits may be less relevant in consideration of the shorter duration of use). Recently, an overview of 53 trials (8400 patients) estimated the incidence of bleeding complications among surgical patients who received antiplatelet drugs (mainly aspirin).[55] Two fatal bleedings were observed in the antiplatelet group (0.05%) and none among the controls. Major nonfatal complications (e.g., need for transfusion) were experienced

TABLE 13.2   Pharmacokinetic Parameters and Dosage Recommendation of NSAIDs

NSAID	Half-life (hrs)	Recommended Dose for Mild to Moderate Pain	Recommended Dose for Rheumatoid Arthritis	Analgesic Action Onset (hrs)	Analgesic Action Duration (hrs)	Antirheumatic Action Onset (days)	Antirheumatic Action Plateau (weeks)
Half life < 12 hours							
Acetylsalicylic acid	(2–3)[1] (6–12)[2]	300–650 mg, 4–6 times daily	300–1000 mg, 4–6 times daily	0.5	3–4	few days	1–2
Diclofenac sodium	1–2	50 mg, 2–3 times daily	50 mg, 2–3 times daily	—	—	—	—
Etodolac	7.3	200–400 mg, 2 times daily	200–400 mg, 2 times daily	0.5	4–12	—	—
Ibuprofen	1.8–2.5	400 mg, 3–4 times daily	400–800 mg, 3–4 times daily	0.5	4–6	≤7	1–2
Indomethacin	4.5 / SR: 4.5–6	25–50 mg, 2–3 times daily	25–50 mg, 2–3 times daily / 75 mg SR, once daily	0.5	4–6	≤7	1–2
Ketoprofen	2–4	25–50 mg, 4 times daily	50 mg, 4 times daily	—	—	—	—
Ketorolac	2.4–8.6	10 mg, 3 times daily[3] (only post-operative pain)	NO	IM:10 min	IM: ≤6	—	—
Meclofenamate	2 (3.3)[4]	50–100 mg, 3–4 times daily	50–100 mg, 3–4 times daily	—	—	few days	2–3
Sulindac	7.8 (16.4)[5]	—	150–200 mg, 1–2 times daily	—	—	≤7	2–3
Half life ≥ 12 hours							
Nabumetone[6]	22.5 (30)[5]	—	500–1000 mg, 1–2 times daily	—	—	—	—
Naproxen	12–15	250 mg, 2–3 times daily	250–500 mg, 2 times daily	1	≤7	≤14	2–4
Piroxicam	30–86	—	10–20 mg, once daily	1	48–72	7–12	2–3

Modified from Golshahr V.E., 1996[64]

[1] Half-life of salicylic acid, low doses.
[2] Half-life of salicylic acid, antiinflammatory doses.
[3] According to the European Agency for the Evaluation of Medicinal Products the initial dose should be 10 mg and the maximum daily dose should not exceed 60 mg in the elderly and 90 mg in other patients; the maximum duration of parenteral administration should be two days.
[4] Half-life with multiple doses.
[5] Half-life of active metabolite.
[6] The active metabolite of nabumetone is an acetic acid.

by 0.7% in the antiplatelet and 0.4% in the control group, with an absolute increase of 3 events per 1000 treated patients. Other complications (e.g., reoperation, wound hematoma) were observed in 7.8% of the antiplatelet and 5.6% of the control group, with an absolute increase of 22 events per 1000 treated patients. Despite these adverse events, a global analysis clearly shows that benefits far exceed risks. Among patients with antiplatelet prophylaxis, a reduction in the incidence of deep vein thrombosis (88 per 1000), pulmonary embolism (17 per 1000), and fatal pulmonary embolism (6 per 1000) was observed.

In conclusion, in short-term treatment of postoperative pain, the adverse events of NSAID use (mainly related to gastrointestinal damages) need to be balanced with those of opioids (mainly represented by respiratory depression). Unfortunately, since the frequency of both events is associated with age, as is the rate of intervention in the population, no clear trade off between the two treatments is available.

13.3.4 Other Potential Indications

Several studies have brought evidence of a protective role of NSAIDs in the development of colorectal cancer.[56] In animal studies NSAIDs inhibited the growth of intestinal tumors induced by chemical carcinogens. In a randomized placebo-controlled trial, patients with familial polyposis treated with sulindac had a decreased number and size of colorectal adenomas.[57] After 9 months of treatment, the number and diameter of polyps decreased to 44% and 35% of the baseline levels. Different nonexperimental studies suggest a protective effect of NSAID use in the development of colon or rectum cancer. In a large cohort study, a reduction in the relative risk (RR = 0.56) of colorectal cancer was observed among women who used aspirin regularly for more than 20 years.[58] There was also a slight reduction in the relative risk (RR = 0.7) among women who took aspirin for 10 to 20 years. A beneficial effect was observed also in a case-control study aimed at evaluating the etiologic role of drugs in the occurrence of acute leukemia.[59] When focusing the analysis on NSAID use, the odds ratios of acute leukemia were 1.2 for any use, 1.0 among users of high doses (duration of use greater than the median observed in the control group), and 0.4 among very high users (duration of use greater than 180 days). Though highly suggestive, further data are needed to confirm the protective role of NSAIDs in the occurrence of cancer, mainly to avoid the raising of false expectations.

13.4   Conclusions

Among risks, gastrointestinal toxicity carries the greatest implications for both individual patient and public health point of view. In this regard, the use of minimum effective analgesic (as opposed to antiinflammatory) doses, together with the use of less gastrotoxic NSAIDs (e.g., ibuprofen), may prevent a considerable proportion of gastroduodenal ulcers in the population. Other risks are considerably less frequent. The findings are relatively reassuring for hepatotoxicity, since, with the possible exception of sulindac (other than the well-defined case of paracetamol), only modest if any increase in the risk for individual patients is present for the most frequently used NSAIDs. Issues such as nephrotoxicity and effects on blood pressure require further studies to estimate not only the incidence of adverse events among users, but also the role, among the others of age, comorbidities, and cotherapy as concomitant risk factors.

The benefit/risk profile of NSAID use varies among clinical indications, and no clear-cut definition is available. Even when valid estimates of the risks are present, several prognostic factors, inherent to characteristics of the users (mainly age and anamnesis of gastroduodenal lesions) and patterns of NSAID use (mainly dose and duration of use), greatly affect the estimates. Moreover, individuals with similar benefit/risk profile from the physician's perspective, may judge very differently their own profile. In this context, physicians are required to discuss pros and cons in order to give the patient a real chance to make, as far as possible, an informed decision.

Glossary of Abbreviations

ARe   attributable risk among exposed
COX 1   cyclo-oxygenase, isoform 1
COX 2   cyclo-oxygenase, isoform 2
DDD   defined daily dose
HCV   hepatitis-C virus
MAP   mean arterial pressure
NSAID   nonsteroidal antiinflammatory drug
OR   odds ratio
RR   relative risk

References

1. Bakke, O. M., Manocchia, M., deAbajo, F., Kaitin, K. I., Lasagna, L., Drug safety discontinuations in the United Kingdom, the United States and Spain from 1974 through 1993: a regulatory perspective, *Clin Pharmacol Ther*, 58, 108, 1995.

2. Ahonen, R., Enlund, H., Klaukka, T., Martikainen, J., Consumption of analgesics and antiinflammatory drugs in the Nordic countries between 1978–1988, *Eur J Clin Pharmacol*, 4, 37, 1991.

3. Nordenstam, I., Wennberg, M., Kristoferson, K., *Svensk läkemedelsstatistik 1992*, Apoteksbolaget, Stockholm, 1993.

4. Ray, W. A., Griffin, M. R., Avorn, J., Evaluating drugs after their approval for clinical use, *N Engl J Med*, 329, 2029, 1993.

5. Brooks, P. M., Day, R. O., Nonsteroidal antiinflammatory drugs — differences and similarities, *N Engl J Med*, 324, 1716, 1991.

6. Raskin, J. B., White, R. H., Jackson, J. E., Weaver, A. L., Tindall, E. A., Lies, R. B., Stanton, D. S., Misoprostol dosage in the prevention of nonsteroidal antiinflammatory drug-induced gastric and duodenal ulcers, a comparison of three regimens, *Ann Intern Med*, 123, 344, 1995.

7. Taha, A. S., Hudson, N., Hawkey, C. J., Swannell, A. J., Trye, P. M., Cottrell, J., Mann, S. G., Simon, T. J., Sturrock, R. D., Russell, R. I., Famotidine for the prevention of gastric and duodenal ulcers caused by nonsteroidal antiinflammatory drugs, *N Engl J Med*, 334, 1435, 1996.

8. Stalnikowicz, R., Rachmilewitz, D., NSAID-induced gastroduodenal damage: is prevention needed? A review and metaanalysis, *J Clin Gastroenterol*, 17, 238, 1993.

9. Hawkey, C. J., Non-steroidal antiinflammatory drugs and peptic ulcer: facts and figures multiply, but do they add up? *Br Med J*, 300, 278, 1990.

10. Bollini, P., García Rodríguez, L. A., Pérez-Gutthann, S., Walker, A. M., The impact of research quality and study design on epidemiologic estimates of the effect of nonsteroidal antiinflammatory drugs on upper gastrointestinal tract disease, *Arch Intern Med*, 152, 1289, 1992.

11. Gabriel, S. E., Jaakkimainen, L., Bombardier, C., Risk for serious gastrointestinal complications related to use of nonsteroidal antiinflammatory drugs, *Ann Intern Med*, 115, 787, 1991.

12. Silverstein, F. E., Graham, D. Y., Senior, J. R., Wyn Davies, H., Struthers, B. J., Bittman, R. M., Geis, G. S., Misoprostol reduces serious gastrointestinal complications in patients with rheumatoid arthritis receiving nonsteroidal antiinflammatory drugs, *Ann Intern Med*, 123, 241, 1995.

13. UK-TIA Study Group, The United Kingdom transient ischaemic attack (UK-TIA) aspirin trial: final results, *J Neurol Neurosurg Psychiatr*, 54, 1044, 1991.

14. Dutch TIA Trial Study Group, A comparison of two doses of aspirin (30 mg vs. 283 mg a day) in patients after a transient ischaemic attack or minor ischaemic stroke, *N Engl J Med*, 325, 1261, 1991.

15. Patrono, C., Aspirin as an antiplatelet drug, *N Engl J Med*, 330, 1287, 1994.

16. Langman, M. J. S., Weil, J., Wainwright, P., Lawson, D. H., Rawlins, M. D., Logan, R. F. A., Murphy, M., Vessey, M. P., Colin-Jones, D. G., Risks of bleeding peptic ulcer with individual non-steroidal antiinflammatory drugs, *Lancet*, 343, 1075, 1994.

17. Smalley, W. E., Ray, W. A., Daugherty, J. R., Griffin, M. R., Nonsteroidal antiinflammatory drugs and the incidence of hospitalizations for peptic ulcer disease in elderly persons, *Am J Epidemiol*, 141, 539, 1995.
18. Shorrock, C. J., Langman, M. J. S., Nonsteroidal antiinflammatory drug-induced gastric damage: epidemiology, *Dig Dis*, 13(Suppl 1), 3, 1995.
19. Miettinen, O. S., Caro, J. J., Principles of nonexperimental assessment of excess risk, with special reference to adverse drug reactions, *J Clin Epidemiol*, 42, 325, 1989.
20. Moride, Y., Abenhaim, L., Evidence of depletion of susceptibles effect in nonexperimental pharmacoepidemiologic research, *J Clin Epidemiol*, 47, 731, 1994.
21. Lanza, L. L., Walker, A. M., Bortnichak, E. A., Dreyer, N. A., Peptic ulcer and gastrointestinal hemorrhage associated with nonsteroidal antiinflammatory drug use in patients younger than 65 years. A large Health Maintenance Organization cohort study, *Arch Intern Med*, 155, 1371, 1995.
22. Koch, M., Capurso, L., Dezi, A., Ferrario, F., Scarpignato, C., Prevention of NSAID-induced gastroduodenal mucosal injury: meta-analysis of clinical trials with misoprostol and H_2-receptor antagonists, *Dig Dis*, 13 (Suppl 1), 62, 1995.
23. Levine, J. S., Misoprostol and nonsteroidal antiinflammatory drugs: a tale of effects, outcomes and costs, *Ann Intern Med*, 123, 309, 1995.
24. Hardman, J. G., Goodman Gilman, A., Limbird, L. E., *Goodman & Gilman's The pharmacological basis of therapeutics*, 9th ed., McGraw-Hill, New York, 1996.
25. Weil, J., Colin-Jones, D., Langman, M., Lawson, D., Logan, R., Murphy, M., Rawlins, M., Vessey, M., Wainwright, P., Prophylactic aspirin and risk of peptic ulcer bleeding, *Br Med J*, 310, 827, 1995.
26. Henry, D., Lim, L. L. Y., García Rodríguez, L. A., Pérez-Gutthann, S., Carson, J. L., Griffin, M., Savage, R., Logan, R., Moride, Y., Hawkey, C., Hill, S., Fries, J. T., Variability in risk of gastrointestinal complications with individual non-steroidal antiinflammatory drugs: results of a collaborative meta-analysis, *Br Med J*, 312, 1996.
27. Traversa, G., Walker, A. M., Menniti Ippolito, F., Caffari, B., Capurso, L., Dezi, A., Koch, M., Maggini, M., Spila Alegiani, S., Raschetti, R., Gastroduodenal toxicity of different nonsteroidal antiinflammatory drugs, *Epidemiology*, 6, 49, 1995.
28. Pérez-Gutthann, S., García Rodríguez, L. A., The increased risk of hospitalizations for acute liver injury in a population with exposure to multiple drugs, *Epidemiology*, 4, 496, 1993.
29. Mele, A., Stroffolini, T., Pasquini, P., *SEIEVA Integrated Epidemiological System for Acute Viral Hepatitis. Report 1985–1994*, Istituto Superiore di Sanità, Roma, Rapporti ISTISAN 96/3, 1996.
30. García Rodríguez, L. A., Williams, R., Derby, L. E., Dean A. D., Jick, H., Acute liver injury associated with non-steroidal antiinflammatory drugs and the role of risk factors, *Arch Intern Med*, 154, 311, 1994.
31. Walker, A. M., (personal communication).
32. Lanza, L. L., Walker, A. M., Bortnichak, E. A., Gause, D. O., Dreyer, N. A., Incidence of symptomatic liver function abnormalities in a cohort of NSAID users, *Pharmacoepidemiology and Drug Safety*, 4, 231, 1995.
33. Friis, H., Andreasen, P. B., Drug-induced hepatic injury: an analysis of 1100 cases reported to The Danish Committee on Adverse Drug Reactions between 1978 and 1987, *J Intern Med*, 232, 133, 1992.
34. Bray, G. P., Mowt, C., Muir. D. F., Tredgar J. M., Williams R., The effect of chronic alcohol intake on the prognosis and outcome in paracetamol overdose, *Human and Experimental Toxicology*, 10, 435, 1991.
35. Perneger, T. V., Whelton, P. K., Klag, M. J., Risk of kidney failure associated with the use of acetaminophen, aspirin and nonsteroidal antiinflammatory drugs, *N Engl J Med*, 331, 1675, 1994.
36. Sandler, D. P., Smith, J. C., Weinberg, C. R., Buckalew, V. M., Dennis, V. W., Blyphe, W. B., Burges, W. P., Analgesic use and chronic renal disease, *N Engl J Med*, 320, 1238, 1989.
37. Sandler, D. P., Burr, F. R., Weinberg, C. R., Nonsteroidal antiinflammatory drugs and the risk for chronic renal disease, *Ann Intern Med*, 115, 165, 1991.

38. Pierucci, A., Patrono, C., NSAIDs in renal impairment and dialysis, in *Therapeutic Applications of NSAIDs*, Famaey, J. P. and Paulus, H. E., Eds., Marcel Dekker, New York, 1992, chap. 11.

39. Fox, D. A., Jick, H., Nonsteroidal antiinflammatory drugs and renal disease, *J Am Med Assoc*, 251, 1299, 1984.

40. Gurwitz, J. H., Avorn, J., Ross-Degan, D., Lipsitz, L. A., Nonsteroidal antiinflammatory drug-associated azotemia in the very old, *J Am Med Assoc*, 264, 471, 1990.

41. Pérez-Gutthann, S., García Rodríguez, L. A., Raiford, D. S., Duque Oliart, A., Non-steroidal anti-inflammatory drugs and the risk of hospitalization for acute renal failure in Saskatchewan: a nested case-control study, *Pharmacoepidemiology and Drug Safety*, 4(S) S53, 1995.

42. Pope, J. E., Anderson, J. J., Felson, D. T., A meta-analysis of the effects of nonsteroidal antiinflammatory drugs on blood pressure, *Arch Intern Med*, 153, 477, 1993.

43. Johnson, A. G., Nguyen, T. V., Day, R. O., Do nonsteroidal antiinflammatory drugs affect blood pressure? A meta-analysis, *Ann Intern Med*, 121, 289, 1994.

44. International Agranulocytosis and Aplastic Anemia Study, Risk of agranulocytosis and aplastic anemia, *J Am Med Assoc*, 256, 1749, 1986.

45. Strom, B. L., Carson, J. L., Shinnar, R., Snyder, E. S., Shaw, M., Lindin, F. E., Jr., Nonsteroidal antiinflammatory drugs and neutropenia, *Arch Intern Med*, 153, 2119, 1993.

46. Mackowiak, P. A., Fever: blessing or curse? A unifying hypothesis, *Ann Intern Med*, 120, 1037, 1994.

47. Saper, C. B., Breder, C. D., The neurologic bases of fever, *N Engl J Med*, 330, 1880, 1994.

48. Giusti, M. P., Marchetti, F., Tognoni, G., Bonati, M., Uso dei FANS in pediatria per il trattamento delle infezioni respiratorie acute e della febbre. Una revisione critica della letteratura. Ministero della sanità, *Bollettino d'informazione sui farmaci*, 6, 3, 1995.

49. Lesko, S., Mitchell, A. A., An assessment of the safety of pediatric ibuprofen. A practitioner-based randomized clinical trial, *J Am Med Assoc*, 273, 929, 1995.

50. Bradley, J. D., Brandt, K. D., Katz, B. P., Kalasinki, L. A., Ryan, S. I., Comparison of an antiinflammatory dose of ibuprofen, and analgesic dose of ibuprofen, and acetaminophen in the treatment of patients with osteoarthritis of the knee, *N Engl J Med*, 325, 87, 1991.

51. Del Favero, A., Anti-inflammatory analgesics and drugs used in gout, in *Side effects of drugs annual 18*, Aronson, J. K., van Boxtel, C.J., Eds., Elsevier, Amsterdam, 1995.

52. March, L., Irwig, L., Schwarz, J., Simpson, J., Chock, C., Brooks, P., n of 1 trials comparing a nonsteroidal anti-inflammatory drug with paracetamol in osteoarthritis, *Br Med J*, 309, 1041, 1994.

53. Parr, G., Darekar, B., Fletcher, A., Bulpitt, C. J., Joint pain and quality of life: results of a randomized trial, *Br J Clin Pharmacol*, 27, 235, 1989.

54. Golshahr, V. E., Neubauer, D. J., Reinert, A. E., Sery, M. R., Sullivan, B., Threlkeld, D. S., *Drug facts and comparisons, loose-leaf drug information service*, Facts and Comparisons Inc., St. Louis, 1996.

55. Antiplatelet Trialists' Collaborative Group, Collaborative overview of randomized trials of antiplatelet therapy. Part III: reduction in venous thrombosis and pulmonary embolism observed with antiplatelet prophylaxis among surgical and medical patients, *Br Med J*, 308, 235, 1994.

56. Marcus, A. J., Aspirin as prophylaxis against colorectal cancer, *N Engl J Med*, 333, 656, 1995.

57. Giardiello, F. M., Hamilton, S. R., Krush, A. J., Piantadosi, S., Hylind, L. M., Celano, P., Booker, S. V., Robinson, C. R., Offerhaus, G. J., Treatment of colonic and rectal adenomas with sulindac in familial adenomatous polyposis, *N Engl J Med*, 328, 1313, 1993.

58. Giovannucci, E., Egan, K. E., Hunter, D. J., Stampfer, M. J., Colditz, G. A., Willett, W. C., Speizer, F. E., Aspirin and the risk of colorectal cancer in women, *N Engl J Med*, 333, 609, 1995.

59. Traversa, G., Menniti Ippolito, F., Da Cas, R., Mele, A., Pulsoni, A., Mandelli, F., Drug use and acute leukemia, *Pharmacoepidemiology and Drug Safety*, 1998 (in press).

14

Toxicity Profile of Liposomal Anthracyclines

Alberto A. Gabizon
Hadassah Hebrew University
Medical Center

14.1 Introduction

Recently, two liposomal anthracycline formulations have been approved by the U.S. Food and Drug Administration for clinical use (Doxil, DaunoXome).[1-3] Another formulation (D-99)[4] is in advanced clinical trials. As these products become available to an enlarged circle of clinicians and nurses, the need to recognize their toxicity profile becomes more urgent. When a drug is presented in a liposome formulation, the medical teams are faced with a new entity that does not fit the usual definition of a new drug. This new entity is more complex than a simple drug since it consists of a carrier, the liposome, and a drug, which may be an approved and well-known drug (e.g., doxorubicin or daunorubicin) or a newly designed drug (e.g., annamycin[5]). Some clinicians tend to look at liposomal doxorubicin or liposomal daunorubicin as another drug analog among the many (epirubicin, mitoxantrone, idarubicin) that have become available in recent years. Moreover, many clinicians will refer to liposomal anthracyclines as a homogeneous group of formulations with relatively similar pharmacological and biological effects. These are dangerous oversimplifications. A learning process is required to grasp the novelty and complexity of liposome-based drug delivery systems.

In considering a liposome formulation, we should be aware that there are basically three variables that may affect its biological activity and toxicity profile: the lipid vesicle, the drug, and the form of interaction between drug and lipid vesicle. Let us briefly examine each of these variables.

1. The liposome (lipid vesicle) is a supramolecular assembly whereby amphipathic lipids, generally phospholipids, form a closed bilayer creating a vesicle with an entrapped water phase separated physically from the external medium.[6] The liposome serves as the drug carrier and may contain additional lipid ingredients, such as cholesterol, intercalated in the bilayer. Liposomes may differ in composition (e.g., fatty acid and headgroup composition of phospholipid, modification of lipids with special chemical groups) and physical properties (number of concentric bilayers, vesicle diameter, surface charge, surface hydrophilicity). The liposome chemical composition and physical properties will affect significantly its pharmacokinetic properties and as such may be tailored in the desired direction. Liposomes are generally nontoxic inasmuch as their components are biocompatible and biodegradable. In fact, one basic principle of the liposome pharmaceutics field is that none

of the manipulations done to the chemical and physical properties of liposomes should result in a toxic preparation. Thus, the aim is to engineer a carrier essentially devoid of inherent toxicity. In addition, liposomes are not expected to have any pharmacologic activity interfering with the active principle of the preparation.

2. The drug is the active principle of the formulation. One should realize that as long as the drug is still encapsulated in the liposome the drug is not bioavailable. To become bioavailable the drug has to be released from the vesicles either by a process of drug leakage with preservation of the liposome integrity, or following vesicle breakdown as in the case of phagocytosis and lysosomal enzymatic attack.[7] In most cases, the drug is a compound approved already in humans for administration as free drug, as in the case of doxorubicin, daunorubicin, and amphotericin B. A baseline toxicity profile is therefore known for the free drug. The unknown is how liposome encapsulation will modify toxicity. From a regulatory standpoint, the safest approach would probably be to consider a liposome formulation of an old drug as a new drug. However, the free drug should always be used as the baseline control against which a balanced evaluation of the toxicology of the liposome formulation is done.

3. Interaction of drug with carrier: This will determine the control the liposome has over the bioavailability and the biodistribution of the drug. Liposomes may function as dual-purpose carriers: providing a slow-release system in the intravascular compartment and tissue of distribution and/or redirecting the tissue distribution of the drug according to the liposome profile of biodistribution.[8] Drugs can be entrapped either in the water phase or in the lipid bilayer of the vesicle. Entrapment may be a passive process or driven actively by a chemical or electrochemical gradient. The method of entrapment will affect the rate of release of the drug with implications on the pharmacological activity.[9] Thus, it is possible to have two liposome formulations of the same drug with the same chemical characterization, but displaying a different pharmacologic and toxicologic profile because of the way the drug has been encapsulated.[10] Another important result of liposome encapsulation is a reduction of the drug renal clearance.[11] This is due to the fact that a small molecular weight drug is encapsulated in a nanoparticle that is far beyond the threshold of glomerular filtration. Thus, no renal excretion of the drug is possible as long as the drug has not been released from the liposomes.

With regard to doxorubicin, there were a handful of reasons that motivated investigators to search for liposomal formulations. This drug is a potent anticancer compound widely used in a broad spectrum of cancer types.[12] However, the toxicity of doxorubicin is substantial and imposes serious limits on our ability to use the drug repeatedly. In addition to acute toxicity consisting mainly of myelosuppression, doxorubicin also causes an irreversible, cumulative dose-dependent myocardial damage. As a result, the maximal cumulative dose of doxorubicin is generally limited to 550 mg/m^2.[13] Besides myelosuppression, other toxic effects related to administration of doxorubicin include nausea and vomiting, transient ECG disturbances, stomatitis, alopecia, local phlebitis, and extravasation necrosis. Thus, a number of investigators have tried to formulate doxorubicin and other anthracyclines in liposomes to buffer its toxicity while preserving its antitumor activity. A vast amount of preclinical literature on this topic is available,[14] from which several pharmaceutical-grade formulations have emerged and are currently being tested in clinical trials (Table 14.1). These formulations differ widely in their pharmacokinetic parameters, a factor that may explain the variations in toxicity profile.[14] In this report, we will concentrate on Doxil, a formulation of liposomal doxorubicin characterized by prolonged circulation time and high stability,[14] and with which we have extensive and direct experience at the clinical level.

14.2   Clinical Toxicity

Doxil (Sequus Pharm. Inc., Menlo Park, CA) is a liposomal formulation of doxorubicin presented in liquid form and stored at 5°C. Each vial contains a 10 ml liposome suspension. The concentration of doxorubicin is 2 mg/ml. The drug is encapsulated in the water phase of the vesicles by an ammonium sulfate–generated proton gradient.[15] Because of its high intraliposomal concentration, the drug appears

TABLE 14.1 Liposomal Anthracylines–Formulations Currently in Clinical Testing[a]

Name	Composition	Vesicle Size	Max. Tolerated Dose	Pharmacokinetics[e]
Doxil[b]	HPC-Ch-DSPE/MPEG	80–120 nm	50–60 mg/m² q4wk	$t_{1/2}$, 84 min/46 hr; Cl, 0.09 L/hr; V_{ss}, 5 L
DaunoXome[c]	DSPC/Ch	50–80 nm	60 mg/m² q2wk	$t_{1/2}$ (single), 4–8 hr; Cl, 0.6 L/hr; V_{ss}, 4 L
D-99[d]	PC/Ch	180 nm	75–90 mg/m² q3wk	$t_{1/2}$, 12 min/17 hr; Cl, 19 L/hr; V_{ss},160 L

Abbreviations:

HPC = hydrogenated phosphatidyl-choline
Ch   = cholesterol
DSPE = distearoylphosphatidyl-ethanolamine
MPEG = methoxy-polyethylene-glycol
DSPC = distearoylphosphatidyl-choline
PC   = phosphatidyl-choline

[a] Based on References 2–4, and 24. Doxil and DaunoXome have been approved for the treatment of AIDS-related Karposi's sarcoma.
[b] From SEQUUS Pharmaceuticals (Menlo Park, CA), active ingredient: doxorubicin.
[c] From NEXSTAR Pharmaceuticals (San Dimas, CA), active ingredient: daunorubicin.
[d] From The Liposome Company (Princeton, New Jersey), active ingredient: doxorubicin.
[e] Median values for dose levels of 50 mg/m² (Doxil), 40–60 mg/m² (DaunoXome), and 90 mg/m² (D-99).

to undergo a reversible gelification process, which may help to stabilize its retention in the vesicles.[16] One critical component of Doxil is a polyethylene-glycol derivatized phospholipid that accounts for the long circulating properties of these liposomes.[17]

We will divide the toxicity observations into three categories according to standard criteria: acute, subacute, and chronic.

14.2.1 Acute Side Effects

Acute reactions to Doxil infusion have been reported at an incidence of 10% of treated patients during the course of a phase I study.[2] These reactions are of variable severity and are characterized by facial and neck flushing, breathing difficulties, and, in some cases, back pain, cyanosis, and transient drop in blood pressure. They resolve usually within a few minutes after discontinuing the infusion. These reactions appear to be related to a fast rate of infusion. In many cases, resuming the infusion at a slower rate does not result in toxicity. Acute back pain during infusion has also been reported for another formulation of liposomes containing a nonanthracycline drug.[18] The mechanism for this acute type of reaction to liposome infusion is still unclear, but it is probably related to an interaction of some blood component with the liposome surface. One possibility is activation of the alternate pathway of the complement system as described in a rat model.[19]

In our institution, we currently administer Doxil by intravenous drip after prior dilution in 250 ml 5% dextrose bags, and the infusion rate is typically 1 to 1.5 mg Doxil per minute. When a patient is treated with Doxil for the first time, it is advisable to start with a test dose of Doxil at 0.1 to 0.2 mg/min for 10 minutes to check for a possible acute reaction. If the patient feels normal, the infusion rate can then be raised to 1.0 to 1.5 mg/ml. We do not premedicate patients, except for those who have had an acute reaction, in which case we administer 50 mg hydrocortisone i.v. immediately prior to Doxil infusion. With these precautions, we have seen only one acute reaction in our last series of 25 patients (author's unpublished results).

Toxicity to the site of injection in the form of local pain, phlebitis, or vein sclerosis has not been observed with Doxil and other liposomal anthracycline formulations. Regarding the vesicant damage to the skin that follows extravasation at the site of injection, there seems to be a protective effect of liposomal encapsulation. In fact, several episodes of Doxil extravasation that did not result in skin damage have been reported.[20] In a mouse study in which Doxil was injected intradermally, we found a mild inflammatory reaction with no ulceration or skin necrosis,[21] suggesting that the liposomes are removed from the site of injection by scavenger cells and lymph drainage before any significant drug leakage from the vesicles takes place. This is in contrast to free doxorubicin, which causes severe skin damage with necrosis and ulceration.

Acute nausea and vomiting is seldom seen with Doxil treatment. Altogether, gastrointestinal toxicity appears to be reduced with liposomal anthracyclines. As a rule, we do not use any intravenous antiemetics as premedication before Doxil administration unless a particular patient is extremely susceptible.

Fever with or without chills has been observed at a significant incidence in patients receiving D-99,[4] but not in the case of Doxil,[2] a finding that may be related to the difference in reticulo-endothelial system (R.E.S.) uptake between these two formulations.[14]

14.2.2 Subacute Side Effects

The most important subacute toxicity of Doxil is mucositis or stomatitis. This is also the dose-limiting toxicity with regard to the maximal tolerated dose (M.T.D.) of a single dose of Doxil. Out of 9 patients receiving 70 to 80 mg/m^2 Doxil, 3 developed grade 3 or 4 stomatitis, which lasted in some cases for 7 days or more.[2] Therefore, the maximal recommended dose of Doxil should not exceed 70 mg/m^2. Stomatitis is also a significant side effect of free doxorubicin, especially when administered as prolonged infusion or when dose-intensive regimens with hematopoietic growth factors are used.[22] However, for conventional bolus administration of doxorubicin at the standard 3-week schedule and without growth factor support, myelosuppression (leucopenia and granulocytopenia) is generally the dose-limiting toxicity.[12] In contrast, Doxil causes mild myelosuppression. Grade 4 leucopenia/granulocytopenia was observed in only one case out of 105 courses in 35 patients treated with a dose range of 60 to 80 mg/m^2. Mild leucopenia is commonly seen at doses of 40 mg/m^2 and greater, but it is generally uneventful and not dose-limiting.[2] As to the timing of leucopenia, there seems to be a slight delay as compared to free doxorubicin. When weekly blood counts are done, the nadir is usually seen at 2 weeks after injection, but in a substantial number of patients we have observed the nadir 3 weeks after treatment (author's unpublished data). With other liposomal anthracycline formulations, granulocytopenia is a significant form of toxicity.[14] Bone marrow is a tissue rich in macrophages and with a fenestrated capillary system that generally results in a high uptake of liposomes. The fact that some of the longer circulating liposomes partly avoid the R.E.S. or are removed by the R.E.S. at a very slow pace may account for the differences in severity of myelosuppression between Doxil, a long-circulating formulation, and other formulations with shorter circulation times.

A unique feature of Doxil is the high incidence of cutaneous toxicity,[2] especially in the form of hand–foot syndrome.[22] Skin toxicity is seldom seen after a single course of Doxil. It becomes more frequent with repeated treatments, especially when short intervals (less than 4 weeks) are allowed. In fact, for multiple treatments with Doxil, skin toxicity is the dose-limiting factor. As a result a schedule of 4 weeks, rather than the standard 3-weeks, is becoming the recommended schedule for doses of Doxil greater than 50 mg/m^2, although more experience is needed before definitive statements can be made. Clearly, there is a cumulative damage in the skin from repeated treatments with Doxil. This appears to be caused by cytotoxic damage to proliferating keratinocytes of the basal layers of the epidermis. Skin repair may be a process too slow to enable retreatment at standard 3-week intervals. This, however, does not hold for low doses (20 mg/m^2) as in the case of AIDS-related Kaposi's sarcoma.[1] Clinically, skin toxicity manifests generally by painful erythema and edema, followed by desquamation, and finally by reepithelization of the damaged areas. Generally, it develops between 1 to 3 weeks after a second or further injection of Doxil, affecting primarily contact pressure areas such as palms of hands and soles of feet. Its severity may range from a minimal erythema to major skin breakdown, temporarily crippling the patient. Recovery lasts between 2 to 3 weeks, thus sometimes delaying further treatment with Doxil for up to 6 weeks. Interestingly, other liposomal anthracycline formulations have not been reported to cause this type of toxicity, despite administration of greater nominal doses of the drug. With regard to free doxorubicin, hand–foot syndrome has been observed only in a small fraction of patients receiving dose intensive regimes such as prolonged infusions (>15 days) or high-dose treatments.[23] This special feature of Doxil may have to do with its extremely long circulation time and small vesicle size (distribution half-life of ~2 days),[24] which may enable liposomes to localize in the skin, depositing a substantial fraction of the drug payload.

Despite the skin toxicity problem, Doxil treatment (dose range: 40 to 80 mg/m²) has a very low incidence of alopecia (~10%),[2] as compared to free doxorubicin, which causes universal alopecia. In the case of D99 and DaunoXomes, the reported incidence of alopecia is ~40% and 5% respectively.[3,4]

14.2.3 Chronic Toxicity

By chronic toxicity of anticancer drugs, we refer to the permanent damage caused by these drugs to a specific tissue. This is the case of the anthracycline-induced cardiomyopathy. Free radical damage caused by doxorubicin and analogs to the mitochondriae of cardiac myocytes appears to account for this severe, irreversible, and cumulative dose-related toxic effect.[12] Although electron microscopy can reveal the increasing damage of repeated doses of doxorubicin relatively soon after treatment,[25] the clinical manifestations may develop only many years later. Because scoring of cardiac damage with electron microscopy is not widely available, clinical studies with liposomal anthracyclines published so far have relied for detection of cardiotoxicity on cardiac function measurements. In a phase I study of Doxil, 14 patients received cumulative doses equal to or greater than 450 mg/m² without any significant loss of cardiac function as determined by the left ventricle ejection fraction.[2] This is encouraging, because it points at a possible reduction of cardiotoxicity when Doxil is used to deliver doxorubicin. However, caution is needed because of the small patient numbers and the relatively short follow-up of patients treated with liposomal anthracyclines.

14.2.4 Other Potential Toxicities

One interesting observation from the reported clinical studies with liposomal anthracyclines is the lack of hepatic toxicity. Because of the high hepatic uptake of most types of liposomes, it was feared that liver toxicity may ensue liposome-based therapy. This has not been the case so far, at least with regard to liposomal anthracyclines. Another source of concern is the possibility of damage to Kupffer cells.[26] Although there are no direct clinical laboratory tests to assess Kupffer cell function, the clinical experience with Doxil and other liposomal anthracyclines does not point at a higher sensitivity of patients to infections. Even if we are overlooking a subtle damage to the R.E.S., this may still be an acceptable toxicity price when the clinical outlook of patients with metastatic cancer is considered.

14.3 Concluding Remarks

The potential of liposomal anthracyclines to buffer a number of undesirable side effects of doxorubicin is a clinical fact. Toxicity buffering stems mainly from the slow release of drug from liposomes and from the tissue distribution changes. There are, however, significant differences among the various liposome formulations that may have a major clinical weight. The unique toxicity of Doxil to skin underscores the contention that drastic pharmacokinetic changes may result in a compound with a qualitatively different toxicity profile. Ultimately, the extended use of liposomal drugs will depend on an assessment of their therapeutic index, i.e., the benefits/risks odds equation. For AIDS-related Kaposi's sarcoma, Doxil at 20 mg/m²[1] and DaunoXome[3] at 50 mg/m² offer an improved therapeutic index, judged to be sufficient by regulatory authorities for their approval. In the case of common forms of cancer, promising antitumor activity has been observed with Doxil in phase I-II trials,[2,27] but we still have to wait for the results of definitive phase III clinical trials that will determine the benefit/risk ratio, in reference to the standard forms of therapy currently available.

References

1. Harrison M., Tomlinson D., and Stewart S., Liposomal-entrapped doxorubicin, an active agent in AIDS-related Kaposi's sarcoma, *J. Clin. Oncol.*, 13, 914, 1995.
2. Uziely B., Jeffers S., Isacson R., Kutsch K., Wei-Tsao D., Yehoshua Z., Muggia F.M., and Gabizon A., Liposomal doxorubicin antitumor activity and unique toxicities during two complementary phase I studies, *J. Clin. Oncol.*, 13, 1777, 1995.
3. Gill P. S., Espina B. M., Muggia F., Cabriales S., Tulpule A., Esplin J. A., Liebman H. A., Forssen E., Ross M. E., and Levine A. M., Phase I/II clinical and pharmacokinetic evaluation of liposomal daunorubicin, *J. Clin. Oncol.*, 13, 996, 1995.
4. Cowens J. W., Creaven P. J., Greco W. R., Brenner D. E., Yung T., Ostro M., Pilkiewicz F., Ginsberg R., and Petrelli N., Initial clinical (phase I) trial of TLC D-99 (doxorubicin encapsulated in liposomes), *Cancer Res.*, 53, 2796-2802, 1993.
5. Zou Y., Priebe W., Stephens L. C., and Perez-Soler R., Preclinical toxicity of liposome-incorporated annamycin, selective bone marrow toxicity with lack of cardiotoxicity, *Clin. Cancer Res.* 1, 1369, 1995.
6. Lasic D. D., *Liposomes, From Physics to Applications.* Elsevier, Amsterdam, 1993.
7. Gregoriadis G., *Liposomes as Drug Carriers, Recent Trends and Progress.* John Wiley & Sons, Chichester, 1988.
8. Gabizon A., Liposomes as a drug delivery system in cancer chemotherapy. In: *Drug Carrier Systems. Biochemical and Biophysical Basis and Medical Prospects.* Roerdink F. H. and Kroon A. M. (Eds.), *Horizons in Biochemistry and Biophysics.* John Wiley & Sons, Chichester, 9, 185–211, 1989.
9. Barenholz Y., and Cohen R., Rational design of amphiphile-based drug carriers and sterically stabilized carriers, *J. Liposome Res.*, 5, 905, 1995.
10. Goren D., Gabizon A., and Barenholz Y., The influence of physical characteristics of liposomes containing doxorubicin on their pharmacological behavior, *Biochim. Biophys. Acta*, 1029, 285, 1990.
11. Gabizon A., Liposome circulation time and tumor targeting, implications for cancer chemotherapy, *Adv. Drug Deliv. Rev.*, 16, 285, 1995.
12. Young R. C., Ozols, R. F., and Myers C.E., The anthracycline antineoplastic drugs, *N. Engl. J. Med.*, 305, 139, 1981.
13. Von Hoff D. D., Layard M. W., Basa P., Davis H. L., Von Hoff A. L., Rozencweig M., and Muggia F. M., Risk factors for doxorubicin-induced congestive heart failure, *Ann. Int. Med.*, 91, 710, 1979.
14. Gabizon A., Liposomal anthracyclines, *Hematol. Oncol. Clin. North Am.*, 8, 431, 1994.
15. Haran G., Cohen R., Bar L. K., and Barenholz Y., Transmembrane ammonium sulfate gradients in liposomes produce efficient and stable entrapment of amphipathic weak bases, *Biochim. Biophys. Acta*, 1025, 143, 1990.
16. Lasic D. D., Frederik P.M., Stuart M. C., Barenholz Y., and McIntosh T. J., Gelation of liposome interior — a novel method for drug encapsulation. *FEBS Lett.*, 312, 255, 1992.
17. Woodle M. C., and Lasic D. D., Sterically stabilized liposomes, *Biochim. Biophys. Acta*, 11, 171, 1992.
18. Sculier J. P., Coune A., Brassinne C., Laduron C., Atassi G., Ruysschaert J. M., and Fruhling J., Intravenous infusion of high doses of liposomes containing NSC 251635, a water-insoluble cytostatic agent. A pilot study with pharmacokinetic data. *J. Clin. Oncol.*, 4, 789, 1986.
19. Szebeni J., Wassef N. M., Spielberg H., Rudolph A. S., and Alving C. R., Complement activation in rats by liposomes and liposome-encapsulated hemoglobin, evidence for anti-lipid antibodies and alternative pathway activation, *Biochem. Biophys. Res. Com.*, 205, 255, 1994.
20. Madhavan S., and Northfelt D. W., Lack of vesicant injury following extravasation of liposomal doxorubicin, *J. Natl. Cancer Inst.*, 87, 1556, 1995.
21. Gabizon A., Pappo O., Goren D., Chemla M., Tzemach D., and Horowitz A. T., Preclinical studies with doxorubicin encapsulated in polyethyleneglycol-coated liposomes, *J. Liposome Res.*, 3, 517, 1993.

22. Lokich J. J., and Moore C., Chemotherapy-associated palmar-plantar erythrodysesthesia syndrome, *Ann. Int. Med.*, 101, 798, 1984.
23. DeSpain J. D., Dermatologic toxicity. In: *The Chemotherapy Source Book,* Perry, M. C., (ed.) Williams & Wilkins, Baltimore, 531–547, 1992.
24. Gabizon A., Catane R., Uziely B., Kaufman B., Safra T., Cohen R., Martin F., Huang A., and Barenholz Y., Prolonged circulation time and enhanced accumulation in malignant exudates of doxorubicin encapsulated in polyethylene-glycol coated liposomes, *Cancer Res.*, 54, 987, 1994.
25. Billingham M. E., Mason J. W., Bristow M. R., and Daniels J. R., Anthracycline cardiomyopathy monitored by morphologic changes, *Cancer Treat. Rep.*, 62, 865, 1978.
26. Daemen T., Hofstede G., Ten Kate M.T., Bakker-Woudenberg I. A., and Scherphol G. L., Liposomal doxorubicin-induced toxicity, depletion and impairment of phagocytic activity of liver macrophages, *Int. J. Cancer,* 61, 716, 1995.
27. Muggia F., Hainsworth J., Jeffers S., Groshen S., Tan M., and Greco F. A., Liposomal doxorubicin (Doxil) is active against refractory ovarian cancer, *Proc. Am. Soc. Clin. Oncol.*, 15, 287, 1996.

15

Antibacterial Agents[1]

Rodolphe Garraffo
University Hospital, Nice

15.1 Introduction

The right choice of antibiotic is one of the the most important factors, though not the only one, in determining the success or failure of an antimicrobial therapy. The drug chosen should combine the highest *in vivo* efficacy, i.e., the best chance of success, with the least likelihood of side effects. The principal objective in giving an antibiotic is to achieve a concentration at the site of infection that is sufficient to kill or inhibit the growth of the bacteria present. At the same time, the antibiotic(s) and the dosage regimen chosen have to be as safe as possible for the patient with regard to the possible occurrence of adverse effects. Thus, to achieve clinical efficacy associated with safety, the prescriber has to deal with several important variables, depending on the patient (age, severity of illness, renal and/or hepatic function, immune status, hypersensitivity to antibiotics), on the infection (site and severity), on the causative organism (level of sensitivity or resistance) and on the antibiotic (pharmacokinetics, pharmacodynamics, or resistance). Unexpected treatment failure and/or severe adverse effects can occur if these factors are not considered. Toxicity avoidance and enhancement of clinical efficacy are the best approach to optimize the benefit/risk ratio of therapy with antibacterial drugs.

15.2 Avoidance of Antibiotic Toxicity

No antimicrobial agent is totally free of unwanted side effects; most of them are trivial, some merely inconvenient, and only a few are life-threatening or fatal. Various factors are responsible for drug toxicity, but two major ways of serious toxicity are usually described for antibiotics: one refers to immuno–allergic side effects, while the other is dose/concentration-related. In the first case the prevention of severe toxicity generally consists in the withdrawal of the responsible antibiotic and/or the choice of an alternative antibacterial, while in the second case, a decrease of the doses under control by therapeutic drug monitoring may be a suitable response.

---

[1] We are very grateful to Dr. Eveline Bernard for her kind help in reviewing this manuscript.

15.2.1   Hypersensitivity, an Example of Dose-Independent Toxicity

Among the antibiotics, the β-lactam compounds have the greatest potential to produce hypersensitivity reactions. Because of the similar structure of the penicillins, hypersensitivity to one agent usually means hypersensitivity to the whole group. Moreover, the structural similarities between cephalosporin antibiotics and the penicillins are accompanied by a degree of cross reactivity between the two groups. The effects are variable and include different types of rash, urticaria, eosinophilia, edema, fever, conjunctivitis, cutaneous photosensitivity reactions, and immunological abnormalities in the blood. They can occur very early in the treatment (immediate hypersensitivity) when the patient is already allergic, and produce vomiting, nausea, laryngeal edema, and cardiovascular collapse requiring adrenaline and control of integrity of the airways. Other hypersensitivity reactions tend to be delayed, and these late reactions include fever, erythema nodosum, and serum sickness-like syndrome. Rashes are very frequent, and a maculopapular eruption is very often found in patients suffering from infectious mononucleosis. Anaphylactic shock, which can be fatal, is the most serious allergic response. The prediction of hypersensitivity is difficult, and thus prevention of this side effect requires a close questioning of the patient to identify those with a strong family history of drug allergy (or allergic diseases like asthma or eczema) and to learn about any previous episode of hypersensitivity. Skin testing is only occasionally carried out because it is unreliable; penicillin desensitization is rarely employed and not easy to perform. Sulfonamides and cotrimoxazole also cause hypersensitivity reactions, and allergic reactions are not uncommon with vancomycin or nitrofurans.

Other toxicities that are not clearly or only partially related to dosage can occur with antibiotics. These are gastrointestinal effects, lupus syndrome (sulfonamides, penicillins), hepatitis, and/or cholestasis (isoniazid, rifampicin, erythromycin), dental staining (tetracyclines), and hematological toxicity (β-lactam antibiotics, sulfonamides).With these drugs there are major difficulties in preventing adverse drug reactions, because patients frequently respond idiosyncratically to antimicrobial agents, especially with rare side effects that are generally unpredictable. Some precautions, e.g., avoiding potentially hepatotoxic antibiotics (erythromycin, isoniazid) in subjects with preexisting liver disease, could restrict risks. When toxic effects develop or are suspected, the decision has to be made as to whether to stop or to change the patient's treatment.

15.2.2   Dose-Dependent Toxicity

The problem is quite different when adverse effects of the antibiotics are dose- or concentration-related. Indeed, in such a situation the antibiotic may frequently be continued providing the dose is adjusted either by reducing each individual dose or prolonging the interval between doses. Routine therapeutic drug level monitoring (TDM) and the consecutive dosage adjustment have been developed during the last decades in order to facilitate the prevention and management of concentration-dependent toxicity of antibiotics.

As stated before, the goal of optimum antimicrobial therapy is to achieve the best possible action against the target microorganism with a minimum risk of unwanted side effects. For this purpose, drug level measurements to individualize the dosage gained much interest during the last decades. This approach, however, was at first intended for drugs with a low therapeutic index (i.e., when the difference between therapeutic and toxic concentrations is small, exhibiting a narrow therapeutic margin) and a poor predictability of serum concentrations. Two other basic requirements have also to be fulfilled for the antibiotics considered: a direct relationship between drug levels and therapeutic and/or toxic effects and a wide variation of serum concentrations among patients under standardized dosage.[41] For several years the approach to the efficacy remained relatively simple, and serum measurements of an antibiotic had just to assure that the minimal concentration achieved in serum, or better, at the infection site, should be approximately equal or superior to the minimal inhibitory concentration (MIC) of the infecting organism. As mentioned above, the majority of antibiotics, including penicillins and cephalosporins, showed little dose-dependent toxicity and consequently can safely be applied well above their MIC. Much attention was paid to the few antibiotics responsible for concentration-dependent toxicity, at least in some patient populations. These are aminoglycosides, vancomycin, and to a smaller extent (because rarely prescribed) chloramphenicol.

Aminoglycosides represent an important class of antibiotics in the management of life-threatening gram-negative infections. Due to their narrow therapeutic range and the poor predictability of serum concentrations, aminoglycosides are among the most frequently selected agents for TDM. A wide inter-patient variation in distribution and elimination characteristics is the main reason for this unpredict-ability. Indeed, Zaske et al.[42] observed in 1640 patients with normal serum creatinine values gentamicin half-lives of 0.4 to 32.7 hours. Moreover, the aminoglycoside pharmacokinetics are best described by a three-compartment open model[17] where the slow terminal half-life reflects elimination from the so-called "deep tissue compartment." The accumulation in this compartment, which includes kidney and inner ear, is probably an important factor in their potential toxicity for these organs.[34] Aminogylcoside-induced injury of the kidney and the inner ear is the major limiting factor in the use of these potent antimicrobials. Controversial opinions about the influence of serum aminoglycoside concentrations on the development of nephro- and ototoxicity have existed for several years. Nevertheless, it appears to me that sufficient data exist to conclude that accumulations at the site of the unwanted effect is related to a clinical manifestation of nephrotoxicity and/or ototoxicity, particularly at the earlier period of treatment.[26,33]

Therefore, determination of aminoglycoside blood levels appears to be the most useful and relevant approach for detecting a drug accumulation caused by nephrotoxicity that may precede changes in other correlates of renal function (e.g., serum creatinine). This allows the prevention of more severe dysfunc-tions of the kidney and cochlear or vestibular damages frequently following kidney injury.

In contrast to this, the existence of a saturable mechanism for the cortical uptake of aminoglycosides in the human kidney can explain the lack of correlations between "peak" serum concentrations and nephro- or ototoxicity.[8] On the other hand, several experimental[11,40] and clinical data[11,18,29,30] demonstrate that the bactericidal activity of aminoglycosides is concentration-dependent, so that clinical efficacy correlates with peak level (see pharmacodynamic section).

The clinical toxicity of aminoglycosides can be minimized by appropriate patient selection and mon-itoring, and taking account for influencing risk factors associated with toxicity (i.e., age, kidney function, duration of treatment, dosage regimen). TDM has to be considered when starting the treatment, partic-ularly within the first few days. Indeed, not only could this allow the reduction of the risk of toxicity (trough level) but also enhance the efficacy by avoiding subtherapeutic levels (peak level) in life-threat-ening infections and/or in specific patient populations (neonates, patients from intensive care units, neutropenic, burned, or old patients).[28] Moreover, serum concentrations may be used in several dosing methods that are now available to help the physician in choosing a new dosage regimen.[14] These include predictive algorithm and nomograms, pharmacokinetics-based dosing methods, and methods that incor-porate Bayesian forecasting. Bayesian methods using the appropriate population-based parameters and non-steady-state serum concentration feedback produce optimal results when compared to other meth-ods for initial individualization of aminoglycoside therapy. Finally, there are now several cost-effectiveness studies demonstrating that such efforts result in cost saving, better therapeutic efficacy, less toxicity, and shorter mean durations of hospital stay.[9]

Vancomycin is a glycopeptide antibiotic highly effective with staphylococcal and enterococcal infec-tions. Except for some allergic reactions like "red man syndrome," the major side effects observed during treatment are nephrotoxicity and ototoxicity. So far a well-defined therapeutic range has not yet been established, but, as with aminoglycosides, accumulation and thus nephro- and ototoxicity should be prevented by TDM, particularly in special populations such as patients with impaired renal function. Moreover, recent data suggest that clinical efficacy could be optimized by maintaining trough values over MIC (see pharmacodynamic section). Thus, TDM should be of particular interest for patients with impaired organ function and with severe infection.[6]

Some other examples could be evoked such as chloramphenicol, 5-fluorocytosine or amphotericin B (antifungal), ganciclovir or aciclovir (antiviral), where TDM should be a valuable approach in preventing toxicity, at least for patients with a high risk of side effects. The rapid development of reproducible and accurate assays of drug concentrations in biological fluids and the possibilities of individualizing the dosage regimen have allowed an increasing role of optimization of the benefit/risk ratio for drugs with a narrow therapeutic index. During the past decades TDM has been essentially devoted to the prevention of the toxicity of antibacterial treatment. Since there are four variables relevant to drug treatment of infections (pharmacokinetics, toxicity, resistance, and pharmacodynamics), and only pharmacodynamics and toxicity have been studied in any detail in patients, there has recently been an increasing attention to the pharmacodynamic relationship between drugs and bacteria, and to the integration of these factors into treatment regimens.

15.3   Enhancement of Clinical Efficacy

The aim of antibiotic therapy is to eradicate the pathogens within the infected site and to avoid relapse. Administration of an antibiotic treatment requires consideration of various parameters: the patient, the bacterium and its level of sensitivity, the infection site, and the potential toxicity of the drug. In fact, clinical failure often results from lack of optimal conditions for interactions between antibiotic and bacteria within the infection site. Moreover, the clinician is now frequently faced with situations requiring an immediate, maximally efficacious and minimally toxic therapy, particularly with patients from intensive care units or neutropenics (or those with depressed immunity). In addition, involved pathogens are most often of nosocomial origin and have acquired high levels of resistance, limiting the therapeutic options. For several years the dosage regimen of antibiotics has been established according to pharmacokinetic data (volume of distribution, total body clearance, elimination half-life, bioavailability) and *in vitro* parameters of microbiological susceptibility tests, mostly MIC and MBC (minimum bactericidal concentration). The general rule is the maintenance of serum concentrations above MIC of the infective bacterium. This method of dosage determination has its limitations, as the dynamics of the bacteria/drug interaction, the specific factors related to the host (immune status, underlying pathology, drug interactions), or those linked with the infection site (accessibility of the antibiotic to the site, presence of foreign material, local inflammation) are not taken into account. Although this approach can determine the conditions of antibacterial activity, it fails to provide sufficient guaranties of *in vivo* efficacy.

As antibiotic therapy is concerned more and more with high-risk patients with severe infections sometimes caused by poorly sensitive bacteria, the clinician may raise the following questions:

- Why do drugs sometimes fail when MIC testing indicates probable efficacy?
- Are there other relevant parameters able to predict *in vivo* efficacy of antibiotics?
- If, yes, how can we further optimize the efficacy of available antibiotics?

To answer these questions, several authors have focused their attention on the pharmacokinetics/pharmacodynamics (PK-PD) relationships and their ability in describing, or better, in predicting the efficacy of various antimicrobial agents. It appeared that, in order for the clinician to make a rational choice of antibiotic for treatment of a given bacterial infection, a specific and precisely defined relationship between antimicrobial pharmacokinetics and the pharmacodynamic interaction between the antimicrobial agent and its bacterial target must be achieved. This specific relationship is a surrogate marker for outcomes such as bacterial eradication and/or clinical cure.[22] This section will address the PK-PD surrogate markers that may influence antibiotic response and clinical outcome, and then examine the applications that have been proposed to rationalize the choice of the antibiotic treatment.

Pharmacodynamic parameters of antimicrobials relate drug concentration and the desired effect. In most cases, these pharmacodynamic parameters are determined *in vitro*, then their relations with pharmacokinetics and clinical outcome are evaluated in animal models, and finally extrapolated to the clinical setting. Some of the more common *in vitro* methods include minimum inhibitory (MIC) and minimum bacterial (MBC) concentrations testing, serum bactericidal titer (SBT), and performance of time-kill curves.

MIC is the minimum antibiotic concentration that prevents growth, i.e., zero net change in the number of organisms over time. It is the most commonly used measure of antibiotic activity. Generally, the lower the MIC is, the greater is the susceptibility of the bacteria. Each antibiotic has an established MIC "breakpoint" that is designed to relate *in vitro* MIC to antibiotic concentrations achievable *in vivo* by means of a complicated and highly subjective process. This breakpoint is used to classify susceptible vs. resistant isolates. Indeed, bacterial strains with a MIC exceeding the breakpoint are considered "resistant" to the antibiotic, whereas those with MICs less than the breakpoint are "susceptible."

MBC is the minimum antibiotic concentration that kills 99.9% of the original inoculum, i.e., the number of initial bacteria is reduced by 3 log 10 units. MBC reflects the ability of an antibiotic to substantially reduce the number of bacteria. For some antibiotics such as β-lactams, aminoglycosides, or fluoroquinolones, the MBC is often similar in magnitude to the MIC. These agents are referred to as "bactericidal" antibiotics in contrast to those with a higher MBC/MIC ratio, which are considered as "bacteriostatic."

SBT is derived from MBC but uses the serum of a treated patient instead of a broth with diluted antibiotic. It is proposed to assess antimicrobial activity in clinical situations where host immune factors are less helpful in eradicating pathogens in conditions such as endocarditis, meningitis, or infections in neutropenic patients. An SBT ≥ 1:8 is usually required to obtain a favorable outcome.[24]

MIC, MBC, and SBT are time-honored measures of antimicrobial activity but the information being provided by them is limited. Several reasons can be pointed out to explain this fact.

MIC, MBC, and SBT are determined after a fixed incubation period and therefore only reflect a specific point of time; thus they are not able to provide any information about the course of antimicrobial activity.

The inoculum usually used for these measurements is relatively low (10^4 to 10^5 CFU/ml), whereas infectious diseases commonly involve much greater inocula (10^6 to 10^8 CFU/ml or even more), requiring much higher antibiotic concentrations than MIC to inhibit growth. Moreover, some antibiotics like β-lactams show a reduced activity when inoculum becomes larger.

In several *in vivo* situations bacteria causing infection at certain sites (cardiac vegetations, intracellular environment, abscesses, cerebrospinal fluid) may only grow slowly. Some antibiotics (β-lactams, for example) exhibit a reduced activity against slowly growing or nongrowing bacteria, requiring antibiotic concentrations markedly over MIC to be effective.

The above described methods use constant antibiotic concentrations over the incubation period, which is very different from the *in vivo* setting where antibiotic concentrations at the site of infection vary widely throughout the administration interval due to drug elimination, thus leading to successive periods of inhibition or killing and regrowth.

Finally, host factors (complement, antibodies) or protein binding either enhancing or reducing *in vivo* antimicrobial activity are not taken into account with MIC and MBC determinations, while this is the case with SBT determinations.

In summary, while MICs and MBCs provide useful information about the intrinsic activity of antimicrobial agents against bacteria, this information is not sufficient for designing dosing regimens aimed at optimizing drug efficacy *in vivo*.

Bactericidal rate: As stated before, MIC and MBC do not provide a full description of drug concentrations time courses of antimicrobial activity against a pathogen. An additional dimension of antibacterial activity has been found by analyzing the kinetics of time-killing curves and regrowth patterns of a specific antibiotic against a specific strain (Figure 15.1). This approach allows us to describe the relationship between bactericidal rate, antibiotic concentrations, and duration of exposure. Using this approach, Vogeleman et al.[40] have pointed out that antibiotics may be categorized into the following three groups:

- Antibiotics, like aminoglycosides or fluoroquinolones, that exhibit a rapid and marked concentration-dependent killing rate (Figure 15.1A) and a degree of suppression of bacterial growth after limited exposure (i.e., the so-called postantibiotic effect).

- Antibiotics with a weak concentration-dependent bactericidal rate over a specific range; when concentrations are increased further, the killing rate becomes maximal and is no longer influenced by the increase of drug concentrations (Figure 15.1B). The killing rate is remarkably slow and correlates with the duration of time of contact between antibiotic and bacteria. These agents, e.g., β-lactams (except carbapenems) or glycopeptids, have no or very short postantibiotic effects.

- Antibiotics that are predominantly bacteriostatic at any concentration normally achieved *in vivo*.

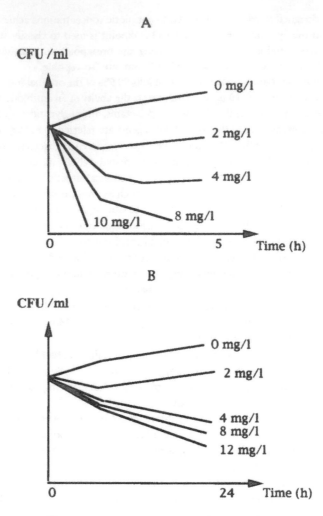

FIGURE 15.1   Typical *in vitro* killing curves with a concentration-dependent (A) and with a time-dependent (B) bactericidal antibiotic.

Postantibiotic effect: The pharmacodynamic parameters discussed so far describe the fate of bacteria during the time at which they are exposed to an antibiotic. However, in the 1940s investigators working with staphylococci observed that these bacteria did not immediately resume growth after transient exposure to penicillin.[32] In the 1970s Mc Donald et al.[27] proceeded with this observation and described a phenomenon actually known as the postantibiotic effect (PAE). It is defined as the delay in bacterial regrowth that occurs as a result of a transient antibiotic exposure after the removal of the antibiotic. Such an effect may be observed with nearly any antimicrobial, but not all antimicrobials produce a postantibiotic effect in all organisms. Briefly, antimicrobials that act by inhibiting DNA, RNA, or protein synthesis (essentially aminoglycosides and quinolones), i.e., concentration-dependent bactericidal antibiotics, exert a postantibiotic effect against most bacteria, particularly gram-negative bacilli. In contrast to this, inhibitors of cell wall synthesis (β-lactams and glycopeptides) routinely produce a postantibiotic effect with staphylococci, but less so with streptococci and not significantly with gram-negative bacteria.[13] The duration of the postantibiotic effect varies from drug to drug but also between different bacterial species and sometimes from strain to strain. The postantibiotic effect could be observed either *in vitro* or *in vivo* but is usually longer *in vivo*. For a given drug/bacteria pair, the postantibiotic effect may be lengthened by increasing the drug concentration during exposure or the duration of drug exposure prior

TABLE 15.1 Main Pharmacodynamic Characteristics of the Most Frequently Prescribed Bactericidal Antibiotics

Main Parameters/ Type of Antibiotic	Bactericidal Rate	Postantibiotic Effect	Resistance During Therapy	Class of Antibiotics
Concentration-dependent antibiotics	Rapid, intense concentration-dependent killing	Yes (depending of period and concentration of exposure)	Possible if C_{max}/MIC too small	Aminoglycosides Fluoroquinolones
Time-dependent antibiotics	Slow concentration independent over a threshold Time-dependent killing	No (except with Staph.)	Possible when T> MIC too short in dosing interval	β-Lactams Glycopeptides

to removal.[15] Moreover, *in vivo* experiments using the neutropenic mice with thigh infections showed that the postantibiotic effect appears to be increased by the presence of leucocytes, as demonstrated with the gentamicin/K. pneumoniae pair. In such a situation the postantibiotic effect in neutropenic mice ranges from 2.6 to 5.0 h (depending on the dose) compared to 5.8 to 9.5 h in nonneutropenic animals.[27]

These data suggest that the postantibiotic effect could be an important pharmacodynamic parameter in designing a dosing regimen. As a matter of fact its presence, for an antibiotic, implies that larger dosing intervals resulting in temporary subinhibitory serum concentrations are possible without compromising efficacy. In addition, less frequent applications of the same daily dose would allow for a higher peak serum level, which has been proved to be an important parameter for a favorable outcome when treating with concentration-dependent bactericidal antibiotics (e.g., aminoglycosides). In contrast to this, for drugs that do not exhibit a postantibiotic effect (e.g., β-lactams and gram-negative bacteria), dosing regimens that do not allow excessive periods with subinhibitory concentrations would be expected to be more efficacious. For a given total daily dose, smaller fractions dosed at frequent intervals would therefore be recommended.

Thus, a great deal of information became available regarding the question as to which variables attendant to drug administration are most closely linked to the killing of bacteria in infected animal models and their relations with the *in vitro* characteristics of bacterial growth inhibition or bactericidal activity of various classes of antibiotics (Table 15.1). On the basis of the aforementioned *in vitro* and animal models, the available data would indicate that the total exposure to aminoglycoside antibiotics as indexed by the area under the serum concentration time curve (AUC or logAUC) and the peak concentration achieved (C_{max}) are most closely linked to the outcome of infections. Likewise, they indicate that the period of time the β-lactam serum concentrations exceed the MIC (T>MIC) of the infecting pathogens is the most important parameter involved in the outcome of infections.[10,11] An important consequence of these results is to be able to decide about the optimal dose regimen for antimicrobials, and how the course of antimicrobial activity should be modified to make the best use of the pharmacodynamic features of a specific antibiotic. Therefore, animal models of infections have also been extensively used to describe pharmacokinetic/pharmacodynamic surrogate relationships and to assess antimicrobial activity at changing concentrations of various classes of antimicrobial agents. These surrogates relate various pharmacokinetic parameters to a measure of the pharmacodynamic interaction (mostly MIC). Three surrogate relationships have been examined in particular (Figure 15.2):

- The C_{max}/MIC ratio, also called inhibitory ratio (or index), which is the peak serum concentration (C_{max}) related to the MIC value of the infecting bacteria.
- The AUC/MIC ratio, also referred to as AUIC, which can be calculated as the serum concentration time curve divided by the MIC, or as the area under the inhibitory serum concentration time curve expressed as the reciprocal of the serum inhibitory titer. The relationship between AUC,

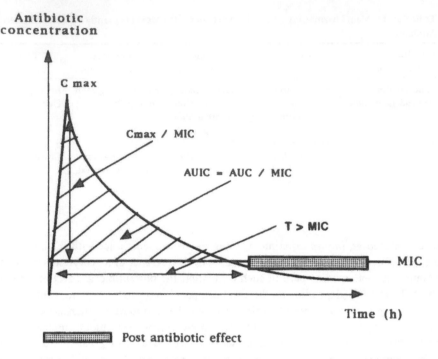

FIGURE 15.2   Most important pharmacokinetic/pharmacodynamic parameters to be considered for enhancement of antimicrobial therapy.

MIC, and effect has become useful for evaluating the efficacy of antibiotic therapy. Nevertheless, use of this method as discussed so far does not address the role of the dosing interval on the AUC/MIC relationship. Indeed, in a clinical setting, provided the interval is not changed, increasing the AUC/MIC ratio by increasing the dose may result rather in increases of T>MIC than in C_{max}/MIC ratio.

• The T>MIC defined as the time the serum concentration of a given antibiotic exceeds the MIC.

As protein binding has to be considered when evaluating therapeutic concentrations of highly protein-bound (>90%) antimicrobial agents, similar considerations apply to the evaluation of pharmacokinetic surrogates. Indeed, for these molecules the free drug concentration could sometimes be more relevant, and thus the pharmakokinetic markers should be related to their free concentration. Drusano[11] has particularly underlined the role of the duration of time the concentration of free β-lactam antiobiotics exceeds the minimum inhibitory concentration of 90% of bacterial strains (MIC_{90}) relative to their respective dosing intervals in explaining or predicting success or failure of a therapy particularly with *P. aeruginosa* and *S. aureus*.

15.4   Prevention of Resistance

The eradication of the most susceptible pathogens (i.e., with low MIC) by antimicrobial therapy may in some cases result in the regrowth of a less susceptible subpopulation (with higher MIC) as commonly observed *in vitro* or in animal models of infections. It is well known that any inoculum of bacteria will usually be constituted of a normally distributed population of organisms with regard to their antibiotic susceptibility. The possibility of emergence of resistance depends on the proportion, the virulence, and the propensity to the development of the less susceptible subpopulation; the ability of host immune factors to cope with it; and the magnitude of the antibiotic concentration (concentration-dependent bactericidal antibiotics) or the duration of presence of a concentration greater than MIC (time-dependent bactericidal antibiotics) at the site of infection. Therefore, it has been demonstrated that with aminogly-cosides and quinolones higher antibiotic concentrations are more likely to eradicate the entire population of an inoculum, and the likelihood of resistance is reduced when the C_{max}/MIC ratio increases. Experimental data suggest that C_{max} has to meet or exceed the MIC to prevent resistance.[2,12] For these agents less frequent administration of a given daily dose would allow for greater peak concentrations and enable concentrations to exceed, at least temporarily, the MIC by eight- to tenfold. In this same way the rate and intensity of bacterial killing are enhanced by higher peak concentrations. This approach has particularly been developed for aminoglycosides, which can prevent the occurrence of *P. aeruginosa* adaptive resistance by increasing the first dose exposure.[7] The conditions required for the prevention of resistance to aminoglycosides along with the concentration-dependent bactericidal rate these agents exhibit imply that less frequent dosing would achieve at least equal but probably greater efficacy to that obtained with frequent dosing, and would probably lessen the occurrence of resistance provided the total daily dose is unchanged.

The aim for time-dependent bactericidal antibiotics is quite different. Their bactericidal rate is less dependent on the magnitude of concentrations, and maximal antibacterial activity can be reached by extending the period of time the concentration exceeds the MIC. Recent data obtained with β-lactams and S. pneumoniae strongly suggest that the likelihood of resistance increases when the period with subinhibitory antibiotic concentrations during the interval of administration is increased.[31] These findings have led to the evaluation of the possible benefits of continuous infusion, which allows the maintenance of antibiotic concentrations over the MIC during the entire interval of administration.

Although it is very difficult to extrapolate the results obtained *in vitro* or with specific animal models of infections to human therapy, it appears that some general rules or indications could be held for clinical purposes. Thus, all PK-PD parameters discussed in the previous sections proved to be helpful for describing efficacy and resistance characteristics of the most often prescribed classes of antibiotics (Table 15.2).

The major problem that has to be resolved is how these relationships can be established between *in vitro* models, animal model studies, and human clinical trials. In fact, when differences in pharmacokinetics between animals and humans are controlled, animal model studies can be very useful for predicting the human clinical response. In order to show similarities between *in vitro* data, animal models, and human exposure, Hyatt et al.[22] recently proposed the $AUIC_{24}$ (i.e., AUC_{24}/MIC) as the general PK-PD surrogate marker able to assume interrelations between these different experimental conditions. Even if the pharmacokinetics (AUC) and the pharmacodynamics (MIC) are taken into consideration in this approach, the shape of AUC, which may be of significance, is neglected; thus it can fail to reflect the influence of the fluctuations of concentrations when the dosing interval is enlarged. Obviously, the C_{max}/MIC ratio and the T>MIC have to be distinguished from $AUIC_{24}$, except when the antibiotic exhibits a short elimination half-life and is administered within small intervals. Nonetheless, these investigators have demonstrated that $AUIC_{24}$ appears to be a relevant parameter for describing the potential of clinical efficacy of an antibiotic (particularly β-lactams and fluoroquinolones), whereas it seems necessary to refer to an adequate C_{max}/MIC ratio to prevent selection of resistant bacteria when using aminoglycosides or fluoroquinolones.

15.5   Application of PK-PD Correlates to the Clinical Setting

Before any dosage regimen resulting from *in vitro* or animal studies becomes clinical practice, it is essential to evaluate it in controlled clinical studies. Investigators, however, usually have little control over many variables that may influence clinical outcome. In contrast to *in vitro* or *in vivo* animal models, dose ranging studies addressing possible side effects of different dosing regimens are not permitted for ethical reasons. Therefore, several limitations exist for comparative studies, making it very difficult to establish that one dosing regimen is superior to another as far as efficacy and/or safety are concerned. For these reasons the knowledge of PK-PD surrogate relationships proved to be useful for describing and comparing antimicrobial efficacy for various agents. They have also to be considered for choosing an effective dosing

TABLE 15.2   Relations between PK-PD Profiles of some Classes of Antibiotics and the Possible Modifications of their Dosage Regimen Aimed at Increasing Efficacy and, in some Cases, Reducing Toxicity

Parameters/ Class of Antibiotic	Bactericidal Rate	Optimization of Efficacy	Postantibiotic Effect	Influence of Dosing Regimen	How to Prevent Resistance	Optimization of Dosing Regimens
Aminoglycosides	Concentration-dependent	↗ C_{max}/MIC ↗ AUC/MIC	Yes, significant	↗ Interval[a]	↗ C_{max}/MIC	"Once a day" dosage regimen
Fluoroquinolones	Concentration-dependent	↗ C_{max}/MIC ↗ AUC/MIC	Yes, significant	↗ Interval[a]	↗ C_{max}/MIC	Increase total dose
β-lactams	Time-dependent	↗ T > MIC ↗ (AUC/MIC) to 125–250	Yes or no	↘ Interval	↗ T > MIC	Continuous infusion or frequent dosing
Glycopeptides	Time-dependent	↗ T > MIC ↗ (AUC/MIC) to 125–250	Yes or no	↘ Interval	↗ T > MIC	Continuous infusion or frequent dosing

Note: ↗ increase, ↘ decrease.

[a] If the patient is not neutropenic.

regimen because they are able to provide information on parameters specific for bacteria and to combine them with pharmacokinetics specific for the patient.[19,20] Moreover, several retrospective studies and a few prospective trials have led us to characterize and quantify PK-PD relationships in the clinical setting. One of the most important applications of these findings is that they serve as a guide for designing appropriate dosage regimens rather for phase I/II clinical studies of newer agents than to optimize therapy with antibiotics that are already used in patients. Interesting approaches have been provided for the most frequently used classes of antibiotics, i.e., aminoglycosides, fluoroquinolones, β-lactams, and glycopeptides.

Aminoglycosides are concentration-dependent bactericidal antibiotics requiring a high C_{max} and/or AUC to MIC ratio to enhance their bactericidal rate and intensity. They exhibit a significant *in vivo* concentration-dependent postantibiotic effect (2 to 10 hours), which indicates that dosage intervals should probably, in a reasonable way, be longer than the period of time during which the concentrations remain above the MIC. Moreover, the first dose exposure effect, which may occur with some gram-negative bacilli, should be controlled by increasing the C_{max}/MIC ratio over a value of 8 to 10, providing optimal conditions for a rapid return to baseline of the MIC and reducing the risk of adaptive resistance. Retrospective clinical studies have confirmed the benefit of administering larger doses with less frequent dosing intervals, suggesting that the administration of the total daily dose as "once a day" dosing regimen will provide the same or a better efficacy and less toxicity than more frequent administrations (2 or 3 administrations per day) with the same daily dose. In a retrospective study, Highet et al.[21] recently asserted that a 24-hour AUIC breakpoint of 100 (as the sum of piperacillin and tobramycin $AUIC_{24}$ values) correlates with clinical cure. Clinical studies documenting once daily dosing of aminoglycosides have primarily involved a combination of antibiotics and generally moderately severe infections. Several clinical trials with monotherapy and more severe infections, however, are actually going on with encouraging results.[1,38]

Fluoroquinolones have demonstrated an ability to exert a concentration-dependent killing. *In vitro* studies suggest that a maximum rate of killing is achieved at concentrations between 50 to 60 times the MIC.[37] By using animal models of infections it has been shown that a C_{max}/MIC ratio of ≥8–10 was able not only to provide a 99% reduction of the initial inoculum within 4 hours but also to prevent the regrowth of more resistant bacteria at 24 hours.[2] This was confirmed in another study by Drusano et al.[12] that concluded that C_{max}/MIC and AUC/MIC ratios were the most predictive parameters of clinical outcome. In an *ex vivo* model with healthy human volunteers it was proved that it is possible to maintain a satisfying efficacy of ciprofloxacin against some gram-negative bacteria with reduced sensitivity by a twofold increase of the given dose.[20] In a clinical study with ciprofloxacin, Forrest et al.[16] finally showed that an $AUIC_{24}$ of 125 to 250 induced slow bacterial killing with bacterial eradication in about 7 days, while with $AUIC_{24}$ ≥ 250, the bacterial killing rate became extremely fast, with eradication obtained in only 1.9 days. In the near future further clinical studies will probably provide valuable guidelines to enhance the treatment strategies with fluoroquinolones.

In contrast to aminoglycosides and fluoroquinolones, β-lactams primarily demonstrated time-dependent killing. Animal models of infections clearly showed that T>MIC is the most important parameter influencing the clinical outcome. In studies utilizing more resistant organisms or larger inoculum, however, $AUIC_{24}$ appears to be another important parameter to be considered for the clinical issue. Thus to optimize the T>MIC, continuous infusion of β-lactams was suggested as a valuable alternative to the usual dosing regimen. There are only a few clinical trials in humans comparing intermittent vs. continuous dosing of β-lactam antibiotics. In a large population of patients with cancer and severe neutropenia Bodey et al.[3] demonstrated the advantage of the continuous infusion of β-lactams as compared to intermittent bolus administration. In a study with cefmenoxime, Schentag et al.[35,36] more recently showed that the choice of a dosing regimen allowing the achievment of a T>MIC of 100% of the dosing interval resulted in more rapid eradication of bacterial pathogens involved in nosocomial pneumonia. Analyzing the same data, AUC/MIC was also found to correlate with microbiological response, and prospective individualized dose adjustment to achieve a target AUC produced earlier bacterial eradication, allowing a reduction of the duration of treatment. Consequently, these studies suggest that bactericidal activity is optimized by maximizing the period of time during which concentrations are above the MIC. The magnitude by which concentrations must exceed the MIC for this prolonged period of time, however, still has to be clearly determined.

Glycopeptides: Adequately characterized PK-PD surrogates for vancomycin and teicoplanin are actually lacking. *In vitro*, these glycopeptides demonstrated relatively slow and concentration-independent killing. As these substances, like β-lactams, are cell-wall-active agents it was presumed that T>MIC is possibly related to efficacy. Two animal model studies confirmed that maintaining plasma levels above the MIC will enhance the clinical issue.[4,5] Some clinical studies with a small number of patients and a glycopeptide frequently given in combination with another antibiotic confirmed that patients with a favorable clinical issue have a higher trough plasma concentration (2 to 10 times the MIC) than those with unsatisfactory outcomes.[25] In a retrospective study it was finally demonstrated that a dosage regimen allowing an AUIC > 125 may optimize outcome.[22] Additional studies are necessary to determine how to increase efficacy of glycopeptides regarding the trough plasma level in the context of a PK-PD approach.

15.6   Conclusions

In recent years the pharmacokinetic and pharmacodynamic properties of antibiotics have received increased attention, and their influence on the clinical outcome and their benefit/risk ratio are currently considered. The outcome of antibiotic therapy actually depends on a number of important variables. Unexpected treatment failures can certainly occur if PK-PD relations are not optimized. On the other hand, both therapeutic drug monitoring and pharmacodynamic-derived administration schedules can contribute to providing less toxicity for the treatment with low therapeutic index antibiotics. The ability to characterize and quantify pharmacodynamics of antibiotics *ex vivo* offers a unique opportunity for clinicians to link these parameters with individualized pharmacokinetics for each patient to propose an optimal dosage regimen. With the aid of *in vitro* and animal infection models overcoming the ethical problems in humans, appropriate PK-PD surrogate markers have been determined. Although it seems that antimicrobial agents should be a class of drugs that is well suited for pharmacodynamic studies, the data available to date, however, are still somehow insufficient. It is extremely important to develop consistent PK-PD modeling for antimicrobials that guide preclinical assessment of any antibiotic, as well as for dosing recommendations in the clinical setting to ensure effective and nontoxic therapy.

Glossary of Abbreviations

AUC	Area under the concentration curve
AUC_{24}	Area under the concentration curve (0 to 24 h)
AUIC	AUC/MIC ratio
$AUIC_{24}$	AUC/MIC ratio (0 to 24 h)
CFU	Colony forming units
C_{max}	Peak concentration
DNA	Deoxyribonucleic acid
K.	Klebsiella
MBC	Minimum bactericidal concentration
MIC	Minimum inhibitory concentration
MIC_{90}	Minimum inhibitory concentration of 90% of bacterial strains
P.	Pseudomonas
PAE	Post-antibiotic effect
PK-PD	Pharmacokinetics/pharmacodynamics
RNA	Ribonucleic acid
S.	Streptococcus
SBT	Serum bactericidal titer
T>MIC	Period of time of concentrations exceeding the MIC
TDM	Therapeutic drug (level) monitoring

References

1. Beaucaire, G., Leroy, O., Beuscart, C., Karp, P., Chidiac, C., Caillaux, M., and the study group, Clinical and bacteriological efficacy and practical aspects of amikacin given once daily for severe infections, *J. Antimicrob. Chemother.*, 27 (Suppl. 7), 91, 1991.

2. Blaser, J., Stone, B. B., Groner, M. C., and Zinner, S. H., Comparative study with enoxacin and netilmicin in a pharmacodynamic model to determine importance of ratio antibiotic peak concentration to MIC for bactericidal activity and emergence of resistance, *Antimicrob. Agents Chemother.*, 31, 1054, 1987.

3. Bodey, G. P., Ketchel, S. J., and Rodriguez, V., A randomized study of carbenicillin plus cefamandole or tobramycin in the treatment of febrile episodes in cancer patients, *Am. J. Med.*, 67, 608, 1979.

4. Cantoni, L., Glaucer, M. P., and Bille, J., Comparative efficacy of daptomycin, vancomycin and cloxacillin for the treatment of S. aureus endocarditis in rats and the role of test conditions in this determination, *Antimicrob. Agents Chemother.*, 34, 2348, 1990.

5. Chambers, H. F., and Kennedy, S., Effects of dosage, peak and trough concentrations in serum, protein binding and bactericidal rate on efficacy of teicoplanin in a rabbit model of endocarditis, *Antimicrob. Agents Chemother.*, 34, 510, 1990.

6. Collin, D., Quintiliani, R., and Nightingale, C. H., Vancomycin therapeutic drug monitoring: is it necessary? *Ann. Pharmacother.*, 27, 594, 1993.

7. Daikos, G. L., Lolans, V. T., and Jackson, G. G., First-exposure adaptative resistance to aminoglycoside antibiotics *in vivo* with meaning for optimal clinical use, *Antimicrob. Agents Chemother.*, 35, 117, 1991.

8. De Broe, M. F., Paulus, G. J., Verpooten, G. A., Roels, F., Buyessens, N., and Weesen, R., Early effects of gentamicin, tobramycin and amikacin on the human kidney, *Kidney Int.*, 25, 643, 1984.

9. Destache, C. J., Meyer, S. K., and Rowley, K. M., Does accepting pharmakokinetic recommendations impact hospitalization? A cost-benefit analysis, *Ther. Drug Monitoring*, 12, 274, 1990.

10. Drusano, G. I., Ryan, P. A., Standiford, H. C., Moody, M. R., and Schimpff, S. C., Integration of selected pharmacologic and microbiologic properties of three new β-lactam antibiotics: a hypothesis for rational comparison, *Rev. Infect. Dis.*, 7, 357, 1984.

11. Drusano, G. L., Human pharmacodynamics of β-lactams, aminogylcosides and their combination, *Scand. J. Infect. Dis.* Suppl. 74, 235, 1990.

12. Drusano, G. L., Johnson, D. E., Rosen, M., and Standiford, H. C., Pharmacodynamics of a fluoroquinolone antimicrobial agent in a neutropenic rat model of Pseudomonas sepsis, *Antimicrob. Agents Chemother.*, 37, 483, 1993.

13. Ebert, S. C., and Craig, W. A., Pharmacodynamic properties of antibiotics: application to drug monitoring and dosage regimen design, *Infect. Control. Hosp. Epidemiol.*, 11, 319, 1990.

14. Erdman, S. M., Rodvold, K. A., and Pryka, R. D., An updated comparison of drug dosing methods, Part III: Aminoglycoside antibiotics, *Clin. Pharmacokin.*, 20 (5), 374, 1991.

15. Fantin, B., Ebert, S., Legett, J., Vogelman, B. and Craig, W. A., Factors affecting duration of *in vivo* post-antibiotic effect for aminoglycosides against gram-negative bacteria, *J. Antimicrob. Chemother.*, 27, 829, 1990.

16. Forrest, A., Nix, D. E., and Ballow, C. H., Pharmacodynamics of intravenous ciprofloxacin in seriously ill patients, *Antimicrob. Agents Chemother.*, 37, 1073, 1993.

17. French, M. A., Cerra, F. B., Plaut, M. E., and Schentag, J. J., Amikacin and gentamicin accumulation pharmacokinetics and nephrotoxicity in critically ill patients, *Antimicrob. Agents Chemother.*, 19, 147, 1981.

18. Garraffo, R., Drugeon, H. B., Dellamonica, P., Bernard, E., and Lapalus, P., Determination of the optimal dosage regimen for amikacin in healthy volunteers by study of pharmacokinetics and bactericidal activity, *Antimicrob. Agents Chemother.*, 34, 614, 1990.

19. Garraffo, R., Optimal adaptative control of pharmacodynamic effects with aminoglycoside antibiotics: a required approach for the future, *Int. J. Biochem. Comput.*, 36, 43, 1994.

20. Garraffo, R., and Drugeon, H. B., Comparative assessment of the pharmacokinetics and pharmacodynamics of ciprofloxacin after single IV doses of 200 and 400 mg, *Drugs*, 49 (Suppl. 2), 317, 1995.

21. Highet, V. S., Ballow, C. H., and Forrest, A., Population-derived AUIC is predictive of efficacy [abstract], 95th Meeting American Society for Clinical Pharmacology and Therapeutics, March, 30–April, 1, New Orleans, 1994.

22. Hyatt, M. J., McKinnon, P. S., Zimmer, G. S., and Schentag, J. J., The importance of pharmacokinetic/pharmacodynamic surrogate markers to outcome. Focus on antibacterial agents, *Clin. Pharmacokinet.*, 23 (2), 143, 1995.

23. Jackson, G. G., Diakos, G. K., and Lolans, V. T., First exposure effect of netilmicin on bacterial susceptibility as a basis for modifying the dosage regimen of aminoglycoside antibiotics, *J. Drug Dev.*, 1 (Suppl. 3), 49, 1988.

24. Klastersky, J. Daneau, D., Swings, G., Weerts, D., Antibacterial activity in serum and urine as a therapeutic guide in bacterial infections, *J. Infect. Dis.*, 129, 187, 1974.

25. Klepser, M. E., Kang, S. L., and MacGrath, B. J., Influence of vancomycin serum concentration on the outcome of gram-positive infections [abstract], American College of Clinical Pharmacy Annual Winter Meeting, Feb, 6–9, San Diego, 1994.

26. Lietman, P. S., Aminoglycosides and spectinomycin: aminocyclitols, in *Principles and Practice of Infectious Diseases*, Mandell, G. L., Douglas, R. G. and Benett, J. E., Eds., Wiley, New York, 192, 1985.

27. McDonald, P. J., Wetherall, B. L., and Prull, H., Post-antibiotic leukocyte enhancement: increased susceptibility of bacteria pretreated with antibiotics to activity of leukocytes, *Rev. Infect. Dis.*, 3, 38, 1981.

28. Mitchell, M. D., Wilson, J., Dodek, P, and Russel, J., Volume of distribution in patients with gram-negative sepsis in the ICU, *Anesthesiology*, 67, A 126, 1987.

29. Moore, R. D., Lietman, P. S., and Smith C. R., Clinical response to aminoglycoside therapy: importance of the ratio of peak concentration to minimal inhibitory concentration, *J. Infect. Dis.*, 155, 93, 1987.

30. Moore, R. D., Smith, C. R., and Lietman, P. S., The association of aminoglycoside plasma levels with mortality in patients with gram-negative bacteremia, *J. Infect. Dis.*, 149, 443, 1984.

31. Moreillon, P., and Tomasz, A., Penicillin resistance and defective lysis in clinical isolates of pneumococci: evidence for two kinds of antibiotic pressure operating in the clinical environment, *J. Infect. Dis.*, 157 (6), 1150, 1988.

32. Parker, R. F., and Luse, S., The action of penicillin on staphylococcus: further observations on the effect of a short exposure, *J. Bacteriol.*, 56, 75, 1984.

33. Sawyers, C. L., Moore, R. D., Lerner, S. A., and Smith, C. R., A model for predicting nephrotoxicity in patients treated with aminoglycosides. *J. Infect. Dis.*, 153, 1062, 1986.

34. Schentag, J. J., Jusko, W. J., Vance, J. W., Cumbo, T. J., Abrutyn, E., Delattre, M., and Gerbracht, L. M., Gentamicin disposition and tissue accumulation on multiple dosing, *J. Pharmacokin. Biopharm.*, 5, 559, 1977.

35. Schentag, J. J., Correlation of pharmacokinetic parameters to efficacy of antibiotics: relationship between serum concentrations, MIC values and bacterial eradication in patients with gram-negative pneumonia, *Scand. J. Infect. Dis.*, 74 (Suppl.) 218, 1990.

36. Schentag, J. J., Nix, D. E., and Adelman, H. H., Mathematical examination of dual individual principles (I): Relationships between AUC above MIC and area under inhibitory curve for cefmenoxin, ciprofloxacin and tobramycin, *DICP, Ann. Pharmacother.*, 25, 1050, 1991.

37. Smith, J. T., Awakening the slumbering potential of the 4-quinolone antibacterials, *Pharm. J.*, 233, 299, 1984.

38. Van der Auwera, P., Pharmacodynamic evaluation of single daily doses of amikacin, *J. Antimicrob. Chemother.*, 27 (Suppl. 7), 63, 1991.

39. Vogelman, B. S., and Craig, W. A., Post-antibiotic effects, *J. Antimicrob. Chemother.*, 15 (Suppl. A), 37, 1985.

40. Vogelman, B. S., Gudmundsson, S., Leggett, J., Ebert, S., and Craig, W. A., Correlation of antimicrobial pharmacokinetic parameters with therapeutic efficacy in an animal model, *J. Infect. Dis.*, 158, 831, 1988.
41. Wenk, M., Vozeh, S., and Follath, F., Serum level monitoring of antibacterial drugs. A review. *Clin. Pharmacokin.*, 9, 475, 1984.
42. Zaske, D. E., Cipolle, R. J., Rotschafer, J. C., Solem, L. D., Mosier, N. R., and Strate, R. G., Gentamicin pharmacokinetics in 164 patients: method for control of serum concentrations, *Antimicrob. Agents Chemother.*, 21, 407, 1982.

16

Antiviral Agents

J. Michael Kilby
*University of Alabama at
Birmingham*

16.1 Introduction

Although viral infections are encountered frequently in clinical practice, the antiviral armamentarium has been limited when compared with other pharmaceutical categories. Many of the currently available antiviral agents were developed within the last two decades. Ongoing research is producing new compounds at a faster rate than ever before. One challenge has been the difficulty of recognizing and diagnosing viral diseases; even as our diagnostic capabilities improve, it is common for a specific viral diagnosis to be confirmed only after the infection has either run its course or caused irreversible damage. Because viruses are intracellular parasites that are dependent on the host's cellular machinery to reproduce, another challenge has been to target the pathogens without causing cellular toxicity.

Agents used to prevent or treat viral infections can be broadly divided into three categories: virucidal compounds that directly inactivate viruses, antiviral drugs that inhibit viral replication, and immuno-modulators that modify the host's response to viral infection. While detergents and chemicals are used to control viruses in the environment, the virucidal category has not proven useful for the treatment of systemic infections because of toxicity to human tissues. Local manifestations of viral infection, however, such as genital warts, respond to nonspecific virucidal approaches such as ultraviolet light or laser therapy.

The commonly used systemic antiviral agents have been designed to inhibit viral replication at the cellular level; the majority of these compounds are nucleoside analogs that inhibit viral nucleic acid transcription. Although these compounds were developed to inhibit viral transcription as selectively as possible, there is potential for adverse effects involving host tissues with high cellular turnover (such as bone marrow and mucosal cells).

In terms of agents that strengthen the host immune response, vaccines against primary infections with certain viruses (including smallpox, measles, mumps, rubella, and poliovirus) have been very successful; the human immunodeficiency virus (HIV) and several of the herpesvirus group are exceptions for which intensive efforts are underway to better describe the correlates of immune protection. There are still many questions to be answered about the complex interplay of cytokines and host immune responses before immunomodulatory therapies become a routine part of clinical practice.

The drugs designed to treat systemic viral infections must achieve intracellular concentrations in affected tissues. Several antiviral drugs accomplish this by undergoing metabolic activation after administration (e.g., phosphorylation of acyclovir and zidovudine). Some compounds that have potent antiviral activity *in vitro* do not achieve the predicted effectiveness *in vivo* due to poor bioavailability, protein binding, or rapid metabolism in serum. For some antivirals, it has been challenging to develop oral compounds that approach the potency of corresponding parenteral formulations (e.g., ganciclovir and foscarnet). With the recent growth in clinical use of antiviral drugs, the recognition of antiviral resistance has increased in parallel. One or more point mutations in critical locations of the viral genome can lead to changes in viral target proteins, resulting in decreased drug susceptibility. This resistance may arise spontaneously from random mutations in the viral genome, but it is more likely that resistant viral variants are selected from a preexisting pool of viral strains (where they initially represent only a very small minority) within an infected patient due to replicative advantages in the presence of a drug. While antiviral drug susceptibility assays have not been adequately standardized, knowledge of drug sensitivity plays a role in the selection of therapy for such common infections as HIV, herpes simplex virus (HSV), and cytomegalovirus (CMV) infection. There is evidence that combinations of antiviral agents may be effective in helping to overcome the limitations of currently available therapies, particularly because of the potential to influence the development of drug resistance.

16.2   General Antiviral Compounds

16.2.1   Acyclovir and Related Compounds

A guanosine analog, acyclovir (9-[2-hydroxyethoxymethyl]guanine), is the most widely used antiviral drug in clinical practice. Acyclovir was developed in the late 1970s, when its effect on Herpesviruses was first recognized. In its triphosphorylated form, acyclovir competitively inhibits viral DNA polymerase and leads to termination of the growing viral DNA strand. Acyclovir is a "prodrug" that must be phosphorylated, first by a viral thymidine kinase and then in two additional steps by cellular enzymes, to achieve its active form within the host. The viral DNA polymerases of herpes simplex (HSV-1 and HSV-2) and varicella zoster (VZV) viruses have a much higher affinity for acyclovir triphosphate than does cellular DNA polymerase. The Epstein-Barr virus (EBV) is inhibited *in vitro* at only very high concentrations. Cytomegalovirus (CMV) is resistant because it is not able to phosphorylate the acyclovir prodrug.[1]

Acyclovir has limited oral bioavailability (15% to 30%); the intravenous formulation can be used to achieve higher serum concentrations. The drug penetrates most body tissues effectively, including the central nervous system. Acyclovir is minimally metabolized and approximately 85% of administered drug is excreted unchanged in the urine. The half-life of acyclovir in the setting of normal renal function is 2 to 3 hours. The dose must be adjusted in patients with impaired renal function. There is also a topical form of acyclovir that has limited usefulness.[1]

Acyclovir is effective for primary mucocutaneous HSV infections, resulting in decreased viral shedding, time to crusting of lesions, and duration of symptoms.[2] While the intravenous form has a slight advantage, most cases of primary HSV infection can be treated with oral acyclovir (200 mg five times daily or 400 mg three times a day; Table 16.1), which is more practical and less expensive. For severe mucocutaneous disease, particularly in immunocompromised hosts, the intravenous fomulation of the drug (5 mg/kg every 8 hours) is indicated.[3] Short-term use of acyclovir has not been shown to decrease the risk of recurrences.

Occasional recurrences of genital HSV can be treated with oral acyclovir, which decreases viral shedding and has an effect on symptoms. The drug must be started as early in the episode as possible, and many patients initiate acyclovir before there are visible lesions when they experience prodromal tingling or pain. Recent evidence confirms that viral shedding is common between symptomatic episodes, and that acyclovir treatment decreases but does not eradicate the amount of infectious virus produced before, during, and after symptomatic outbreaks.[4] Patients with frequent recurrences (for example, more than five or six episodes per year) can suppress HSV episodes by 80% to 90% with chronic oral acyclovir (often at 400 mg twice a day, but the dose can be titrated on an individual patient basis).[5]

TABLE 16.1   Indications and Recommended Adult Dosages for Commonly Used Antivirals

Drug	Indications	Dosages
Acyclovir	Primary HSV	200 mg po five times a day or 400 mg po three times a day
	Frequently recurrent HSV	400 mg po twice a day
	HSV encephalitis	10 mg/kg iv three times a day
	HSV in immunocompromised host	200–400 mg po five times a day or 5 mg/kg iv three times a day
	Chickenpox (VZV)	800 mg po five times a day
	Herpes zoster (VZV)	800 mg po five times a day
	VZV in immunocompromised host	800 mg po five times a day or 10 mg/kg iv three times a day
Valaciclovir	VZV	1 g po three times a day
	Recurrent HSV	500 mg po twice a day
Famciclovir	VZV	500 mg po three times a day
	Recurrent HSV	125 mg po two times a day
Ganciclovir	CMV retinitis (or other end-organ disease)	10 mg/kg iv twice a day (induction); 5 mg/kg iv daily (maintenance)
	CMV retinitis maintenance or prophylaxis in immunocompromised host	1 gm po three times a day
Foscarnet	CMV retinitis (or other end-organ disease)	90 mg/kg iv twice a day (induction); 120 mg/kg iv daily (maintenance)
	Resistant HSV or VZV	40 mg/kg iv three times a day
Amantadine or	Influenza A	100 mg po twice a day
Rimantadine	Influenza A prophylaxis	100 mg po daily

See text for further discussion. HSV = herpes simplex virus; VZV = varicella zoster virus; CMV = cytomegalovirus; iv = intravenously; po = orally.

Intravenous acyclovir for at least 10 days is indicated for the treatment of HSV encephalitis. Mortality is lower among patients treated with acyclovir (19% to 28%) than among those treated with vidarabine (50% to 54%).[6]

Oral acyclovir has modest clinical effects in the setting of uncomplicated primary varicella zoster infection (chickenpox).[7] Acyclovir decreases the duration and intensity of pain due to herpes zoster (shingles) if higher-dose therapy (800 mg five times a day) is initiated within 72 hours of the onset of symptoms.[8] Therapy may decrease the severity of prolonged zoster-associated pain ("postherpetic neuralgia") but studies are inconclusive.

Acyclovir is generally well tolerated. Reversible renal impairment has been reported uncommonly (<5%) among patients given high-dose intravenous acyclovir, but is not associated with routine doses of oral acyclovir. The risk of renal toxicity, which may be related to crystalluria in the presence of high concentrations of the drug, can be further decreased by avoiding rapid acyclovir infusions and ensuring adequate prior hydration. Intravenous acyclovir has been rarely associated with central nervous system toxicity, including mental status changes, tremor, and disorientation. Other reports of adverse effects have been quite rare, even with prolonged continuous administration.[1]

Acyclovir resistance has been an increasing concern in the treatment of both HSV and VZV. The most common mechanism of this resistance is alteration or complete deficiency of thymidine kinase, the viral enzyme necessary for the phosphorylation of acyclovir. Alterations in viral DNA polymerases have also been described. Thus far acyclovir-resistant HSV has primarily been seen among patients with severe immunosuppressive illnesses, especially AIDS.[9] VZV with altered or deficient thymidine kinase has also been isolated from patients with AIDS and other immunocompromising conditions.[10]

New agents related to acyclovir have recently become clinically available. Valaciclovir is a prodrug of acyclovir that is more bioavailable (50% to 60%) than the parent drug, resulting in higher serum concentrations of acyclovir.[11] Valaciclovir is approved for therapy of herpes zoster and herpes simplex virus infections (see Table 16.1). There have been no statistically significant advantages with this new compound compared with acyclovir for the treatment of herpetic infections in the limited clinical trial experience thus far. Famciclovir is the prodrug of penciclovir, a guanosine analog closely related to acyclovir. The oral bioavailability is high (70% to 85%) and the serum half-life of penciclovir is 2 to 3 hours. Famciclovir provides clinical benefits for uncomplicated VZV as compared to placebo.[12] The drug is also approved for the treatment of recurrent HSV. Information comparing the effectiveness of famciclovir with acyclovir is not currently available.

In summary, acyclovir is a well-tolerated antiviral compound that is effective for several manifestations of Herpesvirus infections. The benefit of oral acyclovir is proven in the setting of primary genital HSV and for the suppression of frequently recurring genital HSV. For routine treatment of sporadic labial or genital HSV recurrences in immunocompetent hosts, the indication for oral acyclovir is less clear. In immunocompromised patients, who more frequently suffer from severe, progressive mucocutaneous HSV, oral and in some cases intravenous acyclovir is indicated. Cases that are refractory to high-dose acyclovir therapy may be due to resistant viral strains, and switching to another agent with a different mechanism of action (such as foscarnet) is often necessary. HSV encephalitis, which is associated with significant morbidity and mortality regardless of premorbid conditions, is always treated with intravenous acyclovir.

The use of oral acyclovir for uncomplicated chickenpox in children has been debated recently. If acyclovir is started very early in the disease course, the duration and severity of this generally benign infection can be altered slightly. It has been argued that the cost and (minimal) potential risks of acyclovir for this indication are often outweighed by the potential lost wages when parents are required to stay at home for an additional day or so with their children. The benefit/risk ratio of oral acyclovir administration would appear to be much more favorable when older children and adults develop primary varicella, because of the increased risk of complications (including pneumonitis and encephalitis). Acyclovir therapy is not always justified in routine cases of herpes zoster (shingles), especially in younger adults. However, acyclovir use in elderly patients or for particularly severe VZV cases in young adults likely imparts significant benefits if treatment can be initiated within 72 hours of onset. Because of the significant risk of blindness, intravenous acyclovir is usually recommended for ophthalmic herpes zoster. Immunocompromised patients, in particular transplant recipients and patients with AIDS, should be treated with high-dose oral or intravenous acyclovir because of the possibility of disseminated or multidermatomal VZV disease. There is no proven role for acyclovir in the treatment of mononucleosis or cytomegalovirus infections. Newer agents related to acyclovir, which may have pharmacokinetic advantages, deserve further study.

16.2.2   Ganciclovir

Ganciclovir (9-[1,3-dihydroxy-2-propoxymethyl]guanine) was first reported to have potent anti-CMV effects as well as greater antiviral activity than acyclovir against several other Herpesviruses in 1982. The compound is comparable to acyclovir in that it is a guanosine analog that undergoes phosphorylation in order to attain its active form. However, the initial phosphorylation involving viral-specific thymidine kinase, an obligatory step in the activation of acyclovir, apparently can be carried out by the induction of other kinases in the case of ganciclovir. Although an oral formulation has recently been approved for selected clinical indications, oral bioavailability has been a limiting factor. The serum half-life of the iv formulation is 3 to 4 hours; like acyclovir, the half-life is greatly prolonged in the setting of renal impairment.

Ganciclovir therapy reduces morbidity and mortality due to CMV pneumonitis, a common complication of the immunosuppressive regimens used in the setting of solid organ and bone marrow transplants.[13] The combination of ganciclovir and intravenous immunoglobulin may be beneficial in bone marrow transplant recipients with CMV pneumonitis.[14] HIV-infected patients, for unclear reasons, are less likely to develop CMV pneumonitis than patients receiving transplants, but CMV retinitis and other manifestations of disseminated CMV disease are common. Ganciclovir induction therapy (5 mg/kg iv twice daily) slows progression of CMV retinitis, which frequently progresses to blindness if untreated, in 80% to 95% of patients.[15] However, virtually all treated patients relapse if therapy is discontinued. Maintenance ganciclovir (5 mg/kg iv daily) significantly prolongs disease-free intervals, although breakthrough retinitis or other CMV organ disease eventually occurs in many patients.[15] There is also growing experience with the successful use of ganciclovir therapy for other CMV manifestations in immunocompromised patients, particularly gastrointestinal (esophagitis, colitis, cholangitis) and neurologic (meningoencephalitis, polyradiculopathy) disease.

Unlike acyclovir, ganciclovir has been associated with adverse effects that are relatively common and can be severe. Approximately one third of HIV-infected patients on intravenous ganciclovir develop neutropenia.[15,16] This can make therapeutic decisions complicated as many patients with AIDS already have leukopenia at the time of CMV diagnosis and may be on other medications, particularly zidovudine, that have the potential to cause bone marrow toxicity. Thrombocytopenia occurs in 6% to 19% of patients, while other side effects including nausea, vomiting, and headache, are less frequent.

Most CMV infections in normal hosts are self-limited or asymptomatic, and no therapy is indicated. Ganciclovir increases survival when it is given to immunocompromised patients with life-threatening CMV disease. Ganciclovir is effective in the treatment of CMV retinitis both as an induction and maintenance therapy. A new oral formulation of ganciclovir (1.0 g three times daily) does not have the bioavailability necessary for induction therapy, but it has recently been approved for both maintenance therapy and prophylaxis of CMV retinitis in patients with advanced HIV infection. Although there may be slightly more progression on this therapy compared with daily iv ganciclovir, there are advantages of convenience and the avoidance of line-related complications. Neutropenia and thrombocytopenia are less commonly associated with the oral formulation.[17] Many factors, including expense, adverse effects, overlapping toxicities with other medications, and antiviral resistance, need to be further studied before a consensus can be reached regarding long-term maintenance or prophylaxis with this new formulation. Although ganciclovir-resistant CMV isolates are well-described in patients on therapy, not all breakthrough disease is due to *in vitro* resistance.[18]

16.2.3 Foscarnet

Foscarnet (trisodium phosphonoformate) is a pyrophosphate analog that inhibits viral DNA polymerases including reverse transcriptase (RNA-dependent DNA polymerase). Thus, there is potential for effectiveness against Herpesviruses and HIV infection. Foscarnet does not require a phosphorylation step and provides an alternative therapy for viruses that are resistant to acyclovir and ganciclovir due to altered or deficient thymidine kinase. The drug is not orally absorbed. Intravenous foscarnet has a half-life of 3 to 6 hours and complex pharmacokinetics.

Foscarnet has been successfully used to treat acyclovir-resistant HSV and VZV infections in immunocompromised patients (varying dosages have been used, but 40 mg/kg every 8 hours has been successful).[19,20] Foscarnet resistance has been described and appears to be mediated by alterations in viral DNA polymerase.[21] A large clinical trial of ganciclovir vs. foscarnet for induction therapy of CMV retinitis revealed no difference in disease progression, suggesting the two drugs are equally effective in this setting.[22] However, the trial was suspended after 19 months because of a survival advantage (approximately 8 months vs. 12 months) in the group receiving foscarnet. While this difference may be partially explained by the fact that fewer individuals in the ganciclovir group received concurrent antiretroviral therapy, it is conceivable that a survival advantage is related to antiretroviral effects of foscarnet. Different dosages of foscarnet have been used for the treatment of CMV retinitis in clinical trials — induction therapy with 90 mg/kg every 12 hours and maintenance therapy with 120 mg/kg daily have generally been efficacious.

Foscarnet is associated with significant toxicities, but the adverse effect profile is distinctly different from that of ganciclovir. The most frequently encountered toxicity is renal insufficiency, which is more common when patients are not adequately hydrated prior to infusion. Electrolyte disturbances including depletion of potassium, magnesium, and phosphorous are also common.[16] Cases of painful genital ulceration associated with foscarnet administration have been described.

Foscarnet is indicated for the treatment of immunocompromised hosts with HSV, VZV, and CMV infections that are resistant to acyclovir or ganciclovir. It appears to be as effective as ganciclovir for CMV retinitis, and may provide a modest survival advantage when compared with ganciclovir. While foscarnet has significant toxicities that lead to discontinuation of therapy more often than ganciclovir,[16] the toxicities encountered do not overlap significantly with zidovudine or ganciclovir. The choice of ganciclovir vs. foscarnet in the setting of CMV retinitis must be individualized, based on potential side effects that may be limiting when certain combinations of drugs are necessary (anemia due to zidovudine and ganciclovir, for example, or renal and electrolyte disturbances due to foscarnet and amphotericin B). The relative risks associated with foscarnet must be weighed against the possible survival benefit; the benefit/risk ratios would differ based on preexisiting conditions such as severe anemia or renal failure. It is unlikely that foscarnet will be used in routine clinical practice for the sole indication of antiretroviral therapy because of its toxicities, cost, and difficulty of administration.

Ganciclovir and foscarnet are synergistic against CMV *in vitro* and preliminary reports from trials suggest that combination therapy may provide some clinical advantages. Further experience will be necessary to answer questions about the benefit/risk ratios of such combination therapy. Combining antivirals against the Herpesviruses may become a necessary strategy in the face of increasing levels of drug resistance.

16.2.4   Amantadine/Rimantadine

Amantadine was first approved for prophylaxis of influenza A over 20 years ago although its mechanism of action remains incompletely understood. Amantadine is capable of interfering with influenza virus function at several steps in the life cycle, particularly the uncoating of the invading virus, which is necessary for release of viral RNA into the cell. The drug is typically dosed at 100 mg twice a day, but doses must be adjusted for impaired renal function. Amantadine is generally 70% to 80% effective as prophylaxis for influenza A virus. For influenza prevention, 100 mg once a day may also be effective. The drug is also effective for the treatment of influenza A infection, resulting in a significant reduction of symptoms and of viral shedding.[23] Rimantadine is a closely related compound with identical dosages and indications to amantadine, but it is absorbed more slowly, resulting in lower peak plasma levels. While rimantadine is slightly more potent when tested *in vitro*, the two compounds had comparable effectiveness as prophylaxis in one large clinical trial.[24]

Mild adverse effects occur in 5% to 10% of patients taking amantadine, primarily mild central nervous system complaints such as confusion, insomnia, and anxiety. Anorexia and nausea are also occasionally reported. These toxicities may be less common with rimantadine.[23,24]

Amantadine or rimantadine are recommended for prevention of influenza A for at risk patients who have not received the influenza vaccine or in conjunction with vaccination for patients exposed to an active influenza case. The degree of protection is similar to that of vaccination and, like the vaccine, protection is specific for influenza A and not for other respiratory viruses. Amantadine or rimantadine must be initiated within 48 hours of influenza A symptoms to result in a clinically significant benefit. This is sometimes impractical, as patients frequently delay seeing a physician until symptoms are prolonged or severe, and there is currently no reliable, rapid diagnostic test that aids in efficiently targeting therapy to those who will derive benefit. Nonetheless, there are specific instances in which these drugs are very useful, especially when there is an outbreak in an institutional setting such as a nursing home or dormitory. Unfortunately, central nervous system toxicities are more common in the elderly, the patients who theoretically would benefit the most from aggressive preventitive and preemptive therapy. Thus, lower doses of amantadine or preferably rimantadine (100 mg daily for both drugs) are appropriate in circumstances involving older patients in which the risk of influenza A is clear. Resistance to both amantadine and rimantadine among influenza A isolates has been well-documented, and this is likely to become an increasing concern as these compounds are used more frequently.[25]

16.2.5   Others

Several other antiviral compounds are not encountered as frequently in clinical practice and will not be discussed in detail. Aerosolized ribavirin is used in carefully selected infants and children with severe respiratory syncytial virus (RSV) infections. Because of teratogenic potential, precautions are recommended for female patients or health care workers who are of childbearing age. Vidarabine is a purine nucleoside analog with activity against several members of the herpes virus group, but because of toxicity and the availability of alternative agents it is now rarely used except as a topical compound.

Several immunomodulatory compounds are under study or are already used for certain indications. Interferon-α is approved for intralesional injections of condyloma acuminatum (human papillomavirus) and Kaposi's sarcoma (which may be caused by a newly recognized herpes virus). Parenteral interferon-α is used for active hepatitis due to both hepatitis B and hepatitis C virus; initial therapeutic responses are common, but complete responses generally are achieved in a minority of patients and relapses often occur (especially with hepatitis C) when the drug is stopped. Constitutional symptoms, fever, arthralgias, and myalgias are common side effects of systemic therapy. Intravenous gammaglobulin has been used for a diversity of viral infections, but the benefits of this adjunctive therapy are disputed in many cases.

16.3   Antiretroviral Compounds

16.3.1   Nucleoside Analogs — Zidovudine

Zidovudine (ZDV or AZT), approved by the U.S. Food and Drug Administration in 1987, was the first antiretroviral agent licensed for clinical use. Zidovudine and the four medications that follow in this discussion are nucleoside analogs that act as competitive inhibitors of reverse transcriptase, the retroviral enzyme that transcribes viral RNA into the complementary DNA ultimately integrated into the genetic material of the host. By competing with the nucleoside building blocks (thymidine in the case of zidovudine) and causing chain termination of the forming viral DNA strand, these compounds inhibit viral replication and limit the spread of HIV-1 from cell to cell. Zidovudine inhibits retroviral polymerases 50- to 100-fold more efficiently than the major human DNA polymerases; however, the drug does affect the action of human polymerase gamma in mitochondria at levels achievable in serum.[26,27]

Zidovudine is well-absorbed (60% to 65% bioavailability) when taken orally. The drug requires triphosphorylation by cellular enzymes to achieve its active form. The pharmacokinetics are complex, involving a biexponential decay pattern and intracellular dynamics, but the serum half-life is about 1 to 2 hours. The drug is well-distributed in body tissues, including the central nervous system, where levels estimated at 60% of that in serum can be measured. Zidovudine undergoes glucuronidation during first pass through the liver as well as other extensive metabolism. Approximately 10% is excreted unchanged in the urine while the remainder is renally excreted in the form of inactive metabolites.[26,27]

Zidovudine decreases opportunistic infections and mortality when it is given to symptomatic patients with AIDS.[28] When the drug has been studied in mildly symptomatic or asymptomatic patients with higher CD4 counts, there has been evidence of an alteration in disease progression but no mortality benefit.[29] There is no survival difference when two treatment strategies are compared: giving zidovudine monotherapy to all patients with CD4 counts less than 500/mm³ vs. withholding therapy until symptoms or AIDS-defining illnesses develop.[30] Prolonged monotherapy leads to mutations in the viral genome that correlate with clinical resistance, and this likely is part of the explanation for the limited durability of drug benefit.[31]

Severe anemia (24%) and neutropenia (16%) were common when higher doses of zidovudine were used.[32] These hematologic effects are less frequent at the currently suggested dosages of 600 mg daily in two divided doses; however, leukopenia and anemia can still be significant enough in 1% to 2% of patients to prompt discontinuation of the drug or the initiation of colony stimulating factors, especially in late-stage disease (Table 16.2). Nonspecific symptoms (headaches, achiness, malaise, and insomnia) are common, particularly during initiation of therapy. A lower dose of zidovudine (100 mg three times a day) has some

TABLE 16.2   Antiretroviral Compounds, Recommended Oral Doses, and Common Toxicities

Drug	Dosages	Toxicities
Zidovudine	300 mg twice a day	Anemia, leukopenia, myopathy, headache, malaise, insomnia, nausea
Didanosine	200 mg twice a day (125 mg twice a day if weight < 60 kg)	Neuropathy, pancreatitits, diarrhea, abdominal pain
Zalcitabine	0.75 mg three times a day	Neuropathy, mouth ulcers
Stavudine	40 mg twice a day (30 mg twice a day if weight < 60 kg)	Neuropathy, nausea
Lamivudine	150 mg twice a day	?
Saquinavir	600 mg three times a day	Nausea
Soft gelatin saquinavir (Fortovase™)	1200 mg three times a day	Diarrhea, nausea
Ritonavir	600 mg twice a day (following dose escalation)	Nausea, altered taste, circumoral paresthesias
Indinavir	800 mg three times a day	Nephrolithiasis
Nelfinavir	750 mg three times a day	Loose stools

degree of antiretroviral effectiveness, and can be used for patients intolerant of standard doses or those with renal insufficiency.[26,27,33]

There are also adverse effects shared to varying degrees by most or all of the nucleoside analog antiretroviral compounds that are postulated to be due to mitochondrial DNA polymerase toxicity. These diverse complications include myopathy, neuropathy, and pancreatitis. Zidovudine has been associated with a myopathy, characterized by muscle tenderness and elevated creatinine phosphokinase levels, which at times is difficult to distinguish from myopathy caused by HIV-1 itself. Neuropathy and pancreatitis are rarely seen but are more frequently described in association with other nucleoside analog compounds described further below.[26,27]

Because of new information regarding the effectiveness of combination antiretroviral therapy, prolonged monotherapy with zidovudine for HIV-infected patients can no longer be considered as a standard of care. This is at least in part due to the frequent selection of zidovudine-resistant viral strains within individual patients when they are treated with zidovudine alone. The toxicities seen with this antiretroviral agent, especially at higher doses, may also balance many of the short-term benefits when the drug is used alone. This is especially true in individuals who have certain preexisting conditions (severe anemia or myopathy) or who are already taking medications that share toxicity profiles with zidovudine (ganciclovir, for example). However, combinations of agents that include zidovudine (see below) have clear advantages over zidovudine alone without significantly adding to overall drug-related toxicity. Therefore, while at first glance it would appear counterintuitive, the benefit/risk ratio for HIV-infected patients is lower when one is considering ZDV monotherapy as opposed to a potent combination regimen.

While descriptions of the striking viral dynamics observed even in asymptomatic HIV-infected patients would logically lead one to treat as early and as aggressively as possible,[34] there may also be an incentive to withhold the most potent agents until disease progression occurs. The latter approach would theoretically avoid selecting for multiresistant viral strains early on in HIV-1 infection, when treatments have little or no immediate impact on the way patients feel. Recent studies with potent antiretroviral combination regimens suggest that AIDS complications can be significantly delayed if therapy is started early on in the disease course.

16.3.2   Didanosine

The second approved therapy for HIV-1 infection in the United States was didanosine (ddI), a purine nucleoside analog reverse transcriptase inhibitor. Oral bioavailability of the drug is variable and absorption is most effective in the fasting state. It must be buffered to prevent inactivation by acidic enzymes during digestion. Didanosine is available as a chewable tablet, in a pediatric liquid formulation, and as a powder that must be suspended in antacid. The buffered compound can interfere with the absorption of other medications and so drug regimens often have to be coordinated carefully as does the timing of meals. The plasma half-life of didanosine is less than 1 hour but prolonged intracellular levels allow twice a day dosing. (Once a day dosing is likely to be approved for clinical use in the near future). The recommended dose is 200 mg twice daily, which should be reduced to 125 mg twice daily in patients who weigh less than 60 kg. Sixty percent of the drug is excreted unchanged in the urine, and patients with severe renal insufficiency require an adjustment in dose. The drug does not penetrate into cerebrospinal fluid as well as zidovudine.[27,33]

In patients who have previously received zidovudine, switching to didanosine provides an advantage in terms of disease progression over remaining on zidovudine, although the optimum time for switching therapies remains unclear.[35] Results from two large clinical trials suggest advantages for patients who are placed on didanosine alone or on a zidovudine/didanosine combination rather than zidovudine monotherapy; certain subgroups may have a survival advantage when given regimens containing didanosine.[36,37]

Some patients find didanosine unpalatable. If an individual is able to comply with the drug appropriately in regard to meals and other drugs, this antiretroviral agent has a favorable therapeutic index at currently recommended doses. The most important dose-limiting side effect is painful peripheral neuropathy, which develops in approximately 15% to 20% of patients. A minority of individuals develops asymptomatic elevations in amylase or lipase, which is of questionable significance, and 5% to 10% of all patients on didanosine develop clinical pancreatitis. Headache and diarrhea are also frequent complaints.[27,38] Like all available antiretroviral agents, the development of drug resistance is a problem with didanosine, although the degree of resistance observed after prolonged therapy appears to be of a lesser degree than that selected for by zidovudine monotherapy. There is *in vitro* evidence that one didanosine resistance-conferring mutation diminishes zidovudine resistance, providing a potential explanation for the efficacy of combined therapy.[31]

16.3.3   Zalcitabine

Zalcitabine (didesoxycytidine, ddC) is a cytosine analog that was approved for use in 1992. The drug has high bioavailability (87%) and is more palatable than didanosine. Like didanosine and zidovudine, the drug has a very short plasma half-life but can be dosed at longer intervals (0.75 mg three times a day) due to prolonged intracellular concentrations. Zalcitabine is primarily excreted unchanged in the urine. Painful peripheral neuropathy, which is occasionally irreversible, is more commonly encountered than with didanosine (>20%). On the other hand, pancreatitis is also associated with zalcitabine, but less frequently than with didanosine. Other reported adverse events include rashes, stomatitis, and mucosal ulcerations.[27,33,38]

Didanosine or zalcitabine may be equally efficacious choices if the decision is made to switch from monotherapy with zidovudine to another single agent; the choice should largely be based on the side effect profiles of the drugs.[38] While initiating combination regimens (including zidovudine/zalcitabine) may be beneficial, there is no evidence to support the efficacy of zalcitabine monotherapy at this time.

16.3.4   Stavudine

Stavudine (d4T) is a pyrimidine analog reverse transcriptase inhibitor that was recently approved for use in patients who are intolerant of other available agents or whose disease is progressing despite the use of other agents. The drug has a consistently high oral absorption.[33] The recommended adult dose is 40 mg twice daily, reduced to 30 mg twice daily for patients who weigh less than 60 kg. Stavudine is excreted by both renal and nonrenal routes. Among patients not previously treated with antiretroviral agents, the drug results in modestly increased CD4 counts that persist for longer than 1 year.[39] Patients previously treated with zidovudine switched to stavudine monotherapy have benefits in terms of disease progression and survival.[40] Peripheral neuropathy is common (approximately 20%) but is typically reversible and dose-related. Increases in hepatic enzymes and pancreatitis are reported infrequently.

Studies are underway to evaluate stavudine in combination with other agents. There is evidence of adverse metabolic interactions between zidovudine and stavudine, and these agents should not be combined.

16.3.5   Lamivudine

Lamivudine (3TC) is the (−)-enantiomer of a cytosine analog that is unique in that it has activity against HIV-1 as well as other viruses including HIV-2 and hepatitis B virus. The drug has ~82% oral bioavailability and has the longest serum half-life (~2.5 hrs) of all the available reverse transcriptase inhibitor compounds. About 70% of the drug is excreted unchanged in the urine.[33] There is *in vitro* evidence that a resistance-conferring mutation rapidly selected for by lamivudine therapy "resensitizes" zidovudine-resistant HIV-1 strains to zidovudine[41] and may also slow the development of HIV-1 resistance to a number of antiretroviral compounds.[42] This may explain why patients on combined zidovudine and lamivudine therapy had clear advantages over patients on either drug alone in terms of CD4 counts and disease progression. These patients experienced synergistic benefits without added toxicity.[43] While no adverse effects were directly attributable to lamivudine therapy in several studies, anecdotal evidence suggests that side effects may occasionally occur. Because of rapid selection for resistance in many patients, lamuvidine monotherapy cannot be recommended. AZT (300 mg) and 3TC (150 mg) have recently been marketed as a single combined medication taken twice daily, which should improve patient compliance.

16.3.6   Protease Inhibitors

A diverse group of new compounds has recently gained a great deal of attention because of unprecedented antiretroviral potency without severe side effects. These compounds specifically target the HIV-1 protease enzyme, which normally cleaves viral polyprotein precursors to form mature, infectious virions. At the time of this writing, five protease inhibitors have been approved by the FDA for use in combination with nucleoside analog agents. Similar compounds are likely to be approved in the near future.

Saquinavir is the protease inhibitor for which the most clinical experience is available thus far. Oral absorption of the drug is very poor (~4%) and some clinicians have argued that the recommended dose of 600 mg three times a day is not sufficient for saquinavir to exert its full potential. Saquinavir plus zidovudine is more effective in antiretroviral-naive patients than either compound alone.[44] Triple combination therapy with zidovudine/zalcitabine/saquinavir was more beneficial in terms of CD4 counts and viral load compared with various two-drug combinations.[45] The drug is generally well-tolerated. Gastrointestinal symptoms including nausea are sometimes seen. Important drug interactions must be considered before instituting therapy. When the drug is used alone, resistance develops in many patients; there does not seem to be considerable cross-resistance to the two other available protease inhibitors. A new soft gelatin formulation of saquinavir has recently been approved which has higher bioavailability and thus greater antiviral potency.

Ritonavir is a more potent protease inhibitor that has better bioavailability. When used as monotherapy, significant elevations in CD4 counts and decreases in viral load are seen.[46] When added to regimens of nucleoside agents, survival advantages were evident in less than 6 months in patients with advanced HIV infection.[47] Triple combination therapy (zidovudine/zalcitabine/ritonavir) produced up to 1000-fold (3 $\log_{10}$) reductions in plasma viremia in many patients, such that some had decreases in viral RNA below the detection level with currently used assays.[48] A liquid formulation of the drug was unpalatable for some patients, but a new formulation is available. Non-life-threatening but bothersome adverse effects are common. Many patients experience disturbing circumoral paresthesias soon after taking the drug. Nausea, vomiting, and diarrhea are also reported, although these may be less common with the new pill formulation. Some of the toxicity can be avoided by escalating the dose over a 10-day period in order to avoid high serum drug peaks prior to the induction of metabolic enzymes. The recommended dose following this stepwise escalation is 600 mg twice daily. Metabolism by liver enzymes also results in a number of significant drug interactions and many commonly used medications must be avoided when ritonavir is administered. Drug resistance develops and may lead to cross-resistance to all available protease inhibitors.

Indinavir and nelfinavir are the protease inhibitors with perhaps the most favorable therapeutic indices at this time. When used as monotherapy, indinavir has impressive potency[49] but can select for viral strains that are resistant and cross-resistant to other protease inhibitor compounds. In a preliminary report from one trial, >80% of previously treated patients given zidovudine/lamivudine/indinavir combination therapy had plasma viremia reduced to undetectable levels (over 1000-fold or 3 log_{10} decreases in some instances).[50] Several other trials demonstrate preliminary evidence of similar potency when indinavir is combined with other agents, including didanosine and stavudine. Indinavir is generally well-tolerated. About 10% of patients experience an apparently harmless minor elevation in serum bilirubin. Less than 5% of patients develop kidney stones, which likely represent concretions of the drug or its metabolites, and drinking several glasses of water daily may aid in avoiding this problem. Indinavir also has potential problems with drug interactions, although not as many as ritonavir. Nelfinavir is another promising protease inhibitor with a drug interaction profile similar to indinavir. Nelfinavir can be taken with food, and the most common adverse experience reported is diarrhea.

Thus, early evidence suggests an unprecedented antiretroviral effect, likely conferring survival advantages, when protease inhibitors are combined with nucleoside agents. These potent combinations are likely to significantly alter the natural history of HIV-1 infection and slow the development of clinically significant resistance that occurs with prolonged monotherapy. Unfortunately, each of the available protease inhibitors costs substantially more than the most expensive of the reverse transcriptase inhibitors and triple combination therapy is prohibitively expensive for most HIV-infected patients. Strategies that take advantage of cytochrome enzyme induction to boost the effects of these compounds without compromising safety are under development. More data are needed to assess how long the benefits of these combination regimens can be maintained and how early in the course of HIV infection such regimens should be offered to patients. Another class of antiretroviral drugs, the nonnucleoside reverse transcriptase inhibitors (nevirapine and delavirdine), has also recently become clinically available. Further experience with these compounds is needed to establish how they fit into current combination treatment strategies. As demonstrated throughout this chapter, benefit/risk considerations are complex when considering treatment strategies that involve sequentially administered agents or combinations of several agents. However, it is clear that the benefits of two- and three-drug antiretroviral regimens, when carefully selected, outweigh the potential risks for most HIV-infected patients.

Glossary of Abbreviations

AIDS   Acquired immunodeficiency syndrome
CD4   Cluster of differentiation 4 (lymphocyte subtype)
CMV   Cytomegalovirus
DNA   Deoxyribonucleic acid
HSV   Herpes simplex virus
iv   Intravenous
po   Per os (oral)

References

1. Whitley R. J. and Gnann J. W. Jr. Acyclovir: A decade later. *N Engl J Med* 327:782, 1992.
2. Bryson Y. J., Dillon M., Lovett M., Acuna G., Taylor., Cherry J. D., Johnson L., Wiesmeier E., Growdon W., Creagh-Kirk T., and Keeney R. Treatment of first episodes of genital herpes simplex virus infection with oral acyclovir. *N Engl J Med* 308:916, 1983.
3. Meyers J. D., Wade J. C., Mitchell C. D., Saral R., Lietman P. S., Durack D. T., Levin M. J., Segreti A. C., and Balfour H. H. Multicenter collaborative trial of intravenous acyclovir for treatment of mucocutaneous herpes simplex virus infection in the immunocompromised host. *Am J Med* 73 (Suppl 1A): 229, 1982.

4. Wald A., Zeh J., Barnum G., Davis L. G., and Corey L. Suppression of subclinical shedding of herpes simplex virus type 2 with acyclovir. *Ann Intern Med* 124:8, 1996.

5. Douglas J. M., Critchlow C., Benedetti J., Mertz G. J., Connor J. D., Hintz M. A., Fahnlander A., Remington M., Winter W., and Corey L. A double-blind study of oral acyclovir for suppression of recurrences of genital herpes simplex virus infection. *N Engl J Med* 310:1551, 1984.

6. Whitley R., Arvin A., Prober C., Burchett S., Corey L., Powell D., Plotkin S., Starr S., Alford C., Connor J., Jacobs R., Nahmias A., Soong S.-J., and the National Institute of Allergy and Infectious Diseases Collaborative Antiviral Study Group. A controlled trial comparing vidarabine with acyclovir in neonatal herpes simplex virus infection. *N Engl J Med* 324:144, 1986.

7. Dunkle L. M., Arvin A. M., Whitley R. J., Rotbart H. A., Feder H. M., Feldman S., Gershon A. A., Levy M. L., Hayden G. F., McGuirt P. V., Harris J., and Balfour H. H. A controlled trial of acyclovir for chickenpox in normal children. *N Engl J Med* 325:1539, 1991.

8. Wood M. J., Ogan P. H., McKendrick M. W., Care C. D., McGill J. I., and Webb E. M. Efficacy of oral acyclovir treatment of acute herpes zoster. *Am J Med* 85 (Suppl 2A):79, 1988.

9. Erlich K. S., Mills J., Chatis P., Mertz G. J., Busch D. F., Follansbee S. E., Grant R. M., and Crumpacker C. S. Acyclovir-resistant herpes simplex infections in patients with the acquired immunodeficiency syndrome. *N Engl J Med* 320:293, 1989.

10. Jacobson M. A., Berger T. G., Fikrig S., Becherer P., Moohr J. W., Stanat S. C., and Biron K. K. Acyclovir-resistant varicella-zoster virus infection after chronic acyclovir therapy in patients with the acquired immunodeficiency syndrome. *Ann Intern Med* 112:187, 1990.

11. Jacobson M. A., Gallant J., Wang L. H., Coakley D., Weller S., Gary D., Squires L., Smiley M. L., Blum M. R., and Feinberg J. Phase I trial of valaciclovir, the L-valyl ester of acyclovir, in patients with advanced human immunodeficiency virus disease. *Antimicrob Agents Chemother* 38:1534, 1994.

12. Tyring S., Barbarash R. A., Nahlik J. E., Cunningham A., Marley J., Heng M., Jones T., Rea T., Boon R., Saltzman R., and the Collaborative Famciclovir Herpes Zoster Study Group. Famciclovir for the treatment of acute herpes zoster: effects on acute disease and postherpetic neuralgia. *Ann Intern Med* 123:89, 1995.

13. Sydman D. R. Ganciclovir therapy for cytomegalovirus disease associated with renal transplants. *Rev Infect Dis* 10:554, 1988.

14. Emanuel D., Cunningham I., Jules-Elysee K., Brochstein J. A., Kernan N. A., Laver J., Stover D., White D. A., Fels A., Polsky B., Castro-Malaspina H., Peppard J. R., Bartus P., Hammerling U., and O'Reilly R. J. Cytomegalovirus pneumonia after bone marrow transplantation successfully treated with the combination of ganciclovir and high-dose intravenous immune globulin. *Ann Intern Med* 109:777, 1988.

15. Holland G. N., Sidikaro Y., Kreiger A. E., Hardy D., Sakamoto M. J., Frenkel L. M., Winston D. J., Gottlieb M. S., Bryson Y. J., and Champlin R. E. Treatment of CMV retinopathy with ganciclovir. *Ophthalmology* 94:815, 1987.

16. ACTG. Morbidity and toxic effects associated with ganciclovir or foscarnet therapy in randomized CMV retinitis trials. Studies of the ocular complications of AIDS research group in collaboration with the ACTG. *Arch Intern Med* 155:65, 1995.

17. Crumpacker C. Oral vs. intravenous ganciclovir as maintenance treatment of newly diagnosed cytomegalovirus retinitis in AIDS. In: Abstracts of the 1st National Conference on Human Retroviruses, Washington, D.C., 1993.

18. Drew W. L., Miner R. C., Busch D. F., Follansbee S. E., Gullett J., Mehalko S. G., Gordon S. M., Owen W. F., Matthews T. R., Buhles W. G., and DeArmond B. Prevalence of resistance in patients receiving ganciclovir for serious CMV infection. *J Infect Dis* 163:716, 1991.

19. Safrin S., Crumpacker C., Chatis P., Davis R., Hafner R., Rush J., Kessler H., Landry B., Mills J., and other members of the AIDS Clinical Trials Group. A controlled trial comparing foscarnet with vidarabine for acyclovir-resistant mucocutaneous herpes simplex in the acquired immunodeficiency syndrome. *N Engl J Med* 325:551, 1991.

20. Safrin S., Berger T. G., Gilson I., Wolfe P. R., Wofsy C. B., Mills J., and Biron K. K. Foscarnet therapy in five patients with AIDS and acyclovir-resistant varicella-zoster virus infection. *Ann Intern Med* 115:19, 1991.

21. Birch C. J., Tachedjian G., Doherty R. R., Hayes K., and Gust I. D. Altered sensitivity to antiviral drugs of herpes simplex virus isolates from a patient with the acquired immunodeficiency syndrome. *J Infect Dis* 162:731, 1990.

22. Studies of Ocular Complications of AIDS (SOCA) Research Group in collaboration with the ACTG. Mortality in patients with AIDS treated with either foscarnet or ganciclovir for CMV retinitis. *N Engl J Med* 326:213, 1992.

23. Douglas R. G. Jr. Prophylaxis and treatment of influenza. *N Engl J Med* 322:443, 1990.

24. Dolin R., Reichman R. C., Madore H. P., Maynard R., Linton P. N., and Webber-Jones J. A controlled trial of amantadine and rimantadine on the prophylaxis of influenza A infection. *N Engl J Med* 307:580, 1982.

25. Monto A. S. and Arden N. H. Implications of viral resistance to amantadine in control of influenza A. *Clin Infect Dis* 15:362, 1992.

26. McLeod G. X. and Hammer S. M. Zidovudine: five years later. *N Engl J Med* 117:487, 1992.

27. Hirsch M. S. and D'Aquila R. T. Therapy for human immunodeficiency virus infection. *N Engl J Med* 328:1686, 1993.

28. Fischl M. A., Richman D. D., Grieco M. H., Gottlieb M. S., Volberding P. A., Laskin O. L., Leedom J. M., Groopman J. E., Mildvan D., Schooley R. T., Jackson G. G., Durack D. T., King D., and the AZT Collaborative Working Group. The efficacy of azidothymidine (AZT) in the treatment of patients with AIDS and AIDS-related complex. *N Engl J Med* 317:185, 1987.

29. Volberding P. A., Lagakos S. W., Koch M. A., Pettinelli C., Myers M. W., Booth D. K., Balfour H. H., Reichman R. C., Bartlett J. A., Hirsch M. S., Murphy R. L., Hardy W. D. Soeiro R., Fischl M. A., Bartlett J. G., Merigan T. C., Hyslop N. E., Richman D. D., Valentine F. T., Corey L., and the AIDS Clinical Trials Group of the National Institute of Allergy and Infectious Diseases. Zidovudine in asymptomatic human immunodeficiency virus infection. *N Engl J Med* 322:941, 1990.

30. Concorde Coordinating Committee. Concorde: MRC/ANRS randomized double-blind controlled trial of immediate and deferred zidovudine in symptom-free HIV infection. *Lancet* 343:871, 1994.

31. Richman D. D. Resistance of clinical isolates of HIV to antiretroviral agents. *Antimicrob Agents Chemother* 37:1207, 1993.

32. Richman D. D., Fischl M. A., Grieco M. H., Gottlieb M. S., Volberding P. A., Laskin O. K., Leedom J. M., Groopman J. E., Mildvan D., Hirsch M. S., Jackson, G. J., Durack D. T., Nusinoff-Lehrman S., and the AZT Collaborative Working Group. The toxicity of azidothymidine (AZT) in the treatment of patients with AIDS and AIDS-related complex. *N Engl J Med* 317:192, 1987.

33. Dudley M. N. Clinical pharmacokinetics of nucleoside antiretroviral agents. *J Infect Dis* 171(Suppl. 2): S99, 1995.

34. Wei X., Ghosh S. K., Taylor M. E., Johnson V. A., Emini E. A., Deutsch P., Lifson J. D., Bonhoeffer S., Nowak M. A., Hahn B. H., Saag M. S., and Shaw G. M. Viral dynamics in human immunodeficiency virus type I infection. *Nature* 373:117, 1995.

35. Kahn J. O., Lagakos S. W., Richman D. D., Cross A., Pettinelli C., Liou S-H, Brown M., Volberding P. A., Crumpacker C. S., Beall G., Sacks H. S., Merigan T. C., Beltangady M., Smaldone L., Dolin R., and the NIAID AIDS Clinical Trials Group. A controlled trial comparing continued zidovudine with didanosine in human immunodeficiency virus infection. *N Engl J Med* 327:581, 1992.

36. Hammer S., Katzenstein D., Hughes M., Hirsch M. S., and Merigan T. C. Virologic markers and outcome in ACTG 175 [Abstract S24] 3rd Conference on Retroviruses and Opportunistic Infections, Washington, D.C., 1996.

37. The European Delta Trial. [Abstract] 5th European Conference on Clinical Aspects and Treatment of AIDS, 1995.

38. Abrams D. I., Goldman A. I., Launer C., Korvick J. A., Neaton J. D., Crane L. R., Grodesky M., Wakefield S., Muth K., Kornegay S., Cohn D. L., Harris A., Luskin-Hawk R., Markowitz N., Sampson J. H., Thompson M., Deyton L., and the Terry Beirn Community Programs for Clinical Research on AIDS. A comparative trial of didanosine or zalcitabine after treatment with zidovudine in patients with human immunodeficiency virus infection. *N Engl J Med* 330:657, 1994.

39. Adler M. H., Anderson R. E., Rutkiewicz V., Cross A. P., Dellert K. A., and Dunkle L. Clinical course of long-term therapy with stavudine (D4T). [Abstract M14] 34th Interscience Conference on Antimicrobial Agents and Chemotherapy, Orlando, FL, 1994.

40. Pavia A. T., Gathe J., BMS-019 Study Group Investigators, Grosso R., Dunkle L. M., Cross A. P., Mohanty S., Messina M., and Smaldone L. Clinical efficacy of stavudine (d4T) compared to zidovudine (ZDV) in ZDV-pretreated HIV positive patients. [Abstract I169] 35th Interscience Conference on Antimicrobial Agents and Chemotherapy, San Francisco, CA, 1995.

41. Larder B. A., Kemp S. D., and Harrigan P. R. Potential mechanism for sustained antiretroviral efficacy of AZT-3TC combination therapy. *Science* 269:696, 1995.

42. Wainberg M. A., Drosopoulos W. C., Salomon H., Hsu M., Borkow G., Parniak M. A., Gu Z., Song Q., Manne J., Islam S., Castriota G, and Prasad V. R. Enhanced fidelity of 3TC-selected mutant HIV-1 reverse transcriptase. *Science* 271:1282, 1996.

43. Eron J. J., Benoit S. L., Jemsek J., MacArthur R. D., Santana J., Quinn J. B., Kuritzkes D. R., Fallon M. A., and Rubin M. for the North American HIV Working Party. Treatment with lamivudine, zidovudine, or both in HIV-positive patients with 200 to 500 CD4+ cells per cubic millimeter. *N Engl J Med* 333:1662, 1995.

44. Vella S. HIV therapy advances. Update on a proteinase inhibitor. *AIDS* 8(Supp 3):S25, 1994.

45. Collier A. C., Coombs R. W., Schoenfield D. A., Bassett R. L., Timpone J., Baruch A., Jones M., Facey K., Whitacre C., McAuliffe V. J., Friedman H. M., Merigan T. C., Reichman R. C., Hooper C., and Corey L. for the AIDS Clinical Trials Group. Treatment of human immunodeficiency virus infection with saquinavir, zidovudine, and zalcitabine. *N Engl J Med* 334:1011, 1996.

46. Markowitz M., Saag M., Powderly W. G., Hurley A. M., Hsu A., Valdes J. M., Henry D., Sattler F., La Marca A., Leonard J. M., and D. D. Ho. A preliminary study of ritonavir, an inhibitor of HIV-1 protease, to treat HIV-1 infection. *N Engl J Med* 333:1534, 1995.

47. Cameron B., Heath-Chiozzi M., Kravcik S., Mills R., Potthoff A., Henry D., the Advanced HIV Ritonavir Study Group and Leonard J. Prolongation of life and prevention of AIDS in advanced HIV immunodeficiency with ritonavir. [Abstract LB6a] 3rd Conference on Retroviruses and Opportunistic Infections, Washington, D.C., 1996.

48. Mathez D., De Truchis P., Gorin I., Katlama C., Pialoux G. Saimot A. G., Tubiana R., Chauvin J. P., Bagnarelli P., Clementi M., and Leibowitch J. Ritonavir, AZT, DDC, as a triple combination in AIDS patients. [Abstract 285] 3rd Conference on Retroviruses and Opportunistic Infections, Washington, D.C., 1996.

49. Steigbigel R. T., Berry P., Mellors J., McMahon D., Teppler H., Stein D., Drusano G., Deutsch P., Yeh K., Hilderand C., Nessly M., Emini E., and Chodakewitz J. Efficacy and safety of the HIV protease inhibitor indinavir sulfate (MK 639) at escalating dose. [Abstract 146] 3rd Conference on Retroviruses and Opportunistic Infections, Washington, D.C., 1996.

50. Gulick R., Mellors J., Havlir D., Eron J., Gonzalez C., McMahon D., Richman D., Valentine F., Jonas L., Meibohm A., Chiou R., Deutsch P., Emini E., and Chodakewitz J. Potent and sustained antiretroviral activity of indinavir in combination with zidovudine and lamivudine (3TC). [Abstract LB7] 3rd Conference on Retroviruses and Opportunistic Infections, Washington, D.C., 1996.

17

An Overview and Assessment of the Use of the Antifungal Agents Itraconazole, Terbinafine, and Fluconazole in Dermatology

Aditya K. Gupta
University of Toronto

Piet De Doncker
Janssen Research Foundation

Michel Heenen
Université Libre de Bruxelles

17.1   Introduction

In 1958, griseofulvin became the first significant oral antifungal agent available to treat cutaneous fungal infections.[1-3] Anderson[1] found that about 50% to 60% of patients with tinea corporis, tinea manuum, and tinea pedis were clinically cured with griseofulvin. With pedal onychomycosis however, cure rates were much lower, typical values being 3% to 38%.[2] High relapse rates of 40% to 60% have been reported.[4,5]

Ketoconazole, introduced in the late 1970s was the next important antifungal agent.[6-8] This imidazole was an effective treatment for chronic superficial candidiasis and chronic dermatophytosis.[9] A few years after the introduction of ketoconazole several cases of symptomatic hepatitis were reported, estimated to be from 1 in 10,000 to 1 in 15,000 in frequency and occasionally fatal.[10,11] The reactions were felt to be an idiosyncratic drug-induced hepatitis, not necessarily associated with high doses of ketoconazole or duration of therapy.[12] As a consequence of these results the benefit/risk ratio of oral antifungal agents for the treatment of dermatomycoses was clearly influenced. In dermatology, the use of oral ketoconazole is now restricted to the treatment of chronic mucocutaneous candidiasis and dermatoses requiring short courses of therapy, for example, pityriasis versicolor and vaginal candidiasis.

The new generation of antifungal agents available for the treatment of dermatomycoses are itraconazole, fluconazole, and terbinafine. In this chapter we will present an overview of these agents, and examine their efficacy and adverse-effect profile.

17.2   Itraconazole

Itraconazole is a triazole antifungal agent synthesized in 1980[13] and thus following ketoconazole the first orally active broad spectrum antifungal agent.[14,15] The imidazoles, including miconazole and ketoconazole, have 2 nitrogen atoms in the azole ring. The triazoles include itraconazole, fluconazole, and terconazole, with 3 nitrogen atoms in the 5-membered azole ring. As compared to the imidazoles, the triazole ring and its long lipophilic tail result in longer half-life, increased tissue penetration, improved efficacy, and lower toxicity.[16]

The principal mechanism of action of the imidazoles and triazoles is to inhibit the enzyme lanosterol 14 α-demethylase, a cytochrome P-450 dependent enzyme, that is important for the synthesis of ergosterol. The latter is an essential membrane lipid of most fungi.[17] Host cells, in contrast, can use exogenous cholesterol in the diet. Furthermore, the triazoles, in particular, are markedly more selective for the fungal than the human cytochrome P-450 enzymes.[18] Azoles may also inhibit fungal cytochrome C oxidative and peroxidative enzymes, resulting in an intracellular accumulation of hydrogen peroxide.[19]

17.2.1   Pharmacodynamics

Itraconazole has a broad spectrum of activity *in vitro*. Using the brain-heart infusion broth and agar dilution methods, 94% of dermatophytes (*Microsporum, Trichophyton*, and *Epidermophyton* species) have a minimum inhibitory concentration (MIC) of ≤0.1 µg/mL.[20-23] Itraconazole demonstrates good activity against *Candida* species with 98% of strains being sensitive to itraconazole (1 µg/mL).[21] With *Pityrosporum ovale*, using the Dixon broth test medium, virtually all strains (100%) are sensitive ≤0.1 µg/mM). *In vitro* results, however, may vary, depending upon the culture medium, inoculum size, conditions of incubation, etc.[24-27]

At high concentrations, some azoles may exhibit a fungicidal effect *in vitro* due to direct disruption of the fungal cell phospholipid bilayer.[28,29] This mechanism of action has been demonstrated with imidazoles; whether this also occurs with triazoles such as itraconazole is not certain.[30]

Given the variability in *in vitro* data, information from *in vivo* models may be a better predictor of the efficacy of itraconazole.[31] In *in vivo* animal models, itraconazole is effective in the treatment of experimental infections caused by dermatophyte species, *Candida* species and several pathogens causing systemic mycoses.[20,21,32] For example, *Microsporum canis* infection in guinea pigs results in a 92% cure with itraconazole 2.5 mg/kg/day given for 7 days.[21] In cutaneous candidiasis, itraconazole 5 mg/kg/day for 14 days results in cure in all guinea pigs.[21]

17.2.2   Pharmacokinetics

Itraconazole is an extremely weak base (pK$_a$ = 3.7).[33] The solubility of itraconazole is increased in an acidic environment, with a much larger amount of drug dissolved at pH 1 compared to a higher pH.[34] The normal range of gastric pH is 1 to 3.5, with the lowest values occurring following food intake.[35,36] Poor oral absorption may be associated with hypochlorhydria, for example, in AIDS patients. Similarly, caution should be exercised in a patient receiving antacids, H$_2$-receptor blockers or proton pump inhibitors (e.g., omeprazole) (Table 17.1).

TABLE 17.1   Drug Interactions with Oral Antifungal Agents

	Griseofulvin	Ketoconazole	Fluconazole	Itraconazole	Terbinafine
Alcohol	R1	R1			
Antacids		▼ Keto		▼ Itra	
Aspirin	▼ Aspr				
Astemizole		R2	R2	R2	
Caffeine			▲ Caffeine		▲ Caffeine
Calcium channel blockers (Dihydropyridine)				▲ C. blocker	
Carbamazepine				▼ Itra	
Chlordiazepoxide		▲ Chlor			
Cimetidine		▼ Keto		▼ Itra	▲ Terb
Cisapride		R2		R2	
Corticosteroids		▲ Cor		▲ Cor	
Coumadin (Coumarin or indanedione derivative anticoagulants)	▼ Cou	▲ Cou	▲ Cou	▲ Cou	
Cyclosporine	▼ CsA	▲ CsA	▲ CsA	▲ CsA	▼ CsA
Didanosine (ddl)		▼ Keto		▼ Itra	
Digoxin				▲ Dig	
H₂-receptor antagonists		▼ Keto		▼ Itra	
Hydrochlorothiazide (HCTZ)			▲ Flu		
Insulin		▼ ▲ Ins			
Isoniazid		▼ Keto		▼ Itra	
Loratadine		▲ Lor			
Midazolam				▲ Mid	
Nortriptyline			▲ No		
Omeprazole		▼ Keto		▼ Itra	
Oral hypoglycemic agent		▲ OH	▲ OH	▲ OH	
Phenytoin		▲ Ph ▼ Keto	▲ Ph	▲ Ph ▼ Itra	
Quinidine		▲ Qui		▲ Qui	
Rifampin		▼ Keto	▼ Flu	▼ Itra	▼ Terb
Sucralfate		▼ Keto		▼ Itra	
Tacrolimus		▲ Tac	▲ Tac	▲ Tac	
Terfenadine		R2	R2	R2	▲ Ter
Theophylline		▲ ▼ Th	▲ Th		
Triazolam				▲ Tri	
Vincristine				▲ Vin	
Zidovudine			▲ Zido		

▲ = Drug level may increase;   ▼ = Drug level may decrease
R1 = Disulfram-like reaction;   R2 = Idiosyncratic reaction (arrhythmia, including torsades de pointes)

With itraconazole, dose-dependent pharmacokinetics occur following both single and multiple dosing.[37] Following daily dosing, steady state plasma levels develop after 10 to 14 days.[37,38] However, steady state plasma levels cannot be predicted from the initial oral dose.[30]

Itraconazole is highly lipophilic and keratophilic; the protein-binding is 99.8%. The drug concentrates in fatty tissues such as adipose tissue, omentum, endometrium, liver, and kidneys.[39] Aqueous fluids, for example, the cerebrospinal fluid (CSF), aqueous humor, and saliva contain low concentrations of itraconazole.

The drug reaches therapeutic levels in the skin and nails. Itraconazole is detected in the sweat within 24 hours of intake.[40] Incorporation of drug into sebum takes longer and is a major route by which itraconazole reaches glabrous skin. Incorporation into the basal layer takes place slowly, but drug delivery by this route results in drug levels remaining in the skin for 3 to 4 weeks following discontinuation of itraconazole. At various anatomic sites, the three routes of drug delivery are important to different extents. Following incorporation into epidermis, hair, and nails, itraconazole does not redistribute back into the systemic circulation.[40] Itraconazole reaches the nail plate by diffusion through both the nail bed and the nail matrix.[41] The drug has been detected in the distal part of the fingernails 7 days after starting therapy.[40] Therapeutic concentrations of itraconazole have been found in fingernails and toenails for 6 and 9 months, respectively, following discontinuation of continuous therapy 200 mg/day for 3 months or pulse therapy 200 mg twice daily for 1 week per month for 3 consecutive months.[42]

Itraconazole undergoes extensive hepatic metabolism, with 30 metabolites having been identified. These are excreted mainly in the bile.[43] Most metabolites lack significant antifungal activity; hydroxy-itraconazole, however, has a similar *in vitro* spectrum of activity as compared to the parent compound but with an efficacy two- to fourfold lower in experimental mycoses in rodents.[44,45] This metabolite, being less lipophilic than the parent compound, may be better able to penetrate body fluids.

The serum half-life of ($t_{1/2}$) itraconazole depends on both the dose and duration of therapy.[30] Following a single dose of 100 to 400 mg, the $t_{1/2}$ is 15 to 25 hours. Once the steady state is achieved, the $t_{1/2}$ is 30 to 40 hours.[37] Elimination of drug occurs in the feces (54%), urine (34% of the administered dose after 1 week as inactive metabolites and 3% to 18% of the drug remains active).[39]

In special situations, no differences are found with respect to race, sex, or age of patients.[44,45] In patients with hepatic cirrhosis, a fasting state may reduce the absorption of itraconazole. Itraconazole pharmacokinetics may not be significantly affected with renal dysfunction or hemodialysis.[46] However, peritoneal dialysis may reduce peak plasma levels if antacids agents are being used on a regular basis.[45]

17.2.3   Therapeutic Efficacy

17.2.3.1   Tinea Corporis/Cruris[47-55]

Itraconazole was initially prescribed at a dose of 100 mg/day for 4 weeks to treat tinea corporis/cruris. Since then, based on the pharmacokinetic profile of the drug, the shorter schedule of 200 mg/day for 7 days has been found to be effective.[50] An open study compared oral itraconazole 200 mg/day for 7 days with the lower dose of 100 mg/day for 15 days. At the end of the 4-week follow-up period, clinical response was observed in 100% cases in both groups (83/83 and 27/27, respectively, in the two groups). The mycologic cure at follow-up in the 200 mg/day and 100 mg/day groups was 73 of 84 patients (87%) and 33 of 40 patients (83%), respectively. In a double-blind comparative study,[54] itraconazole 100 mg/day for 15 days was more effective than griseofulvin ultramicronized 500 mg/day for 15 days.

17.2.3.2   Tinea Pedis[56-69]

Initially, itraconazole 100 mg/day for 30 days was the dosage regimen used in many countries. Based on the pharmacokinetics of the drug, good efficacy rates were subsequently obtained at higher doses of itraconazole used for shorter durations of time. Itraconazole 400 mg/day given for 1 week, for example, results in cure rates of 85%, and these are similar, if not higher, than the cure rates of 51% to 84% observed with the 100 mg/day and 200 mg/day dosages given for 2 to 4 weeks.[69] The higher dose of 400 mg/day results in therapeutic concentrations of drug being achieved more rapidly in the thicker stratum corneum, with measurable drug levels in the stratum corneum for 3 to 4 weeks following discontinuation of itraconazole.

17.2.3.3   Pityriasis Versicolor[70,71]

The total dose of itraconazole for effective treatment is ≥1000 mg, given as 200 mg/day for 5 or 7 days.[70] Improvement continues after therapy has been discontinued and results should be assessed clinically and mycologically at 3 to 4 weeks after treatment. Studies have demonstrated that itraconazole is superior to placebo, and as effective as selenium sulfide, clotrimazole, and ciclopirox olamine.[70] It is better tolerated than selenium sulfide.

17.2.3.4   Tinea Capitis[72-75]

In a double-blind study, Lopez-Gomez et al.[72] compared the efficacy of itraconazole 100 mg/day with ultramicronized griseofulvin 500 mg/day each given for 6 weeks. Eight weeks following discontinuation of therapy, 15 of 17 patients (88%) predominantly with *M. canis* infections were cured in each of the itraconazole and griseofulvin groups. The optimum dose in children is 4–5 mg/kg/day.[73] Studies are currently carried out to determine the efficacy of pulse therapy with itraconazole.

17.2.3.5   Onychomycosis[76-85]

In the earlier studies, the dosing of itraconazole was 50–100 mg/day given as continuous dosing.[76] Subsequently, an improved understanding of the pharmacokinetics resulted in the dosing schedule of 200 mg/day for 3 months. In the European studies, using the 3-month schedule for toenails, the clinical response and mycologic cure were 87% and 74%, respectively.[76] For fingernails, the corresponding figures read 93% and 91%. The pulse therapy regimen appears to offer advantages of improved efficacy, fewer adverse effects, better patient compliance, and favorable cost-effectiveness.[85] The use of the three pulse regimens in the treatment of dermatophyte onychomycosis results in a clinical response and a mycological cure in 84% and 64% patients, respectively, at follow-up 1 year after starting therapy.[79]

Itraconazole is also effective in the treatment of *Candida* onychomycosis and *Candida* paronychia.[86,87] De Doncker et al.[88] have reported on their experience in the treatment of nondermatophyte mold onychomycosis with itraconazole continuous and pulse-therapy regimens. Patients with toenail infections caused by *Aspergillus* species, *Fusarium* species, *Scopulariopsis brevicaulis*, and *Alternaria* species were effectively treated with itraconazole.

17.2.4   Drug Interactions

The drugs not to be used with itraconazole include the antihistamines terfenadine and astemizole, the gastrointestinal prokinetic agent cisapride, and the benzodiazepines midazolam and triazolam (Table 17.1). In each case, itraconazole may inhibit the biotransformation system responsible for the metabolism of the drug, resulting in elevation of the above-mentioned drug. For the other drugs it is not an absolute requirement to discontinue them. In some instances it may be possible to reduce the dosage of the object drug or to monitor the dosage according to the side effects (e.g., edema or tinnitus). A substantial body of experience is not available with the use of these drugs with 1 week itraconazole pulse therapy. Many of the drug interactions with itraconazole can be explained on the basis that it inhibits the metabolism of drugs that undergo biotransformation by the cytochrome P-450 3A4 enzyme. Consequently, this may increase and/or prolong the effects of the object drug. Drugs that induce hepatic drug-metabolizing enzymes (e.g., rifampicin, carbamazepine, phenytoin, and phenobarbital) may accelerate the metabolism of itraconazole with resultant reduction in its levels.[89,90]

17.2.5   Adverse Effects

Itraconazole is well tolerated and almost without any serious adverse effects. In a prescription-event monitoring study of more than 13,600 patients, mostly females receiving the drug for the treatment of vaginal candidiasis, itraconazole was found to be virtually free from recognizable adverse effects.[91] Adverse experiences during short-term therapy with itraconazole have been reported in 7.5% of patients.[92] The most common adverse effects are: gastrointestinal (4.4%, nausea), cutaneous (0.7%, rash, pruritus), central nervous system (2%, headache), respiratory system (< 1%), liver and biliary system (0.9%), miscellaneous (0.6%, edema). During long-term therapy in patients, most of whom had underlying pathology and received multiple concomitant treatments, the incidence of adverse experiences was 20.6%.[92]

In dermatologic use, the incidence of asymptomatic and reversible elevation of liver function tests (LFTs) is 1% to 4%.[85] There are isolated reports of significant hepatobiliary dysfunction.[92-96] When considering the treatment of dermatomycosis, the incidence of hepatic injury is rare, occurring in less than 1 in 500,000 patients.[97] Gastrointestinal disorders (e.g., nausea, vomiting, diarrhea, abdominal pain) occurred in 4% of patients in U.S. clinical trials.[34]

Cutaneous eruptions in 3% of patients resulted in a temporary or permanent discontinuation of treatment in U.S. clinical trials.[34] There is a report of a patient on itraconazole who developed acute generalized exanthemic pustulosis.[98] The eruption resolved within 48 hours of discontinuing itraconazole.

At therapeutic doses of itraconazole 100–400 mg/day basal and adrenocorticotropic hormone-stimulated endocrine studies demonstrate no significant influence on adrenocortical function.[99-100] The higher dose of itraconazole 600 mg/day is not used in dermatology; it is more likely to be associated with adverse events such as hypokalemia, hypertension, and gynecomastia.[101]

17.3   Fluconazole

Fluconazole is a bis-triazole that has a difluorophenyl moiety instead of the dichlorophenyl moiety in ketoconazole.[102] Structural modifications such as these have resulted in a greater selectivity, higher metabolic stability and an enhanced water solubility of fluconazole compared to ketoconazole derived from the hydroxy group.

17.3.1   Pharmacodynamics

Until recently, the minimum inhibitory concentrations (MICs) obtained using standard media culture for *in vitro* susceptibility testing was not necessarily of clinical relevance.[103] Use of the broth macrodilution assay[104] and the broth microdilution assay[105] may provide data more predictive of the *in vitro* response.[106] Most *Candida* species are susceptible to fluconazole but vary in their susceptibility. In fact, fluconazole is relatively more active against *C. albicans, C. tropicalis,* and *C. parapsilosis;* less active against *C. guilliermondii;* and inactive against *C. krusei.*[107] Fluconazole exhibits variable activity against *C. glabrata.*[108,109] *In vitro,* fluconazole exhibits less potency against dermatophytes and other filamentous fungi.[110]

In vivo studies have demonstrated that fluconazole is effective in the treatment of dermatophytosis in guinea pigs.[111] Treatment of infection was initiated 6 days following inoculation, and duration of therapy amounted to 10 days. At doses of 2.5, 5, and 10 mg/kg/day, skin and hair were sterile in 38%, 88%, and 100% of animals.

In immunocompetent animal models fluconazole reduces the fungal burden and improves survival in infections caused by *Candida* (primarily *C. albicans*), *Cryptococcus,* or *Histoplasma* species.[112] In murine models, fluconazole was not effective against infections caused by *C. krusei*[113] and *Aspergillus flavus.*[114]

Fluconazole has been widely used, with over 15 million patients being treated since 1988.[115,116] The majority of reports on fluconazole resistance deals with AIDS patients, particularly those with oropharyngeal candidiasis.[115] Failures with fluconazole have involved the use of relatively low doses of fluconazole (100–200 mg/day), failure of recovery from neutropenia, and infection with resistant species such as *C. krusei.* In patients with fungemia, omitting a complete catheter exchange may be associated with the persistence of candidemia.[115] Long-term prophylaxis with fluconazole < 200 mg/day should be avoided. Thus, ideal infections such as oropharyngeal *Candidiasis* should be treated with short courses of fluconazole. When the infective *C. albicans* strains have low MICs, larger doses for shorter periods of time may be preferable, thereby reducing the chance that the organism mutates to a resistant form. Patients known to be colonized with the more resistant forms should be treated with larger doses of fluconazole or with alternative therapies.[115]

17.3.2   Pharmacokinetics

The bioavailability of orally administered fluconazole exceeds 90% in healthy volunteers and patients with AIDS who do not have gastroenteritis.[117] After a single oral dose of 400 mg, peak plasma levels of 6–7 μg/ml occur within 1 to 2 hours. For single doses of 50–400 mg, the peak plasma concentrations and AUC increase dose-proportionally. Fluconazole absorption is not significantly reduced by food,[117] gastric pH modifiers such as antacids,[118] or H_2-receptor antagonist.[119,120] The oral bioavailability of the solution and capsule form is about the same.[121] Multiple dosing results in an increase in the peak plasma concentration approximately 2.5 times that achieved following a single dose.[112] Fluconazole is a small molecule with relatively low lipophilicity (compared to the highly lipophilic nature of ketoconazole and itraconazole) and a low protein binding of 11% (compared to 99% for ketoconazole and itraconazole). These properties result in fluconazole being widely distributed to body tissues and fluids, including cerebrospinal fluid (CSF) and urine. In the latter, fluconazole concentrations exceed plasma levels approximately tenfold.[121]

The pharmacokinetics of fluconazole in skin and nail has been evaluated by several investigators.[123-127] In the study by Hay,[124] healthy volunteers received oral fluconazole 50 mg/day for 14 days. Mean plasma concentrations on day 1 and day 14 were 0.76 µg/ml, respectively. The respective concentrations in skin were 11.7 µg/g and 24.2 µg/g, those in nails 1.3 µg/g and 1.8 µg/g. Therefore, fluconazole is detected in skin and nails within 1 day of starting therapy.[125] Fluconazole is delivered to the stratum corneum by direct diffusion from the capillaries, through the sweat and probably also in the sebum.[123,126,127] Drug concentration in the stratum corneum was 73 µg/g, a level 40-fold higher than in the serum, following the last out of twelve 50 mg oral doses administered daily to healthy subjects.[126,127] In another study (200 mg/day given for 5 days) the concentration of fluconazole in the stratum corneum was 127 µg/g. The elimination of fluconazole from the stratum corneum occurred with a half-life of 60 to 90 hours, which is two to three times slower than the elimination from plasma.[127] Two 150 mg fluconazole doses given at 7-day interval resulted in 7.1 µg/g in the stratum corneum after another 7 days, indicating that once a week dosing may be effective in the management of dermatomycosis.[126] Fluconazole's half-life approximates 30 hours (compared to 8 hours for ketoconazole).[122] It may be increased in patients with renal impairment.[128] Fluconazole is relatively stable to metabolic conversion, with the primary route of excretion being the kidneys.[112] Approximately 80% of the administered dose appears unchanged in the urine and 11% as metabolites.[122,129]

In renal dysfunction, fluconazole excretion is delayed.[130] When the creatinine clearance is 20–51 ml/min, the area under the curve (AUC) and half-life are increased as compared to patients with normal renal function, necessitating a dosage reduction.

Elimination of fluconazole is also impaired in severe liver disease,[131] leading to a reduced total plasma clearance and an increase in the mean residence time. Ruhnke et al.[131] emphasized the need for caution in the treatment with fluconazole of patients with severe liver disease.

17.3.3 Therapeutic Efficacy

17.3.3.1 Tinea Corporis/Tinea Cruris

The most common dosage regimen was fluconazole 150 mg once a week for one to four once-weekly doses.[132-135] A mean of 2.6 doses was administered in the study by Suchil et al.[132] At follow-up 4 weeks after the last dose, 80 of 91 patients (88%) were assessed as being clinically cured, three patients (3%) were improved, and eight (9%) failed. Mycological failure was present in 1 of 86 patients.

17.3.3.2 Tinea Pedis

The dosage schedule used was 150 mg once weekly.[134,136] Del Aguila et al.[136] found that a mean of three doses was administered; patients with *Candida* infection required an average of two doses compared to three to four doses in patients with dermatophyte infections. At follow-up 4 weeks after the last dose was administered, clinical cure was present in 46 of 60 patients (77%), improvement in 13 patients (22%) and failure in 1 patient. Mycological evidence of disease was present in 13 of 59 patients (22%).

17.3.3.3 Onychomycosis

Kuokkanen and Alva[137] treated *Trichophyton rubrum* onychomycosis in 20 patients (43 nails) with fluconazole 150 mg once a week. The mean duration of therapy was 9.3 months. At follow-up, 6 months after the end of therapy, the cure rate in the fingernails and toenails was 100% and 83%, respectively.[138-140]

17.3.3.4 Pityriasis Versicolor

Köse[141] treated pityriasis versicolor with fluconazole 300 mg twice daily for 15 days. At the end of therapy, clinical and mycologic cures were 80% and 88%, respectively. At follow-up, 12 weeks after the end of therapy, the relapse rate was 14%. Following a single 400 mg dose, 17 of 23 patients (74%) were free of lesions 3 weeks after treatment and no recurrences were observed 6 weeks following therapy.[141-142]

17.3.4   Drug Interactions

Fluconazole exhibits a far greater specificity as an inhibitor of fungal cytochrome P-450 compared to mammalian cytochrome P-450 mediated systems[143] as compared to ketoconazole; therefore, fluconazole has less potential for drug interactions.[144] There are, however, some most relevant interactions. While fluconazole 100 mg/day may not significantly affect cyclosporine pharmacokinetics, higher doses of fluconazole can result in elevated cyclosporine plasma concentrations.[110,145] Also interaction with coumadin appears more pronounced at higher doses of fluconazole or in renal impairment.[146] Fluconazole may interfere with FK506 (tacrolimus) metabolism, thereby elevating circulating levels of FK506 (Table 17.1).[147]

17.3.5   Adverse Reactions

In over 4000 patients treated with fluconazole in clinical trials lasting ≥7 days, adverse events were reported in 16%.[110] Side effects occurred more frequently in patients with HIV (21%) as compared to non-HIV-infected individuals (13%). In both groups, the pattern of adverse effects was similar. Adverse effects observed in ≥1% of patients were: gastrointestinal (nausea 3.7%, vomiting 1.7%, abdominal pain 1.7%, diarrhea 1.5%), headache 1.9%, and rash 1.8%. Most serious adverse effects associated with fluconazole are hepatic dysfunction and exfoliative skin disorders. Treatment was discontinued in 1.5% patients because of unwanted clinical reactions and in 1.3% patients due to abnormalities in laboratory values.[110]

Because many of these patients received fluconazole for systemic disease and some were on a variety of concomitant medications, it is difficult to determine if an adverse event is attributable to fluconazole. We are aware of one report of an anaphylactic reaction to fluconazole.[148] There is also a report of a fixed drug eruption,[149] of a Stevens-Johnson syndrome,[150] and of angioedema associated with fluconazole.[151] Alopecia may more likely occur with higher doses, 400 mg/day, given for 2 months or longer.[152,153] Moreover, a case of adrenal insufficiency has been reported in a patient with AIDS.[154]

Also blood dyscrasias have been reported in association with fluconazole therapy: agranulocytosis[156] and thrombocytopenia.[157-162] Not all of these patients had AIDS or were immunocompromised. Although these are rare events with fluconazole therapy,[163] some authors have suggested hematologic monitoring.[155-157] Clinically meaningful deviations from baseline in hematologic values possibly related to fluconazole have been reported with respect to hemoglobin (0.5%), white blood count (0.5%) and total platelet count (0.6%).[110]

There have been reports of hepatic dysfunction associated with fluconazole therapy.[164-170] The paucity of reports suggests that fluconazole-induced liver damage may be rare.[166] Hay[97] reports that approximately 5.1% patients treated for superficial infection may have one or more abnormal liver enzymes. This is comparable to the frequency of abnormal liver enzymes following treatment with topical antifungals or placebo.

With regard to renal function abnormalities, clinically significant elevations of serum urea and creatinine have been reported in 0.4% and 0.3% patients, respectively.[110]

17.4   Terbinafine

Terbinafine is a synthetic allylamine that was discovered in 1978. This was a chance discovery during a chemical research program for the synthesis of compounds for the central nervous system.[171] The first allylamine to be discovered was naftifine.[172] In contrast to terbinafine, which possesses both topical and oral activity, naftifine was active only topically. The allylamines inhibit the enzyme squalene epoxidase and therefore prevent the synthesis of ergosterol, resulting in squalene accumulation.[173] The former is an essential component of most fungal cells and is required for membrane integrity and growth. The ergosterol deficiency may be responsible for the fungistatic action, whereas accumulation of squalene could contribute to the fungicidal action *in vitro.*[174]

17.4.1   Pharmacodynamics

Terbinafine demonstrates *in vitro* activity against a wide range of fungi including dermatophytes, filamentous and dimorphic fungi, pathogenic yeasts, and dematiaceous fungi.[175] MIC values for most dermatophytes range from 0.001 to 0.01 mg/L. The activity of terbinafine against yeasts *in vitro* is species dependent and the drug is more effective against *Candida parapsilosis* compared to *Candida albicans*. The filamentous fungi susceptible to terbinafine *in vitro* include *Aspergillus* species, Fusarium spp., *Scopulariopsis brevicaulis*, and *Scytalidium* spp.

17.4.2   Pharmacokinetics

Terbinafine is well absorbed (> 70%) following oral administration with maximal plasma concentrations of 0.8–1.5 µg/ml 2 hours after a single 250 mg oral dose.[176] In humans, the distribution of terbinafine is large, as evidenced by a large volume of distribution. Terbinafine is extensively metabolized in the liver with at least 15 metabolites having been identified.[177] When terbinafine 125 mg twice daily is taken for 28 days, the maximal plasma concentrations are 0.8 µg/ml.[176]

In hepatic disease, dose adjustments should not be necessary except in cases of extreme hepatic dysfunction because of a large therapeutic index in humans. Although terbinafine is not excreted by the kidneys unchanged, in patients with renal disease the terminal elimination half-life exceeds that of healthy volunteers (24 vs. 16 hours). The absorption and distribution is similar in both groups. In those with renal disease the metabolism of terbinafine may be impaired. The decreased elimination of terbinafine in patients with renal dysfunction could result from a change in the metabolism of terbinafine in this patient group and/or because of a decrease in liver function secondary to renal insufficiency.

Comparing elderly subjects (67 to 73 years old) to healthy young volunteers, no differences in the pharmacokinetics of terbinafine were observed following a single oral dose of 500 mg.[176] The total plasma clearance and volume of distribution of terbinafine in elderly subjects are comparable to those of the volunteers. Thus, in elderly subjects dose adjustments of terbinafine do not appear to be indicated.

Terbinafine has been detected in the stratum corneum as early as 24 hours after commencing therapy.[178] The drug is initially detected in the deeper layers of the stratum corneum following oral administration, suggesting that the route of delivery to the stratum corneum is initially by epidermal diffusion. The detection of terbinafine at all levels within the stratum corneum by the third day suggests that movement of the drug through the stratum corneum is by diffusion and not simply by incorporation into the outward moving corneocytes.[178] When terbinafine 250 mg/day is given to volunteers for 12 days, high levels of the drug are reached in the skin within hours, gradually increasing over several days to 9 µg/g of tissue.[179] Terbinafine is highly concentrated in the sebum, reaching concentrations of 40 µg/ml. This process occurs more slowly than the diffusion of drug from deeper blood vessels to the superficial stratum corneum. The elimination half-life from the sebum and stratum corneum is 3 to 5 days. When terbinafine 250 mg/day is given for 7 days, the levels of drug in the skin are 10 to 100 times the MIC for most dermatophytes 54 days following discontinuation of therapy.[179,180] The elimination kinetics of terbinafine and three metabolites has been found to be multiphasic, being faster initially.[181] The mean terminal elimination half-lives of the parent drug and the metabolites are 18 to 28 days. When terbinafine 250 mg/day is administered for 7 days, the drug is detected in peripheral nail clippings after 7 days at a concentration of 0.5 µg/g; at 90 days after stopping therapy the concentration still is 0.2 µg/g.[180]

17.4.3   Therapeutic Efficacy

17.4.3.1   Tinea Corporis/Cruris

Terbinafine is effective for the treatment of tinea corporis/cruris with recommended duration of therapy being 1 to 2 weeks.[182-185] Farag et al.[185] treated 22 patients who had mycologically confirmed tinea corporis/cruris with terbinafine 250 mg/day for 1 week. At follow-up, 6 weeks after completion of therapy, all patients had been successfully treated and there were no relapses.

17.4.3.2   Tinea Pedis

Terbinafine has been shown to be effective in the treatment of tinea pedis in placebo-controlled[186,187] or comparative trials (vs. griseofulvin[188,189] and vs. itraconazole[64,65]). The recommended duration of terbinafine therapy is 2 to 4 weeks. In the study performed by Hay et al.,[64] terbinafine 250 mg/day given for 2 weeks was found to be as effective as itraconazole 100 mg/day given for 4 weeks, but with fewer long-term relapses, in the treatment of plantar-type tinea pedis.

17.4.3.3   Tinea Capitis

There have been case-reports[190-193] and studies[194-196] where terbinafine has been effectively used to treat tinea capitis. Based on pharmacokinetic studies[195] the optimal dose of terbinafine in children is 125 mg/day when the weight is 20 to 40 kg and 62.5 mg/day for children weighing less than 20 kg.[194] In children whose weight exceeds 40 kg, the adult dose of 250 mg/day is appropriate. With terbinafine, *Microsporum canis* infections causing tinea capitis may be more difficult to eradicate than those caused by *Trichophyton* species.[197] Also, immunocompromised patients may require long-term treatment to prevent relapse.

17.4.3.4   Onychomycosis

The current treatment regimen for pedal and fingernail onychomycosis is continuous therapy with terbinafine 250 mg/day for 12 and 6 weeks, respectively. Open studies,[198-203] double-blind placebo-controlled[204] and comparative studies[205-211] have confirmed the efficacy in onychomycosis. The duration-finding study by Van der Schroeff et al.[203] demonstrated that for toenail onychomycosis, a 12 week treatment period was sufficient.

Terbinafine is more effective than griseofulvin for the treatment of fingernail[205] and toenail onychomycosis.[206,207] Looking at the comparative studies between terbinafine and itraconazole for pedal onychomycosis, in two studies[208,211] there was no significant difference between the two drugs; two other trials were in favor of terbinafine.[209,210] It is important to note that there is an overlap in the clinical response and mycologic cure rates for these two agents in the treatment of onychomycosis.[87]

17.4.4   Drug Interactions

Terbinafine demonstrates negligible potential to inhibit or induce the clearance of drugs metabolized by the hepatic cytochrome P-450 system.[212] This allylamine does not bind as an inhibitor to hepatic enzymes, but it acts as a substrate for some subtypes of the hepatic cytochromes P-450 by which it is degraded to less lipophilic metabolites.[213] The plasma clearance of terbinafine is accelerated by drugs that induce cytochrome P-450 (e.g., rifampicin) and is reduced by drugs that inhibit cytochrome P-450 (e.g., cimetidine).[214] In some instances, menstrual irregularities have been observed in patients on the oral contraceptive pill and terbinafine.[212] The incidence of this disorder, however, is within the background incidence of menstrual irregularities observed in subjects taking the oral contraceptive alone.

17.4.5   Adverse Effects

In general, oral terbinafine is well tolerated with side effects being mild to moderate in nature and transient. The most commonly observed adverse effects are: gastrointestinal (5%, nausea and/or vomiting, diarrhea and/or cramps, dyspepsia, gastritis or gastrointestinal irritation, feeling of fullness of stomach, sickness); cutaneous (3%, erythema or rash, urticaria, eczema, pruritus); central nervous system (1%, headache, change in concentration); other effects (1%, fatigue, pain, change of taste or dry mouth).[212]

Taste disorders may occur in 1 in 800 cases.[215] The taste disturbance usually returns to normal within weeks of discontinuing terbinafine. There have been rare case reports of erythema multiforme,[216-218] toxic epidermal necrolysis,[219] and Stevens-Johnson syndrome.[220] Where sufficient information is available, the mean onset of lesions has been 3 weeks after commencing terbinafine therapy.[116] When a serious cutaneous reaction occurs, the drug should be discontinued. Resolution occurred in most instances within 1 week of stopping terbinafine, with or without initiating prednisone therapy. There has also been a report of a serum sickness-like reaction developing in an 81-year-old man following 6 weeks of therapy with terbinafine[221] and of fixed drug eruption.[222] The exanthematous rash, fever, myalgias, and arthralgias also improved following discontinuation of the terbinafine. In another instance a patient developed severe erythema annulare centrifugum-like psoriatic drug eruption after 5 days of terbinafine therapy.[223]

Isolated cases of significant hepatobiliary dysfunction have been reported.[224,225] Terbinafine should be discontinued if there are symptoms or signs of hepatobiliary dysfunction. The estimated reporting incidence of clinically significant hepatobiliary dysfunction is 1 in 54,000.[212]

Blood dyscrasias have also been reported in association with terbinafine. Kovacs et al.[227] reported a case of severe neutropenia and one of pancytopenia (with severe neutropenia) in association with terbinafine therapy. As of September 1994, there have been 8 cases of blood dyscrasias reported in Canada, in particular agranulocytosis, pancytopenia, and thrombocytopenia, which are possibly or probably related to terbinafine.[226] One case of thrombotic thrombocytopenic purpura (TTP) has been reported; the role of terbinafine in the development of TTP in this case, however, cannot be established.

17.5   Discussion

The availability of the newer antifungal agents has, however, a major impact on the management of dermatomycoses, in particular onychomycoses. In most cases itraconazole, terbinafine, or fluconazole will be preferred over griseofulvin or ketoconazole when treating a cutaneous fungal infection. It should be kept in mind that many localized, superficial fungal infections are amenable to topical therapies and these should be considered initially, when appropriate. In certain cases, for example, tinea capitis, Majocchi's granuloma, widespread tinea infection, chronic tinea pedis, and infections in immunocompromised or immunodeficient individuals, oral therapy may be the first-line treatment. We feel that griseofulvin is still an effective therapy for tinea capitis, with efficacy rates comparable to the newer antifungal agents.[228-230] Oral ketoconazole continues to be an option for the treatment of pityriasis versicolor and vaginal candidiasis requiring short-term dosing. It is also used in the management of chronic mucocutaneous candidiasis.

There are several factors that may enter into the complicated equation that finally results in both the patients and physicians deciding to proceed with oral antifungal therapy (Table 17.2). For the patient and physician issues that may need to be addressed are the perception of disease, its severity, the need to treat it, and the perceived benefits and risks of available therapies. Onychomycosis, for example, can be perceived to be a cosmetic, asymptomatic condition by some patients and/or the physician. Familiarity with an established drug may result in the physician being more likely to use that drug with less attention paid to monitoring for adverse effects; with newer agents there may be some hesitation in prescribing the drug and more regard to potential adverse effects and monitoring issues. The availability of the various drugs, cost of therapy to the patient and health care provider, efficacy of therapies, drugs that the patient is receiving, physician attitude, knowledge and preference, and peer prescribing habits all play a role in the final decision. For both onychomycoses and the other dermatomycoses, it is important that a realistic expectation of efficacy is conveyed to the patient. For example, in onychomycosis the cure rates may be typically 40% to 70% with a relapse of 10% to 20% in the cured group at follow-up 12 months later.

In onychomycosis, particularly where a large portion of the nail plate is diseased, or when the outer part of the nail matrix is involved, oral therapy is necessary. In cases where only a small portion of the nail plate/bed has onychomycosis, topical antifungal agents such as amorolfine or ciclopirox olamine lacquer may be viable alternatives.[231-233] While the newer antifungal agents have an extremely favorable adverse effects profile, uncommon systemic symptoms and signs occur. On the other hand, the topical antifungal agents have the advantage that side effects are rare and if occurring usually manifest as erythema, irritation, or dermatitis at the application site. The pharmacokinetics of the newer antifungal agents enable itraconazole and terbinafine to be administered over much shorter periods of time as compared to griseofulvin, which has to be taken for 9 to 18 months in toenail onychomycosis while the diseased nail is growing out. Also, with fluconazole and itraconazole pulse dosing schedules result in a

TABLE 17.2 Pros and Cons of Antifungal Agents Used in the Treatment of Onychomycosis and Other Dermatomycoses

Property of Antifungal Agent	Disease	Antifungal Agent					
		Griseofulvin	Ketoconazole	Itraconazole	Terbinafine	Fluconazole	Topicals
Efficacy	Onychomycosis	Low	Medium	High	High	High	Low
	Dermatomycosis	High	High	High	High	High	Medium to high (see text)
Duration of therapy	Onychomycosis	Long	Long	Short	Short	Medium	Long
	Dermatomycosis	Short	Short	Short	Short	Short	Short to medium
Dosing	Onychomycosis	Long-term continuous	Long-term continuous	Pulse, short-term	Short-term	Pulse	Long-term continuous
	Dermatomycosis	Short-term continuous	Short-term continuous	Pulse, short-term	Short-term	Pulse	Short-term continuous
Compliance	Onychomycosis	Low	Low	High	High	High	Low to medium
	Dermatomycosis	High	High	High	High	High	High
Adverse effects (some > few > very few)	Onychomycosis	Some	Some	Few	Few	Few	Very few
Relapse rate	Dermatomycosis	Few	Few	Few	Few	Few	Very few
	Onychomycosis	High	High	Low	Low	Not known	High
	Dermatomycosis	Low	Low	Low	Low	Low	Low
Monitoring	Onychomycosis and Dermatomycosis	CBC, LFTs regularly	LFTs regularly	Continuous therapy longer than 1 month Pulse: None	Longer than 6 weeks (U.S. product monograph)	Not approved	None
Overall cost-effectiveness	Onychomycosis	Low	Low	High	High	High	Low to high (see text)
	Dermatomycosis	High	High	High	High	High	High

reduced duration for which the drug remains in plasma after therapy has been discontinued. Skin levels, however, remain high. Both itraconazole and terbinafine are more cost effective than griseofulvin in the treatment of onychomycosis.

The adverse effects profiles of the newer antifungal agents, itraconazole, fluconazole, and terbinafine suggest that they are generally safe and well-tolerated in the treatment of dermatomycoses. Since these agents have been introduced, there have been reports of blood dyscrasias, hepatobiliary dysfunction, and drug interactions concerning these antimycotics. Once the decision has been made to initiate therapy with a given oral antifungal agent, the patient should be counseled about potential adverse effects. These may include symptoms of hepatobiliary dysfunction (e.g., unusual fatigue, nausea, vomiting, abdominal pain, pale urine, dark stools), cutaneous eruption, and blood dyscrasia (e.g., sore throat, fatigue). The patient should be told to discontinue the antifungal agent if such symptoms develop. Furthermore, the health care team, including the patient, physician, and pharmacist, should pay attention to possible drug interactions, particularly in the extremely sick or elderly, who may be on several drugs. Also, in some instances the antifungal agent may interact with nonprescription (over-the-counter) drugs.

The issue of monitoring while a patient is receiving antifungal therapy remains unresolved. At the moment there are no firm guidelines, and the product monographs do not provide definite guidelines concerning monitoring. Some physicians perform a blood count and determine plasma enzymes in case of continuous itraconazole dosing exceeding 1 month. In the United States, hepatic function tests are recommended in patients receiving the drug for longer than 6 weeks. In any case, if a patient develops symptoms suggestive of hepatobiliary dysfunction or a blood dyscrasia, monitoring should be performed.

The newer antifungal agents (itraconazole and fluconazole) are teratogenic in some animal species. The relevance of this effect in humans is not certain. In pregnancy these agents should be used only if the benefit outweighs the potential risk.

The management of cutaneous mycoses involves careful consideration of each patient. The impact of the newer antifungal agents is substantial. The drugs are generally considerably more effective than the traditional therapies, griseofulvin and ketoconazole. Until now, itraconazole, terbinafine and fluconazole showed few serious adverse effects. Physicians, however, should remain vigilant for possible adverse reactions with appropriate screening and counseling of patients. New dosage regimens and monitoring guidelines may be developed when gaining more experience with these drugs.

References

1. Anderson, D.W., Griseofulvin: biology and clinical usefulness: a review. *Ann Allergy* 23, 103, 1965.
2. Korting, H.C., Schäfer-Korting, M., Is tinea unguium still widely incurable? A review three decades after the introduction of griseofulvin. *Arch Dermatol* 128, 243, 1992.
3. Gupta, A.K., Sauder, D.N., Shear, N.H., Antifungal agents: an overview. Part I. *J Am Acad Dermatol* 30, 677, 1994.
4. Villars, V.V., Jones, T.C., Special features of the clinical use of oral terbinafine in the treatment of fungal diseases. *Br J Dermatol* 126 (Suppl. 39), 61, 1992.
5. Davies, R.R., Everall, J.D., Hamilton, E., Mycological and clinical evaluation of griseofulvin for chronic onychomycosis. *Br Med J* 3, 464, 1967.
6. Drouhet, E., Hay, R.J., Jones, H.E., Restrepo, A., *Ketoconazole in the management of fungal disease*. 1st ed. Adis Press, New York, 1982.
7. Jones, H.E., Simpson, J.G., Artis, W.M., Oral ketoconazole: an effective and safe treatment for dermatophytosis. *Arch Dermatol* 117, 129, 1981.
8. Jones, H.E., Ketoconazole. *Arch Dermatol* 118, 217, 1982.
9. Hay, R.J., Ketoconazole in the treatment of fungal infection. Clinical and laboratory studies. *Am J Med* 74, 16, 1983.
10. Knight, T.E., Shikuma, C.Y., Knight J., Ketoconazole-induced fulminant hepatitis necessitating liver transplantation. *J Am Acad Dermatol* 25, 398, 1991.

11. Lake-Bakaar, G., Scheuer, P.J., Sherlock, S., Hepatic reactions associated with ketoconazole in the United Kingdom. *Br Med J* 294, 419, 1987.

12. Hay, R.J., Ketoconazole: a reappraisal. *Br Med J* 290, 260, 1985.

13. Heeres, J., Backx, L.J.J., Van Cutsem, J., Antimycotic azoles. 7. Synthesis and antifungal properties of a series of novel triazol-3-ones. *J Med Chem* 27, 894, 1984.

14. Heeres, J., Backx, L.J.J., Mostmans, J.H., Van Cutsem, J., Antimycotic imidazoles 4. Synthesis and antifungal activity of ketoconazole, a new potent orally active broad spectrum antifungal agent. *J Med Chem* 22, 1003, 1979.

15. Cauwenbergh, G., De Donker, P., Itraconazole (R 51 211): A clinical review of its antimycotic activity in dermatology. *Drug Dev Res* 8, 317, 1986.

16. Clearly, J.D., Taylor, J.W., Chapman, S.W., Itraconazole in antifungal therapy. *Ann Pharmacother* 26, 502, 1992.

17. Greer, D.L., and Hay, R.J., *Itraconazole Monograph: Dermatology.* Adis International, Chester, England, 1994.

18. Vanden Bossche, H., Marichal, P., Gorrens, J., Coene, M.-C., Biochemical basis for the activity and selectivity of oral antifungals. *Br J Clin Pract* 44 (Suppl. 71), 41, 1990.

19. Borgers, M., Mechanism of action of antifungal drugs, with special reference to the imidazole derivatives. *Rev Infect Dis* 2, 520, 1980.

20. Van Cutsem, J., Van Gerven, F., Janssen, P.A.J., Activity of orally, topically, and parenterally administered itraconazole in the treatment of superficial and deep mycoses: animal models. *Rev Infect Dis* 9 (Suppl. 1), S15, 1987.

21. Van Cutsem, J., Van Gerven, F., Janssen, P.A.J., The *in vitro* and *in vivo* antifungal activity of itraconazole. *Recent Trends in the Discovery, Development and Evaluation of Antifungal Agents,* 177, 1987.

22. Arzeni, D., Barchiesi, F., Ancarani, F., Scalise, G., Fluconazole, itraconazole and ketoconazole *in vitro* activity. A comparative study. *Chemotherapy* 3, 139, 1991.

23. Uchida, K., Hosaka, J., Aoki, K., Yamaguchi, H., *In vitro* antifungal activity of itraconazole, a new triazole antifungal agent, against clinical isolates from patients with dermatomycoses. *Jpn J Antibiot* 44, 571, 1991.

24. Kobayashi, G.S., Spitzer, E.D., Testing of organisms for susceptibility to triazoles: is it justified? *Eur J Clin Microbiol Infect Dis* 8, 387, 1989.

25. Odds, F.C., Antifungal susceptibility testing of candida spp. by relative growth measurement at single concentrations of antifungal agents. *Antimicrob Agents Chemother* 36 (8), 1727, 1992.

26. Saag, M.S., Dismukes, W.E., Azole antifungal agents: emphasis on new triazoles. *Antimicrob Agents Chemother* 32, 1, 1988.

27. Sheehan, D.J., Espinel-Ingroff, A., Moore, L.S., Webb, C.D., Antifungal susceptibility testing of yeasts: a brief overview. *Clin Infect Dis* 17 (Suppl. 2), S494, 1993.

28. Sud, I.J., Feingold, D.S., Mechanisms of action of the antimycotic imidazoles. *J Invest Dermatol* 76, 438, 1981.

29. Sud, I.J., Feingold, D.S., Heterogenicity of action mechanisms among antimycotic imidazoles. *Antimicrob Agents Chemother* 20, 71,1981.

30. Zuckerman, J.M., Tunkel, A.R., Itraconazole: a new triazole antifungal agents. *Infection Control* 16 (6), 397, 1994.

31. Maggon, K.K., Slee, A.M., Demos, C.H., Development of antifungal agents by the pharmaceutical industry. *Drugs of Today* 27, 317, 1991.

32. Van Cutsem, J., Oral and parenteral treatment with itraconazole in various superficial and systemic experimental fungal infections. Comparisons with other antifungals and combination therapy. *Br J Clin Pract* 44 (Suppl. 71), 32, 1990.

33. Heykants, J., Van Peer, A., Van de Velde, V., Van Rooy, P., The clinical pharmacokinetics of itraconazole: an overview. *Mycoses* 32, 67, 1989.

34. Janssen Pharmaceutica Inc., *U.S.A. Product Monograph: Sporonox (Itraconazole capsules).* Janssen Pharmeutica, Titusville, NJ, 1996.
35. Heykants, J., Van Peer, A., Lavrijsen, K., Meuldermans, W., Woestenborghs, R., Cauwenbergh, G., Pharmacokinetics of oral antifungals and their clinical implications. *Br J Clin Pract* 71, 50, 1990.
36. Wishart, J.M., The influence of food on the pharmacokinetics of itraconazole in patients with superficial fungal infection. *J Am Acad Dermatol* 17, 220, 1987.
37. Hardin, T.C., Graybill, J.R., Fetchick, R., Woestenborghs, R., Rinaldi, M.G., Kuhn, J.G., Pharmacokinetics of itraconazole following oral administration to normal volunteers. *Antimicrob Agents Chemother* 32, 1310, 1988.
38. Van Peer, A., Woestenborghs, R., Heykants, J., Gasparini, R., Cauwenbergh, G., The effects of food and dose on the oral systemic availability of itraconazole in healthy subjects. *Eur J Clin Pharmacol* 36, 423, 1989.
39. Grant, S.M., Clissold, S.P., Itraconazole: a review of its pharmacodynamic and pharmacokinetic properties, and therapeutic use in superficial and systemic mycoses. *Drugs* 37, 310, 1989.
40. Cauwenbergh, G., Degreef, H., Heykants, J., Woestenborghs, R., Van Rooy, P., Haeverans, K., Pharmacokinetic profile of orally administered itraconazole in human skin. *J Am Acad Dermatol* 18, 263, 1988.
41. Matthieu, L., De Doncker, P., Cauwenbergh, G., Woestenborghs, R., van de Velde, V., Janssen, P.A.J., Itraconazole penetrates the nail via the nail matrix and the nail bed: an investigation in onychomycosis. *Clin Exp Dermatol* 16, 374, 1991.
42. Willemsen, M., De Doncker, P., Willems, J., Woestenborghs, R., van de Velde, V., Heykants, J., Van Cutsem, J., Posttreatment itraconazole levels in the nail. *J Am Acad Dermatol* 26, 731, 1992.
43. Van Cauteren, H., Heykants, J., De Coster, R., Cauwenbergh, G., Itraconazole: pharmacologic studies in animals and humans. *Rev Infect Dis* 9 (Suppl. 1), S43, 1987.
44. Heykants, J., Michiels, M., Meuldermans, W., Monbaliu, J., Larijsen, K., Van Peer, A., Leveron, J.C., Woestenborghs, R., The pharmacokinetics of itraconazole in animals and man: an overview. In: *Recent Trends in the Discovery, Development and Evaluation of Antifungal Agents.* 1st ed. (Fromtling, R.A.) J.R. Prous Science Publishers, Barcelona, 223, 1987.
45. Negroni, R., Arechavala, A.I., Itraconazole: pharmacokinetics and indications. *Arch Med Res* 24 (4), 387, 1993.
46. Boelaert, J., Schurgers, M., Matthys, E., Daneels, R., Van Peer, A., De Beule, K., Woestenborghs, R., Heykants, J., Itraconazole pharmacokinetics in patients with renal dysfunction. *Antimicrob Agents Chemother* 32, 1595, 1988.
47. Roseeuw, D., Willemsen, M., Kint, R.T., Peremans, W., Mertens, R.L.J., Cutsem, J.V., Itraconazole in the treatment of superficial mycoses: a double-blind study vs. placebo. *Clin Exp Dermatol* 15, 101, 1990.
48. Nuijten, S.T.M. Schuller, J.L., Itraconazole in the treatment of tinea corporis: a pilot study. *Rev Infect Dis* 9, S119, 1987.
49. Saul, A., Bonifaz, A., Arias, I., Itraconazole in the treatment of superficial mycoses: An open trial of 40 cases. *Rev Infect Dis* 9 (Suppl. 1), S100, 1987.
50. Parent, D., Decroix, J., Heenen, M., Clinical experience with short schedules of itraconazole in the treatment of tinea corporis and/or tinea cruris. *Dermatology* 189, 378, 1994.
51. Katsambas, A., Antoniou, C.H., Frangouli, E., Rigopoulos, D., Vlachou, M., Michailidis, D., Stratigos, J., Itraconazole in the treatment of tinea corporis and tinea cruris. *Clin Exp Dermatol* 18, 322, 1993.
52. Pariser, D.M., Pariser, R.J., Ruoff, G., Ray, T.L., Double-blind comparison of itraconazole and placebo in the treatment of tinea corporis and tinea cruris. *J Am Acad Dermatol* 31, 232, 1994.
53. Panagiotidou, D., Kousidou, T., Chaidemenos, G., Karakatsanis, G., Kalogeropoulou, A., Tekentzis, A., A comparison of itraconazole and griseofulvin in the treatment of tinea corporis and tinea cruris: a double-blind study. *J Int Med Res* 20, 392, 1992.

54. Bourlond, A., Lachapelle, J.M., Aussems, J., Boyden, B., Campaert, H., Conincx, S., Decroix, J., Geeraerts, C.H., Ghekiere, L., Double-blind comparison of itraconazole with griseofulvin in the treatment of tinea corporis and tinea cruris. *Int J Dermatol* 28, 410, 1989.

55. Engelhard, D., Or, R., Naparstek, E., Leibovici, V., Treatment with itraconazole of widespread tinea corporis due to *Trichophyton rubrum* in a bone marrow transplant recipient. *Bone Marrow Transplant* 3, 517, 1988.

56. Lachapelle, J.M., De Doncker, P., Tennstedt, D., Cauwenbergh, G., Janssen, P.A.J., Itraconazole compared with griseofulvin in the treatment of tinea corporis/cruris and tinea pedis/manus: an interpretation of the clinical results of all completed double-blind studies with respect to the pharmacokinetic profile. *Dermatology* 184, 45, 1992.

57. Rakosi, T., Gerber, M., Treatment of tinea with itraconazole: an open multicentre study. *J Dermatol Treatm* 6, 35, 1995.

58. Rosseuw, D., Willemsen, M., Kint, R.T., Peremans, W., Mertens, R.L.J., Van Cutsem, J., Itraconazole in the treatment of superficial mycoses: a double-blind study vs. placebo. *Clin Exp Dermatol* 15, 101, 1990.

59. Decroix, J., Tinea pedis (mocassin-type) treated with itraconazole. *Int J Dermatol* 34 (2), 122, 1995.

60. Van Hecke, E., Van Cutsem, J., Double-blind comparison of itraconazole with griseofulvin in the treatment of tinea pedis and tinea manuum. *Mycoses* 31, 641, 1988.

61. Wishart, J.M., A double-blind study of itraconazole vs. griseofulvin in patients with tinea pedis and tinea manus. *N Z Med J* 107, 126, 1994.

62. Degreef, H., Marien, K., De Veylder, H., Duprez, K., Borghys, A., Verhoeve, L., Itraconazole in the treatment of dermatophytoses: A comparison of two daily dosages. *Rev Infect Dis* 9 (Suppl. 1), S104, 1987.

63. Cauwenbergh, G., De Doncker, P., The clinical use of itraconazole in superficial and deep mycoses. In: *Recent Trends in Discovery, Development and Evaluation of Antifungal Agents.* (Fromtling, R.A.) J.R. Prous Science Publishers, Barcelona, 1987, 273.

64. Hay, R.J., McGregor, J.M., Wuite, J., Ryatt, K.S., Ziegler, C., Clayton, Y.M., A comparison of 2 weeks of terbinafine 250 mg/day with 4 weeks of itraconazole 100 mg/day in plantar-type tinea pedis. *Br J Dermatol* 132, 604, 1995.

65. De Keyser, P., De Backer, M., Massart, D.L., Westerlinck, K.J., Two-week oral treatment of tinea pedis, comparing terbinafine (250 mg/day) with itraconazole (100 mg/day): a double-blind, multicentre study. *Br J Dermatol* 130 (Suppl. 43), 22, 1994.

66. Saul, A., Bonifaz, A., Itraconazole in common dermatophyte infections of the skin: fixed treatment schedules. *J Am Acad Dermatol* 23, 554, 1990.

67. Difonzo, E.M., Panconesi, E., and Cilli, P., Itraconazole in dermatophyte infections: clinical experience in Italy. *Proc Symp Fungal Infections in the Nineties* 44 (Suppl. 9), 115, 1990.

68. Hay, R.J., Clayton, Y.M., Moore, M.K., Midgely, G., Itraconazole in the management of chronic dermatophytosis. *J Am Acad Dermatol* 23 (3), 561, 1990.

69. Gupta, A.K., De Doncker, P., Heremans, A., Stoffels, P., Piérard, G.E., Decroix, J., Heenen, M., Degreef, H., Itraconazole for the treatment of tinea pedis: a dose of 400 mg/day given for 1 week is similar in efficacy to 100 or 200 mg/day given for 2 to 4 weeks, *J Am Acad Dermatol* 36, 789, 1997.

70. Delescluse, J., Itraconazole in tinea versicolor: a review. *J Am Acad Dermatol* 23, 551, 1990.

71. Estrada, R.A., Itraconazole in pityriasis versicolor. *Rev Infect Dis* 9 (Suppl. 1), S129, 1987.

72. Lopez-Gomez, S., Cuetara, M.S., Iglesias, L., Rodriguez-Noriega, A., Itraconazole vs. griseofulvin in the treatment of tinea capitis: A double-blind randomized study in children. *Int J Dermatol* 33 (10), 743, 1994.

73. Nolting, S., Gupta, A.K., Deprost, Y., Delescluse, J., Degreef, H., Theissen, U., Wallace, R., Heremans, A., Stoffels, P., Itraconazole for the treatment of dermatomycoses in children, submitted.

74. Legendre, R., Esola-Macre, J., Itraconazole in the treatment of tinea capitis. *J Am Acad Dermatol* 23, 559, 1990.

75. Elewski, B., Tinea capitis: itraconazole in *Trichophyton tonsurans* infection. *J Am Acad Dermatol* 31, 65, 1994.
76. Haneke, E., Delescluse, J., Plinck, E.P.B., Hay, R.H., The use of itraconazole in onychomycosis. *Eur J Dermatol* 6, 7, 1996.
77. Walsoe, I., Stangerup, M., Svejgaard, E., Itraconazole in onychomycosis: open and double-blind studies. *Acta Derm Venereol* (Stockh) 70, 137, 1990.
78. De Doncker, P., Van Lint, J., Dockx, P., Roseeuw, D., Pulse therapy with one-week itraconazole monthly for three or four months in the treatment of onychomycosis. *Cutis* 53 (3), 180, 1995.
79. De Doncker, P., Decroix, J., Pierard, G.E., Roeland, D., Woestenborghs, R., Jacqmin, P., Odds, F., Heremans, A., Dockx, P., Roseeuw, D., Itraconazole pulse therapy is effective in the treatment of onychomycosis: A pharmacokinetic/pharmacodynamic and clinical evaluation. *Arch Dermatol* 132, 34, 1995.
80. Arenas, R., Fernandez, G., Dominguez, L., Onychomycosis treated with itraconazole or griseofulvin alone with and without a topical antimycotic or keratolytic agent. *Int J Dermatol* 30 (8), 586, 1991.
81. Korting, H.C., Schäfer-Korting, M., Zienicke, H., Georgii, A., Ollert, M.W., Treatment of tinea unguium with medium and high doses of ultramicrosize griseofulvin compared with that with itraconazole. *Antimicrob Agents Chemother* 37 (10), 2064, 1993.
82. Piepponen, T., Blomqvist, K., Brandt, H., Havu, V., Hollmen, A., Kohtamaki, K., Lehtonen, L., Turjanma, K., Efficacy and safety of itraconazole in the long-term treatment of onychomycosis. *J Antimicrob Chemother* 29, 195, 1992.
83. Rongioletti, F., Robert, E., Tripodi, S., Persi, A., Treatment of onychomycosis with itraconazole. *J Dermatol Treatm* 2, 145, 1992.
84. Hay, R.J., Clayton, Y.M., Moore, M.K., Midgely, G., An evaluation of itraconazole in the management of onychomycosis. *Br J Dermatol* 119, 359, 1988.
85. Gupta, A.K., De Doncker, P., Itraconazole for the treatment of onychomycosis: an overview, submitted.
86. Kagawa, S., A clinical evaluation of itraconazole, a new oral antifungal agent, in the treatment of tinea unguin, Candida onychomycosis and Candida onychia/paronychia. *Clin Rep* 25, 433, 1991.
87. Gupta, A.K., Scher, R.K., De Doncker, P., Current management of onychomycosis: an overview. *Dermatol Clin,* in press.
88. De Doncker, P.R.G., Scher, R.K., Baran, R.C., Decroix, J., Degreef, H.J., Roseeuw, D.I., Havu, V., Rosen, T., Gupta, A.K., Piérard, G.E., Itraconazole therapy is effective for pedal onychomycosis caused by some non-dermatophyte molds and in mixed infection with molds and dermatophytes: a multicenter study with 36 patients. *J Am Acad Dermatol,* in press.
89. Drayton, J., Dickinson, G., Rinaldi, M.G., Coadministration of rifampin and itraconazole leads to undetectable levels of serum itraconazole. *Clin Infect Dis* 18, 266, 1994.
90. Bonay, M., Jonville-Bera, A.P., Diot, P., Lemarie, E., Lavandier, M., Autret, E., Possible interaction between phenobarbital, carbamazepine, and itraconazole. *Drug Safety* 9, 309, 1993.
91. Imman, W., Kiyoshi, K., Pearce, G., Wilton, L., PEM report number 7. Itraconazole. *Pharmacoepidemiol Drug Safety* 2, 423, 1993.
92. Janssen Pharmaceutica Inc., Sporanox — Itraconazole capsules — Canadian product monograph, 1993.
93. Lavrijsen, A.P.M., Balmus, K.J., Nugteren-Huying, W.M., Roldaan, A.C., Van't Wout, J.W., Stricker, B.H., Leverbeschadiging tijdens gebruik van itraconazol (Trisporal). *Ned Tijdsch Geneeskd* 137 (1), 38, 1993.
94. Hann, S.K., Kim, J.B., Im, S., Han, K.H., Park, Y.K., Itraconazole-induced acute hepatitis. *Br J Dermatol* 129 (4), 500, 1993.
95. Gallardo-Quesonda, N., Hepatotoxicity associated with itraconazole. *Int J Dermatol* 34 (8), 589–591, 1995.
96. Lavrijsen, A.P.M., Balmus, K.J., Nugteren-Huying, W.M., Roldman, A.C., Wout, J.W., Stricker, B.H., Hepatic injury associated with itraconazole. *Lancet* 340, 251, 1992.

97. Hay, R.J., Risk/benefit ratio of modern antifungal therapy: focus on hepatic reactions. *J Am Acad Dematol* 29 (1), S50, 1993.

98. Heymann, W.R., Manders, S.M., Itraconazole-induced acute generalized exanthemic pustulosis. *J Am Acad Dermatol* 33, 130, 1995.

99. Van Cauteren, H., Lampo, A., Vandenberghe, J., Vanparys, P., Coussement, W., De Coster, R., Marsboom, R., Safety aspects of oral antifungal agents. *Br J Clin Pract* (Suppl. 71), 47, 1993.

100. De Coster, R., Beerens, D., Haelterman, C., Doolaege, R., Effects of itraconazole on the pituitary-testicular-adrenal axis: an overview of preclinical and clinical studies. In: *Recent Trends in the Discovery, Development and Evaluation of Antifungal Agents.* (Fromtling, R.A.) J.R. Prous Science Publishers, Barcelona, 251, 1987.

101. Perfect, J.R., Lindsay, M.H., Drew, R.H., Adverse drug reactions to systemic antifungals. *Drug Safety* 7, 323, 1992.

102. Kowalsky, S.F., Dixon, D.M., Fluconazole: A new antifungal agent. *Clin Pharm* 10, 179, 1991.

103. Goa, K.L., Barradell, L.B., Fluconazole: An update of its pharmacokinetic properties and therapeutic use in major superficial and systemic mycoses in immunocompromised patients. *Drugs* 50 (4), 658, 1995.

104. Pfaller, M.A., Dupont, B., Kobayashi, G.S., Muller, J., Rinaldi, M.G., Espinel-Ingroff, A., Shadomy, S., Troke, P.F., Walsh, T.J., Warnock, D.W., Standardized susceptibility testing of fluconazole: an international collaborative study. *Antimicrob Agents Chemother* 36, 1805, 1992.

105. Pfaller, M.A., Vu, Q., Lancaster, M., Espinel-Ingroff, A., Fothergill, A., Grant, C., McGinnis, M.R., Pasarell, L., Rinaldi, M.G., Steele-Moore, L., Multisite reproducibility of colorimetric broth microdilution method for antifungal susceptibility testing of yeasts isolates. *J Clin Microbiol* 32, 1625, 1994.

106. Anaisse, E.F., Karyotakis, N.C., Hachem, R., Dignani, M.C., Rex, J.H., Paetznick, Correlation between *in vitro* and *in vivo* activity of antifungal agents against *Candida* species. *J Infect Dis* 170, 384, 1994.

107. Dermoumi, H., *In vitro* susceptibility of fungal isolates of clinically important specimens to itraconazole, fluconazole, and amphotericin B. *Chemotherapy* (Basel) 40, 92, 1994.

108. Slavin, M.A., Osborne, B., Adams, R., Levenstein, M.J., Schoch, H.G., Feldman, A.R., Meyers, J.D., Bowden, R.A., Efficacy and safety of fluconazole prophylaxis for fungal infections after marrow transplantation: a prospective, randomized, double-blind study. *J Infect Dis* 171, 1545, 1995.

109. Wingard, J.R., Merz, W.G., Rinaldi, M.G., Miller, C.B., Karp, J.E., Sara, I.R., Association of *Torulopsis glabrata* infections with fluconazole prophylaxis in neutropenic bone marrow transplant patients. *Antimicrob Agents Chemother* 37, 1847, 1993.

110. Pfizer Canada Inc., Fluconazole Product Monograph, 1995.

111. Richardson, K., Brammer, K.W., Marriott, M.S., Troke, P.F., Activity of UK-49,858, a bis-triazole derivative, against experimental infections with *Candida albicans* and *Trichophyton mentagrophytes.* *Antimicrob Agents Chemother* 27, 832, 1985.

112. Grant, S.M., Clissold, S.P., Fluconazole: a review of its pharmacodynamic and pharmacokinetic properties, and therapeutic potential in superficial and systemic mycoses. *Drugs* 39, 877, 1990.

113. Karyotakis, N.C., Anaissie, E.J., Hachem, R., Dignani, M.C., Samonis, G., Comparison of the efficacy of polyenes and triazoles against hematogenous *Candida krusei* infection in neutropenic mice. *J Infect Dis* 168, 1311, 1993.

114. Cacciapuoti, A., Loebenberg, D., Parmegiani, R., Antonacci, B., Norris, C., Moss, E.L. Jr., Menzol, F. Jr., Yarosh-Tomaine, T., Hare, R.S., Miller, G.H., Comparison of SCH 39304, fluconazole and ketoconazole for treatment of systemic infections in mice. *Antimicrob Agents Chemother* 36, 64, 1992.

115. Rex, J.H., Rinaldi, M.G., Pfaller, M.A., Resistance of Candida species to fluconazole. *Antimicrob Agent Chemother* 39 (1), 1, 1995.

116. Hitchcock, C.A., Resistance of *Candida albicans* to azole antifungal agents. *Biochem Soc Trans* 21, 1039, 1993.

117. Shiba, K., Saito, A., Miyahara, T., Safety and pharmacokinetics of single oral and intravenous doses of fluconazole in healthy subjects. *Clin Ther* 121, 206, 1990.
118. Zimmermann, T., Yeates, R.A., Laufen, H., Pfaff, G., Wildfeuer, A., Influence of concomitant food intake on the oral absorption of two triazole antifungal agents, itraconazole and fluconazole. *Eur J Clin Pharmacol* 46, 147, 1994.
119. Thorpe, J.E., Baker, N., Bromet-Petit, M., Effect of oral antacid administration on the pharmaco-kinetics of oral fluconazole. *Antimicrob Agents Chemother* 34, 2032, 1990.
120. Blum, R.A., D'Andrea, D.T., Florentino, B.M., Wilton, J.H., Hiligoss, D.M., Gardner, M.J., Henry, E.B., Godstein, H., Schentag, J.J., Increased gastric pH and the bioavailability of fluconazole and ketoconazole. *Ann Intern Med* 114, 755, 1991.
121. Lim, S.G., Sawyer, A.M., Hudson, M., Sercombe, J., Pounder, R.E., Short report: The absorption of fluconazole and itraconazole under conditions of low intragastric acidity. *Aliment Pharmacol Ther* 7, 317, 1993.
122. Brammer, K.W., Farrow, P.R., Faulkner, J.K., Pharmacokinetics and tissue penetration of flucon-azole in humans. *Rev Infect Dis* 12 (Suppl. 3), S318, 1990.
123. Faergemann, J., Godleski, J., Laufen, H., Liss, R.H., Intracutaneous transport of orally administered fluconazole to the stratum corneum. *Acta Dermatol Venereol* (Stockh) 75, 361, 1995.
124. Hay, R.J., Pharmacokinetic evaluation of fluconazole in skin and nails. *Int J Dermatol* 31, 6, 1992.
125. Haneke, E., Fluconazole levels in human epidermis and blister fluid. *Br J Dermatol* 123, 273, 1990.
126. Faergemann, J., Laufen, H., Levels of fluconazole in serum, stratum corneum, epidermis-dermis (without stratum corneum) and eccrine sweat. *Clin Exp Dermatol* 18, 102, 1993.
127. Wildfeuer, A., Faergemann, J., Laufen, H., Pfaff, G., Zimmermann, T., Seidl, H.P., Lach P., Bio-availability of fluconazole in the skin after oral medication. *Mycoses* 37, 127, 1994.
128. Toon, S., Ross, C.E., Gokal, R., Rowland, M., An assessment of the effects of impaired renal function and hemodialysis on the pharmacokinetics of fluconazole. *Br J Clin Pharmacol* 29, 221, 1990.
129. Debruyne, D., Ryckelynck, J.P., Clinical pharmacokinetics of fluconazole. *Clin Pharmacokinet* 24, 10, 1993.
130. Dudley, M.N., Clinical pharmacology of fluconazole. *Pharmacotherapy* 10 (Suppl.), 141S, 1990.
131. Ruhnke, M., Yeates, R.A., Pfaff, G., Sarnow, E., Hartmann, A., Trautman, M., Single-dose phar-macokinetics of fluconazole in patients with liver cirrhosis. *J Antimicrob Chemother* 35, 641, 1995.
132. Suchil, P., Montero-Gei, F., Robles, F., Perera-Raminez, A., Welsh, O., Male, O., Once-weekly oral doses of fluconazole 150 mg in the treatment of tinea corporis/cruris and cutaneous candidiasis. *Clin Exp Dermatol* 17, 397, 1992.
133. Stengel, F., Galimberti, R., Suchil, P., Gei, F.M., Robles, M., Perera-Raminez, A., Male, O., Flucon-azole vs. ketoconazole in the treatment of dermatophytoses and cutaneous candidiasis. *Int J Der-matol* 33 (10), 726, 1994.
134. Montero-Gei, F., Perera, A., Therapy with fluconazole for tinea corporis, tinea crusis, and tinea pedis. *Clin Infect Dis* 14 (Suppl. 1), S77, 1992.
135. Halasz, C.L.G., Successful treatment with fluconazole of tinea corporis caused by *Trichophyton verrucosum* (Barn itch). *Cutis* 54, 207, 1994.
136. Del Aguila, R., Montero-Gei, F., Robles, M., et al., Once-weekly oral doses of fluconazole 150 mg in the treatment of tinea pedis. *Clin Exp Dermatol* 17, 402, 1992.
137. Kuokkanen, K., Alva, S., Fluconazole in the treatment of onychomycosis caused by dermatophytes. *J Dermatol Treatm* 3, 115, 1992.
138. Hochman, L.G., Scher, R.K., Meyerson, M.S., Cohen, J.L., Holwell, J.E., The safety and efficacy of oral fluconazole in the treatment of onychomycosis. *J Geriatr Dermatol* 1 (4), 169, 1993.
139. Smith, S.W., Sealy, D.P., Schneider, E., Lackland, D., An evaluation of the safety and efficacy of fluconazole in the treatment of onychomycosis. *South Med J* 87 (12), 1217, 1995.
140. Coldiron, B., Recalcitrant onychomycosis of the toenails successfully treated with fluconazole. *Arch Dermatol* 128, 909, 1992.

141. Köse, O., Fluconazole vs. itraconazole in the treatment of tinea versicolor. *Int J Dermatol* 34, 498, 1995.

142. Faergemann, J., Treatment of pityriasis versicolor with a single dose of fluconazole. *Acta Dermatol Venereol* (Stockh) 72, 74, 1992.

143. Shaw, J.T.B., Tarbit, M.H., and Troke, P.F., Cytochrome P-450 mediated sterol synthesis and metabolism: differences in sensitivity of fluconazole and other azoles. In: *Recent Trends in the Discovery, Development and Evaluation of Antifungal Agents.* (Fromtling, S.A.) J.R. Prous Science Publishers, Barcelona, Spain, 125, 1987.

144. Baciewicz, A.M., Baciewicz, F.A. Jr., Ketoconazole and fluconazole drug interactions. *Arch Intern Med* 153, 1970, 1993.

145. Lopez-Gil, J.A., Fluconazole-cyclosporine interaction: a dose-dependent effect? *Ann Pharmacother* 27, 417, 1993.

146. Gericke, K.R., Possible interactions between warfarin and fluconazole. *Pharmacotherapy* 13 (5), 508, 1993.

147. Assan, R., Fredj, G., Larger, E., Feutren, G., Bismuth, H., FK506/fluconazole interaction enhances FK506 nephrotoxicity. *Diab Metab* 20, 49, 1994.

148. Gussenhoven, M.J.E., Haak, A., Peereboom-Wynia, J.D.R., et al., Stevens-Johnson syndrome after fluconazole. *Lancet* 338, 120, 1991.

149. Neuhaus, G., Pavic, N., Pletscher, M., Anaphylactic reaction after oral fluconazole. *Br Med J* 302, 1341, 1991.

150. Morgan, J.M., Carmichael, A.J., Fixed drug eruption with fluconazole. *Br Med J* 308, 454, 1994.

151. Abbott, M., Hughes, D.L., Patel, R., Kinghorn, G.R., Angio-oedema after fluconazole. *Lancet* 338, 633, 1991.

152. Martinez-Roig, A., Torrez-Rodrigues, J.M., and Bartlett-Coma, A., Double-blind study of ketoconazole and griseofulvin in dermatophytoses. *Ped Infect Dis J* 7, 37, 1988.

153. Pappas, P.G., Kauffman, C.A., Perfect, J., Johnson, P.C., McKinsey, D.S., Bamberger, D.M., Hamill, R., Sharkey, P.J., Chapman, S.W., Sobel, J.D., Alopecia associated with fluconazole therapy. *Ann Intern Med* 123, 354, 1995.

154. Gradon, J.D., Sepkowitz, D.V., Fluconazole-associated acute adrenal insufficiency. *Postgrad Med J* 67, 1084, 1991.

155. Murakami, H., Katahira, H., Matsushima, T., Sakura, T., Tamura, J., Sawamura, M., Tsuchiya, J., Agranulocytosis during treatment with fluconazole. *J Int Med Res* 20, 492, 1992.

156. Chuncharunee, S., Sathapatayavongs, B., Singhasivanon, P., Singhasivanon, V., Fluconazole-induced agranulocytosis. *Therapiewoche* 49 (6), 517, 1994.

157. Agarwal, A., Sakhuja, V., Chugh, K.S., Fluconazole-induced thrombocytopenia. *Ann Intern Med* 113 (11), 899, 1990.

158. Mercurio, M.G., Elewski, B.E., Thrombocytopenia caused by fluconazole therapy. *J Am Acad Dermatol* 32, 525, 1995.

159. Sugar, A.M., Saunders, C., Oral fluconazole as suppressive therapy of disseminated cryptococcosis in patients with acquired immunodeficiency syndrome. *Am J Med* 85, 481, 1988.

160. Larsen, R.A., Leal, M.A., and Chan, L.S., Fluconazole compared with amphotericin B plus flucytosine for cryptococcal meningitis in AIDS. *Ann Intern Med* 113, 183, 1990.

161. Bozette, S.A., Larsen, R.A., Chiu, J., A placebo-controlled trial of maintenance therapy with fluconazole after treatment of cryptococcal meningitis in the acquired immunodeficiency syndrome. *N Engl J Med* 324, 580, 1991.

162. Robinson, P.A., Knirsch, A.K., Joseph, J.A., Fluconazole for life-threatening fungal infections in patients who cannot be treated with conventional antifungal agents. *Rev Infect Dis* 12 (Suppl. 3), S349, 1990.

163. Perfect, J.R., Lindsay, M.H., Drew, R.H., Adverse drug reactions to systemic antifungals. *Drug Experience* 7 (5), 1992.

164. Munoz, P., Moreno, S., Berenguer, J., Bernaldo de Quiros, J.C., Bouza, E., Fluconazole-related hepatotoxicity in patients with acquired immunodeficiency syndrome. *Arch Intern Med* 15, 1020, 1991.

165. Wells, C., Lever, A.M.L., Dose-dependent fluconazole hepatotoxicity proven on biopsy and rechallenge. *J Infect* 24, 111, 1992.

166. Trujillo, M.A., Galgiani, J.N., Sampliner, R.E., Evaluation of hepatic injury arising during fluconazole therapy. *Arch Intern Med* 154 (1), 102, 1994.

167. Jacobson, M.A., Hanks, D.K., Ferrell, L.D., Fatal acute necrosis due to fluconazole. *Am J Med* 96, 188, 1994.

168. Chmel, H., Fatal acute hepatic necrosis due to fluconazole. *Am J Med* 99, 224, 1994.

169. Holmes, J., Clements, D., Jaundice in HIV positive haemophiliac. *Lancet* 1, 1027, 1989.

170. Gearhart, M.O., Worsening of liver function with fluconazole and review of azole antifungal hepatotoxicity. *Ann Pharmacother* 28, 1177, 1994.

171. Birnbaum, J.E., Pharmacology of the allylamines. *J Am Acad Dermatol* 23, 782, 1990.

172. Ryder, N.S., Mechanism of action and biochemical selectivity of allylamine antimycotic agents. *Ann NY Acad Sci* 544, 208, 1988.

173. Ryder, N.S., Terbinafine: mode of action and properties of the squalene epoxidase inhibition. *Br J Dermatol* 126 (Suppl. 39), 2, 1992.

174. Ryder, N.S., The mechanism of action of terbinafine. *Clin Exp Dermatol* 14, 98, 1989.

175. Balfour, J.A., Faulds, D., Terbinafine. A review of its pharmacodynamic and pharmacokinetic properties, and therapeutic potential in superficial mycoses. *Drugs* 43, 259, 1992.

176. Jensen, J.C., Clinical pharmacokinetics of terbinafine (Lamisil). *Clin Exp Dermatol* 14, 110, 1989.

177. Battig, F.A., Nefzer, M., Schultz, G., Major biotransformation routes of some allylamine antimycotics. In: *Recent Trends in the Discovery, Development and Evaluation of Antifungal Agents.* (Fromtling, R.A.) J.R. Prous Science Publishers, Barcelona, 479, 1987.

178. Lever, L.R., Dykes, P.J., Thomas, R., Finlay, A.Y., How orally administered terbinafine reaches the stratum corneum. *J Dermatol Treatm* 1 (Suppl. 2), 23, 1990.

179. Faergemann, J., Zehender, H., Jones, T., Maibach, I., Terbinafine levels in serum, stratum corneum, dermis-epidermis (without stratum corneum), hair, sebum, and ecrine sweat. *Acta Dermatol Venereol* (Stockh) 71, 322, 1990.

180. Faergemann, J., Zehender, H., Millerioux, L., Levels of terbinafine in plasma, stratum corneum, dermis-epidermis (without stratum corneum), sebum, hair and nails during and after 250 mg terbinafine orally once daily for 7 and 14 days. *Clin Exp Dermatol* 19, 121, 1994.

181. Zehender, H., Denouël, J., Faergemann, J., Donatsch, P., Kutz, K., Humbert, H., Elimination kinetics of terbinafine from human plasma and tissues following multiple-dose administration, and comparison with 3 main metabolites. *Drug Invest* 8 (4), 203, 1994.

182. Del Palacio Hernanz, A., Gomez, S.L., Lastra, F.G., A comparative double-blind study of terbinafine (Lamisil) and griseofulvin in tinea corporis and tinea cruris. *Clin Exp Dermatol* 15, 210, 1990.

183. Cole, G.W., Stickin, G., A comparison of a new oral antifungal, terbinafine, with griseofulvin as therapy for tinea corporis. *Arch Dermatol* 125, 1537, 1989.

184. De Wit, R.F.E., A randomized double-blind multicentre comparative study of Lamisil (terbinafine) vs. ketoconazole in tinea corporis. *J Dermatol Treatm* 1 (Suppl. 2), 41, 1990.

185. Farag, A., Taha, M., Halim, S., One-week therapy with oral terbinafine in cases of tinea cruris/corporis. *Br J Dermatol* 131, 648, 1994.

186. White, J.E., Perkins, P.J., Evans, E.G.V., Successful 2-week treatment with terbinafine (Lamisil) for moccasin tinea pedis and tinea manuum. *Br J Dermatol* 125, 60, 1991.

187. Savin, R.C., Treatment of chronic moccasin-type tinea pedis with terbinafine: a double-blind, placebo-controlled trial. *J Am Acad Dermatol* 23, 804, 1990.

188. Savin, R.C., Oral terbinafine vs. griseofulvin in the treatment of moccasin-type tinea pedis. *J Am Acad Dermatol* 23, 807, 1990.

189. Hay, R.J., Logan, R.A., Moore, M.K., A comparative study of terbinafine vs. griseofulvin in "dry type" dermatophyte infections. *J Am Acad Dermatol* 24, 243, 1991.

190. Gip, L., Black piedra: the first case treated with terbinafine (Lamisil). *Br J Dermatol* 130 (Suppl. 43), 26, 1994.

191. Derrick, E.K., Voyce, M.E., Price, M.L., *Trichophyton tonsurans* kerion in an elderly woman. *Br J Dermatol* 683, 1995.

192. Gordon, P.M., Stankler, L., Rapid clearing of kerion ringworm with terbinafine. *Br J Dermatol* 129 (4), 503, 1993.

193. Goulden, V., Goodfield, M.J.D., Treatment of childhood dermatophyte infections with oral terbinafine. *Pediatr Dermatol* 12 (1), 53, 1995.

194. Jones, T.C., Overview of the use of terbinafine (Lamisil) in children. *Br J Dermatol* 132, 683, 1995.

195. Nejjam, F., Zagula, M., Cabiac, M.D., et al., Pilot study of terbinafine in children suffering from tinea capitis: evaluation of efficacy, safety and pharmacokinetics. *Br J Dermatol* 132, 98, 1995.

196. Haroon, T.S., Hussain, I., Mahmood, A., Nagi, A.H., Ahmad, I., Zahid, M., An open clinical pilot study of the efficacy and safety of oral terbinafine in dry non-inflammatory tinea capitis. *Br J Dermatol* 126 (39), 47, 1992.

197. Dragos, V., Podrumas, B., Karalj, B., Bartenjev, I., Efficacy of oral terbinafine treatment in tinea capitis in children caused by *Microsporum canis. J Eur Acad Dermatol Venereol* 5 (Suppl. 1), S171, 1995.

198. Zaias, N., Management of onychomycosis with oral terbinafine. *J Am Acad Dermatol* 23, 810, 1990.

199. Goodfield, M.J.D., Rowell, N.R., Forster, R.A., Evans, E.G.V., Raven, A., Treatment of dermatophyte infection of the finger- and toe-nails with terbinafine (SF 86-327, Lamisil), an orally active fungicidal agent. *Br J Dermatol* 121, 753, 1989.

200. Goodfield, M.J.D., Clinical results with terbinafine in onychomycosis. *J Dermatol Treatm* 1, 55, 1990.

201. Baudraz-Rosselet, F., Rakosi, T., Wili, P.B., Kenzlemann, R., Treatment of onychomycosis with terbinafine. *Br J Dermatol* 126 (Suppl. 39), 40, 1992.

202. Cribier, B., Grosshans, E., Efficacité et tolérance de la terbinafine (Lamisil) dans une série de 50 onychomycoses à dermatophytes. *Ann Dermatol Venereol* 131, 15, 1994.

203. Van der Schroeff, J.G., Cirkel, P.K.S., Crijns, M.B., Van Dijk, T.J.A., A randomized treatment duration-finding study of terbinafine in onychomycosis. *Br J Dermatol* 126 (Suppl. 39), 36, 1992.

204. Goodfield, M.J.D., Andrew, L., Evans, E.G.V., Short term treatment of dermatophyte onychomycosis with terbinafine. *Br Med J* 304, 1151, 1992.

205. Haneke, E., Tausch, I., Bräutigam, M., Weidinger, G., Welzel, D., Short-duration treatment of fingernail dermatophytosis: a randomized, double-blind study with terbinafine and griseofulvin. *J Am Acad Dermatol* 32, 72, 1995.

206. Hofmann, H., Bräutigam, M., Weidinger, G., Zaun, H., Treatment of toenail onychomycosis: A randomized, double-blind study with terbinafine and griseofulvin. *Arch Dermatol* 131, 919, 1995.

207. Faergemann, J., Anderson, C., Hersle, K., Hradil, E., Nordin, P., Kaaman, T., Molin, L., Pettersson, A., Double-blind, parallel-group comparison of terbinafine and griseofulvin in the treatment of toenail onychomycosis. *J Am Acad Dermatol* 32, 750, 1995.

208. Arenas, R., Dominguez-Cherit, J., Fernandez, L.M., Open randomized comparison of itraconazole vs. terbinafine in onychomycosis. *Int J Dermatol* 34 (2), 138, 1993.

209. Bräutigam, M., Nolting, S., Schopf, R.E., Weidinger, G., Randomised double-blind comparison of terbinafine and itraconazole for treatment of toenail tinea infection. *Br Med J* 311, 919, 1995.

210. De Backer, M., van Lierde, M.A., De Keyser, P., De Vroey, C., Lesaffre, E., A 12 weeks treatment for dermatophyte toe-onychomycosis: terbinafine 250 mg/d or itraconazole 200 mg/d? A double-blind comparative trial. *J Eur Acad Dermatol Venereol* 5, S188, 1995.

211. Tosti, A., Piraccini, B.M., Stinchi, C., Venturo, N., Colombo, M.D., Standard of pulse therapy with terbinafine. *J Eur Acad Dermatol Venereol* 5 (1), S188, 1995.

212. Sandoz, Lamisil terbinafine hydrochloride: Product Monograph. Sandoz Canada, Dorval Quebec, 1994.

213. Schuster, I., Metabolic degradatation of terbinafine in liver microsomes from man, guinea pig and rat. In: *Recent Trends in the Discovery, Development and Evaluation of Antifungal Agents.* (Fromtling, S.A.) J.R. Prous Science Publishers, Barcelona, 461, 1987.

214. Breckenridge, A., Clinical significance of interactions with antifungal agents. *Br J Dermatol* 126 (Suppl. 39), 19, 1992.

215. Beutler, M., Hartmann, K., Kuhn, M., Gartmann, J., Taste disorders and terbinafine. *Br Med J* 307, 26, 1993.

216. McGregor, J.M., Rustin, M.H.A., Terbinafine and erythema multiforme. *Br J Dermatol* 131 (4), 587, 1994.

217. Carstens, J., Wendelboe, P., Sogaard, H., Thestrup-Pedersen, K. Toxic epidermal necrolysis and erythema multiforme following therapy with terbinafine. *Acta Dermatol Venereol* (Stockh) 74, 391, 1994.

218. Todd, P., Halpern, S., Munro, D.D., Oral terbinafine and erythema multiforme. *Clin Exp Dermatol* 20, 247, 1995.

219. White, S.I., Bowen-Jones, D., Toxic epidermal necrolysis induced by terbinafine in a patient on long-term anti-epileptics. *Br J Dermatol* 134, 148, 1996.

220. Rzany, B., Mockenhaupt, M., Gehring, W., Schöpf, E., Stevens-Johnson syndrome after terbinafine therapy. *J Am Acad Dermatol* 30 (3), 509, 1994.

221. Kruczynski, K., and Balter, M.S., Serum sickness-like reaction associated with oral terbinafine therapy. *Can J Clin Pharmacol* 2, 129, 1995.

222. Munn, S.E., Russell Jones, R., Terbinafine and fixed drug eruption. *Br J Dermatol* 133, 815, 1995.

223. Wach, F., Stolz, W., Rüdiger, R., Landthaler, M., Severe erythema anulare centrifugum-like psoriatic eruption induced by terbinafine. *Arch Dermatol* 131, 960, 1995.

224. van 't Wout, J., Hermann, W.A., de Vries, R.A., Stricker, B.H., Terbinafine-associated hepatic injury. *J Hepatol* 21, 115, 1994.

225. Lowe, G., Green, C., Jennins, P., Hepatitis associated with terbinafine teratment. *Br Med J* 306, 248, 1993.

226. Health Protection Branch, Canadian Adverse Drug Reaction Newsletter. *Can Med Assoc J* 151(1), 63, 1994.

227. Kovacs, M.J., Alshammari, S., Guenther, L., Bourcier, M., Neutropenia and pancytopenia associated with oral terbinafine. *J Am Acad Dermatol* 31, 806, 1994.

228. Krowchuk, D.P., Lucky, A.W., Primmer, S.I., McGuire, J., Current status of the identification and management of tinea capitis. *Pediatrics* 72(5), 625, 1983.

229. Tanz, R.R., Herbert, A.A., Burton Esterly, N., Treating tinea capitis: Should ketoconazole replace griseofulvin? *J Pediatr* 112, 987, 1988.

230. Gan, V.N., Petruska, M., Ginsburg, C.M., Epidemiology and treatment of tinea capitis: ketoconazole vs. griseofulvin. *Ped Infect Dis J* 6, 46, 1987.

231. Jue, S.G., Dawson, G.W., Brogden, R.N., Ciclopirox olamine 1% cream: a preliminary review of its antimicrobial activity and therapeutic use. *Drugs* 29, 330, 1985.

232. Qadripur, V.S.A., Horn, G., Höhler, T., Zur Lokalwirksamkeit von Ciclopiroxolamin bei Nagelmykosen. *Drug Res* 31(II), 1360, 1981.

233. Zaug, M., Bergstraesser, M., Amorolfine in the treatment of onychomycosis and dermatomycoses (an overview). *Clin Exp Dermatol* 17(Suppl. 1), 61, 1992.

18

Drugs Used in Reproductive Endocrinology and Infertility

Joseph F. Mortola
Harvard Medical School
Beth Israel Hospital

18.1   Introduction

Therapeutic agents used in reproductive endocrinology and infertility are in almost all cases either naturally occurring hormones or agonists and antagonists derived from the native compounds. At present, agonists and antagonists have been developed for the hypothalamic hormone gonadotropin releasing hormone (GnRH) and for each of the major sex steroids produced by the gonads: estrogen, progesterone, and testosterone. Several physiologic principles underlie the clinical use of the these agents. One of these is the requirement of an intact hypothalamic-pituitary-gonadal axis for normal reproductive function. Within this axis, the hypothalamic production of GnRH stimulates the synthesis and secretion of gonadotropins, luteinizing hormone (LH) and follicle stimulating hormone (FSH). LH and FSH in turn stimulate the gonadal production of the major sex steroids. Any alteration of this axis may result in either absent or exaggerated responses within the reproductive system. Perhaps the simplest perturbation of this system occurs when a hormone is given in excess. This is done in the case of gonadotropins for ovulation induction, where exogenous LH and FSH are administered. A second method of perturbing the system is to employ the principle of the negative feedback by sex steroids on gonadotropins and GnRH. This principle can be employed either to suppress gonadal function by supplying an agonist of sex hormone, as in the oral contraceptive pill, or to augment gonadal function by administering an antagonist of the sex steroid as with clomiphene. A third method of perturbing the axis is to make use of the principle of receptor desensitization. This is extensively utilized in reproductive medicine to decrease gonadotropin secretion using potent GnRH agonists. Use of one of these three strategies under-lies the bulk of the rationale for therapies in reproductive endocrinology and infertility. With the notable exception of the treatment of hyperprolactinemia, where use is made of the tonic suppression of the hormone by dopamine, this chapter will address the application of these principles to specific clinical problems.

18.2   Ovulation and Spermatocyte Induction

The use of ovulation induction has achieved a prominent role in the treatment of infertility. The two major indications for the use of ovulation induction are oligo-ovulation, based either on central or peripheral causes, and infertility due to a variety of causes when the woman is regularly ovulatory. In the case of oligo-ovulatory or anovulatory patients, the choice of the ovulation induction agent is dependent on whether the etiology of the anovulation is hypogonadotropism or secondary to ovarian physiologic disturbances as in polycystic ovary syndrome. In the case of women who are infertile for reasons other than oligoovulation, the therapy is employed to increase the fertility chances by inducing several mature oocytes per month. This technique is used either alone or in combination with assisted reproductive technologies such as *in vitro* fertilization or gamete intrafallopian transfer. In the case of assisted reproductive technologies success rates are greatly enhanced with the use of ovulation induction agents to produce up to 12 oocytes as compared to the single oocyte that would be achieved per attempt without these agents.

By analogy to the production of excess mature oocytes, the same agents may be used to increase sperm production, albeit with a much more limited role. In this case, the use of the therapy is most effective for men who have a hormonal etiology for decreased spermatogenesis as occurs in congenital deficiency in GnRH (Kallman's syndrome) or central nervous system lesions that decrease gonadotropins. More limited efficacy is observed in cases where there is a decrease in spermatogenesis of uncertain etiology.

18.2.1   Clomiphene

Clomiphene is a nonsteroidal triphenylethylene derivative[1] that exerts both estrogen agonist and estrogen antagonist activity.[2] It is composed of two isomers, enclomiphene and zuclomiphene,[3] in a racemic mixture that is typically 62% of the former and 38% of the latter. The enclomiphene isomer is responsible for all or the majority of the ovulation inducing effects of the agent.[4] In contrast, the zuclomiphene isomer has a markedly longer half-life, and traces of clomiphene citrate are detectable in the feces up to 6 weeks after administration, suggesting a marked enterohepatic circulation.[5] The majority of the actions of clomiphene are exerted through its interactions with the estrogen receptor. This drug has been shown to occupy the estrogen receptor for several weeks; the most potent native ligand for the estrogen receptor, estradiol, binds for a maximum of only 24 hours.[6] Whether clomiphene acts primarily as an estrogen agonist or antagonist depends both on the circulating levels of more potent estrogens, such as estradiol or estrone, and on the tissue studied. In low estrogen states such as menopause, clomiphene appears to have primarily estrogenic effects in that it reduces the degree of elevation of gonadotropins, which are the hallmark of menopause,[7-9] and has estrogenic effects on vaginal cytology.[10] In contrast, in premeno-pausal women *antiestrogen* actions predominate as evidenced by the drug's propensity to induce hot flashes,[11] decrease cervical mucus,[12] and reverse estrogen-induced changes in endometrial histology.[13,14] With respect to specific tissues, clomiphene tends to act as an antiestrogen at the hypothalamus, and as an estrogen at the pituitary and ovary.[15] The effects in enhancing ovulation are likely to reflect actions at all of these levels.

The usual method of clomiphene administration is to begin with a 50 mg dose for 5 days. In sponta-neously occurring cycles, the starting day is from 3 to 5 after the first day of menstrual bleeding. In anovulatory or oligoovulatory women, the treatment is usually started 3 to 5 days following the start of a progestin withdrawal-induced bleeding episode. Typically, the progestin therapy is a 7 to 10 day course of medroxyprogesterone acetate. There is little doubt that this regimen often results in the ovulation of more than one oocyte, typically two. In cases where ovulation does not occur, the dose is increased by 50 mg for each of the 5 days in successive months up to a maximally effective dose of 250 mg. In order to document ovulation with greater accuracy than the presence of a menstrual period at the expected time, a variety of other methods can be employed, including basal body temperature charts, urinary detection kits for an LH surge, or luteal phase progesterone determination.

The major benefit of clomiphene therapy is that it is highly effective in improving pregnancy rates in well-selected patients. In oligo- or anovulatory patients, more than 80% will ovulate regularly with clomiphene and pregnancy rates of 80% to 90% have been reported.[16,17] Success rates are much poorer when poorly selected patients are included. In particular, patients with normal menstrual cycles and no other cause for infertility have pregnancy rates that are no higher than spontaneously ovulating women.[18] Administration to normal fertile women does not increase the speed at which pregnancy is achieved as compared to untreated women.[19]

The drug has been used empirically in treatment of subfertile men. However, controlled clinical trials have failed to demonstrate its efficacy.[20]

Administration has been associated with a number of undesirable side effects. These include hot flashes in 11% of women.[11] Visual symptoms, particularly scintillating scotomata, occur in approximately 2% of patients. The mechanism of hot flashes is thought to reflect the antiestrogenic action; the mechanism for the visual symptoms, however, is not known. There have been isolated case reports of severe visual disturbance using clomiphene.[21]

Since clomiphene is associated with the increased stimulation of the ovary, probably by increasing FSH levels, it is not surprising that the incidence of twins is five- to tenfold higher than in spontaneous cycles. Seven percent of all pregnancies achieved while on clomiphene are twin gestations.[22]

There have been persistent reports of poor cervical mucus in patients taking this drug.[23-28] Despite the unfavorable effects of poor cervical mucus on pregnancy rates, the overall pregnancy rate is improved on clomiphene in anovulatory women.

Ovarian enlargement occurs in approximately 5% of individuals taking clomiphene. This may be accompanied by abdominal pain, distention, or soreness. Nausea and vomiting (2.2%), headache (1.3%), and dryness or loss of hair (0.3%) occur less frequently. Depression and mood changes have also been reported, although the incidence of these side effects is not well determined. The occurrence of headache is usually considered a contraindication to continuing the medication because of the theoretical risk of stroke that has been associated with administration of estrogenic compounds. Although a pelvic examination is routinely performed by many practitioners at the end of a clomiphene cycle, the incidence of significant ovarian enlargement that does not spontaneously resolve is negligibly small, and this precaution is probably unnecessary.

Considerable attention has focused on the possible teratogenicity of clomiphene. This is of particular concern given the extremely long retention of the drug, particularly in estrogen sensitive tissues, and the well-known serious sequelae of another nonsteroidal estrogen compound, diethylstilbestrol. Although it is virtually certain that clomiphene is still present in the mother's body during early gestation when pregnancy is achieved, there is excellent data that the agent is not teratogenic when used according to the protocol described above. In a study of more than 2600 births, the incidence of birth defects was 2.4% when ovulation induction with clomiphene was associated with pregnancy, as compared to 2.7% in the general population.[29] Nonetheless, there is concern when this agent is inadvertently administered during pregnancy. In the same study, of a total of 58 birth defects observed, 8 occurred in 7 out of 15 mothers who took the medication during pregnancy. Thus it is extremely important that the clinician is certain that the patient is not pregnant before starting the drug or initiation of subsequent cycles of the medication.

The rate of pregnancy loss with clomiphene has been widely studied. It is clear that this is not increased over the general population.[30,31] Although there are reports of increased rates of hydatidiform mole and ectopic pregnancies,[32,33] the association remains unproven.

Recently, there has been extreme concern regarding the increased risk of ovarian cancer in women who have used any type of ovulation induction agent. This association has been reported to be particularly strong, however, with more than 12 cycles of clomiphene. Although current studies are in progress to assess whether a true association exists between ovulation induction agents and ovarian cancer, a metaanalysis of currently available data suggests that the risk of ovarian tumors of borderline malignant potential, rather than the risk of frank ovarian malignancy is increased in infertile women taking these agents. What has not been determined is whether the infertility per se or the use of ovulation induction agents is the pertinent risk factor. Until further clarification is available, it is wise to counsel patients that the use of ovulation induction may be associated with an increased risk of ovarian borderline neoplasms.

Overall, clomiphene is an extremely effective and well tested agent for ovulation induction. The ease of administration by the oral route is a distinct advantage, as is the low rate of serious ovarian hyperstimulation. When used as directed in patients who are determined not to be inadvertently pregnant, extensive data suggest the agent is not teratogenic. The risks are relatively minor and appear to be almost universally reversible.

18.2.2   Human Menopausal Gonadotropins

Human menopausal gonadotropins (hMG) constitute the mainstay of ovulation induction therapy. They are indicated in patients who fail to ovulate on clomiphene, patients who are not candidates for clomiphene (particularly those with central hypogonadotropic hypogonadism), and for controlled ovarian hyperstimulation in patients with endometriosis, male factor infertility, and infertility of unexplained etiology. In controlled ovarian hyperstimulation, the efficacy of the agent is improved if combined with intrauterine sperm insemination.[34] hMG therapy is also used in almost all cases of assisted reproductive technologies including *in vitro* fertilization and gamete intrafallopian transfer.

Two forms of hMG exist, those that contain LH and FSH in approximately equal concentrations, and those that are relatively purified preparations of FSH (>60:1 FSH/LH). Even more highly purified forms of FSH are available, and a recombinant form of FSH that is virtually 100% pure is currently in clinical trials. Despite considerable hope that purified forms of FSH may be superior to standard LH/FSH combinations, particularly in women with polycystic ovary syndrome, where the tendency for excessive ovarian stimulation is high, neither pregnancy rates nor complication rates differ between currently marketed preparations.[35] It is clear, however, based on the success of controlled ovarian hyperstimulation using FSH alone, that LH is not clinically necessary in the vast majority of patients. The only exception to this is patients with a complete lack of endogenous LH as in isolated gonadotropin deficiency or severe hypothalamic-pituitary damage.

The usual protocol for administration of hMG, regardless of the preparation, is to begin with 150 to 225 IU of FSH by intramuscular injection, beginning at the start of a menstrual cycle or after a progestin-induced withdrawal bleed in oligoovulatory patients. The dose is changed after 3 days based on ultrasound examination of the extent of follicular development in the ovaries and serum estradiol. Thereafter, dosage adjustments are made every 1 to 3 days.[36] When at least three follicles have achieved a size on ultrasound greater than or equal to 16 mm, 5,000 to 10,000 units of human chorionic gonadotropin (hCG) are administered to mimic the LH surge and induce ovulation. The luteal phase of the menstrual cycle is then supported with exogenous progesterone. A pregnancy test is performed 12 to 14 days after hCG administration.

Using the standard protocol, pregnancy rates of approximately 20% per cycle are appreciated in patients who were previously anovulatory.[37] Modifications to the standard protocol include the pretreatment with GnRH agonists or use of lower doses, the latter regimen being commonly used in patients with polycystic ovary syndrome. Pretreatment with GnRH agonist has been shown in some,[38,39] but not all,[40] studies to improve cycle fecundity. In patients with polycystic ovary syndrome treated with low-dose hMG, pregnancy rates were similar, although the multiple pregnancy rates and rates of ovarian hyperstimulation were decreased.[41]

The principle side effects of hMG therapy include multiple gestations, ectopic pregnancies, ovarian hyperstimulation syndrome, and the possible risk of increased ovarian cancer. As mentioned previously, (see clomiphene), the association between ovulation induction agents and ovarian cancer has not been proven.

Multiple pregnancies constitute the major detriment to hMG therapy. Of women who become pregnant on hMG therapy, multiple pregnancy rates between 14% and 50% are reported.[37,42-44] The rate of twin pregnancies is reported on average to be between 20% to 25%, that of triplets to be approximately 5%, and higher-order gestations 1% to 2%.[45]

Ectopic pregnancies are also higher using hMG as compared with the spontaneous rate of approximately 1%. Ectopic pregnancy rates in the range of 5% have generally been reported.[46,47] Whether this fivefold increase is truly attributable to hMG or whether the patient population who received hMG was at higher risk for ectopic pregnancy has not been determined.

The most feared complication of hMG use is the development of ovarian hyperstimulation syndrome. It invariably develops after the hMG therapy is discontinued, usually 1 to 2 weeks later. The syndrome is characterized by ovarian enlargement, ascites, weight gain, oliguria, hemoconcentration, and at times pleural effusion. The severe form of the disorder is characterized by ovarian enlargement greater than 12 cm, clinical ascites, decreased creatinine clearance, serum creatinine greater than 1.0 mg/dL, and a hematocrit greater than 45%.[48,49] More recently a life-threatening form of the disorder has been classified that includes a creatinine >1.6 mg/dL, a creatinine clearance <50 mL/min, hematocrit >55%, thromboembolic phenomena, adult respiratory distress syndrome, and renal failure. The pathophysiology of ovarian hyperstimulation syndrome is increased capillary permeability of mesothelial surfaces.[50] Implicated in this regard are the neovascularization that accompanies corpus luteum formation,[51] increased activity of the renin-angiotensin cascade,[52] histamine release,[53] cytokines,[54] prostaglandins,[55] and vascular permeability factors.[56] Despite the attempts to better understand the pathophysiology of the syndrome, specific treatments have not yet been developed.

The predominant risk factor is the presence of a spontaneous or pharmacologic (hCG mimicked) LH surge.[57] Other risk factors include the establishment of a pregnancy during that cycle, serum estradiol concentrations in excess of 4000 pg/ml in assisted reproductive technology cycles, or 1700 pg/ml in ovulation induction cycles, a large number of follicles (>35 in the former and >6 in the latter case), young age (<35), polycystic ovary syndrome, and an asthenic habitus. The higher levels of estradiol and larger number of follicles required to place an individual at significant risk for ovarian hyperstimulation syndrome in assisted reproductive technology as compared to ovulation induction cycles are thought to be due to the deleterious effects on corpus luteum formation that result from surgically induced follicular hemorrhage attendant upon follicular aspiration. The most reliable method of preventing ovarian hyperstimulation syndrome is withholding hCG when risk factors are present.[58]

Treatment of the syndrome includes the consideration of diuretics or a combination of albumin and diuretics, although the benefit of these has not been conclusively demonstrated. Ultrasound guided paracentesis has been shown to be beneficial in severe cases.[59] In life-threatening cases with a severe renal hypoperfusion, dopamine must be considered along with invasive monitoring because of the precarious hemodynamic status. At times, a termination of the pregnancy, if there is one, must be considered.

The use of hMG has revolutionized the field of infertility and has provided the chance for pregnancy when no other reasonable likelihood exists. This is particularly true when combined with assisted reproductive technologies. Nonetheless, the risk of multiple gestations should be carefully considered. Although ovarian hyperstimulation in the severe form is rare, it must be considered a formidable risk.

18.2.3 Pulsatile Gonadotropin Releasing Hormone

Pulsatile gonadotropin releasing hormone is a reasonable alternative to human menopausal gonadotropin therapy in anovulatory women with hypothalamic dysfunction. Patients suitable for this therapy include individuals with low gonadotropins but an *intact* pituitary.[60] Most frequently these patients include those with life-style induced (functional) hypothalamic amenorrhea such as athletic women. It is also suitable for women born with gonadotropin releasing hormone deficiency (called isolated gonadotropin deficiency). Because of the extreme propensity for gonadotropin releasing hormone to downregulate its own receptor (see Section 18.4.2), any long-acting preparation of gonadotropin releasing hormone quickly changes from an agonist to an agent that acts as a virtual antagonist. This is obviously completely unsuitable for ovulation induction. To circumvent this, gonadotropin releasing hormone must be administered in pulsatile fashion that mimics the endogenous profile of hypothalamic release of the hormone.

The use of pulsatile gonadotropin releasing hormone is highly effective for ovulation induction for properly selected patients, where ovulation rates reach 85% to 90% per cycle and pregnancy rates approach normal fecundity.[61] A distinct advantage of the use of this agent is the low risk of multiple gestations as compared to human menopausal gonadotropins. The requirement of a portable pump to deliver GnRH in pulsatile fashion, however, has met with less patient satisfaction than daily human menopausal gonadotropin injection, and it is therefore less widely used.

18.3   Drugs Used for Hyperprolactinemia

Hyperprolactinemia has been observed in up to 30% of women with infertility[62] and 10% of women with primary amenorrhea.[63] As such, it is a relatively common disorder. The causes of hyperprolactinemia are several. In particular, these include a variety of psychotropic and antihypertensive medications, chest wall lesions, hypothyroidism, estrogen excess states such as pregnancy and oral contraceptive pill use, chronic renal failure, cirrhosis, adrenal disease, and central nervous system (CNS) causes. Among the CNS causes are tumors, particularly in the area of the hypothalamus, granulomatous diseases, acromegaly, and most commonly prolactinomas. Several treatments of hyperprolactinemia are available. In general, the principle is to treat the underlying cause when possible. These causes include such entities as CNS tumors, granulomatous disease, endocrine disorders, or pharmacotherapy. In the majority of other cases, the hyperprolactinemia is due to prolactinoma. Hyperprolactinemia need not be treated provided the patient does not have amenorrhea. Hyperprolactinemic women who are amenorrheic are at risk for osteoporosis.[64] Although a subject of rather recent controversy, it has now been shown that hyperprolactinemia in the absence of amenorrhea is not associated with osteoporosis.[65,66] Even if the patient is diagnosed with a microadenoma (<1 cm) the option not to treat should be strongly considered, since the overall progression from microadenoma to macroadenoma is 6.9% over a period of up to 8 years.[67] In the case where treatment is not instituted, surveillance is required.

Surgical therapy employing transsphenoidal hypophysectomy is used for macroadenomas and some microadenomas; however, more than 30% of patients with normal pituitary function prior to surgery develop some degree of pituitary insufficiency after the surgery. Because medical therapy is effective in most cases, regardless of the size of the prolactinoma, it should be the first line of therapy. With regard to medical therapy, the dopamine agonists are clearly the most successful agents. The rationale for this therapy is based on the potent physiologic inhibitory effects of dopamine on prolactin secretion.[68]

18.3.1   Bromocriptine

Bromocriptine is an ergot derivative with potent dopaminergic activity. In a study of more than 400 hyperprolactinemic women, ovulatory menses returned in 90%.[69] In a metaanalysis of 236 patients 77% had a reduction in tumor size during observation periods that lasted 6 weeks to more than 10 years.[70] The most usual response is for the tumor size reduction to continue over many years. In those with visual field abnormalities as a result of compression of the optic chiasm by the prolactinoma, improved visual acuity is observed in 90%, and often precedes discernable decreases in tumor size on imaging studies.[71] In individuals with other pituitary dysfunction as a result of pressure effects from prolactinomas, restoration of normal pituitary function is usually achieved.[72]

Dosages of bromocriptine are usually begun at 1.25 mg at bedtime. The bedtime dose is particularly effective because of minimization of side effects, and because there is a normal sleep-entrained rise of prolactin, which is best eliminated by this regimen. The dose is gradually increased every 7 days until prolactin levels normalize. In general, if the required dose is higher than 1.25 mg, which it is in most cases, a b.i.d. dosing schedule is preferred. A maximum dose of bromocriptine that may be safely used has not been established; rarely doses higher than 7.5 mg/d are required.

The predominant side effect of bromocriptine use is nausea and vomiting. Nausea is common and occurs in up to 45% of patients. Headaches are reported in 19%, dizziness in 17%, fatigue and light-headedness in 5%, and vomiting in 5%. These side effects are minimized by taking the medication with a snack at bedtime. Digital spasm, orthostatic hypotension, and nasal congestion are rare at doses lower than 7.5 mg/day and occur in less than 3% of patients. Psychotic reactions have been reported in less than 1% of patients. Although there are isolated reports of cerebrovascular accidents in women who take bromocriptine for milk reduction postpartum, this has not been reported in other circumstances. The routine use of bromocriptine to prevent milk production in postpartum women who do not wish to breast-feed should be discouraged.

A valuable technique to reduce the nausea associated with bromocriptine is intravaginal administration of the oral tablets. Similar reductions of prolactin levels to those observed following oral intake occur with vaginal administration.[73,74] The drug effect lasts up to 24 hours after a single dose.

Bromocriptine has not been shown to have teratogenic effects in humans.[75] In addition, the rates of multiple pregnancies, ectopic pregnancies, spontaneous abortions, or trophoblastic disease[76] are not increased. Nonetheless, experience with the medication in pregnancy has not been extensive, and it is recommended that the medication be stopped when pregnancy is diagnosed. Despite the increase in hyperprolactinemia invariably seen in pregnancy, women managed in this way have shown remarkably good results with careful monitoring.

Overall, bromocriptine is an extremely effective drug for the treatment of hyperprolactinemia and prolactinomas. It is also remarkably free of serious sequelae. Nonetheless, the side effects of the medication can cause extreme discomfort. This has prompted a search for newer agents of equal efficacy that will be better tolerated.

18.3.2 Newer Agents

Pergolide is a dopamine agonist that has recently been approved by the Food and Drug Administration (FDA) for treatment of Parkinsonism. Although not approved for hyperprolactinemia, there is extensive experience with the drug for this indication. Hyperprolactinemia is well controlled with doses of 50 µg–150 µg/d. Several studies have shown both similar efficacy and a similar side effect profile to bromocriptine.[77-79] In addition, it has the disadvantage of having a higher incidence of cardiac arrhythmias.

Quinagolide is a nonergot dopamine agonist that has been shown in limited studies to have an efficacy rate and side effect profile equal to bromocriptine and pergolide. As with pergolide, there are some patients who tolerate this alternative better than bromocriptine.[80]

Cabergoline has recently been tested in a large study in the United Kingdom[81] in direct comparison with bromocriptine. In this study, stable normoprolactinemia was achieved in significantly more women using cabergoline than bromocriptine (83% vs. 59%). Ovulatory cycles also returned in significantly more of the cabergoline group (72% vs. 52%). Although both drugs had gastrointestinal side effects, the symptoms were much better tolerated in the cabergoline group, where only 3% discontinued therapy as compared to 12% in the bromocriptine group. Overall, cabergoline appears to be a superior medication to the more commonly used bromocriptine. Its use in the United States is limited by the lack of approval.

18.4 Drugs Used for Ovulation Suppression

Certainly, the most commonly used medications to suppress ovulation are oral contraceptive pills. These are available in a wide variety of formulations, the two major categories of which are combined estro-gen/progestin formulations and progestin only type, "the minipill." Of these, the estrogen/progestin combination pills are much more reliable at suppressing ovulation. Nonetheless, the contraceptive efficacy of the progestin only formulation is similar to the combined formulation because of the deleterious effects of chronic unopposed progesterone on both cervical mucus and endometrial histology, both of which inhibit fertilization and implantation. Depot medroxyprogesterone acetate given as a single monthly injection of 150 mg every 3 months, and levonorgestrel in silastic implants changed every 5 years are parenterally administered, long-acting variants of the oral minipill. Depot medoxyprogesterone acetate, however, much more reliably suppresses ovulation than levonorgestrel implants or the progestin-only oral contraceptive.

Oral contraceptives suppress ovulation through administration of progestin, with or without estrogen. Similar effects are achieved with androgenic agents such as danazol. In contrast, newer GnRH analogs differ substantially in their mechanisms of action in that no active sex steroid is used.

In addition to contraception, ovulation suppression is highly desirable in a variety of sex-steroid influenced conditions. These include uterine myomata, which are both estrogen and progesterone sensitive; endometriosis, which is estrogen sensitive; precocious puberty in both boys and girls; and prostate cancer in men. In these conditions, the selection of an agent that does *not* include exogenous sex steroids is very much more desirable. In these cases GnRH agonists have greater efficacy. In the case of endometriosis where progestin inhibits endometrial proliferation, medroxyprogesterone acetate is as effective as GnRH agonists. The oral contraceptive, although less effective than the other agents, is also useful because of the continuous administration of progestin along with estrogen.

18.4.1 Oral Contraceptive Pills

The overall efficacy of oral contraceptive pills in preventing pregnancy is remarkable, with a failure rate of only 2 per 100 years of women use.[82] Thus the efficacy of the method continues to be one of its strongest advantages.

Exhaustive studies of the short- and long-term risks and benefits of oral contraceptives have been conducted.[83,84] Based on these studies there is evidence for an increase in thromboembolic events in women on oral contraceptives.[85] In addition, there appears to be a slightly increased risk of breast cancer in nulliparous women who start the pill early in life (before the age of 25) and continue the pill for durations longer than 10 years.[86] These risks, however, appear to be offset by the decreased risk of ovarian cancer, uterine cancer, and the risks attendant upon pregnancy.[87] It is now apparent that women under the age of 35, and women over the age of 35 who do not smoke, are at no greater risk for cardiovascular death than nonpill users. Insulin resistance seems to be somewhat increased in oral contraceptive users; however, this is only a problem in particularly brittle diabetics.[88]

Many women experience headaches, some weight gain, and fluid retention symptoms on oral contraceptives. Although these are bothersome, they are rarely a cause for concern. Less than 1% of women are unable to tolerate the birth control pill as a result of migraine headaches.

Overall, the risks associated with oral contraceptive use are small compared to the benefits of preventing undesired pregnancy. In addition, there is substantial evidence that it is a safe and well-tolerated treatment for endometriosis.[89]

18.4.2 GnRH Agonists

The use of GnRH agonists for ovulation suppression depends on its ability to cause pituitary desensitization to GnRH. The efficacy of GnRH agonist administration has been demonstrated for endometriosis,[90] uterine myomas[91] prostate cancer[92] precocious puberty[93] and premenstrual syndrome.[94] The desensitization of the pituitary is thought to be the result of internalization of the GnRH receptor.[95,96] Depending on the potency of the agonist, the desensitization phase (termed down-regulation) requires 7 to 21 days. Once down-regulation has been established, it persists for as long as the agonist is administered. During down-regulation, LH and FSH secretion by the pituitary is substantially reduced. As a result, there is insufficient stimulation of the ovary for normal sex steroid production. Circulating estrogen levels are therefore in the postmenopausal range, and progesterone levels are similarly low.

The daily subcutaneous injection form of GnRH agonists as described is cumbersome for the patient. Administration of GnRH analogs may be associated with localized pain and irritation at the injection site. More recently, depot formulations of the compounds have become available. These are administered as monthly intramuscular injections. A nasal spray has also been formulated that is available for two or three times daily use. Although less uncomfortable than subcutaneous administration, absorption may be somewhat more erratic.

In women, the side-effect profile of GnRH agonists is largely the result of hypoestrogenism. Most women on the medication experience significant hot flashes. These are generally classical postmenopausal hot flashes that last on the order of minutes and tend to be more pronounced on the upper torso and face.[97] The hot flash is accompanied by an increased pulse rate and vasomotor instability. Often, the patient will report awakening in the middle of the night drenched in perspiration. These hot flashes tend to be most bothersome at the initiation of the down-regulation phase. In some women, they continue to be highly disturbing, while in others their perceived severity decreases over a period of weeks to months.

The acute menopausal syndrome, in addition to hot flashes includes emotional lability and insomnia.[98] In general, these symptoms tend to be less disturbing than the symptoms of the disease being treated.

The long-term use of GnRH analogs is limited by the effects of chronic hypoestrogenism. The most pronounced of these is osteoporosis.[98] As a result, GnRH analogs are limited to a period of 6 months unless accompanied by serial bone densitometry to demonstrate maintenance of bone integrity. In premarketing trials, bone pain has been reported in some men. The incidence of this symptom in women is not well known.

In addition to osteoporosis, there is concern regarding the long-term consequences of negating the putative protective effect of estrogen on cardiovascular disease in women. Estrogen has been demonstrated to exert beneficial effects on the high-density lipoprotein (HDL) to low-density lipoprotein (LDL) ratio.[99,100] Moreover, there is evidence that estrogen may have additional protective effects on vasculature, independent of alterations in the lipid profile.[101] Large epidemiologic studies will be required to quantify this risk. Other symptoms of menopause, while not posing health risks, are also of concern when prescribing GnRH analogs to women. These include vaginal dryness, an increase in urinary tract symptoms, and decreases in skin collagen content.

Given the association of breast and gynecologic cancers with ovarian function, there is a potential protective effect of long-term GnRH agonist administration on breast, ovarian, and endometrial carcinoma risk. Proof of this protective effect, however, awaits large-scale epidemiologic studies.

In order to reverse the potential side effects of GnRH agonist administration, low-dose estrogen and progestin replacement therapy, similar to that used in postmenopausal women has been advocated. Since almost all of the short- and long-term side effects of the therapy are the result of this hypoestrogenism, this is based on a sound rationale. Preliminary evidence suggests that this "add-back" therapy may maintain the majority of the beneficial effects of GnRH agonists on the symptoms of endometriosis,[102] premenstrual syndrome,[103] and fibroids.[104]

18.4.3 Danazol

Several reports have demonstrated the efficacy of danazol in endometriosis.[105] Its efficacy has also been reported in premenstrual syndrome.[106] This has more recently been shown to be particularly efficacious in the treatment of premenstrual migraines.[107] Danazol is a derivative of the synthetic androgen 17-α-ethinyl testosterone.[108] As such it possesses significant androgenic properties. Administration of danazol results in amenorrhea in the majority of women on the medication. The objective of therapy is to obtain the beneficial effects that occur secondary to this hypoestrogenic amenorrhea. The usual dose required to achieve this is 600–800 mg/d in divided doses. The mechanisms of the amenorrhea are multiple. In particular, danazol suppresses the midcycle of LH and FSH surges required for ovulation, while having relatively few effects on the baseline circulating concentrations of LH and FSH. In addition, danazol is an inhibitor of the majority of the enzymes responsible for the synthesis of androgens and estrogens from cholesterol.[109] As such, it is likely that with danazol therapy sufficient estrogen is not produced to generate the mid-cycle LH surge.

In addition to its effects in eliminating menstrual cyclicity, danazol has significant androgenic effects. Levels of free androgen are increased by binding of danazol to the major androgen binding protein in the peripheral circulation, sex hormone binding globulin. Danazol also exerts androgenic side effects by binding of the androgen receptor and receptor translocation into the nucleus, thereby facilitating androgen-receptor initiated mRNA synthesis.[109]

The side effect profile of danazol is considerable, as is the result of both its androgenic activity and its antiestrogen properties. Acne and weight gain are commonly reported. Decreased breast size is a particularly disturbing complaint for many women. More rarely, overtly masculinizing side effects are noted. These include deepening of the voice and clitoromegaly.[110] Fluid retention on danazol therapy is particularly disturbing to women with premenstrual syndrome.

The antiestrogenic side effects, while better tolerated by most women than the androgen effects, are at times quite bothersome. These are the same side effects as those observed on GnRH analogs and include hot flashes, vaginal dryness, and emotional lability. Although the osteoporosis that accompanies GnRH agonist therapy is less of a concern with danazol, the effects on lipid profiles are more worrisome than with GnRH agonists. This is due to the combined adverse effects of hypoestrogenism and hyperandrogenism. For this reason, the use of danazol should be accompanied by monitoring of lipid profiles.

Danazol is contraindicated during pregnancy because of in-utero female pseudo-hermaphroditism.[111] There have also been reports of hepatotoxicity, manifested by increased liver function tests. Liver function should be monitored periodically in patients taking this drug.

Taken together, approximately 80% of women on danazol will experience side effects. Based on information obtained from the use of danazol in endometriosis, however, only 10% of women discontinued the medication because of side effects. Since some cases of endometriosis are severely disabling, the motivation to continue despite side effects is high.

Based on the benefit/risk ratio, danazol is a less favorable alternative to GnRH agonist and for most indications. At the present time, it should not be considered a first-line agent.

18.5   Drugs Used for Hormone Replacement in Menopause

Epidemiologically, the beneficial effects of estrogen on cardiovascular disease risk in postmenopausal women clearly constitute the most important reason to prescribe hormone replacement therapy. Estrogen replacement has a highly favorable effect on the lipid profile. On average, use of the most commonly prescribed estrogen replacement formulation (conjugated equine estrogens, at doses of 0.625 mg daily) results in a 10% to 15% increase in HDL (largely resulting from increases in the HDL_2 fraction) and a 4% decrease in LDL. Simultaneously increases in triglycerides averaging 20% and modest increase in very low density lipoproteins occur. In most studies where progestin was added to the regimen there has been diminution in the beneficial effects of the estrogen therapy alone. With respect to LDL, some studies have found that the lowering of levels is less than that seen with estrogen alone, although LDL levels on combined therapy are still lower than baseline.[112] Notably, however, this was not the case in the recent carefully controlled National Institute of Health–sponsored Postmenopausal Estrogen and Progestin Intervention (PEPI) Trial.[113] In this study, combined estrogen/progestin regimens were as effective in lowering LDL as estrogen alone, and a far lower risk of endometrial hyperplasia was observed in the combined group. Less favorable results have been reported in controlled clinical trials with regard to HDL. The beneficial increases in HDL and the HDL_2 seen on estrogen alone were markedly negated by the addition of progestin. In addition, the elevation in triglycerides seen in estrogen replacement therapy users appears to be exacerbated by the addition of progestin. Of interest, however, some cross-sectional studies have found no differences in triglycerides, HDL, or LDL in estrogen-only users as compared to combined estrogen/progestin users. Several explanations for the discrepancy between controlled trials and cross-sectional observations have been advanced. One possibility is that compliance with taking the progestin was poor. A second is that the negative effects of progestin on lipid parameters observed in short-term controlled clinical trials is a transient phenomenon that is not observed after several years of use.

In the best controlled epidemiologic study to date, Swedish women taking combined estrogen/progestin therapy had a 50% decrease in their risk for myocardial infarction as compared to untreated women. The reduction in risk was not significantly different from those taking estrogen alone, where the risk decreased by 76%. Similarly in this study, the relative risk for stroke was significantly reduced by 28% in combined estrogen/progestin users as compared to controls, and similar to the protection reported in estrogen-only users.[114]

Overall, the cardioprotective effects of combined estrogen/progestin use appear established with reasonable certainty, although further investigation is required for complete surety. It is therefore most prudent for the clinician to recommend combined therapy for patients who have not had a hysterectomy and do not have contraindications to its use. In addition to its protective effects on cardiovascular status, there is little doubt that combined estrogen/progestin hormone replacement therapy is protective against osteoporosis.[115]

Despite numerous studies, the association of breast cancer with hormone replacement therapy is still a matter of debate. Overall, there does appear to be some increase in breast cancer rates with estrogen use, and a metaanalysis of the data suggests that this increase is in the order of 2% to 3% per year.[116] It is not possible from available information to determine the isolated effect of the progestin component of hormone replacement therapy with regard to breast cancer risk. However, there are no good studies to suggest that progestin administration offers substantial protective effects.

The use of estrogen alone as hormone replacement therapy carries an unacceptably high risk of endometrial hyperplasia and cancer. Therefore, progestin administration is advisable for all women who have a uterus. Both the dose and the duration of progestin administration are important in affording adequate endometrial protection. With respect to medroxyprogesterone acetate, a careful study by Whitehead et al.[117] suggests that endometrial hyperplasia is prevented by 12 to 14 days of a 10 mg daily dose. In the event a 19-nortestosterone derivative is selected for cyclic replacement, 12 to 14 days should also be used. Lower doses of 19-nortestosterone derivatives, equivalent to 2.5 mg of norethindrone, are adequate. The earlier recommendation was that a 10-day course of medroxyprogesterone acetate per month affords substantial protection against endometrial hyperplasia. This regimen was developed by a trial-and-error approach, and is less efficacious than a 12 to 14 day regimen.

The required dose of progestin is somewhat dependent on the estrogen dose used in the combination regimen. The protective effect of a 10-day course of 10 mg of medroxyprogesterone acetate per month is adequate to protect more than 98% of women from endometrial hyperplasia when a 0.625 mg dose of conjugated equine estrogens is used. However, the most complete protection is provided by the 12 to 14 day course, and the latter regimen is effective even with 1.25 mg of conjugated equine estrogens or the equivalent.

More recently, several alternative regimens have been studied that utilize less frequent progestin administration, continuous daily progesterone administration, and alternative forms of progestin therapy. With respect to endometrial protection several of these approaches are largely efficacious, although less well studied than the regimen that uses 14 days of progestin administration each month. Of these, the most popular is the daily administration of 2.5 mg of medroxyprogesterone acetate and 0.625 mg of conjugated equine estrogens. A recent metaanalysis of 42 studies on this regimen has revealed the rate of endometrial hyperplasia to be 1%.[118] In this regard, it can be considered as effective as a 10 mg dose given for 10 days each month, and perhaps only marginally less effective than a 10 mg dose administered for 12 to 14 days. Even more novel are recent studies with administration of this agent restricted to 14 days every 3 months. The incidence of endometrial hyperplasia in these small studies is 1% to 3% after 1 year.[119] Data is still being evaluated on the endometrial protective effects when oral micronized progesterone is used as the progestin. While a 400 mg dose administered for 10 days of each month appears fully protective, it is possible that much lower doses may be sufficient.

While there is substantial evidence that estrogen has beneficial mood effects, particularly in estrogen deficient postmenopausal women, the opposite appears to be true for progesterone. In premenopausal women, premenstrual symptoms have been shown to appear coincident with the luteal phase rise in serum progesterone levels. A similar constellation of premenstrual symptoms has been observed in carefully controlled studies of postmenopausal women taking cyclic estrogen/progestin therapy.[120] Thus, the depressive and premenstrual complaints of women taking cyclic estrogen/progestin therapy are often clinically the most challenging problem in treating these patients. In this context, it should be recalled that the goal of hormone replacement therapy is to improve quality of life as well as longevity. Thus, changes in the progestin regimen, or in selected cases even the complete elimination of the progestin component from the regimen, is required.

The overriding principle is to prescribe a dose of progestin that is high enough to provide protection from endometrial hyperplasia and cancer and low enough to minimize the effects on mood and cardiovascular risk. Since the relief of vasomotor symptoms and the osteoporosis prevention benefits of hormone replacement therapy are largely dependent on the estrogen dose, these are of far lesser consideration in the selection of the progestin component.

There is strong theoretical rationale to favor medroxyprogesterone acetate as the progestin of choice for hormone replacement therapy. C_{21}-derived progestins such as this agent or megestrol acetate have been shown in several model systems to exert fewer adverse effects on lipid profiles than the 19-nortestosterone derivatives such as norethindrone or d,l-norgestrel. Although the data from epidemiologic studies in favor of the C_{21} derivatives is far less convincing than the theoretical evidence, there is little advantage to choosing the 19-nortestosterone derivatives. Of the C_{21} compounds, medroxyprogesterone acetate is chosen because it is the best-studied one. It is likely that this recommendation may change with further study of the parent C_{21} compound progesterone itself given in oral micronized form. At present, several pharmacies in the United States have made oral micronized progesterone available to clinicians. Although not FDA approved for hormone replacement therapy, it has recently become a popular alternative. At the present time, however, until the safe dose with respect to uterine hyperplasia is firmly established, it is prudent to subject patients using this regimen to yearly endometrial biopsies.

Careful review of the literature suggests that medroxyprogesterone acetate administration at doses of 10 mg for 14 days each month remains the most proven regimen for protection from endometrial hyperplasia. In the patient for whom the common occurrence of cyclic monthly bleeding is not bothersome, this is the favored regimen. Much more commonly, patients prefer amenorrhea. In this group, daily continuous administration of 2.5 mg of this agent should be selected. Such patients should be counseled that irregular bleeding occurs in the majority of cases for the first 3 to 6 months of therapy and not uncommonly persists up to 1 year.

The low continuous dose of medroxyprogesterone acetate is also an alternative that will be effective for women who cannot tolerate symptoms similar to premenstrual syndrome effects, which are more pronounced on the 10 mg 14-day regimen. For some of these patients, 14 days of 5 mg is a better alternative with respect to mood symptoms. In the most resistant cases, progestins at any dose may precipitate depression or intolerable premenstrual syndrome symptoms. For these patients, the decision to discontinue hormone replacement therapy or to use estrogen alone with annual endometrial biopsies are equally viable alternatives. The decision between these options should rest on assessment of the osteoporosis and cardiovascular risk profile of the individual patient.

Adequate surveillance for patients on combined estrogen/progestin replacement includes annual assessment of blood pressure, lipid profile, and abnormal bleeding pattern. In the event the lipid profile is not acceptable, consideration should be given to using estrogen alone combined with annual endometrial biopsy. This latter option is unlikely to be effective if the triglycerides are elevated, since hypertriglyceridemia is much more likely to result from estrogen administration.

All cases of abnormal bleeding on hormone replacement therapy should be evaluated with annual endometrial biopsies except in the extenuating circumstance where the patient is unable to tolerate the procedure in the office. In such cases consideration may be given to vaginal probe ultrasonography for measurement of endometrial thickness. An endometrial thickness of less than 4 mm is highly reassuring that the patient does not have endometrial hyperplasia. Only in this circumstance should biopsy be deferred with appropriate informed consent that ultrasonographic measurement of uterine thickness is still an unproven diagnostic modality for the assessment of endometrial abnormalities. In the event that bleeding persists, the endometrial stripe is larger than 4 mm, or the patient is unwilling to accept the risk of possible hyperplasia or cancer, a dilatation and curettage under anesthesia is required.

Abnormal uterine bleeding on hormone replacement therapy can be defined as any bleeding that occurs prior to the ninth day of progestin administration on any regimen where the progestin is given in cyclic monthly fashion. For continuous regimens, any bleeding that persists beyond 12 months of the initiation of the continuous therapy should be considered abnormal.

In the event the patient is on unopposed estrogen hormone replacement therapy, annual endometrial biopsies are warranted regardless of the bleeding pattern because of the high rates of endometrial hyperplasia in this circumstance.

Use of these guidelines should afford almost all postmenopausal women the option of hormone replacement therapy. Attendant upon this choice is the likelihood that greater quality of life will be enjoyed by those who choose to take sex hormones.

18.6   Drugs Used for Hirsutism

18.6.1   Spironolactone

Of the drugs for hirsutism, spironolactone is among the most commonly used. A randomized trial of spironolactone and cyproterone acetate found the two drugs to be equally effective in treating hirsutism.[121] Spironolactone is an aldosterone analog that was developed as an antihypertensive based on aldosterone inhibition. As such it shares considerable structural similarity with steroid hormones. In fact spironolactone has been demonstrated to have very specific antiandrogen effects, which has popularized its use in androgen excess symptoms, particularly hirsutism. Spironolactone acts as an antiandrogen in three ways. First, it inhibits androgen synthesis, albeit to a variable degree, through inhibition of both the 17-hydroxylase and the 17 to 20 desmolase activities of cytochrome P_{450} 17 α-hydroxylase enzyme.[122] Second, it is a potent competitive antagonist for the androgen receptor, with 67% of the affinity of dihydrotestosterone for the receptor. And third, it inhibits 5-α reductase, the enzyme that converts testosterone to its more potent metabolite dihydrotestosterone in peripheral tissues.

The doses required for the treatment of hirsutism are high relative to those used for the treatment of hypertension. The usual starting dose is 50 mg twice daily, although doses as high as 200 mg twice daily may be required. On a dose of 50 mg twice daily, 72% of women observe decreased hair growth within 6 months.[123] Because of the long cycle of hair follicles, effects are rarely observed in shorter than 3 months and the maximal degree of improvement occurs after 6 months.

Because of its aldosterone antagonist activity, spironolactone is a potassium sparing diuretic. Patients should be warned against taking potassium supplements while using this medication. Hypotensive effects are rarely observed in healthy young women, except when the medicine is administered concomitantly with other diuretics.

The most common side effect of spironolactone is menstrual irregularity, which may occur in up to 20% of women. This is completely reversible by the use of oral contraceptives. In the hyperandrogenic women such as those with polycystic ovary syndrome, the combined use of oral contraceptives and spironolactone has been shown to be superior to that of spironolactone alone in decreasing hair growth. Other side effects of spironolactone are rare. Of these, gastrointestinal symptoms are the most common. These include cramping and diarrhea, and in more severe situations, gastric and peptic ulceration and gastrointestinal bleeding. Central nervous system side effects also occasionally occur, including drowsiness, lethargy, headache, and mental confusion. Skin lesions have been noted of both the maculopapular and erythematous variety. Paradoxically, although spironolactone is an antiandrogen, some women have been noted to have androgenic effects including increased hair growth and deepening of the voice. Altogether, however, side effects other than menstrual irregularity occur in less than 5% of women.

The efficacy in reducing hair growth in addition to the very low incidence of side effects makes this a highly useful medication in the treatment of hirsutism.

18.6.2   Cyproterone Acetate

Cyproterone acetate is the drug most commonly used to treat hirsutism in Europe. Like spironolactone it decreases androgen synthesis and inhibits the activity of androgens at the receptor. Unlike spironolactone, however, cyproterone acetate has significant progestational activity. Because of this, it is widely available in Europe as part of a combined oral contraceptive containing 2 mg of cyproterone acetate and either 30 μg or 50 μg of ethinyl estradiol. Although cyproterone acetate containing oral contraceptives are the treatment of choice for hirsutism in Europe, this agent is not approved for use in the United States. In Europe, an alterative method of prescribing cyproterone acetate is to prescribe it for the first 10 to 11 days of a 19-nortestosterone containing oral contraceptive. The efficacy of this treatment has been widely demonstrated.[124] It has been shown when used in this way that a 25 mg daily dose is as effective as higher doses.[125]

The primary side effects of cyproterone acetate are those associated with progestin use. These include weight gain and a bloated sensation. Depression may also rarely occur. The incidence of these combined side effects is less than 10%. Drug-induced hepatitis has been observed with cyproterone acetate. Although rare, it is prudent to follow liver function tests when the drug is prescribed.

18.6.3   Flutamide

Flutamide is an antiandrogen that is approved in the United States for the treatment of prostate cancer. At doses of 250 mg b.i.d., it is an effective treatment for hirsutism. Despite this, its side effect profile makes it far less desirable than spironolactone and cyproterone acetate. These side effects include dry skin (70%), hot flashes (25%), increased appetite (25%), headaches (15%), and fatigue (15%). Nausea, decreased libido and breast tenderness also occur, although each is reported in less than 10% of women. In addition, the costs of flutamide are 5 to 6 times higher that of spironolactone. Since both spironolactone and cyproterone acetate are available, there appears to be little rationale to chose flutamide as a first-line therapy for hirsutism.

18.7   Future Directions

The armamentarium of medications in reproductive endocrinology and infertility is formidable. Through these agents, fertility has become possible for millions of couples. Moreover, medical therapies for a variety of hormonally sensitive disorders have been devised. These include endometriosis, fibroids, hyperandrogenic syndromes, precocious puberty, and prostate cancer. Perhaps most important, there has been a marked decline in the burgeoning population of the Western World as the result of hormonal contraceptives. Most recently, the benefits of the hormone replacement therapy to women after menopause have been realized.

In the future, new compounds such as tibolone will be developed further. Such compounds possess simultaneous estrogenic, androgenic, and progestational activity. Already this has provided a popular form of hormone replacement therapy in Europe. Other compounds are likely to be developed in this category where the relative androgenic, estrogenic, and progestational activity can be uniquely tailored to the specific clinical syndrome.

Perhaps the most exciting advances will be achieved in the development of a variety of new antagonists. GnRH antagonists are currently in clinical trials. These may well offer advantages over GnRH agonists since the upregulation phase of the pituitary, prior to the attainment of downregulation, can be avoided. In particular, this upregulation phase can be extremely painful in cases of prostate cancer and endometriosis.

The development of sex-steroid antagonists is still in its infancy. Compounds such as mifepristone have already provided new reproductive choices as a method of termination of pregnancy and as a postcoital contraceptive. Antiprogestins with lesser glucocorticoid activity than mifepristone are currently in development, and there is an indication that they may be useful as contraceptives and in the treatment of uterine fibroids.

Overall, numerous advances to our current therapies can be expected in the very near future.

Glossary of Abbreviations

GnRh	Gonadotropin releasing hormone
FSH	Follicle stimulating hormone
LH	Luteinizing hormone
hMG	Human menopausal hormone
hCG	Human chorionic gonadotropin
CNS	Central nervous system
mRNA	Messenger ribonucleic acid
HDL	High density lipoprotein
LDL	Low density lipoprotein
FDA	Food and Drug Administration

References

1. Shelton, R.S., Van Campen, M.G. Jr., Meisner, D.F., Parmerter, S.M., Andrews, E.R., Allen, R.E., Wyckoff, K.K., Synthetic estrogens. Halotriphenylethylene derivatives, *J Am Chem Soc*, 75, 5491, 1953.
2. Clark, J.H., Markaverich, B.M., The agonistic-antagonistic 65, 148, properties of clomiphene: a review, *Pharmacol Ther*, 15, 467, 1982.
3. Ernst, S., Hie, G., Cantrell, J.S., Richardson, A. Jr., Benson, H.D., Stereo-chemistry of geometric isomers of clomiphene: a correction of the literature and a reexamination of structure-activity relationships, *J Pharm Sci*, 65, 148, 1976.
4. Glasier, A.F., Irvine, D.S., Wickings, E.J., Hillier, S.G., Baird, D.T., A comparison of the effects on follicular development between clomiphene citrate, its two separate isomers and spontaneous cycles, *Hum Reprod*, 4, 252, 1989.
5. Schreiber, E., Johnson, J.E., Plotz, E.J., Wiener, M., Studies with ¹⁴C-labeled clomiphene citrate, *Clin Res*, 14, 287, 1996.
6. Clark, J.H., Peck, E.J. Jr., Oestrogen receptors and antagonism of steroid hormone action, *Nature* (London), 251, 446, 1974.
7. Czygan, P-J., Schultz, K.D., Studies on the anti-oestrogenic and oestrogen-like action of clomiphene citrate in women, *Gynecol Invest*, 3, 126, 1972.
8. Hashimoto, T., Miyai, K., Izumi, K., Kumahara, Y., Effect of clomiphene citrate on basal and LRH-induced gonadotropin secretion in post-menopausal women, *J Clin Endocrinol Metab*, 42, 593, 1976.
9. Ravid, R., Jedwab, G., Persitz, E., David, M.P., Karni, N., Gil, S., Cordova, T., Harell, A., Ayalon, D., Gonadotropin release in ovariectomized patients. I. Suppression by clomiphene or low doses of ethinyl oestradiol, *Clin Endocrinol* (Oxford) 6, 333, 1977.
10. Natrajan, P.K., Greenblatt, R.B., Clomiphene citrate: induction of ovulation, In: Greenblatt R.B., ed., *Induction of ovulation*, Philadelphia: Lea & Febiger; 35, 1979.
11. Jones, G.S., de Moraes-Ruehsen, M., Induction of ovulation with human gonadotropins and with clomiphene, *Fertil Steril*, 16, 461, 1965.
12. Riley, G.M., Evans, T.N., Effects of clomiphene citrate on anovulatory ovarian function, *Am J Obstet Gynecol*, 88, 1072, 1964.
13. Wall, J.A., Franklin, R.R., Kaufman, R.H., Reversal of benign and malignant endometrial changes with clomiphene, *Am J Obstet Gynecol*, 88, 1072, 1964.
14. Kistner, R.W., Lewis, J.L., Steiner, G.J., Effects of clomiphene citrate on endometrial hyperplasia in the premenopausal female, *Cancer*, 19, 115, 1966.
15. Adashi, E.Y., Clomiphene citrate: Mechanism(s) and site(s) of action — a hypothesis revised, *Fertil Steril*, 42, 331, 1984.
16. Rust, L.A., Israel, R., Mishell, D.R. Jr., An individualized graduated therapeutic regimen for clomiphene citrate, *Am J Obstet Gynecol*, 120, 785, 1974.

17. Drake, T.S., Tredway, D.R., Buchanan, G.C., Continued clinical experience with an increasing dosage regimen of clomiphene citrate administration, *Fertil Steril*, 30, 274, 1978.
18. Gorlitsky, G.A., Kase, N.G., Speroff, L., Ovulation and pregnancy rates with clomiphene citrate, *Obstet Gynecol*, 51, 265, 1978.
19. Hammond, M.G., Halme, I.K., Talbert, L.M., Factors affecting the pregnancy rate in clomiphene citrate induction of ovulation, *Obstet Gynecol*, 62, 196, 1983.
20. Micic, S., Dotlic, R., Evaluation of sperm parameters in a clinical trial of oligospermic men, *J Urol*, 133, 221, 1985.
21. Purvin, V.A., Visual disturbance secondary to clomiphene citrate, *Arch Opthalmol*, 113, 482, 1995.
22. Atlay, R.D., Pennington, G.W., The use of clomiphene citrate and pituitary gonadotropin in successive pregnancies: the Sheffield quadruplets, *Am J Obstet Gynecol*, 109, 402, 1971.
23. Kokia, E., Bider, D., Lunenfeld, B., Blakstein, J., Mashiach, S., Ben-Rafael, Z., Addition of exogenous estrogens to improve cervical mucus following clomiphene citrate medication, *Acta Obstet Gynecol Scand*, 69, 139, 1990.
24. Fedele, L., Brioschi, D., Marchini, M., Dorta, M., Parazzini, F., Enhanced preovulatory progesterone levels in clomiphene citrate-induced cycles, *J Clin Endocrinol Metab*, 69, 681, 1989.
25. Tepper, R., Lunenfeld, B., Shalev, J., Ovadia, J., Bankstein, J., The effect of clomiphene citrate and tamoxifen on the cervical mucus, *Acta Obstet Gynecol Scand*, 67, 311, 1988.
26. Bateman, B., Nunley, W., Kolp, L., Exogenous estrogen therapy for treatment of clomiphene citrate-induced cervical mucus abnormalities: is it effective? *Fertil Steril*, 54, 577, 1990.
27. Van der, J.V., The effect of clomiphene and conjugated estrogens on cervical mucus, *S Afr Med J*, 60, 347, 1981.
28. Maxson, W.S., Pittaway, D.E., Herbert, C.M., Garner, C.H., Wentz, A.C., Antiestrogenic effect of clomiphene citrate: correlation with serum estradiol concentrations, *Fertil Steril*, 42, 356, 1984.
29. Asch, R.H., Greenblatt, R.B., Update on the safety and efficacy of clomiphene citrate as a therapeutic agent, *J Reprod Med*, 17, 175, 1976.
30. Ahlgren, M., Kallen, B., Rannevik, G., Outcome of pregnancy after clomiphene therapy, *Acta Obstet Gynecol Scand*, 53, 371, 1976.
31. Adashi, E.Y., Rock, J.A., Sapp, K.C., Martin, E.J., Wentz, A.C., Jones, G.S., Gestational outcome of clomiphene-related conceptions, *Fertil Steril*, 31, 620, 1979.
32. Mor-Joseph, S., Anteby, S.O., Granat, M., Brzezinksy, A., Evron, S., Recurrent molar pregnancies associated with clomiphene citrate and human gonadotropins, *Am J Obstet Gynecol*, 151, 1085, 1985.
33. Marchbanks, P.A., Coulam, C.B., Annegers, J.F., An association between clomiphene citrate and ectopic pregnancy: a preliminary report, *Fertil Steril*, 44, 268, 1985.
34. Chaffkin, L.M., Nulsen, J.C., Luciano, A.A., Metzger, D.A., A comparative analysis of the cycle fecundity rates associated with combined human menopausal gonadotropin (hMG) and intrauterine insemination (IUI) vs. either hMG or IUI alone, *Fertil Steril*, 55, 252, 1991.
35. Larsen, T., Larsen, J.F., Schioler, V., Bostofte, E., Felding, C., Comparison of urinary human follicle-stimulating hormone and human menopausal gonadotropin for ovarian stimulation in polycystic ovarian syndrome, *Fertil Steril*, 53, 426, 1990.
36. Brown, J.B., Evans, J.H., Adey, F.D., Taft, H.P., Townsend, L., Factors involved in the induction of fertile ovulation with human gonadotrophins, *J Obstet Gynaecol Br Commonw*, 76, 289, 1969.
37. Hull, M.G.R., Gonadotrophin therapy in anovulatory infertility. In: Howles C.M., ed. *Gonadotrophins, gonadotrophin-releasing hormone analogues and growth factors in infertility: future perspectives*, Hove, England: Medi-Fax; 56, 1991.
38. Fleming, R., Haxton, M.J., Hamilton, M.P., Conaghan, C.J., Black, W.P., Yates, R.W., Coutts, J.R., Combined gonadotropin-releasing hormone analog and exogenous gonadotropins for ovulation induction in infertile women: efficacy related to ovarian function assessment, *Am J Obstet Gynecol*, 159, 376, 1988.

39. Dodson, W.C., Hughes, C.L., Yancy, S.E., Haney, A.F., Clinical characteristics of ovulation induction with human menopausal gonadotropins with and without leuprolide acetate in polycystic ovary syndrome, *Fertil Steril*, 52, 915, 1989.
40. Buckler, H.M., Philips, S.E., Kovacs, G.T., Burger, H.G., Healy, D.L., GnRH agonist administration in polycystic ovary syndrome, *Clin Endocrinol*, 31, 151, 1989.
41. Hamilton-Fairley, D., Kiddy, D., Watson, H., Sagle, M., Franks S., Low-dose gonadotrophin therapy for induction of ovulation in 100 women with polycystic ovary syndrome, *Hum Reprod*, 6, 1095, 1991.
42. Wang, C.F., Gemzell, C., The use of human gonadotropins for induction of ovulation in women with polycystic ovarian disease, *Fertil Steril*, 33, 479, 1980.
43. Lunenfeld, B., Therapy with gonadotropins, In: Jacobs H.S., ed., *Advances in gynaecological endocrinology*, London: Royal College of Obstetricians and Gynaecologists, 191, 1978.
44. Hamilton-Fairley, D., Franks, S., Common problems in induction of ovulation, *Balliere Clin Obstet Gynaecol*, 4, 609, 1990.
45. Dodson, W.C., Haney, A.F., Controlled ovarian hyperstimulation and intrauterine insemination for treatment of infertility, *Fertil Steril*, 55, 457, 1991.
46. Keenan, J.A., Moghissi, K.S., Luteal phase support with hCG does not improve fecundity rate in human menopausal gonadotropin-stimulated cycles, *Obstet Gynecol*, 79, 983, 1992.
47. Balasch, J., Carmona, F., Llach, J., Arroya, V., Jove, I., Vanerell, J.A., Acute prerenal failure and liver dysfunction in a patient with severe ovarian hyperstimulation syndrome, *Hum Reprod*, 5, 348, 1990.
48. Rabau, E., Serr, D.M., David, A., Mashiach, S., Lunenfeld, B., Human menopausal gonadotropins for anovulation and sterility, *Am J Obstet Gynecol*, 98, 92, 1967.
49. Schenker, J.G., Weinstein, D., Ovarian hyperstimulation syndrome; a current survey, *Fertil Steril*, 30, 255, 1978.
50. Polishuk, W.Z., Schenker, J.G., Ovarian overstimulation syndrome, *Fertil Steril*, 20, 443, 1969.
51. Jakob, W., Jentsch, K.D., Maursberger, B., Oehme, P., Demonstration of angiogenesis activity in the corpus luteum of cattle, *Exp Pathol*, 13, 231, 1977.
52. Ong, A.C.M., Eisen, V., Rennie, D.P., Homburg, R., Lavchelin, G.C.L., Jacobs, H.S., Slater, J.D.H., The pathogenesis of the ovarian hyperstimulation syndrome (OHS): a possible role for ovarian renin, *Clin Endocrinol*, 34, 43, 1991.
53. Pride, S.M., Yuen, B.H., Moon, Y.S., Clinical, endocrinologic, and intraovarian prostaglandin F response to H-1 receptor blockade in the ovarian hyperstimulation syndrome: studies in the rabbit model, *Am J Obstet Gynecol*, 148, 670, 1984.
54. Lunenfeld, B., Insler, V., Gonadotropins. In: Lunenfeld B., Insler J. eds. *Diagnosis and treatment of functional infertility*, Berlin: Thieme-Verlag, 76, 1978.
55. Schenker, J.G., Polishuk, W.Z., The role of prostaglandins in ovarian hyperstimulation syndrome, *Eur J Obstet Gynecol Reprod Biol*, 6, 47, 1976.
56. Frederick, J.L., Hoa, N., Preston, D.S., Frederick, J.J., Campeau, J.D., Onto, T., diZerega, G.S., Initiation of angiogenesis by porcine follicular fluid, *Am J Obstet Gynecol*, 152, 1073, 1985.
57. Navot, D., Bergh, P.A., Laufer, N., Ovarian hyperstimulation syndrome in novel reproductive technologies; prevention and treatment, *Fertil Steril*, 58, 249, 1992.
58. Smitz, J., Devroey, P., Camus, M., Deschacht, J., Khan, I., Staessen, C., Van Waesberghe, L., Wisanto, A., Van Steirteghem, A.C., The luteal phase and early pregnancy after combined GnRH-agonist/hMG treatment for superovulation in IVF and GIFT, *Hum Reprod*, 3, 585, 1988.
59. Thaler, I., Yoffe, N., Kaftory, J.K., Brandes, J.M., Treatment of ovarian hyperstimulation syndrome: the physiologic basis for a modified approach, *Fertil Steril*, 36, 110, 1981.
60. Valk, T.W., Corley, K.P., Kelch, R.P., Marshall, J.C., Hypogonadotropic hypogonadism: hormonal responses to low dose pulsatile administration of gonadotropin-releasing hormone, *J Clin Endocrinol Metab*, 51, 730, 1980.

61. Bratt, D.D., Schoemaker, R., Schoemaker, J., Life table analysis of fecundity in intravenously gonadotropin-releasing hormone treated patients with normogonadotropic and hypogonadotropic amenorrhea, *Fertil Steril*, 55, 266, 1991.

62. Skrabanek, P., McDonald, D., De Valera, E., Lanigan, O., Powell, D., Plasma prolactin in amenorrhoea, infertility, and other disorders: a retrospective study of 608 patients, *Ir J Med Sci*, 149, 236, 1980.

63. Coulam, C.B., Laws, E.R. Jr., Abboud, C.F., Randall, R.V., Primary amenorrhea and pituitary adenomas, *Fertil Steril* 35, 615, 1981.

64. Klibanski, A., Neer, R.M., Beitins, I.Z., Ridgway, E.C., Zervas, N.T., McArthur, J.W., Decreased bone density in hyperprolactinemic women, *N Engl J Med*, 303, 1511, 1980.

65. Schlechte, J., Walkner, L., Kathol, M., A longitudinal analysis of premenopausal bone loss in healthy women and women with hyperprolactinemia, *J Clin Endocrinol Metab*, 75, 698, 1992.

66. Ciccarelli, E., Savino, L., Carlevatto, V., Bertagna, A., Isaia, G.C., Camanni, F., Vertebral bone density in non-amenorrhoeic hyperprolactinaemic women, *Clin Endocrinol*, 28, 1, 1988.

67. Schlechte, J., Dolan, K., Sherman, B., Chapler, F., Luciano, A., The natural history of untreated hyperprolactinemia: a prospective analysis, *J Clin Endocrinol Metab*, 68, 412, 1989.

68. Leblanc, H., Lachelin, C.L., Abu-Fadil, S., Yen, S.S.C., Effects of dopamine infusion on pituitary hormone secretion in humans, *J Clin Endocrinol Metab*, 43, 668, 1976.

69. Molitch, M.E., Reichlin, S., Hyperprolactinemic disorders, *Dis Month*, 28, 1, 1982.

70. Molitch, M.E., Prolactinomas, In: Melmed S., ed. *The pituitary*, Boston: Blackwell Scientific. In press.

71. Molitch, M.E., Elton, R.L., Blackwell, R.E., Caldwell, B., Chang, R.J., Jaffe, R., Joplin, G., Robbins, R.J., Tyson, J., Thorner, M.O., Bromocriptine as primary therapy for prolactin-secreting macroadenomas: results of a prospective multicenter study, *J Clin Endocrinol Metab*, 60, 698, 1985.

72. Warfield, A., Finkel, D.M., Schatz, N.J., Savino, P.J., Snyder, P.J., Bromocriptine treatment of prolactin-secreting pituitary adenomas may restore pituitary function, *Ann Intern Med*, 101, 783, 1984.

73. Vermesh, M., Fossum, G.T., Kletzky, O.A., Vaginal bromocriptine: pharmacology and effect on serum prolactin in normal women, *Obstet Gynecol*, 72, 693, 1988.

74. Jasonni, V.M., Raffeli, R., de March, A., Frank, G., Flamigni, C., Vaginal bromocriptine in hyperprolactinemic patients and puerperal women, *Acta Obstet Gynecol Scand*, 70, 493, 1991.

75. Krupp, P., Monka, C, Richter, K., The safety aspects of infertility treatments, presented at the 2nd World Congress of Gynecology and Obstetrics, Rio de Janeiro, Brazil, October 1988.

76. Raymond, J.P., Goldstein, E., Konopka, P., Leleu, M.F., Merceron, R.E., Loria, Y., Follow-up of children born of bromocriptine-treated mothers, *Horm Res*, 22, 239, 1985.

77. Franks, S., Lynch, S.S., Horrocks, P.M., Butt, W.R., Treatment of hyperprolactinaemia with pergolide mesylate: acute effects and preliminary evaluation of long-term treatment, *Lancet*, 2, 659, 1981.

78. Blackwell, R.E., Bradley, E.L., Kline, L.B., Duvall, E.R., Vitek, J.J., De Vane, G.W., Chang, R.J., Comparison of dopamine agonists in the treatment of hyperprolactinemic syndromes: a multicenter study, *Fertil Steril*, 39, 744, 1983.

79. Kletzky, O.A., Borenstein, R., Mileikowsky, G.N., Pergolide and bromocriptine for the treatment of patients with hyperprolactinemia, *Am J Obstet Gynecol*, 154, 431, 1986.

80. Vance, M.L., Cragun, J.R., Reimnitz, C., Chang, R.J., Rashef, E., Blackwell, R.E., Miller, M.M., Molitch, M.E., C.V. 205–502 treatment of hyperprolactinemia, *J Clin Endocrinol Metab*, 68, 336, 1989.

81. Webster, J., Piscitelli, G., Poli, A., Ferrari, C.I., Ismail, I., Scanlon, M.F., A comparison of cabergoline and bromocriptine in the treatment of hyperprolactinemic amenorrhea, *N Engl J Med*, 331, 904, 1994.

82. Vessey, M., Doll, R., Peto, R., Johnson, B., Wiggins, P., A long-term follow-up study of women using different methods of contraception, *J Biosoc Sci*, 8, 373, 1976.

83. Greenspan, S.L., Klibanski, A., Rowe, J.W., Age alters pulsatile prolactin release: influence of dopaminergic inhibition, *Am J Physiol,* 258, E799, 1990.

84. Stern, J.M., Reichlin, S., Prolactin circadian rhythm persists throughout lactation in women, *Neuroendocrinology,* 51, 31, 1990.

85. Royal College of General Practitioners Oral Contraceptive Study, Further analysis of mortality in oral contraceptive users, *Lancet,* i, 541, 1981.

86. Olsson, L.L., Landin-Olsson, M., Moller, T.R., Ranstam, J., Holm, P., Oral contraceptive use and breast cancer in young women in Sweden, *Lancet,* 1, 748, 1985.

87. Cancer and Steroid Hormone Study of the CDC and NICHD, The reduction in the risk of ovarian cancer associated with oral contraceptive use, *N Engl J Med,* 316, 650, 1987.

88. Van den Vange, N., Klogsterbuer, H.J., Haspels, A.A., Effect of seven low-dose combined oral contraceptive preparations on carbohydrate metabolism, *Am J Obstet Gynecol,* 156, 918, 1987.

89. Andrews, W.C., Larsen, D.C., Endometriosis: treatment with hormonal pseudopregnancy and/or operation, *Am J Obstet Gynecol,* 118, 643, 1974.

90. Dmowski, W.P., Radwanska, E., Binor, Z., Tumon, I., Pepping, P., GnRH analogues in the management of endometriosis. The results of two randomized trials, In Vickery, B.H., Lunenfeld, B., eds, *GnRH Analogues in Cancer and Human Reproduction,* Boston: Kluwer Academic, 2, 7, 1990.

91. Healy, D.L., Lawons, S.R., Abbott, M., Baird, D.T., Fraser, H.M., Toward removing uterine fibroids without surgery: subcutaneous infusion of a luteinizing hormone-releasing hormone agonist commencing in the luteal phase, *J Clin Endocrinol Metab,* 63, 619, 1986.

92. Parmar, H., Edwards, L., Phillips, R.H., Allen, L., Lightman, S., Orchidectomy vs. long acting D-Trip-6-LHRH in advanced prostatic cancer, *Gynecol Endocrinol,* 2(Suppl), 56, 1988.

93. Kauli, R., Schally, A.V., Laron, Z., Long term experience with a superactive GnRH analog in the treatment of precocious puberty, In: Vickery B.H., Lunenfeld B., eds., *GnRH Analogues in Cancer and Human Reproduction,* Boston: Kluwer Academic: 4-43-52, 1990.

94. Muse, K.N., Cetel, N.S., Futterman, L.A., Yen S.S.C., The premenstrual syndrome; effects of medical ovariectomy, *N Engl J Med,* 311, 1345, 1984.

95. Morel, G., Dihl, F., Albert, M.L., Dubois, P.M., Binding and internalization of native gonadoliberin (GnRH)s by anterior pituitary gonadotrophs of the rat, *Cell Tissue Res,* 248, 541, 1987.

96. Hazum, E., Cuatrecasas, P., Marion, J., Conn, P.M., Receptor-mediated internalization of gonadotropin releasing hormone by pituitary gonadotropins, *Proc Natl Acad Sci U.S.A.,* 77, 6692, 1980.

97. Askel, S., Schomberg, D.W., Tyrey, L., Hammond, C.V., Vasomotor symptoms, serum estrogens and gonadotropin levels in surgical menopause, *Am J Obstet Gynecol,* 126, 165, 1976.

98. Meldrum, D.R., The pathophysiology of postmenopausal symptoms, *Reprod Endocrinol,* 1, 11, 1983.

99. Sorva, R., Kuusi, T., Dankel, L., Taskinen, M-R., Effects of endogenous sex steroids on serum lipoproteins and post heparin plasma lipolytic enzymes, *J Clin Endocrinol Metab,* 66, 408, 1988.

100. Hart, D.M., Farish, E., Fletcher, D.C., Howie, C., Kitchener, H., Ten years postmenopausal hormone replacement therapy — effect on lipoproteins, *Maturitas,* 5, 271, 1984.

101. Stamfer, M.J., Willett, W.C., Colditz, G.A., Rosner, B., Speizer, F.E., Hennekens, C.H., A prospective study of postmenopausal therapy and coronary heart disease, *N Engl J Med,* 313, 1044, 1985.

102. Reid, B.A., Gangar, K.F., Beard, R.W., Severe endometriosis treated with gonadotropin releasing hormone agonist and continuous combined hormone replacement therapy, *Br J Obstet Gynecol,* 99, 344, 1992.

103. Mortola, J.F., Girton, L., Fischer, U., Successful treatment of severe premenstrual syndrome by combined use of gonadotropin-releasing hormone agonist and estrogen/progestin, *J Clin Endocrinol Metab,* 71, 252, 1991.

104. Maheux, R., Lemay, A., Blanchet, P., Fried, J., Pratt, X., Maintained reduction of uterine leiomyoma following addition of hormonal replacement therapy to a monthly luteinizing hormone-releasing hormone agonist implant: a pilot study, *Hum Reprod,* 6, 500, 1991.

105. Bayer, S.R., Seibel, M.M., Medical treatment: Danazol, In: Schenken R.S., ed., *Endometriosis: contemporary concepts in clinical management*, Philadelphia: Lippincott, 169, 1989.
106. Casson, P., Hahn, P.M., Van Vugt, D.A., Reid, R.L., Lasting response to ovariectomy in severe intractable premenstrual syndrome, *Am J Obstet Gynecol*, 162, 99, 1990.
107. Carlton, G.J., Burnett, J.W., Danazol and migraine, *N Engl J Med*, 310, 721, 1984.
108. Dmowski, W.P., Endocrine properties and clinical application of danazol, *Fertil Steril*, 31, 237, 1979.
109. Barbieri, R.L., Hornstein, M.D., Medical therapy for Endometriosis, In: Wilson R.A., ed., *Endometriosis*, Alan R. Liss, Inc., New York, 111, 1987.
110. Warole, P.G., Whitehead, M.I., Mills, R.P., Nonreversible and wide ranging vocal changes after treatment with danazol, *Br Med J*, 287, 946, 1983.
111. Quaglionella, J., Alba Greco, M., Danazol and urogenital sinus formation in pregnancy, *Fertil Steril*, 49, 939, 1985.
112. Nabulsi, A.A., Folsom, A.R., White, A., Pasch, W., Greiss, G., Wu, K.K., Szklo, M., Association of hormone relacement therapy with various cardiovascular risk factor in postmenopausal women, *N Engl J Med*, 328, 1069, 1993.
113. The Writing Group for the PEPI Trial, Effects of estrogen or estrogen/progestin regimens on heart disease risk factors in postmenopausal women, The Postmenopausal Estrogen/Progestin Intervention (PEPI) Trial, *J Am Med Assoc*, 273, 199, 1995.
114. Falkeborn, M., Persson, I., Adami, H.O., Bergstrom, R., Eaker, E., Lithell, H., Mohsen, R., Naessen, T., The risk of acute myocardinal infarction after oestrogen- and oestrogen-progesterone replacement, *Br J Obstet Gynecol*, 99, 821, 1992.
115. Lindsay, R., Hart, D.M., Purdie, D., Ferguson, M.M., Clark, A.S., Kraszewski, A., Comparative effects of oestrogen and a progestogen on bone loss in postmenopausal women, *Clin Sci*, 54, 193, 1978.
116. Pike, M.C., Bernstein, L., Spicer, D.V., Exogenous hormones in breast cancer risk, In Niederhuber J.E., ed., *Current therapy in oncology*, Philadelphia: Dekker, 292, 1993.
117. Whitehead, M.I., Townsend, P.T., Pryse-Davies, J., Ryder, T.A., King, R.J.B., Effects of estrogen and progestins on the biochemistry and morphology of the postmenopausal endometrium, *N Engl J Med*, 305, 1599, 1981.
118. Udoff, L., Langenberg, P., Adashi, E.Y., Combined continuous hormone replacement therapy: a critical review, *Obstet Gynecol*, 86(2), 306, 1995
119. Williams, D.B., Voight, B.J., Fu, Y.S., Schoenfeld, M.J., Judd, H.L., Assessment of less than monthly progestin therapy in postmenopausal women given estrogen replacement, *Obstet Gynecol*, 84, 787, 1994.
120. Magos, A.L., Brewster, E., Singh, R., O'Dowd, T., Brincat, M., Studd, J.W.W., The effects of norethisterone in postmenopausal women on oestrogen therapy: a model for premenstrual syndrome, *Br J Obstet Gynecol*, 93, 1290, 1986.
121. O'Brien, R.C., Cooper, M.E., Murray, R.M.L., Seeman, E., Thomas, A.K., Jerums, G., Comparison of sequential cyproterone acetate/estrogen vs. spironolactone/oral contraceptive in the treatment of hirsutism, *J Clin Endocrinol Metab*, 72, 1008, 1991.
122. Loriaux, D.L., Menard, R., Taylor, A., Pita, J.C., Santen, R., Spironolactone and endocrine dysfunction, *Ann Intern Med*, 85, 630, 1976.
123. Crosby, P.D.A., Rittmaster, R.S., Predictors of clinical response in hirsute women treated with spironolactone, *Fertil Steril*, 55, 1076, 1991.
124. Hammerstein, J., Meckies, J., Leo-Rossberg, I., Moltz, L., Zielske, F., Use of cyproterone acetate in the treatment of acne, hirsutism, and virilism, *J Steroid Biochem*, 6, 827, 1975.
125. Barth, J.H., Cherry, C.A., Wojnarowska, F., Dawber, R.P.R., Cyproterone acetate for severe hirsutism: results of a double-blind dose-ranging study, *Clin Endocrinol* (Oxford), 35, 5, 1991.

19

Nicotine Preparations in Smoking

Neal L. Benowitz
University of California

John Pinney
University of California

19.1 Introduction

Nicotine is the major alkaloid in tobacco and is a minor alkaloid in a number of other plants of the Solanaceae family. Nicotine is a potent insecticide and has been marketed extensively for that purpose. Nicotine is well known to produce human intoxication and even death. Nicotine intoxication consequent to intensive tobacco smoking has been used by South American shamans to induce a state of simulated death from which a miraculous rebirth would occur, establishing the shaman's power over life and death.[1] Nicotine has also been a fatal poison involved in suicides, homicides, and accidental deaths.[2] Thus, nicotine at high doses is poisonous and is widely viewed as one of the major toxins that contribute to tobacco-related disease.

Cigarette smoking is a leading preventable cause of death and disease in the world. Although there has been substantial progress in reducing the prevalence of smoking in developed countries, many millions of individuals worldwide continue to smoke. In the U.S., it has been estimated that up to 70% of smokers would like to quit.[3] Each year in the U.S., about 17 million smokers make an attempt to quit smoking, but only 1.2 million (2.5% of all smokers) become former smokers. For an individual who is a lifelong smoker, the risk of dying from cigarette smoking is one in two or three.[4] Thus, the public health significance of even a small decrease in smoking cessation rates is enormous.

Nicotine is responsible for maintaining tobacco addiction and, as such, is the proximate cause of all tobacco-related diseases.[5] Direct toxicity due to actions of nicotine at levels of exposure related to tobacco use or as relevant to the use of nicotine as a medication has been suspected, but causation still remains unproven. The potential toxicity of nicotine and tobacco as derived from medications will be discussed in detail later in this chapter.

19.2   Benefits of Nicotine as a Medication

Once it became clear that nicotine addiction maintains tobacco use, the idea of using nicotine pharmaceutical products to substitute for the nicotine in tobacco and to aid in smoking cessation was a logical next step. Subsequently, nicotine has been marketed in a variety of pharmaceutical products, including nicotine polacrilex chewing gum, transdermal nicotine (nicotine patches), nicotine nasal spray, and nicotine aerosol. Still other forms of nicotine delivery are under development, including nicotine lozenges, nicotine lollipops, and others.

Nicotine medications to aid smoking cessation, often called nicotine replacement therapy, have been shown in a number of studies to be effective. Nicotine gum, transdermal nicotine, and nicotine nasal spray in clinical trials usually performed in specialized smoking cessation clinics, double smoking cessation rates compared to placebo treatment.[6-9] Typical long-term cessation rates (6 to 12 months) in such a setting are 20% to 30% with nicotine replacement therapy, compared to 10% to 15% on placebo. In physicians' offices and with over-the-counter use, long-term cessation rates with nicotine replacement therapy are typically lower, more often in the range of 7% to 10%. While these cessation rates are low, the potential for widespread use of over-the-counter nicotine replacement therapy medications could have a significant effect on overall cessation.

Despite low quit rates, economic analyses of smoking cessation using nicotine replacement therapy have estimated the dollar cost per year of life saved, and these costs are quite favorable compared to those of other accepted preventive medicine measures. Assuming quit rates of 6.1% with nicotine gum, the cost per year of life saved was estimated to be $4,113 to $9,473. For comparison, the cost per year of life saved is estimated to be $91,000 for primary prevention of hypercholesterolemia and $11,000 to $72,000 for the medical treatment of mild hypertension.[10,11]

Nicotine replacement therapy probably works to aid smoking cessation by several mechanisms. Nicotine reduces withdrawal symptoms, which are believed to contribute to relapse of smoking, particularly in the first few weeks of smoking cessation.[12] Nicotine replacement therapy, particularly nicotine nasal spray, reduces the craving for cigarettes and nicotine replacement may reduce the rewarding or pleasurable effects of cigarette smoking if there is a lapse (that is, smoking a cigarette during a quit attempt).[13] Nicotine replacement therapy to aid smoking cessation is generally used with full doses for 6 to 8 weeks, followed by 4 to 8 weeks of dose tapering. The reason for dose tapering is based on theoretical concepts of reducing the severity of withdrawal symptoms that occur after rapid cessation of an addicting drug, although the necessity of tapering in promoting smoking cessation is unproven.

The doses of nicotine derived from nicotine replacement therapies are generally similar to or lower than those taken by addicted smokers. For example, with nicotine gum use, the systemic dose of nicotine is 1 or 2 mg from each piece of the 2 mg or 4 mg marketed Nicorette products, respectively.[14] This amounts to a daily intake of 10 mg to 20 mg/day in a typical Nicorette user. Transdermal nicotine systems deliver 21 mg or 22 mg per 24 hours or 15 mg per 16 hours in the full-strength systems.[15] Nicotine nasal spray delivers 0.5 mg per dose, or about 15 mg per day with the use of 30 doses per day.[16] In contrast, a typical 20 cigarette per day smoker takes in about 20 mg nicotine per day.[17] Thus, the use of nicotine replacement therapy for smoking cessation exposes smokers to doses of nicotine similar to or lower than those taken in from smoking, and exposes the individual to nicotine for 3 to 6 months as opposed to years if one continues to smoke.

Nicotine is also being considered for use as a medication for a variety of medical disorders. These include ulcerative colitis, Alzheimer's disease, Parkinson's disease, Tourette's syndrome, depression, attention deficit disorder, and others.[5] Nicotine therapy has been shown in clinical trials to increase the chance of regression of uncontrolled ulcerative colitis,[18] and in uncontrolled observations to reduce the severity of Tourette's syndrome in children.[19]

The mechanism of benefit of nicotine for Alzheimer's disease and Parkinsonism is thought to be activation of nicotinic cholinergic receptors with facilitated release of dopamine in the brain. The mechanism for antidepressant effect would similarly be release of neurotransmitters, including dopamine, norepinephrine, and serotonin. The mechanism of benefit for other disorders is unclear. In any case, the unique aspect of nicotine therapy for medical diseases is that nonsmokers may be exposed to nicotine, and the exposure may last for many years. Thus, the risk-benefit considerations differ from those relevant to smokers.

19.3 Risks of Nicotine: General Considerations

Cigarette smoke is a complex mixture of chemicals and includes not only nicotine but also toxic substances such as carbon monoxide, oxidant gases, cyanide, polycyclic aromatic hydrocarbons, and others. The role of nicotine in contributing to or aggravating disease caused by cigarette smoking has not been definitively demonstrated. Several issues need to be considered concerning the relative risks of cigarette smoking and nicotine replacement therapy: What are the possible adverse effects of nicotine per se? How do pharmacokinetic differences in delivery affect the potential toxicity of nicotine? What is the dose-response for nicotine, particularly at higher dose levels? What are the abuse, misuse, and dependence implications of transferring use from one form of nicotine (cigarette smoking) to another form (nicotine replacement therapies)?

19.3.1 Pharmacologic Effects of Nicotine

Some of the known pharmacologic effects of nicotine could potentially aggravate medical diseases. The main concerns are cardiovascular disease and adverse effects on reproductive physiology, resulting from activation of the sympathetic nervous system.[20,21] The consequences include increased heart rate and cardiac contractility, transiently increased blood pressure, constriction of blood vessels including the coronary arteries and the uteroplacental blood vessels, and systemic release of catecholamines. This last effect may contribute to increased body metabolic rate and possibly to adverse effects of nicotine on serum lipids.[22] Nicotine may also affect neuronal development in the fetus, which is discussed in a subsequent section.

19.3.2 Relevant Pharmacokinetics

The intensity of effect of nicotine is determined at least in part by the rate of delivery of nicotine into the circulation, and is related to the peak level of nicotine in arterial blood achieved and the time available for the development of tolerance. Faster delivery of nicotine results in higher arterial levels reaching various body organs, and rapid absorption results in effects occurring rapidly before there is time for development of tolerance.

Cigarette smoking delivers nicotine rapidly through the lung into the central circulation and through the heart and other organs. Arterial levels of nicotine are 6 to 10 times higher compared to venous blood levels during cigarette smoking.[23] Nicotine polacrilex gum results in gradual delivery of nicotine over 30 minutes.[24] Transdermal nicotine preparations deliver nicotine even more slowly, with levels rising gradually over several hours.[15] In this case, arterial and venous levels remain in equilibrium. Nicotine nasal spray results in absorption of nicotine over 10 minutes, which is faster than other nicotine replacement therapies but slower than that from cigarette smoking.[16] The slower the delivery the lower the peak arterial levels for a given dose, and the greater the time for development of tolerance. These phenomena result in general in lesser physiologic effects for the same total dose of nicotine with nicotine replacement products compared to cigarette smoking.

19.3.3 Relevant Pharmacodynamics

Pharmacodynamic factors, including the development of tolerance and nature of the nicotine dose-response curve, are important in understanding the potential risks of nicotine replacement therapy, particularly at high doses and/or with concomitant cigarette smoking. Tolerance develops to many effects of nicotine, including the effects on the sympathetic nervous system.[25] Tolerance is, however, not complete in that some persistent sympathetic activation is seen, but it does result in flattening of the dose–response curve. Thus, the effects of nicotine at higher doses are similar to those at lower doses. This has been shown in two studies. In one, it was shown that heart rate acceleration throughout the day was similar for people smoking 30 cigarettes per day of high vs. low nicotine content, which resulted in fourfold differences in plasma nicotine concentrations.[26] A second study examined the effects of cigarette smoking with and without intravenous nicotine and showed similar cardiovascular effects despite an almost twofold difference in plasma nicotine concentrations.[27] These data suggest that even if levels of nicotine higher than those normally consumed from cigarette smoke are achieved, there will be little or no increased risk related to sympathetic nervous system activation.

19.4   General Safety Profile of Nicotine

Toxicity issues for nicotine can be considered in three categories: Acute systemic toxicity, chronic systemic toxicity, and local toxicity.

Acute systemic toxicity may occur, but such toxicity tends to be mild for most nicotine replacement products in smokers. Effects of acute toxicity commonly include central nervous system effects (headache, dizziness, insomnia, abnormal dreams, and nervousness); GI distress (dry mouth, nausea, dyspepsia, and diarrhea); and musculoskeletal symptoms (arthralgias and myalgias).[12] Interpretation of central nervous system effects of nicotine in smokers who have recently quit smoking is complicated by the potential emergence of nicotine withdrawal symptoms that are similar to those of nicotine intoxication symptoms.

Chronic systemic toxicity is of concern based on pathophysiologic considerations for nicotine-related adverse health effects of cigarette smoking. The main concerns are (1) aggravation of atherosclerotic vascular disease, including coronary heart disease and stroke; (2) aggravation of hypertension; (3) peptic ulcer disease; (4) reproductive toxicity; and (5) delayed wound healing. That nicotine replacement therapy actually causes any of these conditions has not been demonstrated. The considerations for cardiovascular disease and reproductive toxicity will be dealt with in detail in later sections. Delayed wound healing is of concern primarily in postoperative patients and is not likely to be an issue with nicotine replacement therapy.

Local toxicity includes sore mouth, mouth ulcers, and jaw fatigue from chewing nicotine polacrilex gum, cutaneous sweating, itching, burning, and erythema from patch application, and nasal irritation with burning, itching, sneezing, and watery eyes with nicotine nasal spray. The mechanism of these effects is complex, but appears to include activation of local afferent neurons and axon reflexes with release of vasodilators such as bradykinin, substance P, and histamine.[28] Local reactions from skin patches generally resolve within 24 to 48 hours. Nasal irritation with the use of nicotine nasal spray usually resolves with the development of tolerance over 2 to 3 days.

Data on the acceptability of various nicotine replacement products taken from drug registration files indicate that acceptability is good, with fewer than 5% dropping out of therapy because of nicotine product-related side effects.[12]

19.5   Risk Considerations in Special Populations

19.5.1   Cardiovascular Patients

The major impact of cigarette smoking in producing premature morbidity and mortality is cardiovascular disease. Smoking substantially increases the risk of acute myocardial infarction, sudden death, vasospastic angina, carotid artery disease, aortic aneurysm, peripheral arterial disease, and stroke.[29] Cigarette smoking also increases the risk of vascular reocclusion after revascularization procedures such as thrombolysis after acute myocardial infarction, angioplasty, peripheral arterial surgery, and coronary artery bypass grafting.[30-32] The toxins in cigarette smoke that promote cardiovascular disease are not fully determined, but probably include carbon monoxide and nicotine. Naturally, there is concern that nicotine replacement therapy could have injurious effects in patients with cardiovascular disease.

The pharmacologic actions of nicotine most likely to influence cardiovascular disease involve activation of the sympathetic nervous system. The results of sympathetic nervous system activation are acceleration of heart rate, a brief rise in blood pressure, increased cardiac contractility and myocardial oxygen requirement, and constriction of some blood vessels. Coronary blood vessels are constricted to some degree, manifested as impaired vasodilation in response to increased oxygen demand.[33] This results in lesser capacity to supply the heart with adequate oxygen during times of exercise or other stress. Nicotine may also be associated with coronary spasm.[34] Nicotine may also have adverse effects on endothelial cells and lipid metabolism, although these effects have not been clearly demonstrated in humans.

Carbon monoxide from cigarette smoke is believed to be important because it reduces the oxygen-carrying capacity of the blood and impedes the release of oxygen from hemoglobin to tissues.[35] Inhalation of carbon monoxide at levels comparable to those found in cigarette smokers has been shown to reduce exercise tolerance in patients with angina pectoris, to increase exercise-induced ventricular dysfunction as well as the number and complexity of ventricular arrhythmias during exercise.[36-38]

While hemodynamic stresses may contribute, there is much evidence that the precipitant of most acute ischemic cardiac events is rupture of a coronary atherosclerotic placque, with subsequent activation of platelets and local coronary thrombosis. One of the major mechanisms by which cigarette smoking is believed to cause acute coronary events is via production of a hypercoagulable state. The idea that smoking is associated with a hypercoagulable state is supported by observations in patients undergoing coronary angiography after myocardial infarction, indicating that smokers have less severe underlying atherosclerosis compared to nonsmokers.[32] The prognosis is better for smokers compared to nonsmokers following myocardial infarction if smokers are able to quit. The likely explanation for these observations is that smoking produces a hypercoagulable state that results in thrombosis on top of underlying coronary heart disease. When a smoker quits, a major risk factor is reversed, and the prognosis is markedly improved.

Whereas cigarette smoking clearly produces a hypercoagulable state, nicotine does not appear to do so. One human study found that transdermal nicotine had no effect on various measures of coagulation beyond that of placebo patch, whereas cigarette smoking appeared to be associated with activation of platelets and higher fibrinogen levels, both factors promoting hypercoagulability.[39] Clinical trial data support the idea that nicotine replacement therapy does not pose a substantial risk to patients with cardiovascular disease. One placebo-controlled trial of transdermal nicotine in patients with cardiovascular disease has been published.[40] In this study, 156 patients with documented stable coronary artery disease were treated for 5 weeks with 14 mg to 21 mg per day nicotine patches. Cardiac symptoms were recorded and, during the last week of treatment, ambulatory electrocardiography was performed in a subgroup. Of note, the smoking cessation rates were low, so there was much concomitant smoking and patch use in each group. More patients dropped out of the study because of cardiovascular events on placebo than on nicotine patch. The frequency of angina declined in both nicotine and placebo groups, with no differences in arrhythmias or ST segment depression changes in the nicotine vs. the placebo treated group. No significant changes in blood pressure were noted. Thus, in this brief study there was no evidence of aggravation of coronary artery disease by nicotine replacement therapy.

Joseph et al.[40a] recently reported the results of a large Veterans Affairs cooperative study of 584 smokers with cardiovascular disease. Patients received a 10-week course of transdermal nicotine (beginning at 21 mg/day and tapering to 7 mg/day) or placebo. Many participants continued to smoke cigarettes. The incidence of primary endpoints (death, myocardial infarction, cardiac arrest and admission to the hospital for increased severity of angina, arrhythmias or congestive heart failure) was similar in both groups (nicotine group: 5.4%; placebo group: 7.9%). Thus, these two studies have found no evidence of aggravation of coronary disease by nicotine replacement therapy.

Safety aspects of the Lung Health Study have recently been published.[41] This was a large prospective study designed to look at pulmonary function over 5 years in a large group of smokers who at entry had moderately severe obstructive lung disease. A large number of these individuals were treated with nicotine polacrilex gum for as long as 5 years. Hospitalizations for cardiovascular disease were actually less common in the nicotine gum group compared to the no-nicotine replacement therapy group of smokers. Thus, in this very large study where people used nicotine gum for up to 5 years, there is no evidence of adverse effects of nicotine. Also of note in the Lung Health Study is that hospitalization for peptic ulcer disease was no different in people receiving nicotine or not taking nicotine.

Sporadic case reports describing patients with acute cardiovascular events during the use of nicotine replacement therapy have been published. These case reports include descriptions of patients who developed atrial fibrillation, acute myocardial infarction, or stroke.[42-44] Some of these patients were smoking at the same time they were using transdermal nicotine. There was no consistent pattern with respect to how long these individuals had been using nicotine, time of day of the adverse event, or any other factor clearly identifying these events as being related to the pharmacologic effects of nicotine replacement therapy. It must be recognized that cardiovascular disease is common in the age group of smokers undergoing smoking cessation therapy, and that some adverse cardiovascular events are expected to occur by chance in any 1 to 3 month period of time (the time duration of most courses of nicotine replacement therapy). A United States FDA advisory committee reviewed the cases of myocardial infarction in people using nicotine patches in 1992 and judged the events not to be causally related to nicotine patch use.

19.5.2   Pregnant Women

Cigarette smoking during pregnancy increases the risk of low birth weight babies, prematurity, prenatal and perinatal mortality, spontaneous abortion, and sudden infant death syndrome.[21] Furthermore, it is associated with postnatal neurological and behavioral disorders. As discussed for cardiovascular disease, the contribution of nicotine vs. other constituents of tobacco smoke in causing reproductive disorders is unclear. Likewise, the mechanisms by which smoking exerts its harmful effect on the fetus are not well defined.

There are at least four proposed mechanisms by which smoking interferes with fetal development, two of which may involve the effects of nicotine: (1) nicotine-induced vasoconstriction of umbilical blood vessels may decrease the blood supply to the fetus and thus inhibit fetal growth; (2) nicotine may have a direct inhibitory effect on the growth of fetal tissue and/or on brain development; (3) carbon monoxide from cigarette smoke may result in fetal hypoxemia, inhibiting growth, and (4) vitamin B_{12} and essential amino acids are depleted in smokers, and other nutritional deficits may potentially occur in smokers due to the appetite suppressing effects of smoke on the mother.

With respect to the hemodynamic effects of nicotine during pregnancy, the risk considerations are similar to those discussed for cardiovascular disease. Thus, it is likely that any adverse effects of nicotine will be substantially greater during cigarette smoking compared with nicotine replacement therapy. With respect to possible adverse effects of nicotine on *in utero* fetal development, there is less information, but it is also likely that cigarette smoking would be as or more harmful than nicotine replacement therapy. Nicotine is present in the blood and body tissues of the fetus of a mother who smokes cigarettes. In addition to entering the fetus via the uteroplacental circulation, there is also a reservoir of nicotine in the amniotic fluid.[45] If there is a risk to the fetus, it is likely that the total daily dose of nicotine will be the primary determinant of that risk. As noted earlier, for most smokers daily nicotine exposure is less during nicotine replacement therapy than while smoking cigarettes.

Some experimental studies have examined the physiologic effects of nicotine replacement therapy compared to cigarette smoking in pregnant women. In all of these studies, nicotine replacement therapy produced effects on fetal heart rate and breathing that were qualitatively similar to, but generally smaller in magnitude than, those produced by cigarette smoking.[46] The same results have been reported on fetal aortic blood flow and umbilical vein flow. The studies published to date have examined the effects of nicotine polacrilex gum. The effects of transdermal nicotine would be expected to be similar.

19.5.3   Nicotine and Lactation

In smokers, nicotine is found in breast milk and it can be passed through the milk to the infant.[47] This would certainly be the case with nicotine replacement therapy as well. In general, the concentration of nicotine in the milk is so low that milk consumption by an infant would not be expected to produce significant clinical effects. However, there are authorities who believe that even a low concentration of nicotine might be detrimental.

Overall, it would be desirable if there is no maternal or fetal nicotine exposure during pregnancy and breast-feeding. However, it seems that the risk of nicotine replacement therapy to aid smoking cessation in pregnant or postpartum women who cannot stop smoking without such therapy is substantially outweighed by the risks of continued smoking.

19.5.4 Children

There is a natural concern that accidental exposure of children to nicotine replacement products could pose a serious threat to health. Some products, such as transdermal nicotine, still contain substantial amounts of nicotine after removal when it has been worn for the prescribed period of time.[15] However, postmarketing surveillance in the U.S. at least indicates that nicotine exposure in children from nicotine replacement products is not a frequent or a serious problem.[48] Most exposures result in either no effects or minor effects. Intoxication of children consuming tobacco products is much more serious.

19.5.5 Adolescents

There is theoretical concern that nicotine products might encourage adolescents to experiment or develop abusive use patterns. This appears not to have been the case based on postmarketing surveillance data.[48] Several factors are likely to mitigate against the misuse or abuse of nicotine by adolescents. Because of slower delivery of nicotine, most or all nicotine replacement products will not produce pleasurable effects, which would predict a low abuse liability. Another significant factor is price. It is much more expensive to buy nicotine as nicotine medications than to buy nicotine in tobacco.

19.6 Abuse, Misuse, and Dependence

Cigarette smoking is maintained by addiction to nicotine. Therefore, it is reasonable to question whether there is a risk of improper use of nicotine provided by nicotine replacement therapy. Three relevant issues include abuse, misuse, and dependence. Abuse refers to improper or incorrect use, where the individual is attempting to gain a high, euphoric effect or some other psychotropic effect other than that intended to aid smoking cessation. Misuse refers to uses outside those defined in product labeling, but for reasons other than pursuit of a psychotropic effect. Dependence refers to long-term, regular use beyond that recommended for smoking cessation. Of note, with respect to the last, is that some smokers may use nicotine replacement products for long periods of time to prevent relapse or maintain cessation. Thus, it is difficult to distinguish the transference of nicotine dependence from tobacco to a medication source from extended treatment with nicotine products to maintain nonsmoking.

Nicotine from cigarettes maintains addiction from both positive reinforcement (via direct stimulation of nicotinic receptors in the brain) and negative reinforcement (avoidance of nicotine withdrawal symptoms).[5,49] As discussed earlier, the effects of nicotine in the brain, including acute psychotropic effects, are more intense when nicotine is delivered rapidly, as occurs with cigarette smoking. In contrast, most nicotine replacement products have very little acute psychotropic effect because of the gradual introduction into the systemic circulation. This is particularly true for nicotine gum and transdermal nicotine. The rate of absorption from nicotine nasal spray is much slower than from cigarettes, but still is more rapid than from other products. Thus, there has been some concern that nicotine nasal spray may have a greater abuse potential.

The low abuse liability for nicotine polacrilex gum and transdermal nicotine has been shown in studies examining human reactions to nicotine, particularly "liking" ratings.[50] In this study, cigarette smoking, intravenous nicotine, and smokeless tobacco all produced positive changes in drug-liking score. Transdermal nicotine produced no change relative to placebo, and nicotine gum decreased the liking score relative to placebo. Whereas there is a greater potential for acute psychoactivity from nicotine nasal spray, its administration is quite aversive and is disliked by almost all novice users. In one clinical trial of nicotine nasal spray, participants used nasal spray for longer than 6 months, which has raised some concern about the development of dependence.[51] Likewise, about 10% of nicotine gum chewers use gum beyond 1 year, raising the same question.[52] In both cases, it is difficult to disentangle the development of dependence from the use of nicotine replacement therapy to maintain smoking cessation or out of fear of relapse to cigarette smoking. Dependence might be evidenced by escalation of dose over time, but no data with nicotine replacement therapy indicate that this is occurring. Postmarketing surveillance data on nicotine polacrilex gum and transdermal nicotine indicate that abuse, misuse, and dependence are not serious problems.[48]

19.7   Concomitant Smoking and Nicotine Treatment

There is concern that some individuals will continue to smoke cigarettes while using nicotine replacement products, or that different types of nicotine replacement products will be used together (for example, nicotine gum and nicotine patches). Concurrent cigarette smoking and nicotine replacement use is documented in many reports of clinical trials of transdermal nicotine. However, combined cigarette smoking and nicotine replacement therapy does not appear to produce harm to any greater extent than smoking alone. There are several reasons for this assertion.

First, as discussed previously, cigarette smoking is qualitatively more hazardous than nicotine replacement therapy because it delivers many toxins in addition to nicotine, and it delivers nicotine in a manner that produces more intense nicotinic effects that are more likely to be injurious to the body compared to nicotine replacement therapy.

Second, the use of nicotine replacement therapy to assist smoking cessation generally results in less cigarette smoking if people do smoke.[53,54] Suppression of cigarette smoking has been demonstrated in several clinical trials. The result of smoking fewer cigarettes is less exposure to tobacco smoke toxins other than nicotine as compared to smoking before nicotine replacement therapy.

Third, because of the flat dose response curve for nicotine described previously, even if nicotine levels are higher with concurrent cigarette smoking and nicotine replacement therapy compared to smoking alone, the effects of the additional nicotine are not likely to produce any toxicity greater than that produced by cigarette smoking alone. For all of these reasons, it is likely that cigarette smoking while using nicotine replacement therapy in individuals trying to quit smoking will be no more hazardous, and possibly less hazardous, than was cigarette smoking in the same individual prior to quitting.

19.8   Use of Nicotine in Nonsmokers

Most of the discussion above has focused on nicotine use in cigarette smokers, who already have a long history of exposure to nicotine. For this population, it is unequivocal that the risks of nicotine replacement therapy are negligible compared to the risks of continued smoking, and the benefits of quitting are tremendous.

In considering nonsmokers, the question is whether long-term exposure to nicotine poses substantial health hazards, and, if so, what is the nature of the hazard? There are no empirical data to answer this question at this time. The only controlled trials of nicotine therapy in nonsmokers have been in patients with ulcerative colitis. In such trials, individuals were started on low doses of nicotine, 5 mg/16 hour patches, and the doses gradually escalated to maximum doses of 15 mg/16 hours.[18,55] Most individuals tolerated this escalation well, although some did develop some symptoms of nicotine toxicity, primarily nausea, lightheadedness, headache, and sleep disturbance. Only a few withdrew from treatment because of side effects. One study of the effects of transdermal nicotine on cardiovascular risk factors, including white and red cell counts, lipid levels, and markers of endothelial damage and coagulation, has shown no adverse effect in nonsmokers over 12 weeks of treatment.[56]

A concern with long-term nicotine treatment in nonsmokers is whether nicotine accelerates atherogenesis. Some (but not all) studies of nicotine feeding to rabbits on a high cholesterol diet have shown acceleration of the development of atherosclerosis.[57,58] The mechanism is unclear, but could include effects on lipids, activation of endothelial damage — either directly or via induction of hemodynamic stress — and/or coagulation.

A human population that has been used to examine the risks of nicotine without exposure to combustion products is that of people who use smokeless tobacco — that is, snuff or chewing tobacco. Studies of the circadian cardiovascular effects of snuff or chewing tobacco use indicate similar effects on heart rate, blood pressure, and catecholamine release in smokeless tobacco users compared to smokers.[59] In Sweden, the use of oral snuff is widespread among men, and nicotine levels have been shown to be similar with snuff use and cigarette smoking.[60] Measurement of platelet activation, using urinary excretion of metabolites of thromboxane A2, which is released by platelets during activation, indicates that snuff users are similar to nontobacco users, both of whom show less evidence of platelet activation than do smokers.[61] This is reassuring, as thrombosis is a major mechanism of both atherogenesis and acute cardiac events.

Epidemiologic studies of snuff users have shown mixed results. One case control study found no increased risk of myocardial infarction in snuff users.[62] Another cohort study of Swedish construction industry workers showed a significant increase in risk of cardiovascular disease, including hypertension, myocardial infarction, and stroke, in snuff users.[63] Snuff users may have differed from nontobacco users in diet, lipid levels, and other risk factors, but the Swedish study does raise concern that long-term nicotine exposure might contribute to cardiovascular disease. Conceivably, nicotine therapy in nonsmokers could also have an adverse effect on pregnancy and fetal development, as discussed in an earlier section.

19.9 Risks Versus Benefits of Nicotine Therapy

While nicotine is a potential toxin, it appears to be well tolerated during weeks and months of nicotine medication therapy without evidence of adverse health effects, except perhaps in people with active cardiovascular disease and during pregnancy. Compared to cigarette smoking, which exposes an individual to carbon monoxide and many other combustion products as well as nicotine, nicotine replacement therapy is much less hazardous. Since treatment with nicotine medications can promote smoking cessation, and smoking cessation produces tremendous health benefits, nicotine replacement therapy appears to have a positive benefit/risk ratio for any smoker who cannot quit smoking without nicotine therapy.

Some individuals may stop smoking using nicotine therapy, but relapse when they stop the nicotine treatment. In such individuals, the safety of long-term nicotine maintenance therapy has to be considered. Again, because exposure to nicotine during nicotine replacement therapy is generally no greater than that during cigarette smoking, and because there is less exposure to other tobacco toxins, the benefit of nicotine maintenance therapy almost certainly outweighs the risks.

The risk benefit ratio for long-term nicotine use in nonsmokers being treated for medical diseases is less clear. For intermittent or short-term therapy, such as 3 to 6-month treatment for control of symptoms in ulcerative colitis, nicotine seems quite safe. The risk vs. benefit in the use of nicotine for many years, as might be the case in treating depression or attention deficit disorder, remains to be determined. Any analysis of benefit/risk will depend strongly on the magnitude of the beneficial effects of nicotine for particular disorders, which at this time is not fully understood.

Acknowledgments

Preparation of this paper was supported in part by USPHS grants DA02277 and DA01696 (NB) and by Hoechst Marion Roussel (JP) and Alza Corporation (NB).

The authors thank Drs. Bill Byrd, Donna Causey, Charles Gorodetzky, Jane Gorsline, John Hughes, and Saul Shiffman for comments that contributed to the development of this paper, and to Kaye Welch for editorial assistance.

References

1. Wilbert, J., *Tobacco and Shamanism in South America,* Yale University Press, New Haven, 1987.
2. Beeman, J. A., and Hunter, W. C., Fatal nicotine poisoning. A report of twenty-four cases, *Arch. Pathol.,* 24, 481, 1937.
3. Centers for Disease Control, Cigarette smoking among adults — United States, 1993, *J. Am. Med. Assoc.,* 273, 369, 1995.
4. Peto, R., Lopez, A. D., Boreham, J., Thun, M., and Heath, C., Jr., Mortality from tobacco in developed countries: indirect estimation from national vital statistics, *Lancet,* 339, 1268, 1992.
5. Benowitz, N. L., Pharmacology of nicotine: Addiction and therapeutics, *Annu. Rev. Pharmacol. Toxicol.,* 36, 597, 1996.
6. Silagy, C., Mant, D., Fowler, G., and Lodge, M., Meta-analysis on efficacy of nicotine replacement therapies in smoking cessation, *Lancet,* 343, 139, 1994.
7. Fiore, M. C., Smith, S. S., Jorenby, D. E., and Baker, T. B., The effectiveness of the nicotine patch for smoking cessation. A meta-analysis., *J. Am. Med. Assoc.,* 271, 1940, 1994.
8. Lam, W., Sze, P. C., Sacks, H. S., and Chalmers, T. C., Meta-analysis of randomized controlled trials of nicotine chewing-gum, *Lancet,* 2, 27, 1987.
9. Sutherland, G., Stapleton, J. A., Russell, M. A. H., Jarvis, M. J., Hajek, P., Belcher, M., and Feyerabend, C., Randomised controlled trial of nasal nicotine spray in smoking cessation, *Lancet,* 340, 324, 1992.
10. Oster, G., Huse, D. M., Delea, T. E., and Colditz, G. A., Cost-effectiveness of nicotine gum as an adjunct to physician's advice against cigarette smoking, *J. Am. Med. Assoc.,* 256, 1315, 1986.
11. Edelson, J. T., Weinstein, M. C., Tosteson, A. N. A., Williams, L., Lee, T. H., and Goldman, L., Long-term cost-effectiveness of various initial monotherapies for mild to moderate hypertension, *J. Am. Med. Assoc.,* 263, 408, 1990.
12. Palmer, K. J., Buckley, M. M., and Faulds, D., Transdermal nicotine. A review of its pharmacodynamic and pharmacokinetic properties, and therapeutic efficacy as an aid to smoking cessation, *Drugs,* 44, 498, 1992.
13. Levin, E. D., Westman, E. C., Stein, R. M., Carnahan, E., Sanchez, M., Herman, S., Behm, F. M., and Rose, J. E., Nicotine skin patch treatment increases abstinence, decreases withdrawal symptoms, and attenuates rewarding effects of smoking, *J. Clin. Psychopharmacol.,* 14, 41, 1994.
14. Benowitz, N. L., Jacob, P., III, and Savanapridi, C., Determinants of nicotine intake while chewing nicotine polacrilex gum, *Clin. Pharmacol. Ther.,* 41, 467, 1987.
15. Benowitz, N. L., Clinical pharmacology of transdermal nicotine, *Eur. J. Pharm. Biopharm.,* 41, 168, 1995.
16. Johansson, C. J., Olsson, P., Bende, M., Carlsson, T., and Gunnarsson, P. O., Absolute bioavailability of nicotine applied to different nasal regions, *Eur. J. Clin. Pharmacol.,* 41, 585, 1991.
17. Benowitz, N. L., and Jacob, P., III, Daily intake of nicotine during cigarette smoking, *Clin. Pharmacol. Ther.,* 35, 499, 1984.
18. Pullan, R. D., Rhodes, J., Ganesh, S., Mani, V., Morris, J. S., Williams, G. T., Newcomb, R. G., Russell, M. A., Feyerabend, C., and Thomas, G. A., Transdermal nicotine for active ulcerative colitis, *N. Engl. J. Med.,* 330, 811, 1994.
19. Silver, A. A., and Sanberg, P. R., Transdermal nicotine patch and potentiation of haloperidol in Tourette's syndrome (Ltr), *Lancet,* 342, 182, 1993.
20. Benowitz, N. L. and Gourlay, S.G., Cardiovascular toxicity of nicotine: implications for nicotine replacement therapy, *J. Am. Coll. Cardiol.,* 29, 1422, 1997.
21. Benowitz, N. L., Nicotine replacement therapy during pregnancy, *J. Am. Med. Assoc.,* 266, 3174, 1991.
22. Hellerstein, M. K., Benowitz, N. L., Neese, R. A., Schwartz, J., Hoh, R., Jacob, P., III, Hsieh, J., and Faix, D., Effects of cigarette smoking and its cessation on lipid metabolism and energy expenditure in heavy smokers, *J. Clin. Invest.,* 93, 265, 1994.

23. Henningfield, J. E., Stapleton, J. M., Benowitz, N. L., Grayson, R. F., and London, E. D., Higher levels of nicotine in arterial than in venous blood after cigarette smoking, *Drug Alcohol Depend.*, 33, 23, 1993.

24. Benowitz, N. L., Porchet, H., Sheiner, L., and Jacob, P., III, Nicotine absorption and cardiovascular effects with smokeless tobacco use: Comparison with cigarettes and nicotine gum, *Clin. Pharmacol. Ther.*, 44, 23, 1988.

25. Porchet, H. C., Benowitz, N. L., and Sheiner, L. B., Pharmacodynamic model of tolerance: Application to nicotine, *J. Pharmacol. Exp. Ther.*, 244, 231, 1988.

26. Benowitz, N. L., Kuyt, F., and Jacob, P., III, Influence of nicotine on cardiovascular and hormonal effects of cigarette smoking, *Clin. Pharmacol. Ther.*, 36, 74, 1984.

27. Benowitz, N. L., and Jacob, P., III, Intravenous nicotine replacement suppresses nicotine intake from cigarette smoking, *J. Pharmacol. Exp. Ther.*, 254, 1000, 1990.

28. Smith, E. W., Smith, K. A., Maibach, H. I., Andersson, P. O., Cleary, G., and Wilson, D., The local side effects of transdermally absorbed nicotine, *Skin Pharmacol.*, 5, 69, 1992.

29. McBride, P. E., The health consequences of smoking: Cardiovascular diseases, *Med. Clin. North Am.*, 76, 333, 1992.

30. Rivers, J. T., White, H. D., Cross, D. B., Williams, B. F., and Norris, R. M., Reinfarction after thrombolytic therapy for acute myocardial infarction followed by conservative management: Incidence and effect of smoking, *J. Am. Coll. Cardiol.*, 16, 340, 1990.

31. Galan, K. M., Deligonul, U., Kern, M. J., Chaitman, B. R., and Vandormael, M. G., Increased frequency of restenosis in patients continuing to smoke cigarettes after percutaneous transluminal coronary angioplasty, *Am. J. Cardiol.*, 61, 260, 1988.

32. Grines, C. L., Topol, E. J., O'Neill, W. W., George, B. S., Kereiakes, D., Phillips, H. R., Leimberger, J. D., Woodlief, L. H., and Califf, R. M., Effect of cigarette smoking on outcome after thrombolytic therapy for myocardial infarction, *Circulation*, 91, 298, 1995.

33. Kaijser, L., and Berglund B., Effect of nicotine on coronary blood-flow in man, *Clin. Physiol.*, 5, 541, 1985.

34. Maouad, J., Fernandez, F., Hebert, J. L., Zamani, K., Barrillon, A., and Gay, J., Cigarette smoking during coronary angiography: Diffuse or focal narrowing (spasm) of the coronary arteries in 13 patients with angina at rest and normal coronary angiograms, *Cathet. Cardiovasc. Diagn.*, 12, 366, 1986.

35. Rietbrock, N., Kunkel, S., Wörner, W., and Eyer, P., Oxygen-dissociation kinetics in the blood of smokers and non-smokers: Interaction between oxygen and carbon monoxide at the hemoglobin molecule, *Naunyn Schmiedebergs Arch. Pharmacol.*, 345, 123, 1992.

36. Allred, E. N., Bleecker, E. R., Chaitman, B. R., Dahms, T. E., Gottlieb, S. O., Hackney, J. D., Pagano, M., Selvester, R. H., Walden, S. M., and Warren, J., Short-term effects of carbon monoxide exposure on the exercise performance of subjects with coronary artery disease, *N. Engl. J. Med.*, 321, 1426, 1989.

37. Adams, K. F., Koch, G., Chatterjee, B., Goldstein, G. M., O'Neil, J. J., Bromberg P. A., and Sheps D. S., Acute elevation of blood carboxyhemoglobin to 6% impairs exercise performance and aggravates symptoms in patients with ischemic heart disease, *J. Am. Coll. Cardiol.*, 12, 900, 1988.

38. Sheps, D. S., Herbst, M. C., Hinderliter, A. L., Adams, K. F., Ekelund, L. G., O'Neil, J. J., Goldstein, G. M., Bromberg, P. A., Dalton, J. L., Ballenger, M. N., Davis, S. M., and Koch, G. G., Production of arrhythmias by elevated carboxyhemoglobin in patients with coronary artery disease, *Ann. Int. Med.*, 113, 343, 1990.

39. Benowitz, N. L., Fitzgerald, G. A., Wilson, M., and Zhang, Q., Nicotine effects on eicosanoid formation and hemostatic function: Comparison of transdermal nicotine and cigarette smoking, *J. Am. Coll. Cardiol.*, 22, 1159, 1993.

40. Working Group for the Study of Transdermal Nicotine in Patients with Coronary Artery Disease, Nicotine replacement therapy for patients with coronary artery disease, *Arch. Intern. Med.*, 154, 989, 1994.

40a. Joseph, A. N., Norman, S. M., Ferry, L. H. et al., The safety of transdermal nicotine as an aid to smoking cessation in patients with cardiac disease, *N. Engl. J. Med.*, 335, 1792, 1996.

41. Murray, R. P., Bailey, W. C., Daniels, K., Bjornson, W. M., Kurnow, K., Connett J. E., Nides M. A., and Kiley J. P., Safety of nicotine polacrilex gum used by 3,094 participants in the Lung Health Study, *Chest*, 109, 438, 1996.

42. Stewart, P. M., and Catterall, J. R., Chronic nicotine ingestion and atrial fibrillation, *Br. Heart J.*, 54, 222, 1985.

43. Dacosta, A., Guy, J. M., Tardy, B., Gonthier, R., Denis, L., Lamaud, M., Cerisier, A., and Verneyre, H., Myocardial infarction and nicotine patch: A contributing or causative factor? *Eur. Heart J.*, 14, 1709, 1993.

44. Pierce, J. R., Stroke following application of a nicotine patch, *DICP*, 28, 402, 1994.

45. Luck, W., Nau, H., and Steldinger, R., Extent of nicotine and cotinine transfer to the human fetus, placenta and amniotic fluid of smoking mothers, *Dev. Pharmacol. Ther.*, 8, 384, 1985.

46. Oncken, C., Replacement therapy during pregnancy, *Am. J. Health Behavior*, 20, 300, 1996.

47. Luck, W. and Nau, H., Exposure of the fetus, neonate, and nursed infant to nicotine and cotinine from maternal smoking, *N. Engl. J. Med.*, 311, 672, 1984.

48. Marion Merrell Dow (Hoechst Marion Roussel), NDA 18–612, NDA 20-066, Summary of Safety Information, Vol. 1, 19–48, August 20, 1993; Alza Corporation NDA 20-165, Integrated Summary of Safety Data, Vols. 11.51–11.55, August 7, 1995.

49. Benowitz, N. L., Cigarette smoking and nicotine addiction, *Med. Clin. North Am.*, 76, 415, 1992.

50. Henningfield, J. E. and Keenan, R. M., Nicotine delivery kinetics and abuse liability, *J. Consult. Clin. Psychol.*, 61, 743, 1993.

51. Sutherland, G., Russell, M. A. H., Stapleton, J., Feyerabend, C., and Ferno, O., Nasal nicotine spray: A rapid nicotine delivery system, *Psychopharmacology*, 108, 512, 1992.

52. Hughes, J. R., Hatsukami, D. K., and Skoog, K. P., Physical dependence on nicotine in gum, *J. Am. Med. Assoc.*, 255, 3277, 1986.

53. Abelin, T., Buehler, A., Muller, P., Vesanen, K., and Imhof, P. R., Controlled trial of transdermal nicotine patch in tobacco withdrawal, *Lancet*, 1, 7, 1989.

54. Transdermal Nicotine Study Group, Transdermal nicotine for smoking cessation, *J. Am. Med. Assoc.*, 266, 3133, 1991.

55. Thomas, G. A. O., Rhodes, J., Mani, V., Williams, G. T., Newcombe, R. G., Russell, M. A. H., and Feyerabend, C., Transdermal nicotine as maintenance therapy for ulcerative colitis, *N. Engl. J. Med.*, 332, 988, 1995.

56. Thomas, G. A. O., Davies, S. V., Rhodes, J., Russell, M. A. H., Feyerabend, C., and Säwe, U., Is transdermal nicotine associated with cardiovascular risk? *J. Roy. Coll. Phys. (London)*, 29, 392, 1995.

57. Strohschneider, T., Oberhoff, M., Hanke, H., Hannekum, A., and Karsch, K. R., Effect of chronic nicotine delivery on the proliferation rate of endothelial and smooth muscle cells in experimentally induced vascular wall plaques, *Clin. Invest.*, 72, 908, 1994.

58. Stefanovich, V., Gore, I., Kajiyama, G., and Iwanaga, Y., The effect of nicotine on dietary atherogenesis in rabbits, *Exp. Mol. Path.*, 11, 71, 1969.

59. Benowitz, N. L., Jacob, P., III, and Yu, L., Daily use of smokeless tobacco: Systemic effects, *Ann. Int. Med.*, 111, 112, 1989.

60. Holm, H., Jarvis, M. J., Russell, M. A. H., and Feyerabend, C., Nicotine intake and dependence in Swedish snuff takers, *Psychopharmacology*, 108, 507, 1992.

61. Wennmalm, A., Benthin, G., Granström, E. F., Persson, L., Peterson, A., and Winell, S., Relation between tobacco use and urinary excretion of thromboxane A_2 and prostacyclin metabolites in young men, *Circulation*, 83, 1698, 1991.

62. Huhtasaari, F., Asplund, K., Lundberg, V., Stegmayr, B., and Wester, P. O., Tobacco and myocardial infarction: Is snuff less dangerous than cigarettes? *Br. Med. J.*, 305, 1252, 1992.

63. Bolinder, G., Alfredsson, L., Englund, A., and deFaire, U., Smokeless tobacco use and increased cardiovascular mortality among Swedish construction workers, *Am. J. Public Health*, 84, 399, 1994.

20

Systemic Drugs for Skin Diseases

Hélène Bocquet
Hôpital Henri Mondor

Laurence Le Cleach
Hôpital Henri Mondor

Jean-Claude Roujeau
Hôpital Henri Mondor

20.1 Introduction

Many years have elapsed since dermatologic therapy was mainly based on topical paintings with esoteric preparations. The first step in changing therapeutic attitudes was the large use of topical steroids (which are discussed in another section of this book). Nowadays dermatologists frequently prescribe systemic treatments in many dermatologic disorders. Some examples are short-term use of antiinfectious agents (antibacterial, antiviral, antifungal) in the management of skin infections, long-term use of antibacterial agents in the treatment of acne, and use of systemic steroids and immunosuppressive agents in the management of autoimmune blistering disorders. None of these treatments are specific to dermatology and will not be discussed here. In this chapter we will focus on those treatment that are more specific to skin diseases, including antimalarials, dapsone, thalidomide, and drugs used in the management of psoriasis: oral retinoids, methotrexate and cyclosporine.

20.2 Limitation of the Benefit/Risk Analysis in Dermatologic Therapy

A benefit/risk analysis should not be restricted to putting together statements on the efficacy of a therapy and the list of its potential side effects. The quantification of both the benefits and the risks of the treatment should be performed to allow the calculation of a valid ratio. That can easily be done for some treatments. For instance, penicillin reduced to nearly zero the risk of death from pneumococcal pneumonia. The risk of short-term therapy with penicillin is related to anaphylactic shock only, with a risk of death of about 1 in 100,000 courses of treatment. In such a situation a benefit/risk ratio can easily be calculated as the ratio of the risk of death from pneumonia to the risk of death from anaphylactic shock with penicillin.

For most dermatologic diseases we cannot calculate the benefit/risk ratio in the same way because most of these diseases are not lethal. The benefit of therapy is much more difficult to appreciate. This results from two factors, the first is the rate of cure or the percentage of improvement given by the treatment and the second is the impact of the disease on the daily life of the patient.

At present information is available only on the rates of remission provided by some dermatologic treatments. The real impact of chronic dermatologic diseases on the quality of life, on the social, familial, and personal behavioral conduct of patients has not been correctly evaluated. This impact is probably much more important than previously appreciated by most physicians, including dermatologists. For instance, a recent study using quantitative measures of patients' preference (so-called utility approach) has demonstrated that patients with psoriasis considered that widespread forms of their disease had the same severity as chronic renal failure needing hospital dialysis.[1] That means that for patients suffering from severe psoriasis a significant improvement in their disease would provide more benefit than can be suspected by physicians who still consider psoriasis as a benign condition with a small increase in morbidity.

Unfortunately, these utility approaches of dermatologic therapy are only in a very preliminary phase and presently we cannot reliably quantify the benefit of dermatologic treatment in a better way than by appreciating the rates of total or partial remissions provided by treatments.

20.3   Amino-4-Quinoleine–Derived Antimalarials

Chloroquine and hydroxychloroquine are frequently used by dermatologists for their antiinflammatory and photoprotective properties.[2] Accepted indications are: curative treatment of discoid (chronic) lupus erythematosus (DLE) and prophylaxis of relapses of systemic lupus erythematosus (SLE).[3] Antimalarials are also considered useful in the prevention of polymorphous light eruption,[4] as a treatment of porphyria cutanea tarda,[5] and in many other diseases, including cutaneous sarcoidosis, urticarial vasculitis, dermatomyositis, and benign cutaneous lymphocytic infiltrates.[2] In these later indications proof of efficacy is usually lacking.

Antimalarials have been used since 1951 in the management of lupus erythematosus. Controlled studies were never performed in DLE, but large series showed rates of complete remission in the range of 50% to 60% and total rates of partial remission in the range of 70% to 90%.[2,3] Recently a controlled study has indirectly demonstrated the efficacy of antimalarials in SLE by showing that the withdrawal of hydroxychloroquine increased the risk of relapse.[6] In DLE clinical results are usually obtained after 3 to 6 weeks of therapy using 400 mg daily of hydroxychloroquine or 200 mg of chloroquine. The treatment is only suppressive and relapses occur in 50% to 100% of cases after withdrawal of the drugs.

In the management of DLE most dermatologists decrease the daily dosage after remission is obtained and even discontinue treatment in winter to resume it at lower doses in summer.

Lethal complications of chloroquine and hydroxychloroquine treatments are restricted to overdosage, with severe cardiac toxicity occurring after acute ingestion of more than 2 g of chloroquine (20 tablets).[7] Risk of overdosage is a major concern when chloroquine and hydroxychloroquine are used in children, but does not justify a contraindication.[8] At therapeutic doses mild gastrointestinal symptoms are the most frequent complications, occurring in up to 10% of patients. Skin and mucous membrane pigmentations are also frequent. Skin reactions led to discontinuation of treatment in up to 3% of patients. Myopathy occurs in about 1% of patients in long-term treatment.[9]

Of more concern for long-term treatment is the potential occurrence of retinal lesions. Retinopathy is dose-dependent. The risk is negligible for daily doses below 3.5 mg/kg with chloroquine and 6 mg/kg for hydroxychloroquine.[10] Complete remission of skin lesion of DLE often requires daily doses slightly above these "safe" doses. In these cases regular ophthalmological examination is recommended. The risk of occurrence of retinopathy has been estimated to 3% to 10%.[2,10] Early involvement of the retina can be detected by Amsler grid or color vision tests at a time when vision is not yet impaired. Discontinuation of treatment at that stage prevents the occurrence of symptomatic lesions.

Because the occurrence of the most deleterious complications can be prevented by the strict observation of dosage and regular ophthalmological examinations, long-term antimalarial treatment of DLE is generally considered safe and the benefit/risk ratio is acceptable.

20.4 Dapsone

Chemically related to sulfonamides, dapsone (4-4' di-amino-phenyl-sulfone, DDS) was initially prescribed 60 years ago as an antibacterial agent. It is used for the treatment of leprosy and as prophylaxis of HIV-related *pneumocystis carinii* pneumonia. It was first used by dermatologists in the treatment of dermatitis herpetiformis and after that in many other dermatological conditions, mostly those characterized by infiltration of tissues by neutrophils.[11] Those disorders include: erythema elevatum diutinum,[12] subcorneal pustulosis (Sneddon-Wilkinson),[13] linear IgA bullous diseases of children and adults,[14] vesiculo-bullous lupus erythematosus,[15] pemphigus foliaceus,[16] cutaneous vasculitis,[17] pyoderma gangrenosum,[11] and Sweet syndrome.[11]

The most clearly validated use of dapsone is in the treatment of dermatitis herpetiformis, a recurrent pruritic and blistering disorder usually beginning in young adults and lasting for life.[18] The incidence of dermatitis herpetiformis is 10 to 20 per 100,000 in northern Europe. It is much rarer in southern Europe. The disease is characterized by granular deposits of IgA in the dermal papillae. It is usually associated with gluten-sensitive enteropathy, most often asymptomatic. Controlled therapeutic trials of dapsone in the treatment of dermatitis herpetiformis were never performed, but large series showed that 95% to 97% of cases were controlled by dapsone.[18-20] Complete remission is usually obtained with daily doses of 100 to 200 mg. Occasionally, higher dosages may be necessary. Therapeutic response is obtained within 48 h, relapses are equally rapid after the drug is stopped. Most dermatologists prescribe an initial low dose (50 mg per day), increased after a few days in successive stages up to the dosage allowing a total control of the disease. Then the doses are progressively decreased to the minimum dose suppressing the eruption. The more the patient is capable of adhering to a gluten-free diet, the lower are the maintenance doses of dapsone. In most cases with good adherence to diet, the drug can be discontinued without relapse.[21]

Dapsone has a sizeable efficacy, with remissions in more than 50% of cases in a few other diseases, linear IgA bullous diseases of adults and children, erythema elevatum diutinum, and Sneddon-Wilkinson disease. All other potential indications are only based on anecdotal case reports.

The morbidity from treatment is considerable. Side effects leading to withdrawal of the drug occur in 20% to 30% of patients.[20] These side effects are considered severe in 6% of treated patients.[20] Hemolysis and methemoglobinemia are nearly constant and dose-dependent. With a daily dose of 100 mg to 150 mg the hemoglobin level is reduced by 2 g/dl as a mean. That degree of anemia is well tolerated by young patients, but the long-term effects of hemolysis have not been well studied. Hemolytic crises may be life-threatening in patients with congenital deficiency in glucose 6-phosphate dehydrogenase. Severity of hemolysis should be checked by blood count 1 to 2 weeks after the beginning of therapy, 1 month later, and then at 3 months intervals later on. Methemoglobinemia may induce cyanosis and dyspnea when more than 10% of the total amount of hemoglobin have been oxidized, a level often reached with daily doses of dapsone of 200 mg or more.

The most dangerous side effect is agranulocytosis. From data of the adverse drug reactions monitoring system in Sweden, it has been calculated that agranulocytosis occurs in 1 in 240 to 1 in 425 patients treated with dapsone.[22] Because the death rate of agranulocytosis is about 9% to 10%,[23] the risk of death from agranulocytosis can be expected to be 1 in 3000 treated patients.

Many other side effects have been attributed to dapsone. Peripheral neuropathies are considered rare,[24] but in one series of 161 patients with dermatitis herpetiformis, 6 had nerve toxicity.[20] Skin eruptions may be severe, including rare cases of Stevens-Johnson syndrome, toxic epidermal necrolysis, and the so-called "Dapsone syndrome,"[25] a hypersensitivity reaction consisting of fever, papular eruption, lymph node enlargement, eosinophilia, and hepatitis. These severe reactions have also induced some fatalities.

Their rate of incidence is unknown. From that compilation of side effects, it is clear that dapsone has a poor safety profile and should be avoided in those diseases where its efficacy has not been well demonstrated. Even in dermatitis herpetiformis, where its efficacy is very dramatic, dapsone should probably be used only in those patients who are not able to adhere to a gluten-free diet.

When dapsone is prescribed, blood examination should be performed at regular intervals, including complete blood count, methemoglobinemia, and liver function tests at regular intervals at the beginning of therapy.

20.5 Thalidomide

Initially launched as a sedative and hypnotic agent, thalidomide was withdrawn from the drug market in the 1960s after the disaster of birth defects. More than 10,000 babies with phocomelia (abnormally short legs and arms) were born from mothers who had taken thalidomide in the first month of pregnancy.[26,27] These horrible side effects conferred upon thalidomide the reputation of the worst possible benefit/risk ratio that one could expect from a drug. At the same time, thalidomide was discovered by chance to be very effective in improving the skin lesions and constitutional symptoms of patients with erythema nodosum leprosum, a severe lepra reaction. Thalidomide has since then been used successfully in many conditions.[26,27] Controlled trials have confirmed the dramatic effectiveness of thalidomide in erythema nodosum leprosum with a 99% response in a few days.[28] Controlled therapeutic trials were also performed in severe aphthous stomatitis (48% complete response),[29] and in Jessner lymphocytic infiltrate of the skin (75% complete remission).[30] In addition, open studies have suggested that high rates of remission can also be obtained in discoid lupus erythematosus, prurigo nodularis, recurrent erythema multiforme, Langerhans cell histiocytosis, chronic graft-versus-host disease, and mucosal ulcers of patients infected with HIV.[27]

Preclinical studies on thalidomide are now in progress in various conditions, e.g., inflammatory bowel diseases, multiple sclerosis, AIDS, cachexia in patients with HIV infection and in patients with cancer, and Kaposi's sarcoma in patients with HIV infection.

Thalidomide decreases the production of tumor necrosis factor alpha.[27] This effect could contribute to the antiinflammatory effects of the drug in many different disorders.

The mechanisms of birth defects remain unknown. The incidence of these defects is not known precisely, but in the German experience in the 1960s it seemed that almost all women exposed to thalidomide in the third or fourth week of gestation had malformed newborns. The risk is so high that the prescription of thalidomide to fertile women must be restricted to highly controlled situations where patients understand the risks (including the possibility of contraceptive failure) and accept the need to change their behavior to minimize that risk.[31] In such conditions it is possible to reduce to nearly zero the potential of causing birth defects. Then the benefits of the drug in improving severe diseases can be considered as outweighing the risk.

Many other side effects of thalidomide are more difficult to control than the risk of teratogenesis. Sedation is nearly constant at high doses. Hypothyroidism has been occasionally reported; its incidence is not known.

With appropriate measures to prevent teratogenesis, peripheral neuropathy became the most common side effect of treatment with thalidomide. The incidence of thalidomide-related neuropathy is variably estimated from 0.5% to more than 70%.[27] It presents itself as a sensitive axonal neuropathy. The prognosis is characterized by slow regression after drug withdrawal with frequent persistent paresthesia.

The following recommendation can provide the safest possible clinical use of that drug:[31]

- In fertile women thalidomide should be prescribed only for severe conditions and after failure of other treatments.
- Patients should be fully informed of all risks and sign a written consent.
- They should have a negative pregnancy test at initiation of therapy.
- They should accept reliable contraceptive methods during the full length of treatment.

- All patients should have neurological examinations and appropriate electrophysiological measurements before treatment and at at regular intervals.
- Follow-up visits and examinations should be performed at monthly intervals.

Thalidomide is the perfect example that the benefit/risk ratio of a drug may change dramatically with time. In that example the drug indications moved from a mild sedative to a very potent antiinflammatory agent very efficient in several disabling disorders. Moreover, its most dramatic side effect turned out to be amenable to control.

20.6   Isotretinoin

Isotretinoin (13-cis-retinoic acid) is a synthetic isomer of the natural vitamin A derivative retinoic acid. Its efficacy as a systemic treatment of severe acne was demonstrated more than 15 years ago.[32-35] Severe cystic acne is the only accepted indication of isotretinoin in Europe, while other retinoids (etretinate, acitretine) are used in the management of psoriasis and keratinization disorders.

Many controlled studies have demonstrated the efficacy of isotretinoin in severe acne at daily doses ranging from 0.1 to 1 mg/kg/day.[32-35] The most usual dosage is 0.5 to 1 mg/kg/day. It is generally accepted that the total cumulative dose is more important than the level of the daily dose for inducing a long-lasting remission. The usual recommendation is a cumulative dosage of 120 to 150 mg/kg.[36,37] The anti-acneic effect of isotretinoin is mainly mediated through the suppression of sebaceous secretion.

Several double-blind placebo-controlled trials have demonstrated the therapeutic value of isotretinoin in cystic acne. Lesions are improved by 50% to 65% after 12 weeks and 80% to 90% after 16 weeks of treatment.

After complete remission and drug withdrawal, relapses are observed in 20% to 40% of patients. In only half of these patients will the relapses be severe enough to need another course of isotretinoin.[36] The treatment of severe cystic acne with isotretinoin has tremendously decreased the risk of scars and dysmorphic sequelae.[38]

Isotretinoin treatment of acne is associated with numerous side effects. Dry lips and cheilitis are nearly constant. Other mucocutaneous side effects occur in 30% to 60% of patients: dry skin, blepharo-conjunctivitis, facial erythema, and dermatitis. These side effects are dose-dependent. These are usually tolerable at a daily dosage of 0.5 mg/kg but can be difficult to tolerate at higher dosage. Other cutaneous side effects are less frequent: increased risks of staphylococcal infections, transient exacerbation of acne, photosensitivity, alopecia, and folliculitis of the scalp.

Systemic side effects are not rare. Myalgia and arthralgia are often reported by patients; tendinitis may occur. Many patients complain of headache. Rarely, and usually in case of simultaneous therapy with tetracycline antibiotics, intracranial hypertension can occur (pseudotumor cerebri). Like all retinoids isotretinoin impairs lipoprotein metabolism. Moderate increases in the level of blood cholesterol occur in about 10% of patients and moderate rises in triglyceride level in 50%.[38,39] Because isotretinoin is used for a few months only, these alterations in lipid metabolism are generally considered as not able to induce a significant increase in the risk of cardiovascular disorders. Slight elevations of aminotransferases can occur in less than 5% of patients and hepatotoxicity is not a major problem of isotretinoin.

The most dangerous side effect of isotretinoin is teratogenicity. Several studies of human pregnancies with fetal exposure to isotretinoin showed that about 30% of newborns had major malformations.[40] Characteristic abnormalities involve the face (microtia/anotia), the heart, and the central nervous system. These defects are similar to those induced in animals by retinoids or vitamin A.

The risk of embryopathy has prompted extensive programs of information for physicians and patients. Unfortunately, these programs have failed in totally preventing the exposure of pregnant women to isotretinoin. A recent study detected 402 pregnancies among 124,216 women exposed to isotretinoin. That number of pregnancy was 8% of the expected number in the general population of similar age distribution.[41] Such a figure clearly demonstrates the impossibility of reducing the teratogenic risk to zero for a drug that is widely prescribed in a population of teenagers and young adults.

Prescribers should remain aware that isotretinoin, the most efficient drug against severe acne to date, is also a dangerous product. That drug should not be used in women of childbearing age for acne of moderate severity, but only for severe cystic acne and/or after failure of other well-conducted therapies.[42] In women, pregnancy tests should be performed before the prescription of the drug, and effective contraceptive methods should be used for the whole duration of the therapy, beginning 1 month before and lasting 1 month after the treatment.

20.7   Systemic Treatments of Psoriasis

Psoriasis is a common disease, with an estimated prevalence of about 2%. The lesions of psoriasis are characterized by accelerated proliferation and abnormal terminal differentiation of epidermal cells. The disease occurs on a genetic basis that is not yet elucidated. The abnormal proliferation of epidermal cells may be triggered by activated T lymphocytes present in the dermis. Recent investigations on the mechanisms of psoriasis have focused on these T lymphocytes and on the activation of the cytokine network, with the hypothesis of an autoimmune reaction against unknown epidermal cell antigens.

The clinical course of psoriasis is that of a chronic disorder with spontaneous remissions and relapses. In more than 80% of cases psoriasis is a mild disease presenting a few chronic plaques. Death directly related to psoriasis, e.g., exfoliative dermatitis, systemic amyloidosis is very rare. About 10% to 15% of patients suffering from psoriasis have severe alterations in their quality of life in case of widespread lesions, involvement of the face or hands, or psoriatic arthritis. Systemic therapy of psoriasis should be considered in severe forms, when an adequate control of the disease cannot be obtained by topical therapies or by photochemotherapy (PUVA). We want to discuss the three drugs that are often prescribed in the treatment of severe psoriasis: retinoids, methotrexate, and cyclosporine.

Because most dermatologists agree that systemic therapy of psoriasis should be restricted to selected severe cases, it should theoritically be easy to compare the benefits from the different drugs used in these similar indications. Practically that comparison is difficult because these drugs have been proposed as a treatment for psoriasis many years apart. Criteria for inclusion of patients and more importantly criteria used in the evaluation of outcome have varied with time. For example, most results of treatment with methotrexate were appreciated as rates of complete or partial responses whereas studies of more recent treatments have relied on the introduction of a scoring system, the psoriasis area and severity index (PASI).[43] Most therapeutic trials in psoriasis nowadays use the PASI score. Clinical benefits from the treatment are usually expressed as a percentage of reduction from the initial PASI score. Because the PASI score integrates several indexes of inflammation, which often improve or worsen simultaneously, it provides some amplification of any clinical variation. That means that a 50% reduction in the PASI score would probably be appreciated as less than 50% improvement by physicians or a patient. A good correlation has been observed between the PASI score and a psoriasis disability index assessing patients' suffering and disability.[44] In spite of that overall correlation between the PASI and indexes of disability, controlled therapeutic trials should incorporate several of such indexes to better appreciate the benefits from therapy.

20.7.1   Retinoids

Two synthetic retinoids are used in the management of psoriasis, etretinate and its active metabolite acitretin.[45] Both drugs are initially prescribed at low doses (10–20 mg/d), progressively increased up to 0.5–1 mg/kg/d.[45] After remission (usually in 6 weeks to 3 months) doses are progressively tapered to the minimum effective dose.

When used in low doses etretinate and acitretine have a moderate activity in psoriasis.[46] Even with high doses (1 mg/kg/d) the rate of remission in chronic plaque psoriasis does not exceed 60%, and complete remissions are obtained in only 25% of patients.[45,47] Efficacy is better in some severe variants of psoriasis, exfoliative dermatitis and pustular psoriasis (either generalized or palmo-plantar).[45,48]

It has been demonstrated by several controlled studies that the combination of etretinate or acitretin with photochemotherapy increased the efficacy of PUVA and decreased the cumulative dose of UVA needed to obtain a remission.[49,50]

The side effects of etretinate and acitretin are essentially the same as those of isotretinoin.[51] The major difference in appreciating what risks can be accepted is the expected duration of treatment. Isotretinoin is prescribed for a few months in acne, while treatment of psoriasis should be scheduled for the long term.

Teratogenicity is the main problem[40] with an important pharmacodynamics difference with isotretinoin. Because of the accumulation and long elimination half-life of etretinate and because of some biochemical *in vivo* transformation of acitretin to etretinate, pregnancy is contraindicated for 2 years after the discontinuation of treatment even if therapy has been short.[45] Fertile women should be clearly advised of that risk and understand the need for a prolonged contraception.

The second main problem concerns the effects on blood lipids. Etretinate and acitretine increase the triglyceride level and induce a shift of cholesterol from the high-density lipoproteins (HDL) to the low-density lipoproteins (LDL). About 20% to 30% of patients treated with etretinate or acitretine develop hypertriglyceridemia or an increase in the LDL/HDL ratio.[51,52] The direct demonstration that retinoids increase the risk of atherosclerosis is still lacking, but high LDL/HDL ratios are well demonstrated risk factors for the development of ischemic heart disease.[52]

In addition to these two main side effects there are some concerns about hepatotoxicity, which seems uncommon (acute hepatitis and liver cirrhosis each occurred in 1% of 956 patients on long-term therapy with etretinate[53]) and about mucocutaneous side effects, which are very frequent at low doses and nearly constant at high doses but which are most often tolerable and improve with dosage reduction.[51]

Before starting a treatment with acitretine or etretinate, a serum pregnancy test should be performed in women. Blood lipid and aminotransferase levels should be determined in all patients. These measurements should be repeated after 2 weeks and then monthly.

20.7.2 Methotrexate

Methotrexate has been proposed for psoriasis since 1969.[54] At that time the drug was expected to control the proliferation of keratinocytes. Nowadays, the drug is rather considered to improve psoriasis by modulating the immune response. The drug is used in weekly dosage, orally or by intramuscular injections. The treatment begins with low doses, 7.5–10 mg per week and is progressively increased to the usual effective dose 15–25 mg per week.

The efficacy of methotrexate in cutaneous manifestations of psoriasis has never been studied by double-blind, placebo-controlled trials. Several studies on large numbers of patients have showed that the rates of complete remission were in the range of 50%. In addition about 30% of patients had a "good" partial response.[54-56] A randomized double-blind placebo-controlled trial of low-dose methotrexate in psoriatic arthritis has demontrated a mild improvement of doubtful clinical relevance.[57]

Side effects of methotrexate treatment of psoriasis are frequent. In the largest series, 70% of patients reported at least one side effect.[55,56] These side effects resulted in withdrawal of the drug in 20% to 30% of cases. The most frequent side effects are nausea, asthenia, and headache. Hematologic toxicity is not exceptional, and several studies (mainly in patients with rheumatoid arthritis) showed a risk of pancytopenia in the range of 1% to 2%.[58] That risk is more important in case of dehydratation and in case of associated therapy with other antimetabolites, including folic acid antagonists. Hypersensitivity pneumonitis are much more rarely reported in psoriasis than in rheumatoid arthritis treated with methotrexate.

Methotrexate is usually considered noncarcinogenic. Methotrexate alone does not increase the risk of skin carcinoma in patients treated for psoriasis.[59] Patients treated with methotrexate, however, may have an increased risk of developing skin carcinoma from PUVA therapy.[60] Moreover, two cases of Epstein-Barr virus-induced reversible lymphomas have been related to methotrexate.[61]

Hepatotoxicity is the side effect of methotrexate that has been the most extensively discussed in the dermatologic literature; numerous studies provided contradictory results. In one series 25% of patients treated for more than 5 years with methotrexate developed liver cirrhosis; the pretreatment prevalence of cirrhosis was 0.6%.[62] A recent series including 55 patients (treated for 8 years as a mean) found 4% of liver cirrhosis.[56] Another recent study compared the liver biopsies performed in the same patients at several year intervals, the first after a cumulative dose of methotrexate of 2.7 g and the second after a mean cumulative drug dose of 5 g. In that study there were no changes either in the rate of liver fibrosis (22% on the first biopsy, 20% on the second one) nor in the rate of cirrhosis (0% in both).[63]

The risk of cirrhosis induced by methotrexate is probably modulated by cofactors, including obesity, alcohol intake, and exposure to other potentially hepatotoxic drugs. Guidelines have recommended that patients should have a liver biopsy soon after the beginning of treatment, after a cumulative dosage of 1.5 g, and at regular intervals every additional 1.5 g cumulative doses.[64] Such guidelines are based on the observation that important histological alterations can be found even in the absence of alteration in the liver function tests or in noninvasive investigations of the liver. These recommendations have been challenged by some dermatologists, who consider that the risk of cirrhosis is low in patients who do not have additional risk factors and that the morbidity associated with liver biopsy is not negligible. Large series of liver biopsies showed a mortality in the order of 1 per 1000 patients (0.3–1.7 per 1000).[65,66] The risk of dying from repeated liver biopises may rise very close to the risk of dying from liver cirrhosis in the context of methotrexate therapy for patients without additional risk factors.

20.7.3   Cyclosporine

Improvement of psoriasis by cyclosporine was discovered by chance in a patient who had a renal transplant and concomitant psoriasis. Efficacy has been confirmed by several double-blind placebo-controlled therapeutic trials.[46,67-69] These trials and many open studies[70-74] have demonstrated that the efficacy is proportional to the daily dose. The PASI score is decreased by at least 60% in 20% of patients treated with very low doses (1.25 mg/kg/d), in 64% of patients treated with intermediate doses (2.5–3 mg/kg/d) and in 81% of patients receiving at least 5 mg/kg/d. These results are obtained between 10 and 16 weeks, depending on the studies.

The rates of complete remission are not reported in many of these studies. A complete remission was obtained in 3% of patients after 12 weeks, at a daily dosage of 2.5 mg/kg in one study[71] and in 20% of patients using the same dosage for 10 weeks in another study.[46]

The benefits of cyclosporine are purely suppressive. Three months after withdrawal of the drug, 50% to 70% of patients had relapsed.

Side effects of cyclosporine are frequent. In long-term studies (more than 6 months duration) 6% to 18% of patients interrupted the treatment because of side effects.[73,74] Most side effects, however, are mild, headache, paresthesia, hirsutism, and influenza-like symptoms. The overall incidence of side effects reported in large series is 30% to 55%. Three side effects are of real concern, high blood pressure, renal toxicity, and potential promotion of cancer.

High blood pressure was reported in 5% to 26% of patients in short-term trials. Some studies suggested that the occurrence of high blood pressure was more frequent in patients receiving higher doses, but such a relationship was not found in most studies.[73,75] High blood pressure usually responds to calcium channel blockers, mainly nifedipine or isradipine, which do not interfere with cyclosporine metabolism.[75] If blood pressure is not controlled by this treatment, cyclosporine doses should be decreased or discontinued. Blood pressure then returns to normal.

Renal impairment is the most frequent severe side effect. In a recently published long-term study,[73] 46% of patients had at least an increase in serum creatinine of more than 30% above the baseline value. In some instances, in relation to higher doses, the creatinine level was more than 100% above the baseline. In 6% of patients, the creatinine level remained increased after discontinuation of therapy. Renal toxicity from cyclosporine is attributed to a vasoconstrictive effect on kidney arteries. Prolonged vasocontriction may lead to permanent alteration of blood vessels and to chronic fibrosis of the interstitium. Kidney biopsies indicated morphologic alterations in 21% of 192 patients receiving high doses of cyclosporine for autoimmune disorders.[76] Because of the risk of chronic and irreversible alteration of the kidney, it has been recommended that cyclosporine should never be used for more than 2 years in patients with psoriasis.

The last concern is the potential promotion of cancer. It has been well documented that the profound immunosuppression needed for organ transplantation was a risk factor for the development of Kaposi's sarcoma, squamous cell carcinoma of the skin, and lymphoma. In transplant recipients the risk of occurrence of non-Hodgkin's lymphoma has been calculated to be about 30 times higher than in the general population.[77] The extent of the risk seems directly proportional to the degree of immunosuppression. Because it is more potent than conventional immunosuppressive drugs, cyclosporine appeared associated to a higher incidence of proliferative disorders in organ transplant recipients.[78] Most lymphomas occurring after cyclosporine therapy were B cell lymphomas with markers of Epstein-Barr virus infection.[79]

It should be stressed that in organ transplant recipients cyclosporine is usually given in addition to other immunosuppressive agents, and that the doses are much higher than those used in the management of psoriasis. To date only 3 cases of lymphoma have been reported in patients with psoriasis treated with cyclosporine.[80,81] The risk is probably much lower than in transplant recipients. Several cases of transient lymphoproliferation have also been observed.[82,83] Such observations clearly demonstrate that there is a risk of lymphoproliferation with the doses of cyclosporine used in patients with psoriasis, even if that risk is definitely lower than for organ transplant recipients.

References

1. Zug, K. A., Littenberg, B., Baughman, R. D., Kneeland, T., Nease, R., Sumner, W., O'Connor, G., Jones, R., Morrison, E., Cimis, R., Assessing the preferences of patients with psoriasis: a quantitative, utility approach, *Arch Dermatol,* 131, 561, 1995.
2. Weiss, J. S, Antimalarial medications in dermatology, a review, *Dermatol Clin,* 9, 377, 1991.
3. Dubois, E.L., Antimalarials in the management of discoid and systemic lupus erythematosus, *Sem Arthritis Rheum,* 8, 33, 1978.
4. Murphy, G. M., Hawk, J. L. M., Magnus, I. A., Hydroxychloroquine in polymorphic light eruption: a controlled trial with drug and visual sensitivity monitoring, *Br J Dermatol,* 116, 379, 1987.
5. Cainelli, T., Di Padova, C., Marchesi, L., Gori, G., Rovagnati, P., Podenzani, S., Bessone, E., Cantoni, L., Hydroxychloroquine vs. phlebotomy in the treatment of porphyria cutanea tarda, *Br J Dermatol,* 108, 593, 1983.
6. The Canadian Hydroxychloroquine Study Group, A randomized study of the effect of withdrawing hydroxychloroquine sulfate in systemic lupus erythematosus, *N Engl J Med,* 324, 150, 1991.
7. Riou, B., Barriot, P., Rimailho, A., Baud, F. J., Treatment of severe chloroquine poisoning, *N Engl J Med,* 318, 1, 1988.
8. Rasmussen, J. E., Antimalarials, are they safe to use in children?, *Ped Dermatol,* 1, 89, 1983.
9. Estes, M. L., Ewing-Wilson, D., Chou, S. M., Mitsumoto, I., Hanson, M., Shirey, K., Ratliff, A., Chloroquine neuromyotoxicity, clinical and pathological perspective, *Am J Med,* 82, 447, 1987.
10. Cox, N. H., Paterson, W. D., Ocular toxicity of antimalarials in dermatology: a survey of current practice, *Br J Dermatol,* 131, 878, 1994.
11. Stern, R. S., Systemic dapsone, *Arch Dermatol,* 129, 301, 1993.
12. Katz, S., Gallin, J., Hertz, K., Fauci, A., Lawley, T., Erythema elevatum diutinum: skin and systemic manifestations, immunologic studies and successful treatment with dapsone, *Medicine,* 56, 443, 1977.
13. Sneddon, I., Wilkinson, D., Subcorneal pustular dermatosis, *Br J Dermatol,* 100, 61, 1979.
14. Wojnarowska, F., Marsden, R., Bhogal, B., Black, M., Chronic bullous disease of childhood, childhood cicatricial pemphigoid, and linear IgA disease of adults, *J Am Acad Dermatol,* 19, 792, 1988.
15. Hall, R. P., Lawley, T., Smith, H., Katz, S., Bullous eruption of systemic lupus erythematosus, dramatic response to dapsone therapy, *Ann Intern Med,* 97, 165, 1982.
16. Heid, E., Grosshans, E., Basset, A., Traitement du pemphigus seborrheique par les sulfones et la corticothérapie locale, *Bull Soc Fr Derm Syph,* 81, 351, 1974.

17. Fredenberg, M. F., Malkinson, F. D., Sulphone therapy in the treatment of leucocytoclastic vasculitis: a report of three cases, *J Am Acad Dermatol*, 16, 722, 1987.

18. Fry, L., Seach, P. P., Dermatitis herpetiformis: an evaluation of diagnostic criteria, *Br J Dermatol*, 90, 137, 1974.

19. Wyatt, E., Shuster, S., Marks, J., A postal survey of patients with dermatitis herpetiformis, *Br J Dermatol*, 85, 511, 1971.

20. McFadden, J., Leonard, J., Powles, A., Rutman, A., Fry, L., Sulphamethoxypyridazine for dermatitis herpetiformis, linear IgA disease and cicatricial pemphigoid, *Br J Dermatol*, 121, 759, 1989.

21. Garioch, J. J., Lewis, H. M., Sargent, S. A., Leonard, J. N., Fry, L., 25 years' experience of a gluten-free diet in the treatment of dermatitis herpetiformis, *Br J Dermatol*, 131, 541, 1994.

22. Hornsten, P., Keisu, M., Wiholm, B. E., The incidence of agranulocytosis during treatment of dermatitis herpetiformis with dapsone as reported in Sweden, 1972 through 1988, *Arch Dermatol*, 126, 919, 1990.

23. Kaufman, D., Kelly, J., Levy, M., Shapiro, S., *The Drug Etiology of Agranulocytosis and Aplastic Anemia*, Oxford University Press, 148, 1991.

24. Waldinger, T. P., Siegle, R. J., Weber, W., Voorhees, J., Dapsone induced peripheral neuropathy, *Arch Dermatol*, 120, 356, 1984.

25. Johnson, D. A., Cattau, E. L., Kuritsky, J. N., Zimmerman, H., Liver involvement in the sulfone syndrome, *Arch Intern Med*, 146, 875, 1986.

26. D'Arcy, P. F., Griffin, J. P., Thalidomide revisited, *Adverse Drug React Toxicol Rev*, 13, 65, 1994.

27. Ochonisky, S., Revuz, J., Thalidomide use in dermatology, *Eur J Dermatol*, 4, 9, 1994.

28. Sheskin, J., Convit, J., Results of a double-blind study of the influence of thalidomide on the lepra reaction, *Int J Lepr*, 37, 135, 1969.

29. Revuz, J., Guillaume, J. C., Janier, M., Hans, P., Marchand, C., Souteyrand, P., Bonnetblanc, J. M., Claudy, A., Dallac, S., Klene, C., Crickx, B., Sancho-Garnier, H., Chaumeil, J. C., Crossover study of thalidomide vs. placebo in severe recurrent aphthous stomatitis, *Arch Dermatol*, 126, 923, 1990.

30. Guillaume, J. C., Moulin, G., Dieng, M.T., Poli, F., Morel, P., Souteyrand, P., Bonnetblanc, J. M., Claudy, A., Daniel, F., Vaillant, L., Bernard, P., Bouillie, M. C., Bournerias, I., Denoeux, J. P., Lambert, D., Leonard, F., Chaumeil, J. C., Revuz, J., Crossover study of thalidomide vs. placebo in Jessner's lymphocytic infiltration of the skin, *Arch Dermatol*, 131, 1032, 1995.

31. Powell, R. J., Gardner-Medwin, M. M., Guideline for the clinical use and dispensing of thalidomide, *Postgrad Med J*, 70, 901, 1994.

32. Peck, G. L., Olsen, T. G., Yoder, F. W., Prolonged remissions of cystic and conglobate acne with 13-cis-retinoic acid, *N Engl J Med*, 300, 329, 1979.

33. Peck, G. L., Olsen, T. G., Butkus, D., Pandya, M., Arnaud-Battandier, J., Gross, E., Windhorst, D., Cheripko, J., Isotretinoin vs. placebo in the treatment of cystic acne, a randomized double-blind study, *J Am Acad Dermatol*, 6, 735, 1982.

34. Jones, D. H., King, K., Miller, A.J., Cunliffe, W. J., A dose-reponse study of 13-cis-retinoic acid in acne vulgaris, *Br J Dermatol*, 108, 333, 1983.

35. Strauss, J. S., Rapini, R. P., Shalita, A. R., Konecky, E., Pochi, P., Comite, H., Exner, J., Isotretinoin therapy for acne, results of a multicenter dose–response study, *J Am Acad Dermatol*, 10, 490, 1984.

36. Stainforth, J. M., Layton, A. M., Taylor, J. P., Cunliffe, W. J., Isotretinoin for the treatment of acne vulgaris: which factors may predict the need for more than one course, *Br J Dermatol*, 129, 297, 1993.

37. Lehucher-Ceyrac, D., Weber-Buisset, M. J., Isotretinoin and acne in practice, a prospective analysis of 188 cases over 9 years, *Dermatology*, 186, 123, 1993.

38. Layton, A. M., Knaggs, H., Taylor, J., Cunliffe, W. J., Isotretinoin for acne vulgaris 10 years later: a safe and successful treatment, *Br J Dermatol*, 129, 292, 1993.

39. Goulden, V., Layton, A. M., Cunliffe, W. J., Long-term safety of isotretinoin as a treatment for acne vulgaris, *Br J Dermatol*, 131, 360, 1994.

40. Lammer, E. J., Chen, D. T., Hoar, R. M., Agnish, N. D., Benke, P. J., Braun, J. T., Curry, C. J., Fernhoff, P. M., Grix, A. W., Lott, I. T., Richard J. M., Sun, S. C., Retinoic acid embryopathy, *N Engl J Med*, 313, 837, 1985.
41. Mitchell, A. A., Van Bennekom, C. M., Louik, C., A pregnancy prevention program in women of childbearing age receiving isotretinoin, *N Engl J Med*, 333, 101, 1995.
42. Layton, A. M., Cunliffe, W. J., Guidelines for optimal use of isotretinoin in acne, *J Am Acad Dermatol*, 27, S2, 1992.
43. Frederiksson, T., Petterson, V., Severe psoriasis, oral therapy with a new retinoid, *Dermatologica*, 157, 238, 1978.
44. Finlay, A.Y., Khan, G. K., Luscombe, D. K., Salek, M. S., Validation of sickness impact profile and psoriasis disability index in psoriasis, *Br J Dermatol*, 123, 751, 1990.
45. Geiger, J. M., Saurat, J. H., Acitretin and etretinate — how and when they should be used, *Dermatol Clin*, 11, 117, 1993.
46. Mahrle, G., Schulze, H. J., Färber, L., Weidinger, G., Steigleder, G. K., Low-dose short-term cyclosporine vs. etretinate in psoriasis: improvement of skin, nail and joint involvement, *J Am Acad Dermatol*, 32, 78, 1995.
47. Lassus, A., Geiger, J. M., Nyblom, M., Virrankoski, T., Kaartamaa, M., Ingervo, L., Treatment of severe psoriasis with etretin (RO 10-1670), *Br J Dermatol*, 117, 333, 1987.
48. White, S. I., Marks, J. M., Shuster, S., Etretinate in pustular psoriasis of palms and soles, *Br J Dermatol*, 113, 581, 1985.
49. Parker, S., Coburn, P., Lawrence, C., Marks, J., Shuster, S., A randomized double-blind comparison of PUVA-etretinate and PUVA-placebo in the treatment of chronic plaque psoriasis, *Br J Dermatol*, 110, 215, 1984.
50. Saurat, J. H., Geiger, J. M., Amblard, P., Beani, J. C., Boulanger, A., Claudy, A., Frenk, E., Guilhou, J. J., Grosshans, E., Mérot, Y., Meynardier, J., Tapernoux, B., Randomized double-blind multicenter study comparing acitretin-PUVA, etretinate-PUVA and placebo-PUVA in the treatment of severe psoriasis, *Dermatologica*, 177, 218, 1988.
51. Halioua, B., Saurat, J. H., Risk benefit ratio in the treatment of psoriasis with systemic retinoids, *Br J Dermatol*, 36, S135, 1990.
52. Vahlquist, A., Long term safety of retinoid therapy, *J Am Acad Dermatol*, 27, S29, 1992.
53. Stern, R. S., Fitzgerald, E., Ellis, C. N., Lowe, N., Goldfarb, M. T., Baughman, R. D., The safety of etretinate as long-term therapy for psoriasis: results of the etretinate follow-up study, *J Am Acad Dermatol*, 33, 44, 1995.
54. Roenigk, H., Fowler-Bergfeld, W., Curtis, G., Methotrexate for psoriasis in weekly oral dose, *Arch Dermatol*, 99, 86, 1969.
55. Nyford, A., Benefits and adverse drug experiences during long-term methotrexate treatment of 248 psoriatics, *Dan Med Bull*, 25, 208, 1978.
56. Van Dooren-Greebe, R., Kuijpers, A., Mulder, J., De Boo, T., Van De Kerkhof, P., Methotrexate revisited, effects of long-term treatment in psoriasis, *Br J Dermatol*, 130, 204, 1994.
57. Willkens, R., Williams, J., Ward, J., Egger, M., Reading, J., Clements, P., Cathcart, E., Samuelson, C., Solsky, M., Kaplan, S., Guttadauria, M., Halla, J., Weinstein, A., Randomized, double-blind, placebo controlled trial of low-dose pulse methotrexate in psoriatic arthritis, *Arthritis Rheum*, 27, 376, 1984.
58. Gutierrez-Urena, S., Molina, J., Garcia, C., Cuéllar, M., Espinoza, L., Pancytopenia secondary to methotrexate therapy in rheumatoid arthritis, *Arthritis Rheum*, 39, 272, 1996.
59. Stern, R., Zierler, S., Parrish, J., Methotrexate used for psoriasis and risk of noncutaneous or cutaneous malignancy, *Cancer*, 50, 869, 1982.
60. Stern, R. S., Laird, N., The carcinogenic risk of treatments for severe psoriasis, *Cancer*, 73, 2759, 1994.

61. Kamel, O., Van De Rijn, M., Weiss, L., Del Zoppo, G., Hench, K., Robbins, B., Montgomery, P., Warnke, R., Dorfman, R., Brief report: reversible lymphomas associated with Epstein-Barr virus occurring during methotrexate therapy for rheumatoid arthritis and dermatomyositis, *N Engl J Med*, 328, 1317, 1993.
62. Zachariae, H., Kragballe, K., Sogaard, H., Methotrexate induced liver fibrosis. Studies including serial liver biopsies during continued treatment, *Br J Dermatol*, 102, 407, 1980.
63. Boffa, M., Chalmers, R., Haboubi, N., Shomaf, M., Mitchell, D. M., Sequential liver biopsies during long-term methotrexate treatment for psoriasis, a reappraisal, *Br J Derm*, 133, 774, 1995.
64. Roenigk, H., Auerbach, R., Maibach, H., Weinstein, G., Methotrexate in psoriasis: revisited guidelines, *J Am Acad Dermatol*, 19, 145, 1988.
65. McGill, D., Rakela, J., Zinsmeister, A., Ott, B., A 21-year experience with major hemorrhage after percutaneous liver biopsy, *Gastroenterology*, 99, 1396, 1990.
66. Gilmore, I., Murray-Lyon, I., Williams, R., Jenkins, D., Hopkins, A., Indications, methods, and outcomes of percutaneous liver biopsy in England and Wales: an audit by the British Society of Gastroenterology and the Royal College of Physicians of London, *Gut*, 36, 437, 1995.
67. Ellis, C., Gorsulowsky, D., Hamilton, T., Billings, J., Brown, M., Headington, J., Cooper, K., Baadsgaard, O., Duell, E., Annesley, T., Turcotte, J., Voorhees, J., Cyclosporine improves psoriasis in a double-blind study, *J Am Med Assoc*, 256, 3110, 1986.
68. Van Joost, T., Bos, J., Heude, F., Meinardi, M., Low-dose cyclosporin A in severe psoriasis, a double-blind study, *Br J Dermatol*, 118, 183, 1988.
69. Ellis, C., Fradin, M., Messana, J., Brown, M., Siegel, M., Hartley, H., Rocher, L., Wheeler, S., Hamilton, T., Parish, T., Ellis-Madu, M., Duell, E., Annesley, T., Cooper, K., Voorhees, J., Cyclosporine for plaque-type psoriasis, Results of a multidose, double-blind trial, *N Engl J Med*, 324, 277, 1991.
70. Powles, A., Baker, B., Valdimarsson, H., Hulme, B., Fry, L., Four years of experience with cyclosporine A for psoriasis, *Br J Dermatol*, 122, S13, 1990.
71. Timonen, P., Friend, D., Abeywickrama, K., Laburte, C., Von Graffenried, B., Feutren, G., Efficacy of low-dose cyclosporin A in psoriasis: results of dose-finding studies, *Br J Dermatol*, 122, S33, 1990.
72. Christophers, E., Mrowietz, U., Henneicke, H., Färber, L., Welzel, D. and the participants in the German multicenter study, Cyclosporine in psoriasis: a multicenter dose-finding study in severe plaque psoriasis, *J Am Acad Dermatol*, 26, 86, 1992.
73. Laburte, C., Grossman, R., Abi-Rached, J., Abeywickrama, K. H., Dubertret, L., Efficacy and safety of oral cyclosporine A (CyA; Sandimmun) for long-term treatment of chronic severe plaque psoriasis, *Br J Dermatol*, 130, 366, 1994.
74. Mrowietz, U., Färber, L., Henneicke-Von Zepelin, H., Bachmann, H., Welzel, D., Christophers, E., and the participants in the German multicenter study, Long-term maintenance therapy with cyclosporine and posttreatment survey in severe psoriasis: results of a multicenter study, *J Am Acad Dermatol*, 33, 470, 1995.
75. Feutren, G., Abeywickrama, K., Friend, D., Von Graffenried, B., Renal function and blood pressure in psoriatic patients treated with cyclosporin A, *Br J Dermatol*, 122, S57, 1990.
76. Feutren, G., Mihatsch, M., Risk factors for cyclosporine-induced nephropathy in patients with autoimmune diseases, *N Engl J Med*, 326, 1654, 1992.
77. Penn, I., Brunson, M.E., Cancers after cyclosporine therapy, *Transplant Proc*, 20(suppl 3), 885, 1988.
78. Shuttleworth, D., Marks, R., Griffin, P., Risalaman, J., Epidermal dysplasia and cyclosporine therapy in renal transplant patients: a comparison with azathioprine, *Br J Dermatol*, 120, 551, 1989.
79. Beveridge, T., Krupp, P., McKibbin, C., Lymphomas and lymphoproliferative lesions developing under cyclosporin therapy, *Lancet*, i, 788, 1984.
80. Krupp, P., Monka, C., Side-effect profile of cyclosporin A in patients treated for psoriasis, *Br J Dermatol*, 122, S47, 1990.

81. Koo, J., Kadonaga, J., Wintroub, B., Lozada-Nur, F., The development of B-cell lymphoma in a patient with psoriasis treated with cyclosporine, *J Am Acad Dermatol*, 26, 836, 1992.

82. Brown, M., Ellis, C., Billing, J., Cooper, K., Baadsgaard, O., Headington, J., Voorhees, J., Rapid occurrence of nodular cutaneous T-lymphocyte infiltrates with cyclosporine therapy, *Arch Dermatol*, 124, 1097, 1988.

83. Puig, L., Puig, J., De Moragas, J., Transient monoclonal gammapathy associated with cyclosporin treatment of psoriasis, *Br J Dermatol*, 133, 141, 1995.

21

Topical Liposome Drugs

Hans C. Korting
Ludwig-Maximilians Universität

Monika Schäfer-Korting
Freie Universität Berlin

Current vehicles for topical drugs are very inefficient and only marginally effective in delivering their active ingredients to the site of disease.

— Anonymous[2]

21.1 Introduction

During the last few years interest in new methods of drug delivery has increased tremendously, corresponding to an "explosion in research".[42] The reasons for this development are manifold, as stated in Table 21.1.

In principle various approaches to modified drug delivery can be devised, as described in Table 21.2. Advanced drug delivery systems of the vesicle type may be microparticulate or colloidal carriers composed of proteins, carbohydrates, synthetic polymeres, or lipids. If the structures are composed primarily of lipids in water and of globular lamellar configuration they are called liposomes. Liposomes represent the most widely discussed vesicles so far. Properties considered advantageous comprise lack of toxicity, lack of immunogenicity, and degradability. In the past, however, the generally poor stability both during storage and use were considered disadvantageous. Moreover, large-scale production initially proved difficult. Thus it took about a quarter of a century since the inception of the liposome concept by Bangham et al.[3] before the first systemic liposomal drugs were approved. Interest was focused on

TABLE 21.1   Reasons for New Methods of Drug Delivery

Improvement of safety and efficacy of active compounds old and new and permission of new therapies;
Provision of more complex active compounds such as proteins necessitating new technological approaches;
Increasing awareness of the biological relevance of release patterns (continuous vs. pulsatile);
Comparatively low cost of the improvement of a conventional drug as compared to creation of a new compound, in particular facing patent expiration;
Advances in materials science.

Adapted from Langer 1990.[42]

chemotherapeutic agents, both antineoplastic and antimicrobial. For the treatment of Kaposi's sarcoma related to HIV infection liposomal doxorubicine and daunorubicine were developed. However, an increase in the benefit/risk ratio compared to conventional corresponding drugs has not been definitely established.[80] Whereas anthracyclines were considered candidate drugs for liposomal encapsulation in the context of antineoplastic agents, the polyenes were in the context of antimicrobial agents. In particular, a liposomal formulation of amphotericin B has been developed and become the first approved systemic liposomal drug. There are many claims regarding an increase in the benefit/risk ratio; however, this has not yet been considered finally proven.[35] The breakthrough in systemic treatment using liposomal drugs has been brought about by the switch from what is now termed conventional liposomes to sterically stabilized liposomes also known as "stealth" liposomes. Sterically stabilized liposomes are also colloidal particles composed of lipid bilayers encapsulating an aqueous medium; however, glycolipids are added to the lipid fraction or the lipids themselves are conjugated with ethylene glycol. In the future liposomes might prove appropriate to enhance DNA introduction into mammalian cells and thus provide an opportunity for gene therapy.[44] Seemingly, it is feasible to devise liposomes incorporating DNA in an orderly structure, the DNA being absorbed between cationic bilayers forming a single layer of parallel helices.[45] Future trends encompass self-assembling multicompartment liposomes of high encapsulation efficiency, allowing the incorporation of extremely toxic agents such as cytostatic agents into small liposomes incorporated in larger ones and liposomes with a ligand attached to address the receptors of interest specifically.[43]

TABLE 21.2   Approaches to Modified Methods of Drug Delivery

Chemical modification;
Vesicles;
Controlled release systems, in particular for peptides and proteins and transdermal release systems;
Novel degradable polymers and polymeric controlled release systems.

Modified from Langer 1990.[42]

Although liposomal drugs to be applied via the intravenous route clearly attract most of the interest by clinicians so far, it should not be overlooked that topical application in various contexts also looks rewarding. Topical liposome drugs are at the horizon in various fields, including ophthalmology and pulmonology. In the latter field in particular infant respiratory distress syndrome, a disease of the newborn, can be effectively treated by surfactant in an adequate formulation introduced in 1993.[14] In this chapter only the application of topical liposome drugs in dermatology will be dealt with in detail as it is this specialty of medicine in which the largest number of different liposome preparations has reached clinical trials. In fact it also was a cutaneous disease, dermatomycosis, for which the first liposomal drug has been approved, econazole lipogel branded as Pevaryl®Lipogel in Switzerland in 1988.

That topical liposome drugs in dermatology form a buzzing field also becomes obvious from the number of recent review articles.[28,29,73,75] In 1996 a special theme issue on "The skin as a site for drug delivery: The liposome approach and its alternatives" was published in a specialized journal.[33] The skin not only is an interesting site for the application of liposomal drugs under the aspect of skin disease but also under the aspect of providing an entry for xenobiotics destined for more deep-seated organs.[79] While the application of active compounds on the basis of transdermal delivery is far-fetched, topical liposomal

TABLE 21.3 Generally Accepted Candidates for Topical Use in Improved Vehicles

Type of Compound	Problem
Antiinflammatory compounds	Adverse local effects including interference with collagen formation
Local anesthetics	Lack of efficacy of "caine" topicals on intact skin making local skin surgery without injection impossible
Antifungals	Long period of time required for clearance of lesions
Adenine arabinoside	Efficacy on the eye but not on the skin
Anti-cancer compounds including fluorouracil	Difficulty of use
Keratinolytic agents including salicylic acid	Lack of efficacy
Antiacne drugs including antibiotics and retinoic acid	Increase of efficacy and safety due to the delivery of amounts of drugs exactly needed

Modified from Anonymous 1979.[2]

drugs for the treatment of skin disease now looks feasible. The reason they are not only feasible but also desirable is that the skin is meant as a barrier protecting the body against the intrusion of xenobiotics irrespective of their beneficial or deleterious character. According to a consensus reached under the auspices of the Society of Investigative Dermatology of the U.S.A. the "development of new vehicles and delivery systems for topical drugs" belongs to the "dermatological needs in drugs and instrumentation".[2] "Needs for improvement of vehicles" have been identified with the various types of active ingredients laid down in Table 21.3.

As early as 1979 it was stated that drugs could be "microencapsulated". A prolongation of efficacy is expected from such an approach.

In the same context also the use of penetration enhancers is discussed. In fact, percutaneous penetration enhancers have become a major issue, as reflected by a recent monograph.[77] Conventional penetration enhancers or penetration enhancers in a stricter sense, however, are believed to disturb the packaging of the intercellular lipid bilayers and to liquefy intercellular lipid sheets.[48] Hence, it can easily be understood that such enhancers as propylene glycol are usually also well-known irritants. In a wider sense, however, liposomes might also be considered penetration enhancers.[71] While conventional penetration enhancers might compromise the ultrastructure of the stratum corneum, as the morphological substrate of the skin barrier, liposomes reflect both the composition and structure of cutaneous elements, i.e., the membrane-coating granules or Odland bodies formed in the lower strata of the epidermis and disintegrating in the uppermost ones to form lipid bilayers.[69] In fact, application of empty liposomes to the skin surface can contribute to skin surface humidity in a dose-dependent fashion. Skin surface humidity can be quantitated using corneometry. Dry skin, that is skin of low humidity, today is a major problem, in particular in those having developed or being prone to develop atopic eczema.[66]

In the following various fields for application of topical liposome drugs in dermatology and various types of candidate active compounds will be discussed under the aspect of their benefit/risk ratio.

21.2 Liposomes and Liposomal Formulations for Topical Use: From the Biochemist's Workbench to Large-Scale Production

Several decades ago the biochemist Dr. Alec Bangham, working at the Institute of Animal Physiology, Barbraham, England, discovered that "hand shaken phospholipid dispersions" in aqueous phase provide a model for the analysis of physicochemical and physiological properties of biological membranes. Mixing phospholipids with water by shaking leads to the formation of liposomes, or spherulites composed of lipid bilayers separated by water. The reason for the formation of such structures is the amphipathy of phospholipids.[4] What Bangham did, however, was only a first step under pharmaceutical aspects: the so-called lipid hydration step. The liposomes obtained that way are characterized by differing, fairly large

TABLE 21.4　Liposome Classification According to Structural Parameters

MLV	Multilamellar vesicles	> 1 μm
OLV	Oligolamellar vesicles	0.1–1 μm
UV	Unilamellar vesicles	
SUV	Small unilamellar vesicles	20–100 nm
MUV	Medium-sized unilamellar vesicles	
LUV	Large unilamellar vesicles	> 100 nm
GUV	Giant unilamellar vesicles	> 1 μm

Modified from Barenholz 1992.[4]

Table 21.5　Classification of Liposomes According to Mode of Production

REV	Single or oligolamellar vesicles made by reverse-phase evaporation
MLV-REV	Multilamellar vesicles made by the reverse-phase method
SPLV	Stable plurilamellar vesicles
FAT-MLV	Frozen and thawed multilamellar vesicles
MLV-ET	Multilamellar vesicles prepared by extrusion methods
FPV	Vesicles prepared by French press
FUV	Vesicles prepared by fusion
DRV	Dehydration-rehydration vesicles

Modified from Barenholz 1992.[4]

TABLE 21.6　Methods of Hydration and Sizing in Liposome Production and Up-Scaling Feasibility

Hydration		Sizing	
Method	Up-scaling Feasibility	Method	Up-scaling Feasibility
Mechanical shaking (MLV)	Very good	Ultrasonic irradiation giving SUV	Poor
Organic solvent(s) replacement (MLV, OLV, UV)		High pressure extrusion giving mainly SUV (French pressure cell, microfluidizer,	Very good
a) Solvents miscible with water, such as ethanol	Very good	high-pressure homogenizer)	
b) Solvents not miscible with water, such as ether	Feasible but problematic	Low or medium pressure extrusion (up to 2000 psi), through pores of defined sizes,	Very good
Formation of lipid-detergent mixed micelles followed by detergent removal (OLV, UV)	Good for OLV but not SUV	e.g., stainless steel filters. Size and number of lamellae can be determined	

Modified from Barenholz 1992.[4]

size, and large numbers of lamellae. Consequently these are now called multilamellar vesicles (MLV). Table 21.4 gives details on liposome classification based on structural parameters.

Another procedure to manufacture liposomes consisted of ultrasonic irradiation. This led to small unilamellar vesicles (SUV). Ultrasonic irradiation allows defined liposome sizing, thus representing the second step. Several other methods for liposome production were devised later. Apart from structure liposomes can also be named after the way they are produced (Table 21.5). In future development it proved particulary helpful that organic solvents could be used as alternative means for lipid hydration. More recently scaling up of production and broadening the spectrum of lipidic raw materials became major subjects. Table 21.6 gives more detailed information on the critical steps in liposome formation, i.e., hydration and sizing, and on the feasibility of up-scaling.

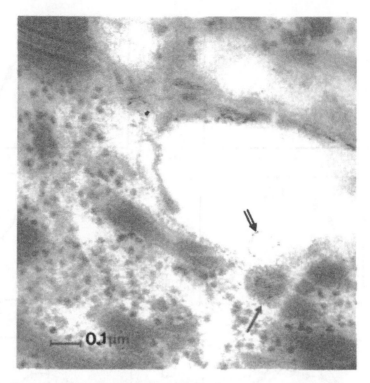

Figure 21.1 Membrane-coating granule (*closed arrow*) belonging to a human keratinocyte grown *in vitro* and a liposome composed of lecithin of similar size (*open arrow*) attached to the keratinocyte. Electron micrograph; × 91,000. From Schmid & Korting 1993 with permission.[72]

Apart from water, classical liposomes are conventionally made of phospholipids, forming one of the major lipid classes in both animal and plant membranes. Phospholipids as a rule are diesters of glycerol esterified with fatty acids of varying length and unsaturation at the sn-1 and sn-2 positions of the glycerol residue, the sn-3 position being esterified with phosphoric acid. Phosphatidyl choline or lecithin is a major congener. Phospholipids are amphiphilic as they carry on the one hand the highly polar head group and on the other the hydrophobic fatty alkyl chains. Correspondingly they spontaneously aggregate when exposed to water while they can be solubilized in mixtures of organic solvents like chloroform-methanol. They can exist in a liquid-crystalline state as found in biological membranes and in a gel state. At a defined temperature called the transition temperature (T_m) there is a phase transition from the gel to the liquid-crystalline state. The temperature depends on the composition of the phospholipid.[7] As described above stratum corneum lipids can be structured as liposomes called membrane-coating granules. Accordingly, major components of stratum corneum lipids, e.g., ceramides, are capable of forming liposomes.[84] Figure 21.1 gives an impression of a membrane-coating granule of a human keratinocyte grown *in vitro* and a liposome composed of lecithin.

The importance of the lipid composition of liposomes on drug action has been demonstrated in a classical experiment by Weiner et al.[83] Guinea pigs were inoculated with herpes simplex virus and treated with conventional and liposomal preparations of interferon-α_2 as well as empty liposomes. For 10 days the lesions were checked clinically for severity of disease using a score system. The scores found for the various treatment modalities investigated are depicted in Figure 21.2.

Not only interferon-α_2 in a conventional vehicle was found inactive but also interferon encapsulated into liposomes composed of phospholipids. The results clearly demonstrate that "liposomes and liposomes can be much different." This gives a clue to the virtual maze of conflicting evidence as to pharmacokinetics and pharmacodynamics related to topical liposome drugs. Nevertheless, skin lipid liposomes so far have not reached clinical application.

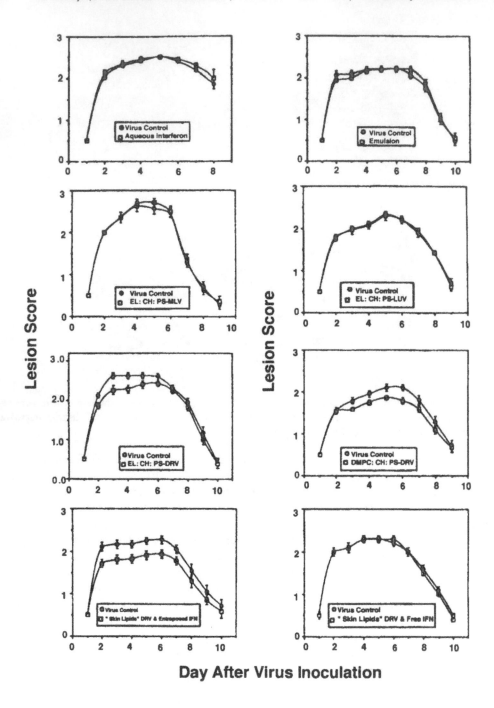

Day After Virus Inoculation

FIGURE 21.2   Effects of various liposomal and conventional preparations of interferon-α_2 as well as control preparations on severity of experimental guinea pig herpes simplex. From Weiner et al. 1989 with permission.[83]

Currently, three types of liposome production deserve our interest most of all: the ethanol injection technique, the detergent dialysis technique and the preliposome system technique. As early as 1973 Batzri and Korn[5] described the production of liposomes by ethanol injection. In 1984 the Swiss Company Cilag of Schaffhausen decided in favor of the new technique to develop a liposome dermatological with econazole as active ingredient. Econazole (base) and not any of the related compounds was chosen due

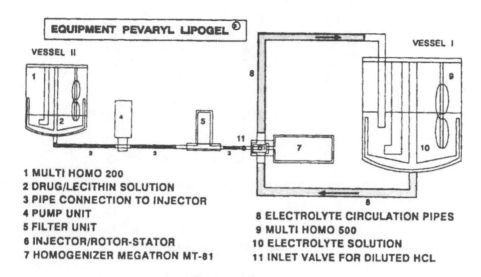

FIGURE 21.3 Scheme of large-scale production of econazole liposome gel. From Kriftner 1992 with permission.[41]

FIGURE 21.4 Freeze-fracture electron microscopic image of fresh econazole liposome gel. From Kriftner 1992 with permission.[41]

to its solubility in ethanol and a mixture of ethanol and lecithin. The process for large scale production is depicted in Figure 21.3.

While previously it had often been doubted that long-term stability might be feasible with a topical liposome drug, this actually was the case. Moreover, dynamic laser slight scattering proved that variability in terms of vesicle size was low from batch to batch, average sizes of particles ranging from about 170 to 210 nm. In the final product 90% to 95% of the active compound were found liposome-bound. Figures 21.4 and 21.5 give freeze-fracture electron microscopic impressions of the fresh final product, the econazole liposome gel, and the same product after 6 months' storage.

The detergent dialysis technique was developed by Weder[81] in the context of topicals. Bilayer forming substances such as phospholipids and solubilizing agents such as sodium cholate form an equilibrium between the aqueous phase and the lipidic phase of the mixed micelles. The equilibrium is primarily determined by the critical micelle concentration of the detergent and the ratio of lipid and detergent on a molar basis. Using phosphatidyl choline from soy bean, lecithin is dispersed in a phosphate buffer containing already dissolved sodium cholate. The active compounds are added. Liposome production

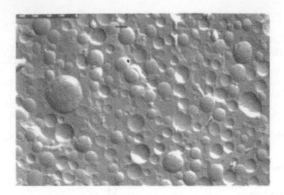

FIGURE 21.5   Freeze-fracture electron microscopic image of econazole liposome gel stored for 6 months. From Kriftner 1992 with permission.[41]

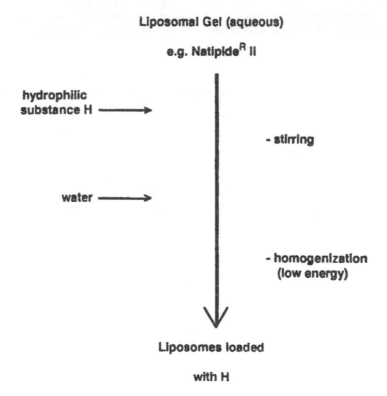

FIGURE 21.6   Preparation of a topical liposome drug from a preliposome system. From Röding 1992.[64]

itself is performed using a special equipment called a Vesisette. An important step in the process is removal of the detergent sodium cholate by diafiltration at a constant volume. The procedure allows the production of vesicles of a diameter of about 80 nm with an incorporation efficiency of about 100%. The preparation does not deteriorate over 60 days at least.

Topical liposome drugs can also be prepared on the basis of ready-made liposome preparations to which the active compound is added after finishing the original production process. The liposomes are composed of lecithin in the case of Natipide II. This preparation consists of densely packed liposomes of about 280 nm diameter representing in a way a gel-like form itself, without the addition of a macromolecule such as carboxymethyl cellulose used for the econazole liposome gel. Figure 21.6 describes the principle of the loading process.

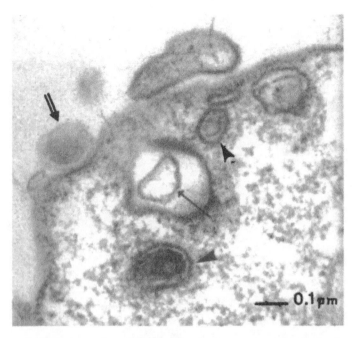

FIGURE 21.7 Electron micrograph of a keratinocyte growing *in vitro* exposed to empty liposomes showing different stages of phagocytosis. Attachment (*double arrow*), invagination of a liposome (*arrow*), and several internalized liposomes within phagoliposomes (*arrowheads*) are demonstrated (x 81,000). From Schaller et al. 1996 with permission.[70]

The preliposome technique with Natipide II is particularly appropriate for the production of drugs incorporating hydrophilic and amphiphilic substances, although encapsulation of lipophilic compounds such as bisabolol is feasible.[64] To incorporate lipophilic substances vegetable oil can be added to liposomes. This has been achieved with a phospholipid solution containing natural glycerides and ethanol called Phosal 75 SA. Mixed with water both true liposomes and monolayer-surrounded oily particles are present.[63]

21.3 Interaction of Liposomes and their Active Ingredients with Human Skin

According to a widely shared belief liposomes can interact with mammalian cells in various ways, by endocytosis, fusion, and lipid exchange.[59] While experimental evidence for this hypothesis was mainly based on investigations with hepatocytes, recently the interaction of liposomes with the major cell of human epidermis, the keratinocyte, has also been investigated. Though Bonnekoh et al.[9] had not been able to visualize intact liposomes within human keratinocytes growing *in vitro*, this has later been possible.[38] Early subcultures of human keratinocytes grown *in vitro* were exposed to large oligolamellar liposomes prepared by the detergent dialysis technique according to Milsmann et al.[58] According to this investigation, liposomes are first attached, then engulfed, later taken up and enclosed by a lysosome, and finally disintegrated. In a more recent investigation this hypothesis was substantiated using liposomes carrying the active compound silver sulfadiazine. Silver acts as an electron-dense marker (Figure 21.7).

This time the liposome encapsulating silver sulfadiazine could be clearly identified as such due to the marker and dispersion of the active compound shown (Figure 21.8).

Most recently it even has been possible to identify the nature of the marker combining electron microscopy and X-ray microanalysis.[71] Uptake of intact liposomes by human keratinocytes *in vitro*, however, does not necessarily mean that uptake of intact liposomes happens if healthy human skin is exposed as postulated by Foldvari et al.[19] This may also be true for mouse skin[93] even though it was questioned until most recently by other investigators.[29]

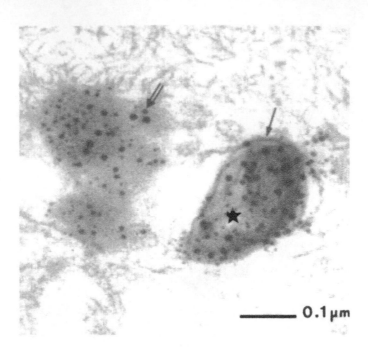

FIGURE 21.8 Electron micrograph showing a keratinocyte with an internalized silver-labeled liposome (*star*) incorporated into a cellular unit membrane (*arrow*). Disintegration of a liposome and release of the encapsulated silver particles (*double arrow*) (× 148,000). From Schaller et al. 1996 with permission.[70]

TABLE 21.7 Phospholipid Concentration in Various Strata of Piglet Skin 180 Min after Application of Liposomes

Skin Compartment	Concentration of Phospholipid (μg/g tissue)
Horny layer	100,000
Epidermis	500
Dermis	28
Subcutis	8

Modified from Röding and Artmann 1992.[65]

Korting et al.[39] investigated the question using human epidermis reconstructed *in vitro*, sometimes also called the living skin equivalent. Again LUV obtained by the detergent dialysis technique were used, the human epidermis was reconstructed according to Regnier and Darmon.[61] Some intact liposomes or their remnants were seen between corneocytes of upper strata. Intact liposomes, however, did not proceed to the living strata or even through the epidermis.

These findings correspond to results of experiments on the lipid kinetics with topical liposomes. Röding and Artmann[65] investigated this subject with tritium-labeled phosphatidyl choline liposomes obtained using the film method in the minipig. At defined points of time after application upper strata of the skin were removed using self-adhesive tape. The bulk of phospholipid was found in the horny layer (Table 21.7).

These findings do not question the fact that liposomal formulations can enhance drug penetration into cutaneous compartments. Even using similar lipids for the control preparation, Egbaria and Weiner[16] were able to demonstrate a much higher uptake of cyclosporin from the liposomal form into the stratum corneum of the hairless mouse. With reconstructed human epidermis an increase of lipidic material has been found between corneocytes. This is reflected by changes in the structure of the stratum corneum observed both by small angle x-ray scattering and freeze fracture electron microscopy.[11] The structural findings correspond to functional ones: using attenuated total reflectance Fourier transform infrared spectroscopy, an increased permeability of the horny layer was observed in the presence of liposomes, which was attributed to the presence of excess lipid in the upper strata.[8] These findings may look contradictory to the ones obtained by corneometry discussed above. Clearly, the beneficial effect of empty liposomes or its opposite needs further clarification.

21.4 Making Drugs Work on the Topical Route: Local Anesthetics and Excision Repair Endonucleases as Examples

There are active compounds that have never worked when put on the skin surface although they clearly are active when applied by a different route, in particular on injection. The reason must be the inadequacy of the vehicle. A classical example is local anesthetics of the "caine" type. In clear contrast to previous experience[13] Hallen and co-workers[25] were able to demonstrate the capability of a particular lignocaine prilocaine cream to relieve pain due to venipuncture. In this preparation the two local anesthetics form a three-dimensional structure, a "eutectic mixture". The original experience in adults was corroborated later in children.[76] However, the preparation needs to be applied in a fairly thick layer and anesthesia might be limited to the upper parts of the skin. At about the same time as Hallen and co-workers Gesztes and Mezei reported on local anesthesia of the skin due to the application of liposome-encapsulated tetracaine.[24] In 1990 this claim was protected by a U.S. patent.[56] Gesztes and Mezei used a 0.5% tetracaine formulation and assessed its efficacy by the pin-prick method. In a later double-blind trial the 0.5% liposomal tetracaine formulation proved superior to placebo despite a small difference in pain perception, which could be explained by the fact that the pain due to venipuncture used as the end point here was perceived deeper in the skin compared to the pain due to pricking.[26] Subsequently, a 2% formulation was tried and found more suitable for venipuncture and other minor surgical procedures related to the skin.[20]

More recently, this 2% liposomal tetracaine formulation has been characterized in detail both under pharmacokinetic and clinical aspects.[18] Multilamellar liposomes were prepared using hydrogenated soy phosphatidyl choline by means of the solvent evaporation method as described by Mezei and Nugent.[53] The penetration of tetracaine from the liposomal preparation as well as a control preparation into human skin from the breast obtained during plastic surgery was investigated in a diffusion cell. For tetracaine a tritium-label was used, for liposomal lipid ^{14}C labeling. On the whole two distinct liposomal preparations and two distinct conventional preparations were investigated. ^{3}H tetracaine penetration was the highest with one of the liposomal preparations (5.3% of total), followed by one conventional preparation (3.3%), the other liposomal preparation (1.7%), and the other conventional preparation (1.2%) that actually represented polyethylene glycol ointment. Correspondingly the two latter preparations had no effect 15 minutes after application in the pin prick test, scores for pain being similar to those for placebo, while the pharmacokinetically optimum preparations both the liposomal and conventional type fared similarly, both being effective ($p < 0.005$). Despite a more long-lasting effect of the better liposomal preparation as compared to the one of the better conventional preparation, both preparations were found about equipotent. The difference in efficacy of the two liposomal preparations was traced back to different pH values. The comparison of penetration of lipid and active compound made the author of the study think that tetracaine might not have been quantitatively conveyed to deeper strata of the skin within intact liposomes (see above).

Certainly, it is obvious that optimizing conventional formulations also can contribute to efficacy of so far inert active compounds. However, a liposomal formulation can clearly modify the uptake of an active compound into the skin at an extent critical for clinical usefulness. The differences found with the two liposomal formulations again make clear that as a rule there is not just one liposomal formulation possible. This should exclude generalized interpretations of findings with a particular liposomal form, especially if no benefit is seen. Figures 21.9 and 21.10 depict the cumulative amount of tetracaine diffused over time as well as the influence on the pain score on pin pricking.

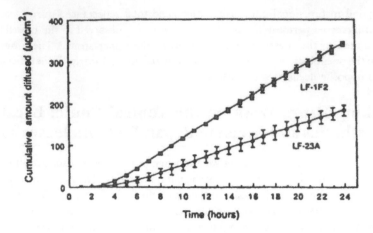

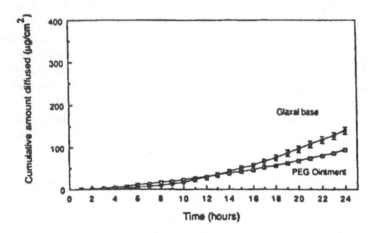

FIGURE 21.9 Cumulative amount of tetracaine diffused over time with four different formulations, two of the liposomal type (LF-1F2 and LF-23A) and two of the conventional type (Glaxal base and PEG ointment). From Foldvari 1994 with permission.[18]

For a long time liposomes have been considered especially in the context of drug delivery with respect to proteins. Relevant proteins in the dermatological context are DNA-repair enzymes. Skin cancer, which is becoming more and more common, is closely linked to UV exposure. DNA damage is considered an initiating event. Often either a pyrimidine dimer is formed by a cyclobutane ring link or 6-4-pyrimidine-pyrimidone. Such photo-products in principle can be removed by excision. The inherited skin disease xeroderma pigmentosum is due to a primary deficiency in excision repair leading to a by order of magnitude higher frequency of skin cancer. A particular enzyme qualifies to contribute to DNA repair in mammalian cells: T4 endonuclease V. This enzyme can be encapsulated into liposomes termed T4N5 liposomes.[90] T4 endonuclease V encapsulation into liposomes was clearly more effective in terms of intracellular drug delivery as compared to other techniques encompassing permeabilization, microinjection, and transfection.[30] The liposomes used are multilamellar vesicles by nature of an approximate diameter of 0.1 μm.[91] The liposomes are particularly pH-sensitive to optimize endocytosis and membrane fusion due to the presence of two lipids in the membrane, phosphatidyl ethanolamine and oleic acid.[92] Cholesterol is also present to control pH-stability. With mice pieces of evidence were forwarded that T4N5 liposomes might enter intact into the cytoplasm of keratinocytes.[88] The efficacy of the liposomes was investigated in hairless mice (SKH-I) irradiated with ¼ minimal erythema dose (MED) UV-B three times per week for 28 weeks. T4N5 liposomes reduced the incidence of skin cancer ($p = 0.0006$) and also

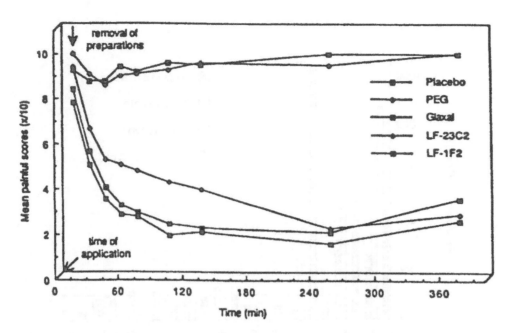

FIGURE 21.10 Local anesthetic effect of various formulations applied topically on the skin before pricking judged by the pain score over time. In this context LF-1F2 is an empty (active compound-free) liposomal formulation. From Foldvari 1994 with permission.[18]

the wrinkle length and improved over-all skin quality as defined by Bissett et al.[6] While in the end all animals in the control group had at least developed one cancerous skin lesion, this was the case in the verum group only in less than 70%.[90] Experience in man still is limited. However, preliminary results from a phase 2 study with T4N5 liposome lotion in xeroderma pigmentosum are available.[31] In the double-blind trial 12 xeroderma pigmentosum patients were treated with T4N5 liposome lotion and a heat-inactivated control lotion immediately and 2 hours after exposition to two minimal erythema doses of solar simulating radiation. Formation of cyclobutane pyrimidine dimers served as the end points. Pyrimidine dimers were found markedly reduced in 9 of 11 patients in the context of verum liposome treatment compared to an untreated area. This was the case only with 6 of 11 with respect to control treatment. Clearly, clinical use awaits further clarification from ongoing trials.

21.5 Lowering Drug Concentration Yet Increasing Efficacy: The Glucocorticoid Paradigm

Glucocorticoids have early been considered as candidates for liposomal encapsulation in the context of optimization of topical treatment of skin disease. There are two main reasons for this:

Topical treatment with conventional medium potent and highly potent topical glucocorticoids is linked to potentially long-standing or permanent unwanted local effects. Among the manifold items on a long list as given by Marks[49] atrophic striae as a major manifestation of skin atrophy are most important.[17] With respect to medium potent topical glucocorticoids the problem can be solved by the use of new chemical entities belonging to a special type termed the nonhalogenated double-ester type encompassing prednicarbate.[79] Apart from the relative frequency of unwanted effects limiting the use in clinical practice the absolute frequency of prescription of topical glucocorticoids is another reason to look for modifications.[21]

As early as 1980 Mezei and Gulasekharam[54] reported on the optimization of the vehicle for triamcinolone acetonide, the particular compound having induced the atrophic striae described by Epstein et al. Liposomally encapsulated triamcinolone acetonide was used in either a lotion or a gel form.[55] In the guinea pig bioavailability was investigated in various compartments of the skin as well as inner organs.

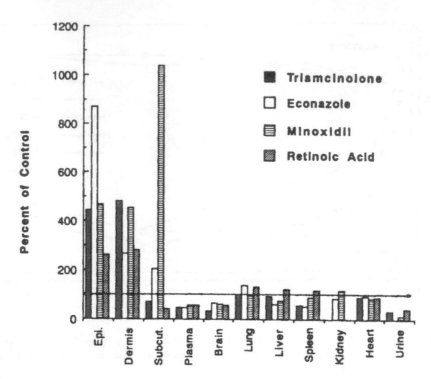

FIGURE 21.11   Relative bioavailability of various compounds used in dermatological treatment in various organs of the guinea pig: relative values with the liposomal form as compared to those with a corresponding conventional vehicle, the values obtained from the latter drug being set at 100%. From Mezei 1992 with permission.[53]

The results with respect to triamcinolone acetonide and other drugs to be considered in the dermatological context are depicted in Figure 21.11.

According to Figure 21.11 drug levels in the epidermis and dermis were severalfold higher with the liposomal as compared with the conventional form whereas other tissue levels including plasma levels were lower. This result deserves particular interest as apart from local unwanted effects there also is a fear of systemic effects, i.e., depression of the hypopituitary adrenal axis upon application of potent topical glucocorticoids to large areas of the skin. The findings related to cutaneous delivery are well in accordance to those found by Wohlrab et al.[88] using an *ex vivo* approach. Wohlrab and Lasch[86] used excised human skin *ex vivo* for comparative investigation of hydrocortisone penetration using a technique described by Schaefer et al.[72] Liposomes were produced of lecithin and hydrocortisone, essentially combining the film and the sonication technique. Thus a liposome suspension could be compared with a conventional emulsion of the w/o-type, both containing hydrocortisone 1%. To trace hydrocortisone tritium labeling was used. After 30 min and 300 min of exposure of the skin specimens to the hydrocortisone preparations, the concentrations of active compound in the stratum corneum were about the same while concentrations in the epidermis and dermis differed by about one order of magnitude in favor of the liposomal preparation. The results at 30 min are depicted in Figure 21.12.

Wohlrab and Lasch[87] investigated the liposome preparation in the guinea pig also, addressing serum concentrations of hydrocortisone and its urinary excretion on topical application to the ear. Serum levels appeared clearly lower following the liposomal formulation; at 24 hours, for example, serum concentrations amounted to 2.19 ± 0.11 and 8.18 ± 2.89 dpm $\times 10^4$/ml. Correspondingly, with the conventional formulation urinary excretion was much higher within the first four days. The authors called their drug delivery system selective and termed it a "drug localizer" to distinguish clearly between such a type of formulation and a formulation incorporating a "penetration enhancer." Wohlrab and Lasch also postulated

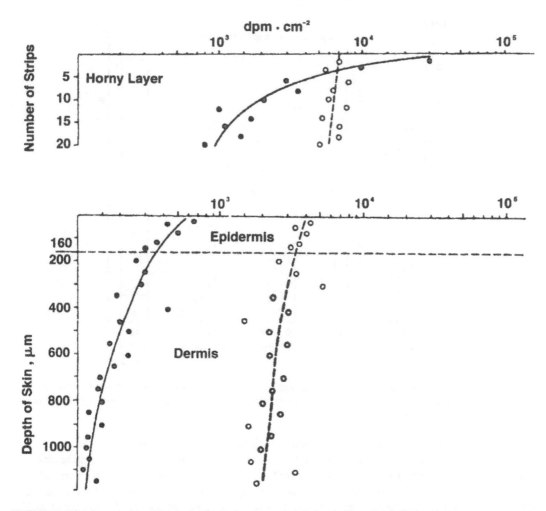

FIGURE 21.12 Penetration of hydrocortisone from conventional ointment (*black dots*) and liposome suspension (*white dots*) into various compartments of excised human skin within 30 min exposition. From Wohlrab and Lasch 1987 with permission.[86]

an increase in the benefit/risk ratio facing a potential for increased therapeutic use at the site of application corresponding to a decrease in the potential for systemic unwanted effects.

It was against this background that we started to analyze the efficacy of a liposomal form incorporating a strong topical glucocorticoid, betamethasone dipropionate, in the two major noncontagious inflammatory skin diseases, atopic eczema and psoriasis vulgaris.[40] The liposomes with a mean diameter of 50 nm were obtained using the detergent dialysis technique by Weder et al.[82] using egg lecithin as a major component of the vehicle. The liposomes encapsulating the active compound were incorporated into a polyacrylate gel to give a final concentration of 0.039%. This concentration corresponded to a concentration of 0.064% with the conventional control preparation, which was commercial betamethasone dipropionate gel with propylene glycol incorporated as a penetration enhancer (Diprosis®Gel from Essex, Munich, D). In a double-blind randomized trial 10 patients, each suffering from atopic eczema or psoriasis vulgaris, were treated with either preparation on one side of the body. The preparations were applied openly once a day for 14 days. Efficacy was determined using a score system. As depicted in Figure 21.13 statistical evaluation obtained gave only one hint at a difference: despite the lower drug concentration the liposomal preparation tended to be superior in terms of scaling after 7 days.

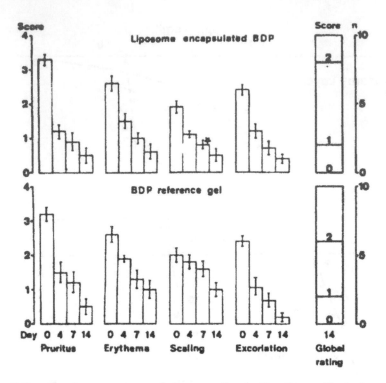

FIGURE 21.13   Rating of various symptoms and signs over time in 10 patients with atopic eczema related to treatment with liposome encapsulated betamethasone dipropionate and a commercial reference gel. From Korting et al. 1990 with permission.[40]

While the much less concentrated formulation of betamethasone dipropionate tended to be more efficacious in atopic eczema, the opposite was the case in psoriasis vulgaris. The results make clear that liposomal encapsulation of a topical glucocorticoid does not only increase tissue levels in critical compartments such as the epidermis but also increases efficacy in at least one major skin disease. Currently, there is no doubt that decreasing the concentration of a highly potent topical glucocorticoid such as betamethasone dipropionate should also decrease local unwanted effects of the atrophy type. Judging efficacy and safety altogether, one can clearly conclude from the findings published so far that liposomal encapsulation of a topical glucocorticoid is capable of increasing the benefit/risk ratio, a particularly important feature of this particular class of topicals (see above).[74] Using the chemical derivatization process in the past only an increase of the benefit/risk ratio with respect to medium potent topical glucocorticoids such as betamethasone 17-valerate and prednicarbate has been demonstrated. It remains to be seen if molecular changes might also finally lead to still more highly potent but safer congeners. So far, however, liposomal formulation appears as a valid alternative. This appears likely as betamethasone dipropionate gel with propylene glycol is considered to be one of the most potent topical glucocorticoids of all.[12]

21.6   Lowering Drug Concentration to Increase Tolerability while Maintaining Efficacy: The Tretinoin Experience

Certainly, an increase in efficacy cannot always be expected from liposomal encapsulation if the concentration of the active compound in the final preparation is decreased. However, lowering the concentration of the active drug might well increase the benefit/risk ratio leading to a decrease in unwanted effects while efficacy is by and large maintained. Already today the experience with tretinoin under this aspect can be called classical.

Since its introduction about two decades ago tretinoin has been considered the mainstay of topical dermatologic treatment of acne vulgaris due to its capability to normalize abnormal follicular keratinization characteristic of acne lesions.[85] Using the well-known 0.025 to 0.05% gel formulations comedones and papules as major noninflammatory and inflammatory lesions of acne are reduced by more than one half.[15] Local irritancy, however, is frequent, occurring in up to three quarters of all patients treated.[23] Initial deterioration described as "flare-up" also is an issue.[57] Such unwanted effects critically limit acceptance of conventional tretinoin preparations in clinical practice.[47] Both the major unwanted effect, i.e., irritancy, and the wanted effect, i.e., comedolytic activity, are dose-dependent. Gels of 0.01% clearly are less irritant but also clearly less efficacious.[60] One approach to an increase in the benefit/risk ratio has been successful already, the addition of erythromycin.[1]

Liposomal encapsulation of tretinoin can provide a valid alternative. In 1985 Meybeck et al.[52] described liposomal encapsulation of tretinoin using liposomes produced of soy bean lecithin and cholesterol (9:1) by means of ultrasonication. These liposomes encapsulating tretinoin can be incorporated into a carboxy-methyl-cellulose gel.[50]

Meybeck[51] investigated liposomal and conventional vitamin A acid formulations in the rhino mouse model, the rhino mouse representing a mutant of the hairless mouse with spontaneously occurring comedones.[32] In the mouse model comedolytic activity is higher if tretinoin is liposomally encapsulated. Corresponding pharmacokinetic data could also be obtained. Using the liposomal formulation commanding a high encapsulation or entrapment efficiency,[10] higher levels of tretinoin were found in deeper skin strata upon application of the liposomal compared to the conventional form.[51] Upon application of the 0.1% liposomal tretinoin gel to the back of hairless rats after 30 hours 59%, 41%, 13% of the tretinoin dose administered were found in the surface, epidermal, and dermal compartment as compared to 57%, 18%, and 8% with the control gel of the conventional type. In this context it is of particular interest that it has also been possible to demonstrate that the liposomes containing tretinoin do not penetrate intact into deeper skin strata.[78] Schäfer-Korting et al.[68] compared a 0.01% liposomal preparation provided by Meybeck and his group in a randomized double-blind trial in acne vulgaris comparing the new formulation to commercial conventional gels containing tretinoin 0.025% and 0.05%, respectively, in 20 patients suffering from uncomplicated acne vulgaris. Within the observation period of 75 days on average the number of inflammatory and noninflammatory acne lesions did not differ in the various treatment groups. While all patients received the liposomal preparation on one side of the face the other was at random either treated with the higher or the lower concentrated conventional form. While efficacy was about alike, tolerability clearly was different. According to patient perception liposome encapsulated tretinoin induced less burning than either reference gel and less erythema than the 0.025% reference gel ($p < 0.05$). Analyzing all data obtained together overall irritancy was less marked with the liposomal form ($p < 0.005$). The cumulative irritancy score for burning, erythema and scaling with the various tretinoin preparations is depicted in Figure 21.14.

Thus up to now there is no definite evidence that liposomal encapsulation also increases efficacy of a tretinoin-containing topical in human disease. However, facing increased tolerability while efficacy is maintained, it can be stated that liposomal encapsulation of tretinoin increases the benefit/risk ratio of a topical drug where this parameter is a major issue in clinical practice.

21.7 The Benefit/Risk Ratio of the First Approved Topical Liposome Drug: Econazole Liposome Gel

The technological possibilities of encapsulating econazole, a major topical antimycotic of the azole type, have already been discussed. Clearly, econazole liposome gel has meant a tremendous breakthrough in terms of pharmaceutical development as the feasibility of a liposome drug has been established in pharmaceutical practice for the first time. Early investigations, however, did not finally define its value in clinical terms. The comparatively higher levels of econazole in upper strata of the skin upon application of a 1% liposomal form as compared to an equally concentrated conventional form (compare above)

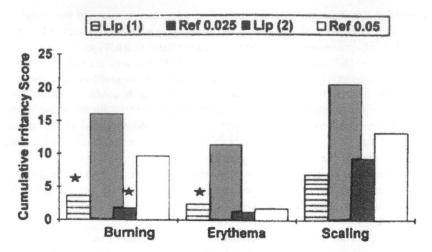

FIGURE 21.14   Cumulative irritancy score following the 0.01% liposomal tretinoin form and the 0.025% and 0.05% conventional control preparations. A star indicates p ≤0.05 with respect to the difference between the liposomal and a conventional form. From Schäfer-Korting et al. 1994 with permission.[68]

has led to the idea that efficacy might be greater clinically and in particular the onset of action earlier.[41] According to the authors of an unpublished report on a comparative clinical trial comparing the econazole 1% liposome gel with conventional 1% naftifine hydrochloride cream such a quicker onset of action was felt (on day 7, on once-daily topical application of either drug) (Fritsch et al., unpublished data, on file at Cilag AG, Schaffhausen, Switzerland, 1987). Although no major difference in terms of efficacy was seen between both treatment groups, the trial could not answer relevant questions on efficacy due to its design.

Therefore a major double-blind comparative trial has recently been performed. Patients suffering from tinea pedis were treated either with the 1% econazole liposome gel once a day for 14 days, with commercial 1% econazole cream or with a generic 1% clotrimazole cream. Reflecting previous experience confirmatory data analysis focused on mycological cure on day 7. While in an intent-to-treat analysis a major difference was not seen at this time, using this criterion another analysis based on evaluable patients on day 28 tended to hint at a higher cure rate in the group treated with the econazole liposome gel. Cure rates amounted to 80.2%, 73.1%, and 69.0% respectively (p = 0.08). Correspondingly, tolerability was classified as very good in 69.7%, 62.6%, and 64.9% of patients. This might hint at an increase in the overall benefit/risk ratio of the liposomal drug (Korting et al., submitted for publication). Recently, liposomal econazole gel and conventional econazole cream have also been compared in a new *ex vivo* model of cutaneous candidosis based on reconstructed human epidermis. In comparative terms the liposomal form reduced Candida albicans–specific alterations of the skin more markedly, while it induced mild morphologic alterations of the blastospores judging on electron microscopic grounds. Intact liposomes could be observed in intercellular spaces of the upper stratum corneum; moreover seemingly liposomal lipid was found attached both to stratum corneum and Candida albicans cells (Schaller et al., in preparation). In summary, today we have various reasons to believe that 1% liposomal gel containing econazole might be slightly superior to a conventional counterpart. Well tolerated and highly efficacious topical azole drugs being available for fungal skin disease, the profit from liposomal encapsulation is less prominent than with glucocorticoids and retinoids that induce more prominent local unwanted effects.

21.8   Liposome Encapsulation Makes Topical Herbal Drugs Work: or, Challenging the Witch Hazel

Large amounts of data have been published on herbal drugs as topical antiinflammatory agents for skin disease.[27] A final evaluation of the benefit/risk ratio, however, currently still appears difficult. Some of

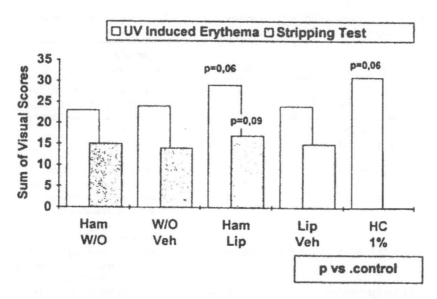

FIGURE 21.15 Activity of conventional hamamelis distillate cream, the corresponding vehicle, hamamelis distillate liposome cream, the corresponding vehicle, and hydrocortisone cream in the UV erythema test and the stripping test according to Wells. Activity is expressed as sum of visual scores. From Korting et al. 1993 with permission.[36]

the chamomile preparations and hamamelis preparations can be considered fairly safe but the frequently used arnica and calendula preparations in particular are linked to not too infrequent cases of allergic contact dermatitis. Data on efficacy being scanty, however, it has long been almost impossible to define the clinical value of chamomile and hamamelis preparations. Recent evidence on activity and/or efficacy in particular has been forwarded in the context of hamamelis preparations. Hamamelis virginiana or witch hazel is a bush growing in several different parts of the world including North America. For pharmaceutical use either distillates or extracts are prepared. Various chemical entities are discussed as potential active ingredients. The leaves in particular contain tannins related to an astringent effect many patients are familiar with. In the bark even higher quantities of tannins can be found. Other potentially active principles repeatedly named comprise catechol derivatives and flavonoids such as quercetin.[62] Among the tannins β-hamamelitannin in particular has recently received widespread interest.[22] β-hama-melitannin, however, is not a major component of the hamamelis preparation currently preferred in dermatology, the distillate.[46] Nevertheless, it was this particular type of plant preparation that had been selected for modern clinical research. While by convention hamamelis distillate had been used in an o/w cream, more recently a modern formulation has also been developed on the basis of a preliposome system as described above (Phosal®). The final preparation of this type contained 5.35% hamamelis distillate or 0.64 mg hamamelis ketone per 100 g (Hametum®Creme, Spitzner, Ettlingen, D). This form was considered in randomized double-blind studies in healthy human volunteers using the UV erythema test and the cellophane tape stripping test.[36] The liposomal form was compared to the conventional one of identical strength previously used clinically. In the UV erythema test the conventional form was not found to be more active than the vehicle while there was a noteworthy difference in favor of the liposomal form. In principle similar observations were made in the stripping test. Results are depicted in Figure 21.15.

Later the hamamelis distillate liposome cream 5.35% was also compared to hydrocortisone cream 0.5% in a double-blind controlled trial in atopic eczema. The corresponding vehicle of the hamamelis preparation was also included. While all treatments reduced itch, erythema, and scaling within 1 week the hydrocortisone preparation was found superior to the hamamelis preparation, which was not superior to its vehicle.[37] Even upon liposomal encapsulation as performed here hamamelis distillate preparations for topical use seemingly are not more efficacious than their corresponding vehicle in atopic eczema. With sunburn, however, the situation may be different. The use of the liposome system increases activity

TABLE 21.8   Main Candidates for Liposome Encapsulation Among Groups of Active
Compounds for Topical Use in Dermatology

Class of Active Compound	Indication	Aspect of Potential Improvement
Glucocorticoids (triamcinolone acetonide 0.1%, hydrocortisone 0.5%, betamethasone valerate 0.01%)	Atopic eczema	Increased activity, reduction of local unwanted effects, in particular less skin atrophy
Interferons (interferon-α)	Genital papilloma virus infection	Improved efficacy
Retinoids (isotretinoin)	Acne	Reduced local irritation and reduced inactivation by oxidation
Methotrexate	Psoriasis	Lower drug concentrations in the blood leading to decreased systemic unwanted effects
Hair growth stimulants (minoxidil)	Alopecia	Due to improved targeting to hair follicles less systemic availability and thus unwanted effects
Antibiotics (gentamycin)	Soft tissue infection	Decreased bacterial counts, i.e., increased efficacy

Modified from Schmid and Korting 1994.[73]
In the table those liposomal drugs already mentioned above in whatever context have not been
included. This definitely is not to say that these drugs do not belong here.

in the UV erythema test to an extent to be considered clinically relevant up to a point. Current data at
least provide a rationale to develop topical herbal drugs for inflammatory skin disease meeting the
expectations of our patients. Good tolerability clearly is an asset. However, liposomal encapsulation only
makes sense if efficacy due to it clearly can be confirmed.

21.9   Topical Liposome Drugs Ahead

To give a clinician an idea of what can be expected in the future from liposomal encapsulation of
conventional active compounds in dermatology a systematic patent search was performed some time
ago.[34] Table 21.8 gives a survey addressing most important groups of active compounds and their indications as well as a statement on potential improvement in benefit and/or risk of the drugs relevant for
benefit/risk ratio assessment.

21.10   Conclusions

While topical liposome drugs represented by econazole liposome gel have come in first in the race for
the first approved liposomal drug of whatever type, at present liposomal drugs for systemic use attract
more interest from the side of the clinician and the pharmaceutical industry. It is in particular the latter
aspect that impedes further progress with respect to topical liposomal drugs. In general prognoses for
the turnover for a topical liposomal drug are considered too limited to justify an engagement in pharmaceutical and clinical development. Nevertheless, the paradigm of topical liposomal econazole gel has
made clear and undebatable that topical liposome drugs are feasible under technological aspects and also
under approval aspects as this particular drug has now been approved in several different European
countries encompassing Switzerland, Italy, Poland, and Germany. Probably, fairly small companies will make
a major contribution to further progress in the field as has been and still is the case with systemic liposome
drugs. Judging from clinical needs in dermatology and possibilities substantiated by experimental work, the

biggest challenge for the near future is to develop a highly potent and well-tolerated topical liposomal drug of the glucocorticoid type. This should be possible within a few years. In clinical practice, an extremely large field of topical treatment, it might be atopic eczema primarily that could be subjected to safer treatment in future. It is good news that today already three technologically different approaches to producing liposomes for topical use have undergone the process of scaling up successfully. In the future perhaps more emphasis is needed on the selection of optimum lipids. It looks particularly challenging to use skin lipids that might be able to increase the benefit/risk ratio of topical antivirals in herpes simplex tremendously.

Acknowledgment

We are grateful to Ms. A. Senf for expert secretarial work.

References

1. Amblard, P., Basex, J., Beurey, J., Beyloth, C., Bonerandi, J.J., Civatte, J., Dupré, A., Garal, J., Grupper, C., Hincky, M., Kalis, B., Lambert, D., Masse, R., Meynadier, J., Moulin, G., Privat, Y., Texier, L., Geniaux, M., Thivolet, J., and Verret, J.L., Antibio-aberel: efficaté, tolérance, *Gazette Med.*, 89, 2193, 1982.

2. Anonymous, Dermatological needs in drugs and instrumentation, *J. Invest. Dermatol.*, 73, 473, 1979.

3. Bangham, A.G., Standish, M.M., and Watkins, J.C., Diffusion of univalent ions across the lamellae of swollen phospholipids, *J. Mol. Biol.* 13, 238, 1965.

4. Barenholz, Y., Liposome production: historic aspects, in Braun-Falco, O., Korting, H.C., and Maibach, H.I., Eds., *Liposome Dermatics*, Springer, Berlin, 1992, 69.

5. Batzri, S., and Korn, E.D., Single bi-layer liposomes prepared without sonication, *Biochim. Biophys. Acta*, 298, 1015, 1973

6. Bissett, D., Chatteryee, R., and Hannon, D., Photoprotective effect of topical anti-inflammatory agents against ultraviolet radiation-induced chronic skin damage in the hairless mouse, *Photodermatol. Photoimmunol. Photomed.*, 7, 153, 1990.

7. Blume, A., Phospholipids as basic ingredients, in Braun-Falco, O., Korting, H.C., and Maibach, H.I., Eds., *Liposome Dermatics*, Springer, Berlin, 1992, 129.

8. Boddé, H.E., Pechtold, L.A.R.M., Subnel, M.T.A., and de Haan, F.H.N., Monitoring *in vivo* skin hydration by liposomes using infrared spectroscopy in conjunction with tape-stripping, in Braun-Falco, O., Korting, H.C., and Maibach, H.I., Eds., *Liposome Dermatics*, Springer, Berlin, 1992, 137.

9. Bonnekoh, B., Röding, J., Krueger, G.R.F., Ghyczy, M., and Mahrle, G., Increase of lipid fluidity and suppression of proliferation resulting from liposome uptake by human keratinocytes *in vitro*, *Br. J. Dermatol.*, 124, 333, 1991.

10. Bonté, F., Chevallier, J.M., and Meybeck, A., Determination of retinoic acid–liposomal association level in a topical formulation, *Drug Dev. Indust. Pharm.*, 20, 2527, 1994.

11. Bouwstra, J.A., Hofland, H.E.J., Spies, F., Gouris, G.S., and Junginger, H.E., Changes in the structure of the human stratum corneum induced by liposomes, in Braun-Falco, O., Korting, H.C., and Maibach, H.I., Eds., *Liposome Dermatics*, Springer, Berlin, 1992, 123.

12. Cornell, R.C., and Stoughton, R.B., Correlation of the vasoconstriction assay and clinical activity in psoriasis, *Arch. Dermatol.* 121, 6367, 1985.

13. Dalili, H., and Adriani, J., The efficacy of local anesthetics in blocking the sensations of itch, burning, and pain in normal and "sun burned" skin, *Clin. Pharmacol. Ther.* 12, 913, 1971.

14. Diederichs, J.E., and Müller, R.H., Liposome in Kosmetika und Arzneimitteln, *Pharm. Ind.* 56, 267, 1994.

15. Eckstein, E., Steiner, H., and Wesenberg, W., Zur externen Behandlung der Akne mit Tretinoin, *Arzneimittel Forsch./Drug. Res.*, 24, 1205, 1974.

16. Egbaria, K., and Weiner, N., Topical delivery of liposomally encapsulated ingredients evaluated by *in vitro* diffusion studies, in Braun-Falco, O., Korting, H.C., and Maibach, H.I., Eds., *Liposome Dermatics*, Springer, Berlin, 1992, 172.

17. Epstein, N.N., Epstein, W.L., and Epstein, J.H., Atrophic striae in patients with inguinal intertrigo, *Arch. Dermatol.*, 87, 450, 1963.

18. Foldvari, M., *In vitro* cutaneous and percutaneous delivery and *in vivo* efficacy of tetracaine from liposomal and conventional vehicles, *Pharm. Res.*, 11, 1593, 1994.

19. Foldvari, M., Gesztes, A., and Mezei, M., Dermal drug delivery by liposome encapsulation: Clinical and electron microscopic studies, *J. Microencapsul.*, 7, 479, 1990.

20. Foldvari, M., Jarvis, B., and Oguejiofor, C.J.N., Topical dosage form of liposomal tetracaine: effect of additives on the *in vitro* release and *in vivo* efficacy, *J. Contr. Rel.*, 27, 193, 1993.

21. Fricke, U., Dermatika, in Schwabe, U., and Paffrath, D., Eds., *Arzneiverordnungs-Report*, Fischer, Stuttgart, 1993, 169.

22. Friedrich, A., and Krüger, N., Neue Untersuchungen über den Hamamelis-Gerbstoff, *Planta Med.*, 25, 138, 1974.

23. Frosch, P.J., Irritative und kontaktallergische Nebenwirkungen von Akne-Externa, *Hautarzt*, 36, Suppl. 7, 179, 1985.

24. Gesztes, A., and Mezei, M., Topical anesthesia of the skin by liposome-encapsulated tetracaine, *Anesth. Analg.*, 67, 1079, 1988.

25. Hallen, B., Carlsson, P., and Uppfeldt, A., Clinical study of lignocaine — prilocaine cream to relieve pain of venipuncture, *Br. J. Anesthesiol.*, 57, 326, 1985.

26. Hansen, L., Reynolds, B., and Foldvari, M., Topical liposomal tetracaine for i.v. cannulation, *Can. J. Anesth.*, 37, 564, 1990.

27. Hörmann, H.P., and Korting, H.C., Evidence for the efficacy and safety of topical herbal drugs in dermatology, Part I, Anti-inflammatory agents, *Phytomedicine*, 1, 161, 1994.

28. Hope, M.J., and Kitson, C.N., Liposomes. A perspective for dermatologists, *Dermatol. Clin.*, 11, 143, 1993.

29. Imbert, B., and Wickett, R.R., Topical delivery with liposomes, *Cosmet. Toiletries*, 110, 32, 1995.

30. Kibitel, J., Yee, V., and Yarosh, D., Enhancement of ultraviolet-DNA repair in denV gene transfectants and T4 endonuclease V-liposome recipients, *Photochem. Photobiol.*, 54, 56, 1991.

31. Klein, J., Chadwick, C., Hawk, J., Potten, C., Proby, C., Young, A., and Yarosh, D.B., Part II, Study of T4 and -5 liposome lotion for the treatment of xeroderma pigmentosum, *J. Invest. Dermatol.*, 104, 689, 1995 (Abstract).

32. Kligman, L.A., and Kligman, A.M., The effect on rhino mouse skin of agents which influence keratinisation and exfoliation, *J. Invest. Dermatol.*, 73, 354, 1979.

33. Korting, H.C., The skin as a site for drug delivery: the liposome approach and its alternatives, *Adv. Drug Deliv. Rev.*, 18, 271, 1996.

34. Korting, H.C., Blecher, P., Schäfer-Korting, M., and Wendel, A., Topical liposome drugs to come: what the patent literature tells us, *J. Am. Acad. Dermatol.*, 25, 1068, 1991.

35. Korting, H.C., and Braun-Falco, O., Liposomes in delivery of antifungals, in *Fungal Disease. Biology, Immunology, and Diagnosis*, Jacobs, P.H., and Nall, L., Eds., Dekker, Basel, 1997, 501.

36. Korting, H.C., Schäfer-Korting, M., Hart, H., Laux, P., and Schmid, M., Anti-inflammatory activity of hamamelis distillate applied topically to the skin. Influence of vehicle and dose, *Eur. J. Clin. Pharmacol.*, 44, 315, 1993.

37. Korting, H.C., Schäfer-Korting, M., Klövekorn, W., Klövekorn, G., Martin, C., and Laux, P., Comparative efficacy of hamamelis distillate and hydrocortisone cream in atopic eczema, *Eur. J. Clin. Pharmacol.*, 48, 461, 1995.

38. Korting, H.C., Schmid, H.-H., Hartinger, A., Maierhofer, G., Stolz, W., and Braun-Falco, O., Evidence for the phagocytosis of intact oligolamellar liposomes by human keratinocytes *in vitro* and consecutive intracellular disintegration, *J. Microencapsul.*, 10, 223, 1993.

39. Korting, H.C., Stolz, W., Schmid, M.-H., and Maierhofer, G., Interaction of liposomes with human epidermis reconstructed *in vitro, Br. J. Dermatol.*, 132, 571, 1995.
40. Korting, H.C., Zienecke, H., Schäfer-Korting, M., and Braun-Falco, O., Liposome encapsulation improves efficacy of betamethasone dipropionate in atopic eczema but not in psoriasis vulgaris, *Eur. J. Clin. Pharmacol.*, 39, 349, 1990.
41. Kriftner, R.W., Liposome production: the ethanol injection technique and the development of the first approved liposome dermatic, in Braun-Falco, O., Korting, H.C., and Maibach, H.I., Eds., *Liposome Dermatics*, Springer, Berlin, 1992, 91.
42. Langer, R., New methods of drug delivery, *Science*, 249, 1527, 1990.
43. Lasic, D.D., Liposomes within liposomes, *Nature*, 387, 26, 1997.
44. Lasic, D.D., and Papahadjopoulos, D., Liposomes revisited, *Science*, 267, 1275, 1995.
45. Lasic, D.D., Strey, H., Stuart, M.C.A., Podgornik, R., and Frederik, P.M., The structure of DNA-liposome complexes, *J. Am. Chem. Soc.*, 119, 832, 1997.
46. Laux, P., and Oschmann, R., Die Zaubernuß — Hamamelis virginiana L., *Z. Phytother.*, 14, 155, 1993.
47. Leyden, J.J., and Shalita, A.R., Rational therapy for acne vulgaris: an update on topical treatment, *J. Am. Acad. Dermatol.*, 15, 907, 1986.
48. Loth, H., Skin permeability methods, *Meth. Find. Exp. Clin. Pharmacol.*, 11, 155, 1989.
49. Marks, R., Adverse side effects from the use of topical steroids, in Maibach, H.I., and Surber, C., Eds., *Topical Corticosteroids*, Karger, Basel, 1992, 170.
50. Masini, V., Bonté, F., Meybeck, A., and Wepierre, J., Cutaneous bioavailability in hairless rats of tretinoin in liposomes or gel, *J. Pharm. Sci.*, 82, 17, 1993.
51. Meybeck, A., Comedolytic activity of a liposomal antiacne drug in an experimental model, in Braun-Falco, O., Korting, H.C., and Maibach, H.I., *Liposome Dermatics*, Springer, Berlin, 1992, 235.
52. Meybeck, A., Michelon, P., Montastier, C., and Redziniak, G., Pharmaceutical composition in particular dermatological or cosmetic, comprising hydrous lipidic lamellar phases or liposomes containing retinoid or structural analogue thereof such as a carotenoid, U.S. Patent No. 5,034,228, 1985.
53. Mezei, M., and Nugent, J.F., Biodisposition of liposome-encapsulated active ingredients applied on the skin, in Braun-Falco, O., Korting, H.C., and Maibach, H.I., Eds., *Liposome Dermatics*, Springer, Berlin, 1992, 206.
54. Mezei, M., and Gulasekharam, V., Liposomes — a selective drug delivery system for the topical route of administration: Lotion dosage form, *Life Sci.*, 26, 1473, 1980.
55. Mezei, M., Gulasekharam, V., Liposomes — a selective drug delivery system for the topical route of administration: a gel dosage form, *J. Pharm. Pharmacol.*, 34, 473, 1982.
56. Mezei, M., Liposomal local anesthetic and analgesic products, U.S. Patent. No. 4, 937, 078, 1990.
57. Mills, O.H. Jr., and Kligman, A.M., Treatment of acne vulgaris with topically applied erythromycin in tretinoin, *Acta Dermato-Venereol.*, 58, 555, 1978.
58. Milsmann, M.H.W., Schwendener, R.A., and Weder, H.-E., The preparation of large single bi-layer liposomes by a fast and controlled dialysis, *Biochim. Biophys. Acta*, 512, 147, 1978.
59. Ostro, M.J., Cullis, P.R., Use of liposomes as injectible drug delivery systems, *Am. J. Hosp. Pharm.*, 46, 1576, 1989.
60. Papa, C.M., The cutaneous safety of topical tretinoin, *Acta Dermato-Venereol.*, Suppl. 74, 128, 1975.
61. Régnier, M., and Darmon, M., Human epidermis reconstructed *in vitro*: a model to study keratinocyte differentiation and its modulation by retinoic acid *in vitro, Cell. Dev. Biol.*, 25, 1000, 1989.
62. Reznik, H., and Egger, K., Myricetin — ein charakteristisches Flavonol der Hamamelidaceae und Arnacadiaceae, *Z. Naturforsch.*, 15, 247, 1960.
63. Röding, J., Natipide®II: a new easy liposome system, *Seifen, Öle, Fette, Wachse*, 14, 509, 1990.
64. Röding, J., Properties in characterisation of pre-liposome systems, in Braun-Falco, O., Korting, H.C., and Maibach, H.I., Eds., *Liposome Dermatics*, Springer, Berlin, 1992, 110.

65. Röding, J., and Artmann, C., The fate of liposomes in animal skin, in Braun-Falco, O., Korting, H.C., and Maibach, H.I., Eds., *Liposome Dermatics*, Springer, Berlin, 1992, 185.

66. Röding, J., and Ghyczy, M., Control of skin humidity with liposomes: stabilisation of skin care oils and lipophilic active substances with liposomes, *Seifen, Öle, Fette, Wachse*, 10, 372, 1991.

67. Schaefer, H., Stüttgen, G., Zesch, A., Schalla, W., and Gazith, J., Quantitative determination of percutaneous absorption of radio-labeled drugs *in vitro* and *in vivo* by human skin, in Maibach, H.I., Ed., *Current Problems in Dermatology*, Vol. 7, Karger, Basel, 1979, 80.

68. Schäfer-Korting, M., Korting, H.C., and Ponce-Pöschl, E., Liposomal tretinoin for uncomplicated acne vulgaris, *Clin. Investigator*, 72, 1086, 1994.

69. Schäfer-Korting, M., Schmid, M.-H., and Korting, H.C., Topical glucocorticoids with improved benefit/risk ratio. Rationale of a new concept, *Drug Safety*, 14, 375, 1996.

70. Schaller, M., Korting, H.C., and Schmid, M.-H., Interaction of cultured human keratinocytes with liposomes encapsulating silver sulfadiazine: proof of the uptake of intact vesicles, *Br. J. Dermatol.*, 134, 445, 1996.

71. Schaller, M., Wurm, R., and Korting, H.C., Direct evidence for uptake of intact liposomes encapsulating silver sulfadiazine by cultured human keratinocytes based on combined transmission electron microscopy and X-ray microanalysis, *Antimicrob. Agents Chemother.*, 41, 717, 1997.

72. Schmid, M.-H., and Korting, H.C., Liposomes for atopic dry skin: the rationale for a promising approach, *Clin. Investigator*, 71, 649, 1993.

73. Schmid, M.-H., and Korting, H.C., Liposomes: a drug carrier system for topical treatment in dermatology, *Crit. Rev. Ther. Drug Carrier Syst.*, 11, 97, 1994.

74. Schmid, M.-H., and Korting, H.C., Liposomes as penetration enhancers and controlled release units, in Smith, E.W., and Maibach, H.I., Eds., *Percutaneous Penetration Enhancers*, CRC Press, Boca Raton, 1995, 323.

75. Schmid, M.-H., and Korting, H.C., Therapeutic progress with topical liposome drugs for skin disease, *Adv. Drug Deliv. Rev.*, 18, 335, 1996.

76. Sims, C., Thickly and thinly applied lignocaine-prilocaine cream prior to venipuncture in children, *Physiol. Intens.*, 19, 343, 1991.

77. Smith, E.W., and Maibach, H.I., Eds., *Percutaneous Penetration Enhancers*, CRC Press, Boca Raton, 1995.

78. Spies, F., Boddé, H.E., Meybeck, A., and Bonté, F., A freeze fracture study of interactions between liposome preparations and human skin *in vitro*, *Proc. Int. Symp. Control. Rel. Bioact. Mater.*, 18, 529, 1991.

79. Touitou, E., Junginger, H.E., Weiner, N.D., Nagai, T., and Mezei, M., Liposomes as carriers for topical and transdermal delivery, *J. Pharm. Sci.*, 83, 1189, 1994.

80. Wagner, D., Kern, W.V., and Kern, P., Liposomal doxorubicin in AIDS-related Kaposi's sarcoma: long-term experiences, *Clin. Investigator*, 73, 417, 1994.

81. Weder, H.G., Liposome production: The sizing-up technology starting from mixed micelles and the scaling-up procedure for the topical glucocorticoid betamethasone-dipropionate and betamethasone, in Braun-Falco, O., Korting, H.C., and Maibach, H.I., Eds., *Liposome Dermatics*, Springer, Berlin, 1992, 101.

82. Weder, H.G., Waldvogel, S., Supersaxo, A., Roos, K., and Zumbühl, O., Präparation homogener Liposomen im Semi-Makromaßstab mittels kontrollierter Detergens-Dialyse, in Schmidt, K.H., Ed., *Liposomes as Drug Carriers*, Thieme, Stuttgart, 1986, 26.

83. Weiner, N., Williams, N., and Bird, G., Topical delivery of liposomal encapsulated interferon evaluated in a cutaneous herpes guinea pig model, *Antimicrob. Agents Chemother.*, 33, 1217, 1989.

84. Wertz, P.W., Abraham, W., Landmann, L., and Downing, D.T., Preparation of liposomes from stratum corneum lipids, *J. Invest. Dermatol.*, 87, 582, 1986.

85. Winston, M.H., and Shalita, A.R., Acne vulgaris. Pathogenesis and treatment, *Pediatr. Clin. North Am.*, 38, 889, 1991.

86. Wohlrab, W., and Lasch, J., Penetration kinetics of liposomal hydrocortisone in human skin, *Dermatologica*, 174, 18, 1987.
87. Wohlrab, W., and Lasch, J., The effect of liposomal incorporation of topically applied hydrocortisone on serum concentration and urinary excretion, *Dermatol. Monatsschr.*, 175, 348, 1989.
88. Wohlrab, W., Lasch, J., Laub, R., Taube, C.M., and Wellner, K., Distribution of liposome-encapsulated ingredients in human skin *ex vivo*, in Braun-Falco, O., Korting, H.C., and Maibach, H.I., Eds., *Liposome Dermatics*, Springer, Berlin, 1992, 215.
89. Yarosh, D., Purification and administration of DNA repair enzymes, Int. Patent Appl. PCT/US 89/02973, 1990.
90. Yarosh, D.B., Liposome-encapsulated enzymes for DNA repair, in Braun-Falco, O., Korting, H.C., and Maibach, H.I., Eds., *Liposome Dermatics*, Springer, Berlin, 1992, 258.
91. Yarosh, D., Bucana, C., Cox, P., Alas, L., Kibitel, J., and Kripke, M., Localisation of liposomes containing a DNA repair enzyme in murine skin, *J. Invest. Dermatol.*, 103, 461, 1994.
92. Yarosh, D.B., Tsimis, J., and Yee, E.V., Enhancement of DNA repair of UV damage in mouse and human skin by liposomes containing a DNA repair enzyme, *J. Soc. Cosmet. Chem.*, 41, 85, 1990.
93. Yarosh, D.B., Tsimis-Kibitel, J., Green, L.A., and Spinowitz, A., Enhanced unscheduled DNA-synthesis in UV-irradiated human skin explants treated with T4- and -5 liposomes, *J. Invest., Dermatol.*, 97, 147, 1991.

22

Topical Glucocorticoids with Improved Benefit/Risk Ratio

Monika Schäfer-Korting
Freie Universität Berlin

Anja Gysler
Freie Universität Berlin

22.1 Introduction

The topical corticosteroid (glucocorticoid) therapy of steroid-sensitive dermatoses such as atopic dermatitis, contact dermatitis, and psoriasis vulgaris aims at a strong antiinflammatory action on the skin and minor (if any) systemic and local side effects. Hypothalamic-pituitary-adrenal axis suppression perceptible by unphysiologically low serum cortisol concentrations may occur after topical application of high-potency but also with low-potency glucocorticoids.[1,2] Partially damaged and diseased skin as well as application on large areas of the body surface for extended periods favor adrenal atrophy.[3-5] Higher amounts of drug are also absorbed from physiologically occluded areas.[6] Special treatment schedules particularly for use in chronic skin diseases and risk groups like children[7,8] and elderly persons[9] may be helpful to minimize adverse effects.

Local side effects, however, appear more relevant as compared to systemic ones. Skin thinning due to the antiproliferative effect is feared most of all.[10,11] Striae formation is the worst and most irreversible form of skin atrophy. Atrophy and teleangiectasia (irreversible dilatation of small blood vessels) are phenomena mainly located in the dermis[12-15] resulting from the loss and degeneration of elastic and collagen fibrils. Clinical experiences with conventional fluorinated topical glucocorticoids indicate that high antiinflammatory and antiproliferative effects run parallel. Treatment schedules using less active drugs at a low application rate have been advocated to reduce the atrophogenic potential. Moreover, much effort was and is still made to develop corticosteroids corresponding to the so-called soft drug principle.[16] This means a proper adjustment of drug structures to obtain a high concentration at the target site and a rapid inactivation in other areas. Rapid metabolism of the potent steroid to inactive or

0-8493-2791-1/99/$0.00+$.50
© 1999 by CRC Press LLC

only slightly active metabolites may dissociate the wanted antipruritic and antiphlogistic effect — produced by the affinity of the parent compound for the glucocorticoid receptor[17-19] — from the unwanted antiproliferative effect in deeper skin layers.

Besides research on therapeutic regimens and molecular modifications, attempts were made to reveal vehicle effects on glucocorticoid release, efficacy, and adverse reactions. The optimization and adaptation of the vehicle to the respective skin disease in combination with agents that stimulate or regulate cell proliferation also appear promising in the treatment of chronic dermatoses. The improvements obtained by any of these measures are reviewed in detail. Moreover, a technique to quantify the benefit/risk ratio of topical glucocorticoid therapy is described.

22.2 Molecular Modifications

Until only recently new steroid compounds ought to be as potent as possible. Considering the patients' increasing awareness of drug side effects, however, research now aims at rather moderate corticosteroids with an improved benefit/risk ratio. In designing better therapeutic agents, it is of fundamental importance to interpret data on the structure-activity relationship[20-22] and to integrate information about the mechanisms of action on the cellular level.[23-25] With respect to dermatological agents for topical use, consequences of type and severity of skin diseases on the barrier and reservoir function of the stratum corneum also have to be kept in mind. Diseased skin leads to different drug absorption rates and thus diminished or increased drug effects, necessitating adjusting steroid potency.[26-29]

Several corticosteroids based on the above-described soft drug principle[16] have been introduced into therapy of steroid-sensitive skin disorders. The soft steroids in general are esters from the less potent glucocorticoids hydrocortisone and prednisolone. These are hydrocortisone aceponate, hydrocortisone buteprate, methylprednisolone aceponate and mometasone furoate, which followed prednicarbate, the first congener of this group of drugs.

22.2.1 Hydrocortisone Esters

Hydrocortisone buteprate (hydrocortisone 17-butyrate, 21-propionate), a double-esterified nonhalogenated corticosteroid differs from halogenated steroids with regard to weaker systemic and atrophogenic effects.[30]

Y. Kitano investigated the enzymatic hydrolysis of hydrocortisone buteprate in human keratinocytes. Hydrolysis at C-21 to hydrocortisone 17-butyrate occurred rather rapidly, followed by nonenzymatic translocation to hydrocortisone 21-butyrate, and then a final hydrolysis to hydrocortisone. The intracellular concentration of hydrocortisone buteprate, however, was 7.7 times higher than that of hydrocortisone.[31] Studying the penetration of corticosteroid-esters through the skin of rats and dogs, Otomo et al.[32] noticed hardly any biotransformation in the horny layer despite a good distribution in this area, but a deesterification to the 17-monoester and unesterified steroid in the viable epidermis. Binding studies by Otomo et al. indicate the affinity of hydrocortisone buteprate to the glucocorticoid receptor to be almost equal to that of hydrocortisone 17-butyrate and much stronger than that of hydrocortisone (Table 22.1). This is rather unique if compared with other halogenated[33] and nonhalogenated[34] glucocorticoid esters. Moreover, Muramatsu et al.[17] evaluated the receptor association and dissociation of [³H]-hydrocortisone and [³H]-hydrocortisone buteprate. The association rate constant was three times higher for hydrocortisone than for the diester, and the dissociation rate constant six to eight times higher. The rather short-lasting but strong affinity of the double-ester to its receptor may moderate the antiproliferative action by means of not interfering with the cell-cycle but keeping the antiphlogistic efficacy.[35]

To investigate the reasons for the low antiproliferative potency combined with a strong antiinflammatory property, Schalla et al. compared hydrocortisone buteprate 0.1% and fluocinolone acetonide 0.025% cream following epicutaneous and intradermal application in healthy volunteers. The increased lipophilicity of the steroid diesters improved the permeation through the horny layer to viable skin. Therefore, with both application routes vasoconstriction following hydrocortisone buteprate had a quick start —

TABLE 22.1   Receptor Binding of Selected Topical Glucocorticoids

	IC_{50} [nMol/l]	K_i	RRA
Dexamethasone	7.5 ± 1.6[17]	5.9 ± 1.3[17]	100[19]
Triamcinolone acetonide	—	—	361[19]
Betamethasone	6.3 ± 1.6[17]	4.9 ± 1.3[17]	—
17-valerate	3.4 ± 0.5[17]	2.6 ± 0.4[17]	—
		0.22 ± 0.016[67]	
Hydrocortisone	66.4 ± 3.2[17]	51.9 ± 2.5[17]	22.7[43]
17-butyrate	11.0 ± 1.3[17]	8.6 ± 1.5[17]	—
buteprate	8.7 ± 2.6[17]	6.8 ± 2.0[17]	—
Prednisolone	—	—	10.4[19]
17-ethylcarbonate	—	—	103[19]
Prednicarbate	—	—	7.3[19]
Methylprednisolone	—	—	125[43]
17-propionate	—	—	143[43]
aceponate	—	—	58.6
Mometasone	—	0.194 ± 0.014[67]	—
17-furoate	—	0.307 ± 0.035[67]	—
6β-hydroxy-17-furoate	—	0.428 ± 0.040[67]	

IC_{50}: Concentration of the competitor to displace 50% of [^{3}H]-dexamethasone[17] or [^{3}H]-triamcinolone acetonide[67] from the glucocorticoid receptor

$$K_i: \quad \frac{IC_{50}}{1 + \dfrac{c}{K_d}}$$

K_d: Dissociation rate constant
RRA: Relative receptor affinity, dexamethasone = 100

followed by a steep decline after reaching maximum effect. Fluocinolone acetonide, however, had a later onset of action following the epicutaneous application as compared to the intradermal route bypassing penetration of the horny layer. There was also a flatter decline with both routes of application compared with the hydrocortisone diester. Thus, the strong antiinflammatory but low antiproliferative effect of hydrocortisone buteprate obviously depends on its molecular structure.[36]

In patients suffering from atopic dermatitis twice daily applications of hydrocortisone buteprate ointment or cream[37] appeared efficacious and safe. This holds true even for such very sensitive skin types as facial or aged skin.[38,39]

Data on the pharmacokinetics of hydrocortisone aceponate (hydrocortisone 21-acetate, 17-propionate) are missing. The drug inhibited [^{3}H]-thymidine incorporation in DNA and thus proliferation of human skin fibroblasts less than the equipotent halogenated corticosteroid betamethasone 17-valerate. Hydrocortisone aceponate also had a more favorable effect on collagen and total protein synthesis compared to betamethasone valerate.[10] Moreover, Korting et al.[40] noticed the less toxic effect of hydrocortisone aceponate (and other nonhalogenated diesters) on human skin fibroblasts and keratinocytes compared to betamethasone valerate in a neutral red release assay. All corticosteroids tested had a more marked toxic effect on keratinocytes than on fibroblasts.

Clinical trials in several forms of eczema showed a strong antiphlogistic efficacy and a low rate of adverse effects. Hydrocortisone aceponate cream or ointment were applied once daily for up to 13 weeks (K. W. Rüping, personal communication). This double ester also appears to be a promising member of the so-called soft steroid family.

22.2.2   Prednisolone Derivatives

Methylprednisolone aceponate (methylprednisolone 21-acetate, 17-propionate) is a diester of the prednisolone type. Pharmacokinetic studies in human cadaver skin indicate a degradation of the highly lipophilic double ester[41] to methylprednisolone 17-propionate by enzymatic hydrolysis in viable epidermis and dermis.[42] As with the hydrocortisone esters the 17-monoester is converted nonenzymatically to its respective 21-ester before skin esterases hydrolyze the compound into the free corticosteroid. Since the 17-monoester shows a three times higher affinity to the glucocorticoid receptor than the parent compound (Table 22.1), the first hydrolysis step in the skin functions as an activation. The nonviable stratum corneum acts as a reservoir tissue and is not influenced in its barrier function by the drug.[43] Zentel and Töpert[44] presume a strong increase of esterase activity in inflamed skin and therefore a faster conversion to the active 17-monoester as a reason for the antiinflammatory potency, as observed in croton oil treated rat ears.

Methylprednisolone aceponate is classified as a potent or very potent topical corticosteroid[45,46]; it is active when applied once daily.[47] Kecskés et al.[48] evaluated the systemic side effects of twice daily applications of methylprednisolone aceponate 0.1% preparations to 60% of the body surface on healthy volunteers. Effects were compared to those following clobetasol propionate 0.05% cream. Plasma cortisol levels and the circadian rhythm of plasma concentrations remained within the normal range after nonoccluded administration of methylprednisolone aceponate cream for 5 days. In contrast to this a significant decrease in endogenous cortisol production was obtained with clobetasol propionate cream. Furthermore, there was even a superiority of methylprednisolone aceponate ointment over clobetasol propionate cream when applied under occlusion, which is well in accordance with a rapid formation of inactive metabolites.[49] Similar results were obtained in patients with atopic dermatitis or psoriasis vulgaris: twice daily applications of methylprednisolone aceponate on skin lesions of at least 40% of the body surface for 7 consecutive days indicated only a minor decrease in plasma cortisol levels; the circadian rhythm of endogenous cortisol release was maintained.[50] Together with its mild local adverse effects, especially the low atrophogenic potential[51,52], methylprednisolone aceponate may also differentiate beneficial actions from antiproliferative effects, as they occur with other prednisolone-derived corticosteroids.[53,54]

The first congener of the soft steroid group, prednicarbate, the 17-ethylcarbonate, 21-propionate diester of prednisolone, exhibits a strong antiinflammatory[55,56] and antipruritic[57] potency. In *in vitro* cultivated fibroblasts a suppression of chemotactic activity was not observed with prednicarbate as there was only minor inhibition of cell proliferation (comparable to that of hydrocortisone and hydrocortisone buteprate).[11]

The biotransformation of prednicarbate during its rather slow penetration through animal skin occurs by a rapid deesterification at C-21 to prednisolone 17-ethylcarbonate, the intrinsic active agent (Table 22.1), followed by a delayed, presumably nonenzymatic, metabolism to mainly inactive prednisolone. Moreover, there is some hydroxylation to 6β-hydroxy-20β-dihydro-prednisolone.[58] Recently pharmacokinetics and pharmacodynamics of prednicarbate have been evaluated in primary cell cultures. Human keratinocytes rapidly cleave the 21-ester forming prednisolone 17-ethylcarbonate which is almost completely metabolized to prednisolone within 24 hours. Fibroblasts, however, degrade prednicarbate at a rate of only 1% per hour.[58a] Pharmacokinetic differences are well reflected by pharmacodynamic ones. A pronounced suppression of interleukin-1α synthesis with prednicarbate is only seen with keratinocytes. Since this cytokine reflects inflammation in keratinocytes and induces fibroblast proliferation, the selective influence of prednicarbate on epidermal cells explains the antiinflammatory activity in dermatitis and suggests a minor atrophogenic potential[58b] Prednisolone 17-ethylcarbonate, however, does not appear selective.[58a,b] Neither prednicarbate nor any metabolite was detected in plasma samples of healthy volunteers following the topical application to the skin.[34,59]

Prednicarbate 0.25% cream applied for 12 h to occluded unbroken skin (approximately 50% of the body surface) did not change cortisol secretion.[34] Besides the lack of systemic side effects, twice daily nonoccluded applications of prednicarbate 0.25% cream induced less skin atrophy in healthy volunteers than betamethasone valerate 0.1% and clobetasol propionate 0.05% cream as assessed by high-frequency ultrasound.[60]

Clinical studies comparing two soft steroids in atopic eczema indicated prednicarbate to be superior to fluocortin butyl ester[61] and equipotent to hydrocortisone buteprate.[57] The atrophogenic potential of prednicarbate, however, turned out to be more pronounced as compared to the hydrocortisone double-ester.[57] It can be speculated whether this results from the additional double bound (ring A) of prednisolone increasing all glucocorticoid effects.

Extensive trials in healthy volunteers implying an increased benefit/risk ratio[62] are discussed in detail below. Further data on tolerance and safety in patients with atopic dermatitis, however, have to be obtained, considering allergic contact dermatitis[63] and topical side effects after prolonged use.[53]

22.2.3 Halogenated Glucocorticoid Esters

Mometasone furoate is the furoate 17-ester of the 16α-methyl analog of beclomethasone containing chlorine atoms in the 9α- and 21-position. In rat dermis, the binding affinities of mometasone-related compounds were considerably increased by means of halogenation of the 9α-position, substitution of the 21-OH by Cl and esterification of the 17-OH with furoate (Table 22.1). Therefore, assuming ester hydrolysis as a biotransformation pathway, the receptor binding will decline during the passage through the skin. Most interestingly mometasone has a lower receptor affinity in dermal cells than in cells of the epidermis.[64] As 6β-hydroxylation, another highly likely biotransformation reaction, also decreases the affinity toward the glucocorticoid receptor (Table 22.1), an increased benefit/risk ratio is suggested.

Mometasone furoate is classified by Niedner as a potent corticosteroid. In various steroid-sensitive dermatoses such as psoriasis vulgaris, atopic and seborrheic dermatitis, and lichen planus once daily applications of mometasone furoate 0.1% preparations appeared almost equipotent to twice daily applications of betamethasone dipropionate 0.05%[65] and superior when compared to betamethasone valerate 0.1% applied twice daily.[66-68] Mometasone furoate was superior as well in reducing UV-B–induced erythema as assessed by laser Doppler blood flowmetry when comparing single applications of the above-mentioned preparations.[69]

Comparative trials vs. hydrocortisone 1% preparations showed a favorable safety profile of mometasone furoate.[70] This holds true with respect both to skin thinning[71-73] and systemic side effects,[74] even with children (7 months to 12 years).[75]

In healthy volunteers, however, mometasone furoate 0.1% appeared inferior to methylprednisolone aceponate 0.1% when applied three times a week under occlusion. There was a significantly higher rate of teleangiectasia and skin atrophy following mometasone furoate after 6 weeks.[74] Thus occlusive dressings for a prolonged time will disturb mometasone furoate tolerability, stressing the importance of the absorption rate, which should not be too high to allow for sufficient degradation to inactive or less active metabolites essential for lower systemic and topical side effects.

A promising new contribution to increased corticosteroid benefit/risk ratio is loteprednol etabonate, a chloromethyl 17α-ethoxycarbonyloxy, 17β–carboxylate derivative of prednisolone, achieving a 4.3 times higher glucocorticoid receptor binding than the fluorinated corticosteroid dexamethasone.[76] The drug permeated hairless mouse skin at rates usually achieved with conventional steroids, and hardly any biotransformation took place in the skin.[76,77] After oral administration to dogs, however, no native substance but metabolites were detected, presumably deesterfied at C-20 to the 17β-carboxylic acid and then hydrolyzed to the analogous cortienic acid.[78] Binding assays with these putative biotransformation products demonstrated poor receptor affinity, indicating a rapid inactivation of this compound as well.[76] In human skin loteprednol etabonate induced skin blanching comparable to betamethasone valerate. Clinical trials are needed to prove loteprednol etabonate a well tolerated and efficacious topical corticosteroid.

Another topical steroid metabolized in the systemic circulation rather than during the penetration through the skin is tipredane, a 9-fluorinated 17α-ethylthio 17β-methylthio derivative of prednisolone. The drug is rapidly metabolized to a variety of compounds in human liver but not in skin homogenates.[79] In the croton oil test, the metabolites were markedly less potent than tipredane. This decrease in antiinflammatory potency and thus glucocorticoid activity suggests tipredane to induce less systemic adverse effects compared to the conventional halogenated glucocorticoids. This, however, has to be proved in human studies. Effort is also made to optimize the tipredane vehicle — in this case a solution for topical application — considering the influence of the vehicle to release of the active agent.[80]

22.2.4   Androstane-17β- and Androstane-16α-Carboxylic Acid Esters

Potent antiinflammatory agents are also found within the group of androstane-17β- and androstane-16α-carboxylic acid esters. Fluticasone 17-propionate, a thioester of the former group, is used for seasonal rhinitis and skin disease. With both groups, receptor affinity is linked to the ester moiety and sharply decreases following hydrolysis.[19,81]

22.3   Influence of the Vehicle

In topical therapy, the vehicle plays an important role in releasing the drug to the target area and also in mediating its own effect. For example, the base preparation moisturizes and lubricates dry, lichenified eczema skin and helps to close and drain weeping areas. In clinical trials with patients suffering from psoriasis vulgaris and various other forms of steroid responsive skin diseases 39% of the psoriasis patients and 66% of the others showed moderate or even excellent results when treated with the steroid-free vehicles exclusively.[82]

Aiming at an optimized interaction between vehicle and drug, Gip and Verjans[83] compared a new hydrocortisone butyrate 0.1% lipocream formulation to betamethasone valerate 0.1% cream in dry severe chronic eczema. The two preparations appeared to be equipotent. The patients favored the lipocream due to its cosmetic characteristics.

Assuming that an ointment influences drug penetration by inducing occlusion and therefore some hydratation of the treated site several betamethasone valerate 0.1% formulations were evaluated by Smith et al.[84] in healthy volunteers. In the nonoccluded application mode, bioavailability (derived from skin blanching) was highest with a scalp preparation, followed by the ointment, the cream, and the lotion. Under occlusion, however, the ranking order reads scalp preparation–lotion–cream–ointment. The absolute blanching activity of the ointment preparation was almost unaffected. Also betamethasone dipropionate was delivered to a higher extent to the skin from an ointment compared to the corresponding cream.[85] Apart from the occlusive effect of the ointment special components of the formulation enhance steroid release: e.g., propylene glycol is a well-known penetration enhancer for glucocorticoids.[86,87] Solubilizing betamethasone dipropionate in high amounts of propylene glycol resulted in an efficiency in psoriasis vulgaris significantly exceeding the one following conventional ointment formulations.[88] In another trial in healthy volunteers cream and ointment formulations of triamcinolone acetonide 0.1% or clobetasol propionate 0.05% were compared concerning their effect on skin thickness.[89] The cream preparations induced a significantly higher degree of skin thinning, indicating an increased risk of skin atrophy. To take the influences of vehicles on glucocorticoid activity into account potency ranking of topical corticosteroids has to consider the respective formulation.

Aside from modulating clinical efficacy and the atrophogenic potential the vehicles also seem to be involved in causing or at least intensifying other topical adverse effects. Korting et al.[90] reported an increase of skin surface roughness when applying a corticosteroid-free cream, whereas the skin surface smoothed after application of the corresponding steroid preparation. This is well in accordance with the observed higher rate of adverse drug reactions following base preparations (about 340 patients) compared to glucocorticoid formulations (about 2600 patients).[82]

22.4   Reduction of Side Effects by the Use of Additional Drugs

To prevent steroid-induced atrophy combinations of glucocorticoids with other drugs, which were suggested to increase viable epidermal thickness (retinoids, ammonium lactate) or which may be lacking in atrophogenic skin, were investigated. In mice 0.025% retinoic acid cream applied for 9 consecutive weekdays (in the afternoon) prevented the reduction of skin thickness by 0.05% to 0.1% dexamethasone (alcohol/propylene glycol) applied in the morning. Histologic examinations verified the antiatrophogenic effect. Antiinflammatory activity, however, was not diminished by the retinoid. This was derived from results in the ear edema model (12-O-tetradecanoylphorbol-13-acetate, TPA) and in croton-oil-induced skin inflammation.[91] Prevention of epidermal thinning may result from retinoic acid, inducing DNA synthesis and thus stimulating cell proliferation. The inhibition of dermal atrophy is explained by the observation of retinoids stimulating glycosaminoglycan and possibly also collagen and proteoglycan synthesis.[92] The results of these studies should be verified in man.

In addition to retinoic acid ammonium lactate was evaluated to prevent steroid atrophy. The influence of topical applications of a 12% solution on the wanted and unwanted effects of potent corticosteroid ointment preparations was evaluated in healthy subjects. Though ammonium lactate did not influence glucocorticoid activity in the skin blanching assay, it clearly diminished the reduction of viable epidermis and glycosaminoglycans.[93] In contrast to these stimulating results, 10% primrose oil failed to influence the reduction of skin thickness induced by betamethasone valerate 0.1% cream.[94] This suggests a deficiency of ω-6 fatty acids to be less important with respect to skin atrophy.

22.5 Treatment Schedule

In recent years numerous therapeutic regimens have been proposed for the treatment of atopic dermatitis. To minimize tachyphylaxis, Woodford et al.[95] assessed vasoconstriction following amcinonide 0.1% cream in healthy volunteers. Following three applications on the first day (loading dose) the drug was reapplied once or twice daily and also on an alternate day base. Already the loading dose caused marked tachyphylaxis. The once daily application appeared more potent than the other application regimens and should be preferred concerning activity.

Tapering the hydrocortisone 17-butyrate concentration (0.1%, 0.05%, 0.03%, 0.015%) down to the respective steroid-free vehicle in patients with atopic dermatitis indicated that adaptation of the glucocorticoid concentration to the course of the disease gives acceptable rates in symptom improvement. There were no signs of rebound or tachyphylaxis even when the treatment was repeated.[96] Another interesting new regimen is the short time exposure of a superpotent topical glucocorticoid. Originally created for dithranol treatment in psoriasis vulgaris,[97,98] short-term therapy may be useful in atopic eczema too. No significant difference in skin blanching following a clobetasol propionate 0.05% cream exposure for 1 and 16 h respectively was observed using the vasoconstriction test.[99]

Besides this, several attempts to moderate adverse effects by means of intermittent or discontinuous treatment have been made assuming the stratum corneum as a corticosteroid reservoir[100,101] slowly providing certain amounts of the steroid to the viable tissue.[102] In patients with atopic dermatitis Høybye et al. monitored the influence of a daily application of mometasone furoate 0.1% fatty cream applied once daily for 3 weeks, followed by an intermittent treatment (three consecutive days a week) for another 3 weeks. Thus optimized treatment schedules with this conventional glucocorticoid are also looked for. The regimen appeared very effective (41 out of 48 patients cleared or markedly improved after 6 weeks) and safe (few side effects with no evidence of skin atrophy).[103] Similar results were obtained with psoriasis vulgaris. Twelve patients applied clobetasol propionate 0.05% cream t.i.d for 4 consecutive days and then once weekly for up to 31 weeks. Serum cortisol levels returned to the normal range after the initial 4 days of treatment. Eight patients remained cleared for an average period of 5 months.[104] The good tolerability of such a regimen is further supported by the results of a study with another potent glucocorticoid. In healthy volunteers the application of 0.1% triamcinolone acetonide ointment once daily for 3 consecutive days a week decreased skin thickness during the first 2 weeks. Then skin thickness returned to almost normal values despite continued application.[105]

Much effort was taken to create the most effective and safe application schedule, depending on type and severity of the disease and on the kind of corticosteroid used. Investigations, however, are not yet completed. New facts (especially on the molecular, genetic, and cellular level) improving our understanding for the course and healing of skin diseases should be taken into consideration. Since steroid structure, vehicle, and treatment schedule contribute to both efficacy and safety further clinical trials have to take all of these points into account.

22.6   Estimation of the Benefit/Risk Ratio

As described above, there are by now various new glucocorticoids reported to induce less adverse effects compared to equipotent conventional counterparts. A technique for the comparative quantification of the benefit/risk ratios is suggested by Schäfer-Korting and co-workers[106] comparing the ratios of drug activity and atrophogenicity. Mean score values for blanching obtained by the UV erythema test or skin-blanching test according to McKenzie and Stoughton are divided by the relative reduction of the skin thickness. Skin thinning is focused upon because of its high relevance in topical glucocorticoid therapy. Glucocorticoid activity should by preference be judged from the skin-blanching test, which is more discriminatory than UV-induced erythema. Reduction of skin thickness can be obtained precisely by high-frequency ultrasound analysis.[107] With these parameters a real quantification of the benefit/risk ratio is possible for the first time.[106]

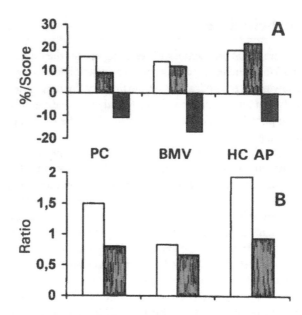

FIGURE 22.1   A: Activity (visual scores) of prednicarbate (PC), hydrocortisone aceponate (HC AP), and betamethasone valerate (BMV) in the skin blanching assay (*open columns*) and UV-B–induced erythema (*striped columns*) as well as the reduction of skin thickness (%, black columns). B: Benefit/risk ratio as derived from skin blanching test (*open columns*) and UV-B–induced erythema test (*striped columns*).

The benefit/risk ratio as described by Schäfer-Korting et al. appears to be better with prednicarbate and hydrocortisone aceponate than with betamethasone valerate (Figure 22.1). The improved ratio may result from pharmacokinetic and/or pharmacodynamic reasons. Pharmacokinetic superiority is due to a fast inactivation in viable skin,[58a] whereas a pharmacodynamic superiority may result from a selectivity for special receptor populations/subtypes, and variations in regulatory processes.[58b]

22.7   Conclusions

Irrespective of the particular mechanism, however, the treatment with better-tolerated glucocorticoids is of great importance for patients with frequently relapsing disease, e.g., atopic dermatitis, to avoid irreversible skin damage. Today there is convincing evidence that a separation of wanted and unwanted glucocorticoid effects is possible. This means a major progress in the treatment of chronic and recurring skin diseases, for example, atopic dermatitis and psoriasis vulgaris. Yet clinical studies comparing two (or more) soft steroids are urgently needed to look for the best congener(s) within this group. Above this more studies evaluating the reason for this separation — combining pharmacokinetic and pharmacodynamic data — are also highly desirable. Risk/benefit analyses of drugs are important for therapeutic decisions.[108,109]

References

1. Gomez, E. C., Kaminester, L., and Frost, P., Topical halcinonide and betamethasone valerate effects on plasma cortisol, *Arch Dermatol* 113, 1196, 1977.
2. Brattsand, R., Thalen, A., Roempke, K., Kallstrom, L., and Gruvstad, E., Influence of 16α,17α-acetal substitution and steroid nucleus fluorination on the topical to systemic activity ratio of glucocorticoids, *J Steroid Biochem* 16, 779, 1982.
3. Turpeinen, M., Influence of age and severity of dermatitis on the percutaneous absorption of hydrocortisone in children, *Br J Dermatol* 118, 517, 1988.
4. Munro, D. D., The effect of percutaneously absorbed steroids on hypothalamic-pituitary-adrenal function after intensive use in in-patients, *Br J Dermatol* 94, 67, 1976.
5. Kirketerp, M., Systemic effects of local treatment with fluocinolone acetonide applied under plastic film, *Acta Derm Venereol* 44, 54, 1964.
6. Feldmann, R. J. and Maibach, H. I., Regional variation in percutaneous penetration of ^{14}C cortisol in man, *J Invest Dermatol* 48, 181, 1967.
7. Sillevis Smitt, J. H. and Winterberg, D. H., Topical corticosteroids in children: local and systemic effects. In H.I. Maibach and C. Surber (Eds.), *Topical Corticosteroids.* Basel: Karger, 1992. pp. 196–209.
8. Turpeinen, M., Lehtokoski-Lehtiniemi, E., Leisti, S., and Salo, O. P., Percutaneous absorption of hydrocortisone during and after the acute phase of dermatitis in children, *Pediatr Dermatol* 5, 276, 1988.
9. Niedner, R., Grundlagen einer rationalen Therapie mit externen Glukokortikosteroiden, *Hautarzt* 42, 337, 1991.
10. Görmar, F. E., Bernd, A., and Holzmann, H., Wirkung von Hydrocortisonaceponat auf Proliferation, Gesamtprotein- und Kollagen-Synthese menschlicher Hautfibroblasten *in vitro*, *Arzneim-Forsch* 40, 192, 1990.
11. Hein, R., Korting, H. C., and Mehring, T., Differential effect of medium potent nonhalogenated double-ester-type and conventional glucocorticoids on proliferation and chemotaxis *in vitro*, *Skin Pharmacol* 7, 300, 1994.
12. Lehmann, P., Zheng, P., Lavker, R., and Kligman, A M., Corticosteroid atrophy in human skin. A study by light, scanning, and transmission electron microscopy, *J Invest Dermatol* 81, 169, 1983.
13. Heng, M. C. Y., Heng, H. L., and Allen, S. G., Basement membrane changes in psoriatic patients on long-term topical corticosteroid therapy, *Clin Exp Dermatol* 15, 83, 1990.
14. Oikarinen, A., Autio, P., Kiistala, U., Risteli, L., and Risteli, J., A new method to measure type I and III collagen synthesis in human skin *in vivo*: demonstration of decreased collagen synthesis after topical glucocorticoid treatment, *J Invest Dermatol* 98, 220, 1992.
15. Siddiqui, A. A., Jaffar, T., Khan, N., Azami, R., and Waqar, M. A., Effect of dexamethasone on collagen synthesis in fibroblasts cultured in collagen lattice, *Biochem Soc Trans* 20, 207, 1992.
16. Bodor, N., The application of soft drug approaches to the design of safer corticosteroids. In E. Christophers, E. Schöpf, A.M. Kligman, and R.B. Stoughton (Eds.), *Topical Corticosteroid Therapy — A Novel Approach to Safer Drugs.* New York: Raven Press, 1988. pp. 13–25.
17. Muramatsu, M., Fujita, A., Tanaka, M., Ishii, Y., and Aihara, H., Enhancement of affinity to receptors in the esterified glucocorticoid, hydrocortisone 17-butyrate, 21-propionate (HBP), in the rat liver, *Biochem Pharmacol* 35, 1933, 1986.
18. Würthwein, G. and Rohdewald, P., Activation of beclomethasone dipropionate by hydrolysis to beclomethasone-17-monopropionate, *Biopharm Drug Disp* 11, 381, 1990.
19. Würthwein, G., Rehder, S., and Rohdewald, P., Lipophilicity and receptor affinity of glucocorticoids, *Pharm Ztg Wiss* 4, 161, 1992.

20. Vogt, H.-J. and Höhler, T., Controlled studies of intraindividual and interindividual design for comparing corticosteroids clinically. In E. Christophers, E. Schöpf, A.M. Kligman, and R.B. Stoughton (Eds.), *Topical Corticosteroid Therapy — A Novel Approach to Safer Drugs.* New York: Raven Press, 1988. pp. 169–179.

21. Täuber, U., Dermatocorticosteroids: structure, activity, pharmacokinetics, *Eur J Dermatol* 4, 419, 1994.

22. Green, M. J., Berkenkopf, J., Fernandez, X., Monahan, M., Shue, H.-J., Tiberi, R. L., and Lutsky, B., Synthesis and structure-activity relationships in a novel series of topically active corticosteroids, *J Steroid Biochem* 11, 61, 1979.

23. Lawlor, F. and Greaves, M., Mode of action of corticosteroid at a molecular level. In E. Christophers, E. Schöpf, A.M. Kligman, and R.B. Stoughton (Eds.), *Topical Corticosteroid Therapy — A Novel Approach to Safer Drugs.* New York: Raven Press, 1988. pp. 137–142.

24. Bamberger, C. M., Bamberger, A.-M., de Castro, M., and Chrousos, G. P., Glucocorticoid receptor β, a potential endogenous inhibitor of glucocorticoid action in humans, *J Clin Invest* 95, 2435, 1995.

25. Woods, M. D., Shipston, M. J., Mullens, E. L., and Antoni, F. A., Pituitary corticotrope tumor (AtT20) cells as a model system for the study of early inhibition by glucocorticoids, *Endocrin* 131, 2873, 1992.

26. Trozak, D. J., Topical corticosteroid therapy in psoriasis vulgaris, *Cutis* 46, 341, 1990.

27. Wester, R. C. and Maibach, H. I., Percutaneous absorption in diseased skin. In H.I. Maibach and C. Surber (Eds.), *Topical Corticosteroids.* Basel: Karger, 1992. pp. 128–141.

28. Marks, R., Aspects of the pharmacology of eczema, *J Dermatol Treatm* 3, 9, 1992.

29. Gianotti, B. and Pimpinelli, N., Topical steroids. Which drug and when?, *Drugs* 44, 65, 1992.

30. Kimura, M., Tarumoto, Y., Nakane, S., and Otomo, S., Comparative toxicity study of hydrocortisone 17-butyrate 21-propionate (HBP) ointment and other topical corticosteroids in rats, *Drugs Exptl Clin Res* 8, 643, 1986.

31. Kitano, Y., Hydrolysis of hydrocortisone 17-butyrate, 21-propionate by cultured human keratinocytes, *Acta Derm Venereol* 66, 98, 1986.

32. Otomo, S., Muramatsu, M., Higuchi, S., Ozawa, Y., and Koyama, I., A potent topical steroid with less systemic side effect — hydrocortisone 17-butyrate, 21-propionate, *Clin Res* 33, 674A, 1985.

33. Ponec, M., Kempenaar, J., Shroot, B., and Caron, J.-C., Glucocorticoids: binding affinity and lipophilicity, *J Pharm Sci* 75, 973, 1986.

34. Barth, J., Lehr, K. H., Derendorf, H., Möllmann, H. W., Höhler, T., and Hochhaus, G., Studies on the pharmacokinetics and metabolism of prednicarbate after cutaneous and oral administration, *Skin Pharmacol* 6, 179, 1993.

35. Holzmann, H. and Hevert, F., Kortikosteroid-Ester als neuer Weg in der Ekzemtherapie, *Dermatologie* 6, 1, 1989.

36. Schalla, W., Busch-Heidger, B., and Hevert, F., Promise of optimization of topical corticoids: some evidence given by the time course of vasoconstriction after epicutaneous and intradermal application, *Z Hautkr* 69, 166, 1994.

37. Wendt, B. and Stähle, H., Drug-Monitoring mit einem neuen Kortikosteroid-Ester, *Dtsch Dermatol* 12, 1472, 1990.

38. Wendt, B., Sind topische Kortikosteroide im Gesicht kontraindiziert? *Hautnah Dermatol* 6, 1, 1991.

39. Wendt, B., Ekzeme der Altershaut — Klinische Erfahrungen mit einem neuen Kortison-Derivat (HBP), *Dermatologie* 7, 24, 1991.

40. Korting, H. C., Hülsebus, E., Kerscher, M. J., Greber, R., and Schäfer-Korting, M., Discrimination of the toxic potential of chemically differing topical glucocorticoids using a neutral red release assay with human keratinocytes and fibroblasts, *Br J Dermatol* 133, 54, 1995.

41. Zaumseil, R.-P., Kecskés, A., Täuber, U., and Töpert, M., Methylprednisolone aceponate (MPA) — a new therapeutic for eczema: a pharmacological overview, *J Dermatol* 3, 3, 1992.

42. Täuber, U. and Rost, K. L., Esterase activity of the skin including species variations. In B. Shroot and H. Schaefer (Eds.), *Pharmacology and the Skin.* Basel: Karger, 1987. pp. 170–183.

43. Täuber, U. and Matthes, H., Percutaneous absorption of methylprednisolone aceponate after single and multiple dermal application as ointment in male volunteers, *Arzneim-Forsch* 42, 1122, 1992.
44. Zentel, H. J. and Töpert, M., Preclinical evaluation of a new topical corticosteroid: methylprednisolone aceponate, *J Am Acad Dermatol* 3, 532, 1994.
45. Kecskés, A., Jahn, P., Kleine-Kuhlmann, R., and Lange, L., Activity of topically applied methylprednisolone aceponate in relation to other topical glucocorticoids in healthy volunteers, *Arzneim-Forsch* 43, 144, 1993.
46. Kecskés, A., Jahn, P., Wendt, H., Lange, L., and Kleine-Kuhlmann, R., Dose-response relationship of topically applied methylprednisolone aceponate (MPA) in healthy volunteers, *Eur J Clin Pharmacol* 43, 157, 1992.
47. Haneke, E., The treatment of atopic dermatitis with methylprednisolone aceponate (MPA), a new topical corticosteroid, *J Dermatol Treatm* 3, 13, 1992.
48. Kecskés, A., Jahn, P., Matthes, H., Kleine-Kuhlmann, R., and Lange, L., Systemic effects of topically applied methylprednisolone aceponate in healthy volunteers, *J Am Acad Dermatol* 28, 789, 1993.
49. Täuber, U., Pharmakokinetik und "Bioaktivierung" von MPA, *Eur J Dermatol Venereol* 4, 1, 1994.
50. Ortonne, J.-P., Safety aspects of topical methylprednisolone aceponate (MPA) treatment, *J Dermatol Treatm* 3, 21, 1992.
51. Fritsch, P., Clinical experience with methylprednisolone aceponate (MPA) in eczema, *J Dermatol Treatm* 3, 17, 1992.
52. Rampini, E., Methylprednisolone aceponate (MPA) — use and clinical experience in children, *J Dermatol Treatm* 3, 27, 1992.
53. Lubach, D. and Platschek, H., Steroidbedingte Gesichtshautschädigungen nach Anwendung von Prednicarbat, *Hautarzt* 41, 43, 1990.
54. Levy, J., Gassmüller, J., Schröder, G., Audring, H., and Sönnichsen, N., Comparison of the effects of calcipotriol, prednicarbate and clobetasol 17-propionate on normal skin assessed by ultrasound measurement of skin thickness, *Skin Pharmacol* 7, 231, 1994.
55. Ulrich, R. and Andresen, I., Treatment of acute episodes of atopic dermatitis. Double-blind comparative study with 0.05% halometasone cream vs. 0.25% prednicarbate cream, *Fortschr Med* 109, 741, 1991.
56. Tafler, R., Herbert, M. K., Schmidt, R. F., and Weis, K. H., Small reduction of capsaicin-induced neurogenic inflammation in human forearm skin by the glucocorticoid prednicarbate, *Agents Actions* 38, 31, 1993.
57. Hevert, F., Schipp, I., Busch, B., and Rozman, T., Kortikosteroid-Ester — ein klinischer Vergleich unter besonderer Berücksichtigung des neuen Hydrocortison-17-butyrat, 21-propionat, *Deutsch Dermatol* 6, 678, 1989.
58. Kim, K. H. and Henderson, N. L., Kinetic studies of skin permeation and biotransformation of prednicarbate. In E. Christophers, E. Schöpf, A.M. Kligman, and R.B. Stoughton (Eds.), *Topical Corticosteroid Therapy — A Novel Approach to Safer Drugs*. New York: Raven Press, 1988. pp. 49–56.
58a. Gysler, A., Lange, K., Korting, H. C., and Schäfer-Korting, M., Prednicarbate biotransformation in human foreskin keratinocytes and fibroblasts, *Pharm Res* 14, (6), 797, 1997.
58b. Lange, K., Gysler, A., Bader, M., Kleuser, B., Korting, H. C., and Schäfer-Korting, M., Prednicarbate versus conventional topical glucocorticoids: pharmacodynamic characterization *in vitro*, *Pharm Res* 14, 793, 1997.
59. Kellner, H.-M., Eckert, H. G., Fehlhaber, H. W., Hornke, I., and Oekonomopulos, R., Pharmacokinetics and biotransformation after topical application of the corticosteroid prednicarbate, *Z Hautkr* 61, 18, 1986.
60. Korting, H. C., Vieluf, D., and Kerscher, M. J., 0.25% prednicarbate cream and the corresponding vehicle induce less skin atrophy than 0.1% betamethasone-17-valerate cream and 0.05% clobetasol-17-propionate cream, *Eur J Clin Pharmacol* 42, 159, 1992.
61. Aliaga, A., Rodríguez, M., Armijo, M., Bravo, J., López Avila, A., Mascaro, J. M., Ferrando, J., Del Rio, R., Lozano, R., and Balaguer, A., Double-blind study of prednicarbate vs. fluocortin butyl ester in atopic dermatitis, *Int J Dermatol* 35, 131, 1996.

62. Kerscher, M. J. and Korting, H. C., Topical glucocorticoids of the non-fluorinated double-ester type, *Acta Derm Venereol* 72, 214, 1992.

63. Senff, H., Kunz, R., Kollner, A., and Kunze, J., Allergic contact dermatitis due to prednicarbate, *Hautarzt* 42, 53, 1991.

64. Isogai, M., Shimizu, H., Esumi, Y., Terasawa, T., Okada, T., and Sugeno, K., Binding affinities of mometasone furoate and related compounds including its metabolites for the glucocorticoid receptor of rat skin tissue, *J Steroid Biochem Molec Biol* 44, 141, 1993.

65. Kelly, J. W., Cains, G. D., Rallings, M., and Gilmore, S. J., Safety and efficacy of mometasone furoate cream in the treatment of steroid responsive dermatoses, *Australas J Dermatol* 32, 85, 1991.

66. Viglioglia, P., Jones, M. L., and Peets, E. A., Once-daily 0.1% mometasone furoate cream vs. twice-daily 0.1% betamethasone valerate cream in the treatment of a variety of dermatoses, *J Int Med Res* 18, 460, 1990.

67. VanderPloeg, D. E., Cornell, R. C., Binder, R., Weintraub, J. S., Jarrat, M., Jones, M. L., and Peets, E. A., Clinical trial in scalp psoriasis — mometasone furoate lotion 0.1% applied once daily vs. betamethasone valerate lotion 0.1% applied twice daily, *Acta Ther* 15, 145, 1989.

68. Wishart, J. M. and Il-Song, L., Mometasone vs. betamethasone creams: a trial in dermatoses, *NZ Med J* 106, 203, 1993.

69. Bjerring, P., Comparison of the bioactivity of mometasone furoate 0.1% fatty cream, betamethasone dipropionate 0.05% cream and betamethasone valerate 0.1% cream in humans, *Skin Pharmacol* 6, 187, 1993.

70. Kirby, J. D. and Munro, D. D., Steroid-induced atrophy in an animal and human model, *Br J Dermatol* 94, 111, 1976.

71. Medansky, R. S., Lepaw, M. I., Shavin, J. S., Zimmerman, E. H., Jones, M. L., Peets, E. A., Samson, C., and Taylor, E., Mometasone furoate cream 0.1% vs. hydrocortisone cream 0.1% in the treatment of seborrhoeic dermatitis, *J Dermatol Treatm* 3, 125, 1992.

72. Katz, H. I., Prawer, S. E., Watson, M. J., Scull, T. A., and Peets, E. A., Mometasone furoate ointment 0.1% vs. hydrocortisone ointment 1.0% in psoriasis, *Int J Dermatol* 28, 342, 1989.

73. Lebwohl, M., Peets, E. A., and Chen, V., Limited application of mometasone furoate on the face and intertriginous areas: analysis of safety and efficacy, *Int J Dermatol* 32, 830, 1993.

74. Kecskés, A., Heger-Mahn, D., Kleine-Kuhlmann, R., and Lange, L., Comparison of the local and systemic side effects of methylprednisolone aceponate and mometasone furoate applied as ointments with equal antiinflammatory activity, *J Am Acad Dermatol* 29, 576, 1993.

75. Vernon, H. J., Lane, A. T., and Weston, W., Comparison of mometasone furoate 0.1% cream and hydrocortisone 1.0% cream in the treatment of childhood atopic dermatitis, *J Am Acad Dermatol* 24, 603, 1991.

76. Druzgala, P., Hochhaus, G., and Bodor, N., Blanching affinity and receptor binding activity of a new type of glucocorticoid: loteprednol etabonate, *J Steroid Biochem Molec Biol* 38, 149, 1991.

77. Bodor, N., Loftsson, T., and Wu, W.-M., Metabolism, distribution, and transdermal permeation of a soft corticosteroid, loteprednol etabonate, *Pharm Res* 9, 1275, 1992.

78. Hochhaus, G., Chen, L.-S., Ratka, A., Druzgala, P., Howes, J., Bodor, N., and Derendorf, H., Pharmacokinetic characterization and tissue distribution of the new glucocorticoid soft drug loteprednol etabonate in rats and dogs, *J Pharm Sci* 81, 1210, 1992.

79. Lan, S. J., Scanlan, L. M., Mitroka, J., Weinstein, S. H., Lutsky, B. N., Free, C. A., Wojnar, R. J., Millonig, R. C., and Migdalof, B. H., Rapid metabolic inactivation of tipredane, a structurally novel topical steroid, *J Steroid Biochem* 31, 825, 1988.

80. Varia, S. A., Faustino, M. M., Thakur, A. B., Clow, C. S., and Serajuddin, A. T., Optimization of cosolvent concentration and excipient composition in a topical corticosteroid solution, *J Pharm Sci* 80, 872, 1991.

81. Yoon, K.-J., Khalil, M. A., Kwon, T., Choi, S.-J., and Lee, H. J., Steroidal anti-inflammatory antedrugs: synthesis and pharmacological evaluation of 16α-alkoxycarbonyl-17-deoxyprednisolone derivates, *Steroids* 60, 445, 1995.

82. Akers, W. A., Risks of unoccluded topical steroids in clinical trials, *Arch Dermatol* 116, 786, 1980.
83. Gip, L. and Verjans, H. L., Hydrocortisone 17-butyrate 0.1% lipocream vs. betamethasone 17-valerate 0.1% cream in the treatment of patients with dry severe chronic eczema, *Curr Ther Res* 41, 258, 1987.
84. Smith, E. W., Meyer, E., and Haigh, J. M., Blanching activities of betamethasone formulations, *Arzneim-Forsch* 40, 618, 1990.
85. Pershing, L. K., Silver, B. S., Krueger, G. G., Shah, V. P., and Skelley, J. P., Feasibility of measuring the bioavailability of topical betamethasone dipropionate in commercial formulations using drug content in skin and a skin blanching assay, *Pharm Res* 9, 45, 1992.
86. Harding, S. M., Sohail, S., and Busse, M. J., Percutaneous absorption of clobetasol propionate from novel ointment and cream formulations, *Clin Exp Dermatol* 10, 13, 1985.
87. Watson, W. S. and Finlay, A. Y., The effect of the vehicle formulation on the stratum corneum penetration characteristics of clobetasol propionate *in vivo*, *Br J Dermatol* 118, 523, 1988.
88. Samson, C., Peets, E. A., Winter-Sperry, R., and Wolkoff, H., Augmented betamethasone dipropionate — Diprolene — Enhancement of topical activity through vehicle formulation. In H.I. Maibach and C. Surber (Eds.), *Topical Corticosteroids*. Basel: Karger, 1990. pp. 302–317.
89. Kerscher, M. J. and Korting, H. C., Comparative atrophogenicity potential of medium and highly potent topical glucocorticoids in cream and ointment according to ultrasound analysis, *Skin Pharmacol* 5, 77, 1992.
90. Korting, H. C., Kerscher, M. J., Vieluf, D., Mehringer, L., Megele, M., and Braun-Falco, O., Commercial glucocorticoid formulations and skin dryness. Could it be caused by the vehicle?, *Acta Derm Venereol* 71, 261, 1991.
91. Lesnik, R. H., Mezick, J. A., Capetola, R., and Kligman, L. H., Topical all-*trans*-retinoic acid prevents corticosteroid-induced skin atophy without abrogating the anti-inflammatory effect, *J Am Acad Dermatol* 21, 186, 1989.
92. Kligman, L. H., Schwartz, E., Lesnik, R. H., and Mezick, J. A., Topical tretinoin prevents corticosteroid-induced atrophy without lessening the anti-inflammatory effect. In H.C. Korting and H.I. Maibach (Eds.), *Topical Glucocorticoids with Increased Benefit/Risk Ratio*. Basel: Karger, 1993. pp. 79–88.
93. Lavker, R. M., Kaidbey, K., and Leyden, J. J., Effects of topical ammonium lactate on cutaneous atrophy from a potent topical corticosteroid, *J Am Acad Dermatol* 26, 535, 1992.
94. Oliwiecki, S., Armstrong, J., Burton, J. L., and Bradfield, J., The effect of essential fatty acids on epidermal atrophy due to topical steroids, *Clin Exp Dermatol* 18, 326, 1993.
95. Woodford, R., Haigh, J. M., and Barry, B. W., Possible dosage regimens for topical steroids, assessed by vasoconstrictor assays using multiple applications, *Dermatologica* 166, 136, 1983.
96. Schulz, H., Hydrocortison-17-butyrat-Verdünnungen (Alfason CreSa°) als Form der Stufentherapie zur topischen Behandlung der Neurodermitis, *Med Cosmetol* 17, 215, 1987.
97. Schaefer, H., Farber, E. M., Goldberg, L., and Schalla, W., Limited application period for dithranol in psoriasis, *Br J Dermatol* 102, 571, 1980.
98. Schalla, W., Bauer, E., Goldberg, L., Farber, E. M., and Schaefer, H., Penetration studies in short-term therapy with dithranol, *Arch Derm Res* 267, 203, 1980.
99. Stoughton, R. B. and Wullich, K., Relation of application time to bioactivity of a potent topical glucocorticoid formulation, *J Am Acad Dermatol* 22, 1038, 1990.
100. Vickers, C. F. H., Existence of reservoir in the stratum corneum, *Arch Dermatol* 88, 20, 1963.
101. Carr, R. D. and Wieland, R. G., Corticosteroid reservoir in the stratum corneum, *Arch Dermatol* 94, 81, 1966.
102. Dupuis, D. and Rougier, A., *In vivo* relationship between horny layer reservoir effect and percutaneous absorption in human, *J Invest Dermatol* 82, 353, 1984.
103. Høybye, S., Møller, S. B., De Cunha Bang, F., Ottevanger, V., and Veien, N. K., Continuous and intermittent treatment of atopic dermatitis in adults with mometasone furoate vs. hydrocortisone 17-butyrate, *Curr Ther Res* 50, 67, 1991.

104. Hradil, E., Lindström, C., and Möller, H., Intermittent treatment of psoriasis with clobetasol propionate, *Acta Dermatol Venereol* 58, 375, 1978.
105. Bensmann, A. and Lubach, D., Untersuchungen über Entstehung und Rückbildung der dermalen Kortikosteroid-Atrophie, *Dermatosen* 34, 20, 1986.
106. Schäfer-Korting, M., Korting, H. C., Kerscher, M. J., and Lenhard, S., Prednicarbate activity and benefit/risk ratio in relation to other topical glucocorticoids, *Clin Pharm Ther* 54, 448, 1993.
107. Korting, H. C., Topical glucocorticoids and thinning of normal skin as to be assessed by ultrasound. In H.C. Korting and H.I. Maibach (Eds.), *Topical Glucocorticoids with Increased Benefit/Risk Ratio.* Basel: S. Karger, 1993. pp. 114–121.
108. Hasford, J. and Victor, N., Risk-benefit analyses of drugs: fundamental considerations and requirements from the point of view of the biometrician. Problems in the assessment of the combination of trimethoprim with sulfamethoxazole, *Infection* 15, 236, 1987.
109. Palminteri, R., Benefit/risk ratio of new drugs: for whom? Discussion paper, *J Roy Soc Med* 81, 155, 1988.

Index

Printed and bound by CPI Group (UK) Ltd, Croydon, CR0 4YY

23/10/2024

01778248-0011